Lecture Notes in Computer Science   10269

Commenced Publication in 1973
Founding and Former Series Editors:
Gerhard Goos, Juris Hartmanis, and Jan van Leeuwen

T0177958

More information about this series at http://www.springer.com/series/7412

Puneet Sharma · Filippo Maria Bianchi (Eds.)

Image Analysis

20th Scandinavian Conference, SCIA 2017
Tromsø, Norway, June 12–14, 2017
Proceedings, Part I

Editors
Puneet Sharma
University of Tromsø
Tromsø
Norway

Filippo Maria Bianchi
University of Tromsø
Tromsø
Norway

ISSN 0302-9743 ISSN 1611-3349 (electronic)
Lecture Notes in Computer Science
ISBN 978-3-319-59125-4 ISBN 978-3-319-59126-1 (eBook)
DOI 10.1007/978-3-319-59126-1

Library of Congress Control Number: 2017940836

LNCS Sublibrary: SL6 – Image Processing, Computer Vision, Pattern Recognition, and Graphics

Printed on acid-free paper

This Springer imprint is published by Springer Nature
The registered company is Springer International Publishing AG
The registered company address is: Gewerbestrasse 11, 6330 Cham, Switzerland

Preface

This book constitutes the refereed proceedings of the 20th Scandinavian Conference on Image Analysis, SCIA 2017, held in Tromsø, Norway, in June 2017.

The 87 revised papers presented were carefully reviewed and selected from 133 submissions. The 87 accepted articles are organized in two volumes, i.e., volumes 1 and 2. Volume 1 comprises topical sections on the history of SCIA, motion analysis and 3D vision, pattern detection and recognition, machine learning, and image processing and applications. Volume 2 is structured in topical sections on remote sensing, medical and biomedical image analysis, feature extraction and segmentation, and face, gesture, and multispectral analysis.

June 2017
Puneet Sharma
Filippo Maria Bianchi

Organizers

Sponsors

Organization

General Chair

Robert Jenssen   University of Tromsø - The Arctic University of Norway

Program Chairs

Puneet Sharma   University of Tromsø - The Arctic University of Norway

Filippo Maria Bianchi   University of Tromsø - The Arctic University of Norway

Program Co-chairs

Arnt Børre Salberg   Norwegian Computing Center, Norway

Jon Yngve Hardeberg   Norwegian University of Science and Technology, Norway

Trym Haavardsholm   Norwegian Defence Research Establishment, Norway

Program Committee

Adrien Bartoli	ISIT – CENTI, France
Anders Heyden	Lund University, Sweden
Anne H. Schistad Solberg	University of Oslo, Norway
Arnt-Børre	Norsk Regnesentral, Norway
Atsuto Maki	Kungliga Tekniska Högskolan, Sweden
Cristina Soguero Ruiz	Rey Juan Carlos University, Spain
Daniele Nardi	Sapienza University, Italy
Domenico Daniele Bloisi	Sapienza University, Italy
Enrico Maiorino	Sapienza University, Italy
Erkki Oja	Aalto University, Finland
Fredrik Kahl	Lund University, Sweden
Gustau Camps-Valls	University of Valencia, Spain
Heikki Kälviäinen	Lappeenranta University of Technology, Finland
Helene Schulerud	Sintef, Norway
Ingela Nyström	Uppsala University, Sweden
Janne Heikkilä	University of Oulu, Finland
Jens Thielemann	SINTEF, Norway
Joni Kämäräinen	Tampere University of Technology, Finland
Karl Øyvind Mikalsen	University of Tromsø, Norway
Kjersti Engan	University of Stavanger, Norway

Machine Learning

Contents – Part II

Feature Extraction and Segmentation

History of SCIA

Image Processing and Its Hardware Support Analysis vs Synthesis - Historical Trends

Ewert Bengtsson[✉]

Department of Information Technology, Centre for Image Analysis,
Uppsala University, Box 337, 75105 Uppsala, Sweden
ewert@cb.uu.se

Abstract. Computers can be used to handle images in two fundamentally different ways. They can be used to analyse images to obtain quantitative data or some image understanding. And they can be used to create images that can be displayed through computer graphics and visualization. For both of these purposes it is of interest to develop efficient ways of representing, compressing and storing the images. While SCIA, the Scandinavia Conference of Image Analysis, according to its name is mainly concerned with the former aspect of images, it is interesting to note that image analysis throughout its history has been strongly influenced also by developments on the visualization side. When the conference series now has reached its 20th milestone it may be worth reflecting on what factors have been important in forming the development of the field. To understand where you are it is good to know where you come from and it may even help you understand where you are going.

Keywords: History · Image processing · Image analysis · Computer graphics · Visualization · Hardware support

1 Introduction

Images offer a very rich way of representing information about our world. This is true for the basic two dimensional monochrome projection image in a classical photograph and an artistic drawing or painting. It is also true for a film or video adding the temporal dimension. For a color image adding the spectral dimension, possible to generalize to multi- or hyper- spectral images. For a 3D image through stereo, depth scanning, tomography or holography registering information about the third dimension. We can talk about 2D, 3D, 4D or 5D images referring to how many of the spatial, temporal and spectral dimensions are represented in the image. In the rest of this paper I will simply use the term "image" to refer to image data of any number of dimensions.

One reason while images are so important is that the highest capacity channel for getting information into our brains is through our visual system. Images would be almost meaningless if no one was there to see them. And it is not only we as humans who have great use of our visual system. There are theories

© Springer International Publishing AG 2017
P. Sharma and F.M. Bianchi (Eds.): SCIA 2017, Part I, LNCS 10269, pp. 3–14, 2017.
DOI: 10.1007/978-3-319-59126-1_1

claiming that the "invention" of visual systems by evolution caused the "Cambrian explosion", the first major dramatic increase in the number and diversity of species [2]. Through the development of image analysis and computer vision the entity "seeing" the image may be a machine.

Images existed long before computers were invented but it is not surprising that images has become a very important data structure in computers. It is likely that images form the greatest part of all data handled in all the computers in the world. We take it for granted today that any computer can handle images, even with high spatial and temporal resolution and in full color. But that has not always been the case. When one of the long standing conference series in the field, the Scandinavian Conference on Image Analysis, SCIA, celebrates the milestone of the 20th conference it may be worth reflecting on what has been driving the development of computer capacity of handling images. That is the topic of this paper.

Computers can be used to handle images in two fundamentally different ways.

1. We may have some data that we want to turn into an image so that we can use our visual system to perceive, enjoy or understand the data. We have the fields of computer graphics and visualization. This can be seen as a forward problem: data to image.
2. We may load an image of some part of the world into the computer and try to analyse it to get some useful quantitative data out of it or to reach some kind of understanding of what is in the image. We have the fields of image analysis, image understanding, computer vision. This is an inverse problem: image to data. From the very complex representation in an image we need to find the underlying information of interest. As most inverse problems it is under determined, there may be many possible solutions and we need some models to choose the best one.

We may combine the two and use visualization of real world images to interactively analyse the images and extract more meaningful data from them than could be done without computer support.

2   The Early History

Working memory and storage capacity was a very limited and expensive resource in early computers. Images are heavy data structures. Even a very low resolution image say $256 \times 256 \times 8$ bit greyscale occupies $64\,kB$, typically the whole working memory of a research computer of the early 1970-is or of the first generation personal computers appearing towards the end of that decade. Since a whole image could not be stored in memory there were no display units that could visualize it. The only output units generally available for the early computers were alphanumeric printers and, a bit later, alphanumeric screens.

The first digital images were obtained by scanning a physical picture or an optical image of some part of the world pixel by pixel, reading the intensity into computer memory. Application areas were mainly remote sensing (mostly military) and medicine. There was also research trying to develop more general computer vision.

2.1   Remote Sensing

The first (civilian) remote sensing satellite, Landsat 1, was launched in 1972. It transmitted images from three different sensors down to earth. The highest resolution one, MSS, had pixels of 68×83 meters over areas covering 185×185 km with four spectral bands [1]. This meant that the images were 2720×2228 or 6 Mega pixels. That was far beyond the primary memory capacity of any computer in those days, it was even more than many computers had as secondary storage capacity. So the images were written to photographic film and analysed visually in the same way as aerial photography images had been analysed since the first world-war. But the availability of these images also triggered research on computerized image analysis, although that had to be done without being able to see the digital images that were being analysed.

The founder of the Swedish IAPR section SSAB and one of the key persons behind the SCIA conference series was Thorleif Orhaug who was head of the research unit at the Swedish Defence Research Institute developing remote sensing techniques. That group also had a drum scanner that could read photographs and produce high resolution digital images and write high resolution images to film or photographic paper. This was to my knowledge the first high resolution digital image input-output device in Scandinavia. So it was possible to input and output digital images of high resolution but no one could see the images while they were in the computer and thus not interact with them in any efficient manner. The progress in actually being able to analyse the satellite images with computers in useful ways was very slow.

2.2   Computed X-Ray Images

In the early 1970-is Mc Cormack and Hounsfield invented computer tomography, CT [3]. This was the first time useful images were the result of computations, not only direct measurements. It revolutionized medicine by providing images of cross sections of the body rather than projections through the body. The first tomographic images were very small some tens of kilobytes still there were no computers around that could store and display them in a useful way so the images were written to film and handled as all other medical X-ray images. Even though the first CT systems only provided single, or a few images, that was the first step towards truly volumetric images, since several CT slices stacked on top of each other formed a 3D volume image. The 3D reconstruction had to be in the head of the radiologist based on a number of 2D slices displayed as a mosaic on X-ray film. Most early image processing research in the radiology field was focused on creating better images faster, not on doing any computerized analysis of the images.

2.3 Automated Cell Image Analysis

Long before the advent of computed tomography, already in the 1950-is, a system for cell image analysis for automated early detection of cervical cancer was developed in the US. The background was that Papanicolaou a decade earlier had shown that precancerous lesions could be detected by visual inspection of cell smears. One distinguishing feature was that cancer cell nuclei were bigger than those of normal cells [4]. But the visual screening was tedious and expensive. So a system was developed that scanned the microscopy samples and "looked" for nuclei larger than 10 microns in diameter. The developed system was based on hard-wired analogue video processing circuits since there were no useful computers in the 1950-is [5].

The system was a failure since it could not tell the difference between a cancer cell and two normal cells on top of each other. But as soon as computers became available in the 60-is and 70-is research projects started to develop image analysis systems for the same purpose. Those systems used image processing to detect more features than the nuclear diameter, such as shape and nuclear texture. They also needed to do robust automated segmentation of the cells. The development was initially done with the line printer of the computer as the only tool for displaying the cell images. A cell image was typically around 128×128 pixels and could be processed in the computer memory. To screen a whole smear many thousand image fields needed to be analysed. To achieve that several slide scanners were developed [6]. All the image analysis had to be done on the fly since there was no way of storing the hundreds of megabyte that comprised a specimen. Around 30 years later memory capacities had caught up with the need and slide scanners that actually stored the images were developed as the first step towards digital pathology, a very active field today.

2.4 Early Computer Vision

In addition to the applied work on remote sensing and medical applications there were also early attempts at using the first generations of computers for developing computer vision systems. In 1966, Seymour Papert at MIT wrote a proposal for building a vision system as a summer project [7]. The abstract of the proposal starts stating a simple goal: "The summer vision project is an attempt to use our summer workers effectively in the construction of a significant part of a visual system". The difficulty in creating a vision system was somewhat underestimated at that time.

2.5 The Birth of Computer Graphics

The first generations of computers generally operated in batch mode. Piles of punched cards which contained both programs and data were handed in at a computer center and a few hours later a printed list with results (or usually error listings) could be collected. In 1962 Sutherland presented Sketchpad, a computer drawing system that was the first graphical user interface [8]. This was really pioneering work but there were no products available that made it possible for people to use the graphical interface. The IBM 2250 Graphics Display Unit was announced in 1964. It cost around USD 280 000 in 1970, equivalent to around USD 2 million in 2017 currency, so it was not widely available [9]. The system was vector oriented, could draw bright lines on a darker background on a display area of around $30 \times 30\,cm$ with $1k \times 1k$ resolution. The Tektronix 4010 series which appeared in the early 1970-is had the same resolution and drawing capability but was based on storage tube technology which meant that you could only erase graphics by flashing the entire screen, erasing everything where after a new graphical image could be written. Since the price-tag was almost 100 times lower, around USD 3000 they became much more widely available. Based on the Tektronix 4010 and similar rather primitive graphical display devices it was possible to create the first interactive image analysis systems. Even the limited display capacity helped increase the effectiveness of algorithm development significantly.

2.6   The Need of Special Image Processing Hardware

The very limited memory and processing capacity of early computers led to much interest in special hardware architectures for image processing. An early noteworthy Scandinavian project was the PICAP [10]. Internationally we had the GLOPR, Cytocomputer, CLIP projects [11]. They all in different ways explored neighbourhood relations and the potentially parallel processing possibilities in images usually through pipe lining architectures although in the CLIP case through a physically parallel architecture. Single or a few units were built but the special architectures never caught on to become successful products.

3   Image Analysis Established as a Research Field - Early SCIA Conferences

3.1   First SCIA Linköping 1980

The first SCIA conference which took place in January 1980 was clearly influenced by the need of increased processing power. There was for instance a paper about the work on PICAP II, a successor of the pioneering PICAP system but with a very different architecture. The need of interactive image analysis systems and the possibilities of creating such systems because of improved display facilities led to a number of papers on image processing software systems. In total 17 papers dealt with system design, more than at any other SCIA conference. The main application fields discussed at the conference were those mentioned above, remote sensing and medical image analysis, in particular microscopy and various industrial applications [12].

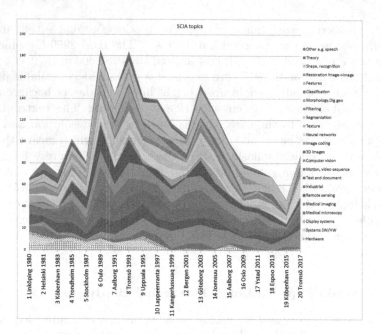

Fig. 1. The number of papers presented for different topics during the SCIA conferences. The topics have been judged from the titles of the papers, a paper may belong to several categories, one has been chosen here based on an estimate of what was the dominant topic. The total number of papers presented at all the SCIA conferences so far including this year is 2148.

3.2   Second SCIA, Helsinki 1981 and Third, Köbenhavn 1983

At the second SCIA in June the following year (the organizer realized it was easier to attract international colleagues to Scandinavia in June than in January) the great interest in developing image analysis systems was illustrated by a survey from VTT which had identified 123 such systems in the literature, many from the US, but Scandinavia was in relation to the population strongly overrepresented. A hardware architecture for GOP, the general operator processor which implemented hierarchical local processing with feed-back between levels was presented. It was similar to convolutional networks, but with hand coded filters coefficients rather than defined by machine learning. There was also strong focus on theoretical foundations of image analysis. KS Fu one of the global pioneers in image analysis gave a keynote lecture on syntactic methods in image analysis. JI Zuravlev from the Soviet Academy of Science gave a most theoretical lecture, spending an hour of intensive writing of equations on a black board, proving the solution to the pattern recognition problem gamma, QED. T Kohonen presented the first results about self-organizing neural networks. The size of the conference and profile of topics was very similar at SCIA 3 two years later [12].

3.3   Image Processing Hardware Developments

During the following years a number of companies were founded in particular in Linköping, based on the idea that it should be possible to use image processing for a number of useful applications, but the general purpose computer architectures were not really suited for processing image data and far too slow. Imtec was based on the PICAP II architecture. It later split in one branch Teragon which developed a first generation desktop publishing system for use at newspapers. The other branch which retained the name Imtec moved to Uppsala to develop medical image analysis based on the research that had been going on there since the early seventies. Sectra was formed to initially develop data coding for secure transmissions but soon also became active in the radiology image processing field. The GOP became the basis of Context Vision, a company that first addressed image analysis and computer vision in a very general way but later specialized in medical image enhancement.

These companies developed graphics subsystem capable of showing high resolution full color images. But they did not come cheaply, the 1024 × 1280 full color display system developed by IMTEC for medical image display cost more than USD 10 000 in manufacturing cost for a single unit. The international companies specializing in computers for research such as Digital Equipment, Sun and Silicon Graphics also offered products with high quality display options for tens of thousands of dollars.

3.4   SCIA 4 Trondheim 1985 and 5 Stockholm 1987

At the 5th SCIA the Scandinavian commercial efforts were at their peak with a large industrial exhibit and much optimism about image analysis finally having come through from an academic curiosity to mainstream commercial products relevant for many application fields. But the number of application papers presented was the lowest for all the SCIA conferences, the application focus had moved to commercial efforts. Instead papers dealing with computer vision problems, understanding 3D scenes and video sequences started appearing at SCIA 4 and were the most common topics at SCIA 5 [12].

3.5   Computers Becoming a Consumer Product

In parallel to these developments computers were moving out of the research labs and becoming a consumer product. The IBM PC was announced in August 1981 about the same time as the second SCIA. It had rudimentary color graphics capability. The CGA-standard graphics card had 16 kB video memory and could display 640 × 200 binary monochrome graphics and up to four colors at the 320 × 200 resolution. But the vendors realized that in order to sell computers to consumers they had to have good graphics capabilities and still be affordable. The potential mass market made it possible to develop such products. So during the 1980-is there were several generations of improving graphic standards. Specialized brands such as Amiga was first with good graphics performance but towards the end of the decade also standard PC:s could display images and videos with reasonable quality and performance.

So in the early 1990-is we had affordable basic image display systems and standard computers had reached a capacity of storing and processing images that made it possible to develop image analysis applications without special hardware. The special image processing hardware developments more or less disappeared. The exponential growth of computer capacity dubbed "Moores law" had led to standard computers having the capacity the special architectures had a decade earlier. The companies that based their business idea on special image processing hardware had either disappeared or reformulated the business to be strongly application oriented using mainly standard hardware. The lesson learned was that image analysis is too narrow a field to support special hardware that keeps ahead of general purpose computing hardware.

3.6   SCIA 6 Oulu 1989, 7 Aalborg 1991 and 8 Tromsø1993

The fact that general purpose computers had reached sufficient capacity to handle process images of useful size and resolution led to a boom in the research in the field. The following three SCIA conferences were the largest of all with attendances of around 250 persons. The relative number of papers on computer architectures and systems had decreased since people no longer had to develop their own tools to be able to do research on image analysis. There were numerous computer vision papers dealing both with static 3D scenes and motion problems. There were also many method oriented papers dealing with mathematical morphology, segmentation, feature extraction, object recognition, classification. As an example, H. Knutsson presented a paper about representation of local structure using tensors which later received rather many citations. The main application fields were still medical and remote sensing but also document analysis was discussed in many papers. But most applications were in rather narrow fields and far from having general impact in society [12].

4   Computer Games Driving Graphics Developments

The development was different for the other, forward, aspect of processing images in computers. Computer games had found mass markets and there were very strong economic motivations for developing powerful image display and realistic real time 3D image rendering systems. Several vendors e.g. ATI, 3Dfx, NVIDIA were competing intensely, developing ever more powerful 3D rendering chips and display systems [13]. At this time the invention of the world wide web had turned internet from being a convenient way for researchers to communicate to become a mainstream communication platform for multimedia, again needing great capacity for image handling and display, but not so much for image analysis.

4.1 SCIA 9 Uppsala 1995, 10 Lappeenranta 1997, 11 Kangerlussuaq 1999 and 12 Bergen 2001

Towards the end of the millennium the size of the SCIA conferences decreased steadily. In particular the number of computer vision related topics decreased, instead there were more application papers reaching around a third of all papers. On the methods side the number of papers dealing with recognition problems increased. Also during these years we saw the first boost in the interest in neural networks, in the first conferences there were a single paper on that topic, now we had between five and ten, but the interest died after the 12th conference, returning to single papers per conference [12,14].

4.2 Image Analysis Lacking a "killer Application"

So the situation was that we had very strong general public impact for image generation and display but not at all corresponding developments for image analysis or computer vision. The markets were very different, there were no "killer applications" like computer games for the image analysis field. The powerful display capabilities were of course useful when doing image analysis research but the processing was done in standard software. Many image analysis applications were actually running slower than they had earlier since the increasing computer capacity had made it possible to write image analysis algorithms in very high level systems such as Matlab and still get them executed at acceptable speeds, although much slower than the same algorithm would have executed if implemented more efficiently.

4.3 SCIA 13 Göteborg 2003, 14 Joensuu 2005, 15 Aalborg 2007, 16 Oslo 2009

SCIA 13 saw a recovery in the number of accepted papers, almost reaching the numbers of the conferences ten years earlier. Application papers decreased in numbers and instead we saw more papers on methods of various kinds, texture, segmentation, feature extraction, shape analysis were popular topics. The first presentations about local binary patterns, LBP, at a SCIA conference was given in 2003, the year after the ground breaking publication in PAMI [15]. They covered an extension to greater neighbourhoods and a combination with neural networks. The LBP methods was later also applied to face detection and analysis, a popular topic at this time [16].

5 Image Processing to Benefit from Graphics Hardware

Around the turn of the millennium it had become possible to do some programming of the graphical processing units. Initially quite difficult low level programming but ten years ago the CUDA software development kit was made public. Later additional packages such as OpenCL offered increased convenience and functionality. So now the hardware developed for generating images became available for helping us analyse images. That's the second major way in which graphics has made a decisive difference for image analysis (the first one was when it became possible to display images on computer screens, remember).

The availability of high performance graphics processors that can be programmed to do general purpose parallel image processing operations at very high speed has been exploited in a large number of image analysis applications significantly speeding up the processing. But the greatest impact has been on convolutional neural networks. As mentioned above, the concept of artificial neural networks is not new but the interest died due to lack of successful applications. But a few years ago something happened, we saw a leap in classification performance for well established, difficult computer vision tasks. And the reason was the availability of huge numbers of images and massive computing power in the GPU:s making it possible to train the networks much more extensively than was possible before. Today deep convolutional neural networks are used for all kinds of image analysis tasks, often with impressive performance. So finally special hardware and software architecture is having an impact on image analysis. But that hardware was not developed for image analysis, it was developed for image synthesis and display for computer games.

5.1   SCIA 17 Ystad 2011, 18 Espoo 2013, 19 Köbenhavn 2015 and 20 Tromsö 2017

During the most recent SCIA conferences the focus was on methods for analysing images and recognizing features and structures. More than half of the presented applications were from the medical field. The numbers of presented papers has been declining, the 19th SCIA was the smallest of all so far. But fortunately this years conference, the 20th SCIA is breaking that trend, with 87 accepted papers we have to go back a decade to find a bigger SCIA conference, it is in the middle size range among all the conferences. As expected a record number of the papers are dealing with neural networks, 17 papers, which is almost 20% of all papers at the conference and also 20% of all neural network papers presented at any SCIA conference. [17].

6   Conclusion and Future Work

So where do we stand today? The forward image processing, generating realistic images in real time is seeing enormous markets, the computer game market is greater than the film market in economic terms. We are rapidly approaching the point where it is impossible to distinguish between computer generated characters and filmed real humans actors. This is likely going to have a huge impact on entertainment. In particular when it is combined with virtual reality so that the spectator can be immersed in the action in the artificially generated world.

The inverse problem, image analysis is also finally finding some mass market applications. Most mobile phones and digital cameras today have face detection, helping to focus on the most important parts of the image, a common research topic at the SCIA conferences a decade ago. We can also do image search on the internet by showing an image to Google receiving similar images in return. Also this a research problem discussed at SCIA a decade ago. But these applications are not at all of the same impact as the gaming industry.

6.1   Augmented Reality

We may see a future high impact image analysis application in augmented reality. If the system can understand the visual neighbourhood of the person taking part in a game the visual and real world can be joined in very powerful ways. We have already seen great public impact of primitive augmented reality in the PokemonGo game. But there the integration with the world was only based on location not vision. In addition to gaming, this technology could find usage in significantly improved user interfaces to smart phones. If the phone can understand what you are saying and see where you are pointing in an improved, heads-up display that is comfortable enough to wear all day we could get away from staring at and manipulating tiny screens.

6.2   Self Driving Cars - a "killer Application"

The "killer application" for image analysis that is beginning to appear is self-driving cars. Even though there are other sensors than passive imaging cameras most of them generate image information that need to be analysed in very strict real time. The car industry is big enough to motivate special hardware development for the needed image analysis capacity so even if the hardware developed for games is currently having a major impact on the development of image analysis for self-driving cars we will most likely see specially developed image analysis hardware solutions in the near future. Self-driving cars could transform our society in very profound ways, even helping solve the climate crisis. But it is crucial that the image analysis works in a really reliable way, otherwise we may have a much too literal killer application.

6.3   Big Brother Is Watching

Another image analysis application that can have major social impact is identification of individuals based on general appearance, gait and facial features. We already have the image analysis performance to identify a person and to follow that person from one camera to another if the cameras are set-up to cover a large area such as a campus or a whole city. This can have major usage in monitoring areas for possible terrorism threats or other unauthorized behaviour. But it really is an implementation of "big brother" that is quite scary in its possibilities for misuse by authoritarian authorities.

6.4   Concluding Remarks

When the SCIA conferences started computers were hardly capable of processing images with meaningful performance. The general development of microelectronics gave us exponential growth of computing capacity, millions of times larger now than at the time of the first SCIA. Consumer markets for graphics gave us even higher performance parallel processing units. For a long time I had the feeling that we were not at all living up to using that capacity to create applications that had any greater impact on society. But now my conclusion is that we are at a point in time where image analysis is beginning to reach real social impact. Perhaps at the 30th SCIA you will be able to look back at a research field that has had a major role in transforming society.

References

1. http://landsat.gsfc.nasa.gov/the-multispectral-scanner-system/
2. Parker, A.: In the Blink of an Eye: How Vision Sparked the Big Bang of Evolution. Perseus Publishing, Cambridge (2003)
3. Hounsfield, G.N.: Computed Medical Imaging. Nobel Lecture (1979)
4. Papanicolaou, G.N., Traut, H.: The diagnostic value of vaginal smears in carcinoma of the uterus. Am. J. Obstet. Gynecol. **42**, 193–206 (1941)
5. Tolles, W.E., Bostrom, R.C.: Automatic screening of cytological smears for cancer: the instrumentation. Ann. N. Y. Acad. Sci. **163**, 1211–1218 (1956)
6. Tucker, J.H., Shippey, G.: Basic performance tests on the CERVIFIP linear array prescreener. Anal. Quant. Cytol. **5**(2), 129–137 (1983)
7. Papert, S.: The summer vision project. MIT AI Memo 100, Massachusetts Institute of Technology, Project Mac (1966)
8. Sutherland, I.E.: Sketchpad a man-machine graphical communication system. Trans. Soc. Comput. Simul. **2**(5), 3 (1964)
9. Pugh, E.W., Johnson, L.R., Palmer, J.H.: IBM's 360 and Early 370 Systems. The MIT Press, Cambridge (1986)
10. Antonsson, D., Gudmundsson, B., Hedblom, T., Kruse, B., Linge, A., Lord, P., Ohlsson, T.: PICAP - a system approach to image processing. IEEE Trans. Comput. **31**(10), 997–1000 (1982)
11. Preston Jr., K., Duff, M.J.B.: Modern Cellular Automata: Theory and Applications. Springer, New York (1984)
12. The first 12 SCIA conferences were published in printed format by the pattern recognition society of the hosting country but are not available digitally
13. http://www.techspot.com/article/650-history-of-the-gpu/
14. Borgefors, G. (ed.): Theory and Applications of Image Analysis. World Scientific, Singapore (1995). A book of selected papers from SCIA 9 is available digitally
15. Ojala, T., Pietikainen, M., Maenpaa, T.: Multiresolution gray-scale and rotation invariant texture classification with local binary patterns. IEEE Trans. Pattern Anal. Mach. Intell. **24**(7), 971–987 (2002)
16. These SCIA conference proceedings were published in the following Springer LNCS editions and are available digitally: 13 - 2749, 14 - 3540, 15 - 4522, 16 - 5575
17. These SCIA conference proceedings were published in the following Springer LNCS editions and are available digitally: 17 - 6688, 18 - 7944, 19 - 9127

Motion Analysis and 3D Vision

Averaging Three-Dimensional Time-Varying Sequences of Rotations: Application to Preprocessing of Motion Capture Data

Tomasz Hachaj[1]($\boxtimes$), Marek R. Ogiela[2], Marcin Piekarczyk[1], and Katarzyna Koptyra[2]

[1] Institute of Computer Science, Pedagogical University of Krakow, 2 Podchorazych Ave, 30-084 Krakow, Poland
tomekhachaj@o2.pl, marcin.piekarczyk@up.krakow.pl
[2] Cryptography and Cognitive Informatics Research Group, AGH University of Science and Technology, 30 Mickiewicza Ave, 30-059 Krakow, Poland
{mogiela,kkoptyra}@agh.edu.pl

Abstract. The aim of this paper is to propose and initially evaluate our novel algorithm which enables averaging of time-varying sequences of rotations with three degrees of freedom described by quaternions. The methodology is based on Dynamic Time Warping barycenter averaging (DBA) with one minus dot product distance function, Markley's quaternions averaging method and Gaussian quaternion signal smoothing. The proposed algorithm was successfully applied to generate single, averaged motion capture recording (MoCap) from ten MoCap of mawashi-geri karate kick of black belt Shorin-Ryu karate master. We have used inverse kinematic model. In our experiment mean DTW normalized distance between averaged signal and original signals varied from $0.713 \cdot 10^{-3}$ for Hips sensor to $6.153 \cdot 10^{-3}$ for LeftForearm sensor, which were very good results. Also the visualization of the averaged MoCap data showed that the proposed method did not introduce unwanted disturbances and may be usable for that task. That type of averaging has many important applications. For example it can be used to calculate and visualize an average performance of an athlete who performs some activity that he wants to optimize during training. The numerical and visual data may be a very important feedback for coach that supervises the training. Also our method is not limited to MoCap data averaging; it can be applied to average any type of quaternion-based time-varying sequences.

Keywords: Signal averaging · Preprocessing · Quaternions · Dynamic Time Warping barycenter averaging · Motion capture · Karate

1 Introduction

In many real-life scenarios time-varying signal analysis requires averaging (or template generation) from samples that come from many measurements [1–4].

© Springer International Publishing AG 2017
P. Sharma and F.M. Bianchi (Eds.): SCIA 2017, Part I, LNCS 10269, pp. 17–28, 2017.
DOI: 10.1007/978-3-319-59126-1_2

There are several state-of-the art methods which can be used averaging. Among most popular is Kalman Filter (KF) [5] that is used for example in kinematic model synthesis [6–8]. Other popular method is DTW barycenter averaging (DBA) [9], which was already used in movements' analysis [10,11]. One can use the smoothing ability of KF to average the signals that came from multiple measurements of the same angle. However, in the situation when signals cannot be wrapped optimally linearly, the nonlinearity between signals will be smoothed by the KF. This smoothing might visually damage the recorded content, because it might treat nonlinearity as noise. Due to this we choose DBA approach.

The motivation of this paper was a need to create the averaged karate action templates for future use in athletes' kinematic analysis. That approach has many important applications. For example it can be applied in computer supervised training to calculate and visualize an average performance of an athlete or to compare it to templates of world class sportsmen's to find the optimize the action's technique. The aim of this paper is to propose and initial evaluate our novel algorithm that enables averaging of time-varying sequences of rotations with three degrees of freedom described by quaternions. To our best knowledge the method of this kind has not yet been reported in scientific papers. The methodology is based on DBA with one minus dot product distance function, Markley's quaternions averaging method [12] and Gaussian quaternion signal smoothing. The proposed algorithm is applied to generate single averaged motion capture (MoCap) recording from ten MoCap of mawashi-geri karate kick of black belt Shorin-Ryu karate master. We have used inverse kinematic model.

2   Materials and Methods

This section presents all algorithms that are used for MoCap signal averaging and smoothing. Due to paper space limitation we did not provide the details of Markley's and DBA algorithm that can be found elsewhere (in [9,12] appropriately).

2.1   Quaternion Averaging

For the Quaternion averaging we have utilized Markley's algorithm [12]. It determines the average norm-preserving quaternion from a set of weighted quaternions. The solution involves an eigenvalue/eigenvector decomposition of a matrix composed of the given quaternions weights matrix.

2.2   Signal Averaging

To average a set of time-varying signals we have used Dynamic Time Warping barycenter averaging (DBA) heuristic algorithm [9] that uses Dynamic Time Warping (DTW) similarity measure. The cost function in DTW for quaternions is defined as:

$$cf(x,y) = 1 - (x \circ y) \tag{1}$$

where x, y are normalized quaternions and $\circ$ is a dot product. Because we are dealing with quaternions, the barycenter averaging is replaced by Markley's algorithm from Sect. 2.1.

Signal Averaging Algorithm

```
Input data:
  A #initially averaged sequence of size L
  S #matrix of sequences, each sequence has length T
Algorithm:
  Associations := empty list of size L
  For each seq in S
    #DTW algorithm returns Cost matrix and Path matrix
    #cf function is defined in Eq. (1)
    [Cost, Path] := DTW (A, seq)
    i := L
    j := T
    While i >= 1 and j >= 1
      #add j-th quaternion from seq to Associations
      Associations[i] := Associations[i] sum seq[j]
      [i, j] := Path[i,j]
    End While
  End For
  For i in 1:T
    Ai := Quaternion averaging(Associations[i]) #See Sect. 2.1
  End For
Return:
  Ai
```

2.3 Smoothing Algorithm

It is possible that DBA algorithm introduces in averaged signals the high frequency noises that are visible as rapid Euler angles hops. This is of course result of the DBA heuristic which does not prevent these situations even if input data does not contain that type of noises. Our smoothing algorithm works similarly to typical discrete linear convolution algorithm with Gaussian kernel, however instead of linear combination of signal samples in kernel window and kernel weights, we use Markley's algorithm from Sect. 2.1 with Gaussian weights.

Smoothing Algorithm

```
Input data:
  Q #quaternion signal of length T which we want to smooth
  windowSize #length of kernel
Algorithm:
  smoothedSignal := Q #rewrtie signal to sequence
```

```
  #that hold output data
startInd := floor(windowSize / 2)
startLoop := startInd
endInd := ceiling(windowSize / 2)
endLoop := T  endInd
G := symmetric vector with Gaussian distribution
   of values with length windowSize
For a in startLoop:endLoop
  i := 1
  sampleToSmooth := vector with length windowSize
  For b in (-startInd+1):endInd
    sampleToSmooth[i] <- Q[a+b]
    i := i + 1
  End For
  smoothedSignal[a] <- Quaternion averaging(sampleToSmooth, G)
End For
Return:
  smoothedSignal
```

2.4   MoCap Averaging

The averaging algorithm for MoCap data works as follows. As an input it takes a set of MoCap recordings of a single activity. They may vary in length, however they have to use the same kinematic model. In our case we use inverse kinematic model with 16 various features that represents the rotation of body joints with three degrees of freedom in Euler angles (see Fig. 1). The Euler angles are recalculated to quaternions. This step remove problems caused by nonlinearity in Euler angles domain: $[-180, 180)$ for rotation towards X axis, $[-90, 90)$ for Y and $[-180, 180)$ for Z. Also quaternions prevent the gimbal lock phenomena. Signal in each feature of each recording is interpolated to the uniform length with nearest neighbor interpolation. It is obvious that signals among single MoCap recording have same number of frames, however different MoCap recordings may have different frames number. The new length equals the longest signal in the input dataset. After averaging all features, each of the averaged signals is smoothed with algorithm from Sect. 2.3. The obtained averaged and smoothed quaternions are recalculated to Euler angles.

3   Results

To evaluate our novel MoCap averaging algorithm, we have used recording of black belt Shorin-Ryu karate master. The activity we wanted to average was a mawashi-geri kick with a right leg. We have recorded ten repetitions of this kick with a Shadow 2.0 wireless motion capture system. The tracking frequency was 100 Hz with 0.5° static accuracy and 2° dynamic accuracy. We have prepared data to be in reverse kinematic model which is presented in Fig. 1. Data was

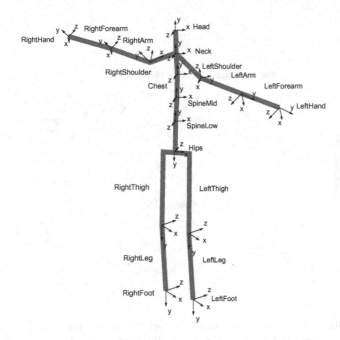

Fig. 1. This figure presents the 16 points over the human body in which three-dimensional rotations in Euler angles are measured. Figure presents also local coordinate systems for each measuring point.

recalculated from Euler angles representation to quaternions and we applied the methodology that was presented in Sect. 2 of this paper. The length of the signals among those ten recordings varied from 226 to 249 frames, due to this the averaged signal had length of 249 frames. We have calculated 100 iterations of DBA method after which results became stable.

In Figs. 2, 3 and 4 we have presented the plots of Euler angles values that describe the rotation in hips measuring point about X, Y and Z axis. The dotted plots are values from input data set (they names begins with Hips in legend of plots). The blue solid line is a result of averaging by algorithm from Sect. 2.2. The red solid line is plot smoothed with algorithm from Sect. 2.3. In Figs. 5, 6 and 7 we have presented the plots of Euler angles values that describe the rotation in RightLeg measuring point about X, Y and Z axis. Dotted plots are values from input data set, solid black lines are results of averaging and red plot are smoothed results from black plot.

We have also prepared visualization of MoCap recordings averaged by our methodology to check if our algorithm did not introduce some visible deviations from expected limbs trajectories. In Fig. 8 we present rendering of important parts of averaged mawashi-geri kick.

To evaluate how similar averaged signals are to the input dataset we have used DTW normalized distance between each time-varying signal (in quaternions) and averaged signal. The normalized distance is defined as a DTW distance divided

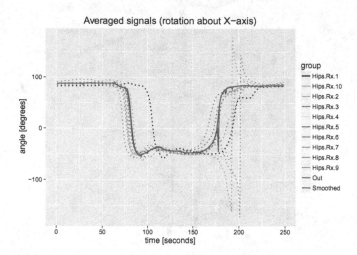

Fig. 2. This figure presents the plots of Euler angles values that describe the rotation in Hips measuring point for about X axis. (Color figure online)

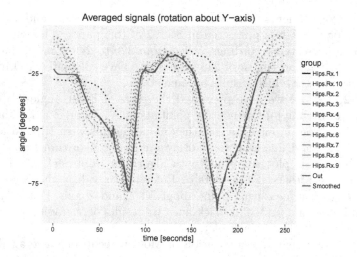

Fig. 3. This figure presents the plots of Euler angles values that describe the rotation in Hips measuring point for about Y axis. (Color figure online)

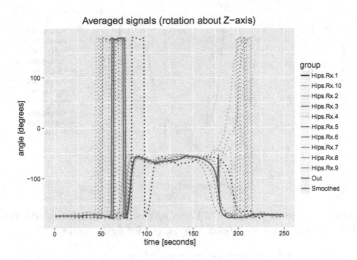

Fig. 4. This figure presents the plots of Euler angles values that describe the rotation in Hips measuring point for about Z axis. (Color figure online)

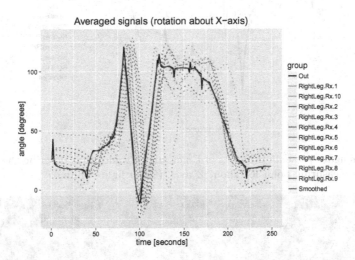

Fig. 5. This figure presents the plots of Euler angles values that describe the rotation in RightLeg measuring point for about X axis. (Color figure online)

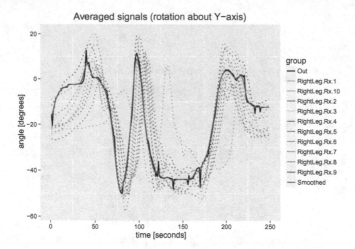

Fig. 6. This figure presents the plots of Euler angles values that describe the rotation in RightLeg measuring point for about Y axis. (Color figure online)

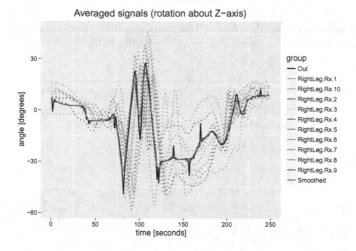

Fig. 7. This figure presents the plots of Euler angles values that describe the rotation in RightLeg measuring point for about Z axis. (Color figure online)

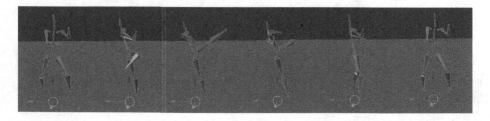

Fig. 8. This figure presents rendering of important parts of averaged mawashi-geri kick done by our algorithm.

Table 1. This table summarizes results from proposed algorithms evaluation.

Feature name	Mean of DTW normalized distance $\cdot 10^{-3}$	Standard deviation of DTW normalized distance $\cdot 10^{-3}$
Hips	0.713	0.181
LeftThigh	0.949	0.602
LeftLeg	0.681	0.419
RightThigh	2.727	2.249
RightLeg	1.996	1.278
SpineLow	0.016	0.011
SpineMid	0.120	0.081
Chest	0.116	0.079
LeftShoulder	2.732	2.447
LeftArm	4.207	2.127
LeftForearm	6.153	5.165
RightShoulder	0.966	0.7039
RightArm	4.097	2.742
RightForearm	3.327	1.392
Neck	1.759	1.231
Head	1.028	0.685

by the sum of length of two signals between which distance is calculated. In Fig. 9 we present those results in the form of heat map with color-coded values. In Table 1 we present means and standard deviations of that comparison grouped by features (measuring points) names.

4 Discussion

As can be seen in Figs. 2 and 4, the proposed method deals very well with non-linearity in rotation description caused by periodicity of Euler angles notation. Also two very similar rotations might be composed of two sets of three rotations about X, Y and Z axis that might have quite different values, for example compare signal 9 and 10 in Figs. 2 and 4 to other signals in those figures. Both those problems are solved thanks to applications of quaternions in DBA averaging. The DBA might introduce some high-frequency noises that are clearly visible in Figs. 2, 3, 4, 5, 6 and 7 as peaks of angle values in relatively smooth angles trajectories. Those noises are unwanted phenomena that should not appear in MoCap recordings which operate in frequency 100 Hz even while dealing with such fast body actions like karate. Those errors are introduced because DBA is a heuristic method. As can be seen on all plots, our smoothing method performed by an

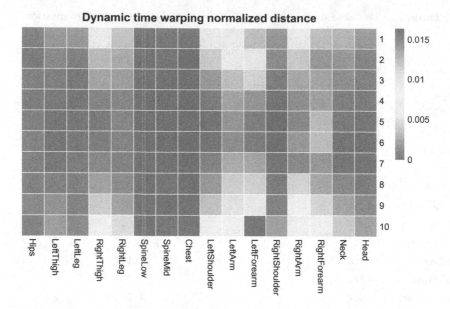

Fig. 9. This figure presents normalized DTW distance between averaged signals and input data in the form of heat map with color-coded values. Each row represents different input recording (there were totally 10 MoCap recordings in our data set) while each column is for different measuring point name.

algorithm described in Sect. 2.3 removes those peaks very well without damaging the overall signal characteristic.

The visualization of MoCap recordings averaged by our methodology was judged by a karate master as very accurate. The averaging did not introduced visible deviations from expected limbs trajectories.

The results presented in Table 1 prove that mean DTW normalized distance between averaged signal and original signals varies from $6.153 \cdot 10^{-3}$ for Left-Forearm sensor to $0.713 \cdot 10^{-3}$ for Hips sensor, which were very good results. The heat map in Fig. 9 shows that the biggest distance between averaged signals and input data is present in measuring points that describe position of hands, especially LeftForearm. This situation is caused by two facts. The first is that arm position is a bit less important than precise lower body movements in this technique and the professional karate athlete might pay a bit less attention to some small variation in his or her hands placement. The second fact was that we have some small MoCap errors introduced during data acquisition in the region of LeftForearm that resulted in less precise angles measurement.

5   Conclusions

In this paper we have presented the algorithm that enables averaging of multiple MoCap signals of the same full body action which is represented as the set of time-varying signals in inverse kinematic. Basing on discussion in previous section, we can conclude that the proposed method seems to be promising for that task. That type of averaging has many important applications. For example it can be used to calculate and visualize an average performance of an athlete who performs some activity that he or she wants to optimize in training. The numerical and visual data may be a very important feedback for the coach that supervises the training [13]. Also our method is not limited to MoCap data averaging; it can be applied to any type of quaternion-based time-varying sequences.

The next step of our researches will be evaluation of our algorithm on significantly larger dataset.

Acknowledgments. This work has been supported by the National Science Centre, Poland, under project number 2015/17/D/ST6/04051.

References

1. Glowacz, A.: Recognition of acoustic signals of synchronous motors with the use of MoFS and selected classifiers. Meas. Sci. Rev. **15**(4), 167–175 (2015). ISSN: 1335–8871. https://doi.org/10.1515/msr-2015-0024
2. Glowacz, A., Glowacz, W., Glowacz, Z.: Recognition of armature current of DC generator depending on rotor speed using FFT, MSAF-1 and LDA. Eksploatacja i Niezawodnosc - Maintenance and Reliability **17**(1), 64–69 (2015). doi:10.17531/ein.2015.1.9
3. Pięta, A., Lupa, M., Chuchro, M., Piórkowski, A., Leśniak, A.: A model of a system for stream data storage and analysis dedicated to sensor networks of embankment monitoring. In: Saeed, K., Snášel, V. (eds.) CISIM 2014. LNCS, vol. 8838, pp. 514–525. Springer, Heidelberg (2014). doi:10.1007/978-3-662-45237-0_47
4. Fabijańska, A., Smurzyński, J., Hatzopoulos, S., Kochanek, K., Bartnik, G., Raj-Koziak, D., Mazzoli, M., Skarżynski, P.H., Jędrzejczak, W.W., Szkiełkowska, A., Skarżyński, H.: The relationship between distortion product otoacoustic emissions and extended high-frequency audiometry in tinnitus patients. Part 1: normally hearing patients with unilateral tinnitus. Med. Sci. Monit. **18**(12), CR765–CRC770 (2012). doi:10.12659/MSM.883606
5. Kalman, R.E.: A new approach to linear filtering and prediction problems. Trans. ASME J. Basic Eng. **82**, 35–45 (1960)
6. Sul, C.W., Jung, S.K., Wohn, K.: Synthesis of human motion using Kalman filter. In: Magnenat-Thalmann, N., Thalmann, D. (eds.) CAPTECH 1998. LNCS (LNAI), vol. 1537, pp. 100–112. Springer, Heidelberg (1998). doi:10.1007/3-540-49384-0_8
7. Burke, M., Lasenby, J.: Estimating missing marker positions using low dimensional Kalman smoothing. J. Biomech. **49**, 1854–1858 (2016)
8. Jin, M., Zhao, J., Yu, J.J.G., Li, W.: The adaptive Kalman filter based on fuzzy logic for inertial motion capture system. Measurement **49**, 196–204 (2014)
9. Petitjean, F., Ketterlin, A., Gançarski, P.: A global averaging method for dynamic time warping, with applications to clustering. Pattern Recognit. **44**(3), 678–693 (2011). http://dx.doi.org/10.1016/j.patcog.2010.09.013

10. Laurent, E., Thomas, D., Maike, B., Gavin, M.: Trajectory box plot: a new pattern to summarize movements. Int. J. Geogr. Inf. Sci. **30**(5), 835–853 (2016). doi:10.1080/13658816.2015.1081205
11. Seto, S., Zhang, W., Zhou, Y.: Multivariate time series classification using dynamic time warping template selection for human activity recognition. In: IEEE symposium series on computational intelligence, SSCI 2015, Cape Town, 7–10 December 2015, pp. 1399–1406 (2015). doi:10.1109/SSCI.2015.199
12. Markley, F.L., Cheng, Y., Crassidis, J.L., Oshman, Y.: Averaging quaternions. J. Guid. Control. Dyn. **30**(4), 1193–1197 (2007). doi:10.2514/1.28949
13. Hachaj, T., Ogiela, M.R., Koptyra, K.: Application of assistive computer vision methods to oyama karate techniques recognition. Symmetry **7**(4), 1670–1698 (2015). doi:10.3390/sym7041670

Plane Refined Structure from Motion

Branislav Micusik[1]([✉]) and Horst Wildenauer[2]

[1] AIT Austrian Institute of Technology, Vienna, Austria
branislav.micusik@ait.ac.at
[2] Zeno Track GmbH, Vienna, Austria
horst.wildenauer@zenotrack.com

Abstract. We present a simple, yet very effective way to enhance the quality of the output of Structure from Motion (SfM) pipelines operating on indoor imagery. Due to the lack of texture and repetitiveness in such environments, the resulting difficulty in frame-to-frame data association presents a challenge and potential error source for state-of-the-art approaches, known to work well for natural scenes. However, in contrast to natural scenery, imagery of man-made structures is often heavily populated by planes, which is a property that can be leveraged to improve overall SfM effectiveness. In our work, we concentrate on the bundle adjustment (BA) stage in the final refinement of structure and motion, where we propose to enforce the planarity constraint of some of the 3D points to obtain improved results. The plane parameters are assumed unknown, and only the association of some of the 3D points to a plane is required. We convert the planarity constraint into the same object space error measure, into the re-projection angular error, allowing for their natural joint treatment. Results on synthetic and real data show significant impact of the proposed constraint for improving both the structure and motion estimates.

Keywords: Structure from Motion · Indoors · Plane · Bundle adjustment

1 Introduction

As a consequence of traits like repetition of texture or general lack thereof, as well as reflections caused by windows and wiry structures, especially indoor environments are acknowledged to be extremely difficult to handle for traditional, point-based SfM systems. The adverse conditions cause image feature detectors and descriptors to perform less reliably, which in turn may introduce detrimental effects into the keystone of SfM pipelines, namely, frame-to-frame correspondence association. Weak correspondences between frames can generate substantial pose errors that are propagated throughout SfM stages, subsequently causing either pose and structure estimates to drift or, even worse, a total failure due to pose loss.

B. Micusik—This research received funding from the Austrian Research Promotion Agencys (FFG) projects LOLOG 840168, LARAH 4586620.

P. Sharma and F.M. Bianchi (Eds.): SCIA 2017, Part I, LNCS 10269, pp. 29–40, 2017.
DOI: 10.1007/978-3-319-59126-1_3

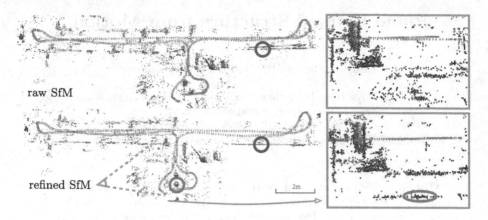

Fig. 1. Raw and the plane refined SfM on the UNIVERSITY sequence. *Left column*: A bird eye view on a 3D reconstruction of an indoor space with in red shown camera poses before and after the refinement. *Right column*: A side view on the 3D model and the estimated camera poses before and after the refinement. The bottom result shows that enforcing some points to lie on a plane, marked by the green circle, improves significantly the original structure and pose estimates. This is seen by the correctly achieved planarity of the estimated camera poses (in red) and the structure (in gray). (Color figure online)

There appears to be a conceptual flaw in the design of traditional SfM pipelines. Indoor and outdoor urban environments exhibit a strongly limited scene structure, which is typically not utilized. The images of indoor scenes show an abundance of planar and perpendicular structures, which constrain the position of the 3D points. Still, with the exceptions of [3,26], the majority of SfM pipelines reconstruct the 3D points without any knowledge about the inherent structure. Like most computer vision problems, SfM is inherently facing a chicken-and-egg dilemma: Knowledge about structure can help to obtain stable poses, and accurate poses lead (in general) to more precise reconstructions [14,26]. For example, the Manhattan [7] or Atlanta world [18] assumption captured by vanishing points of parallel scene lines represent structural constraints utilized for urban navigation [4,8,22]. Similarly, plane-based constraints have been used in the context of SfM, but these are mostly enforced in the dense reconstruction stage [11,15,19]. Only structure is refined accordingly while camera poses are kept fixed. Currently indoor imagery is perceived as being much harder to handle than that of outdoor or natural scenes. However, this is (at least to a certain degree) likely the result of neglecting the prominent scene regularities and the non-use of corresponding constraints.

The fundamental trick to boost accuracy of SfM pipelines is to move on loop like trajectories when mapping the targeted environments and to enforce the loop closure wherever possible. Loop closure eliminates large parts of the error in camera pose estimation that is prevalent in the translational motion component, that is, scale shrinking. Unfortunately, loop closure is often hard to enforce when

Fig. 2. Annotations of the same planes in the images of the UNIVERSITY sequence. Two leftmost images: Points in the manually established green polygons are enforced to lie on the ground plane. Rightmost two: Posters on the opposite sides of a corridor, shown in blue, are to lie on the same vertical plane. None of the parameters of the two planes are known in advance.

dealing with challenging indoor sequences such as office corridors, train station or airport halls. The reason is that the mostly planar patches, which are rich enough in texture and thus usable for interest point detectors, are often seen from widely different vantage points. The difference in geometric appearance is beyond the invariance of the image descriptors, and as result many loop closure events are not recognized. On one side, this causes the measurement matrix to be sparse without loop closed tracks. On the other side, physically identical 3D points are reconstructed more than once, producing so called echoed reconstructed elements. Figure 1 demonstrates where the same plane got reconstructed twice as two close parallel planes. In addition, the camera pose is clearly inaccurate. As a remedy, one could laboriously enforce loop closure by a manual marking of anchor points. The task is however not trivial, since there would need to be sufficient loop closing connections with sufficient strength to pull BA out of a local minimum and towards a more correct solution.

In this paper, we demonstrate that there are simple but strong additional constraints. We propose to force some of the 3D points to lie on the same plane, and thereby provide auxiliary information in addition to automatically detected loop closure correspondence. Annotating a small number of regions lying on a plane in a couple of images, see Fig. 2, is much simpler than establishing loop closure point correspondences, and brings, as we demonstrate, a significant boost in quality of the final structure and motion output. For an example, where multiple echoed reconstructions have been significantly reduced, see Fig. 1. The depiction also indicates how enforcing distant patches to lie on one plane keeps the entire reconstruction consistent. Here, annotating posters on the wall on opposite sides of the long corridor prevented the slight bending and the echoing of the reconstruction.

2   Related Work

Structure from Motion is one of the mature technologies in computer vision. Basically, there are sequential, incremental and global estimation pipelines. Sequential pipelines are closely related to SLAM approaches and build on the assumption that the input sequence is temporally and thus spatially ordered [26]. Incremental pipelines progressively grow a reconstruction by adding a new view at a time, but do not require stringent spatial ordering, see *e.g.* [25]. Global SfM pipelines [21], a more most recent development, aim at building and optimizing the viewing graph from a set of unordered images.

Most approaches are validated on benchmark datasets exhibiting conveniently well-textured scenarios. As the efficiency of conventional SfM approaches are severely hampered in low-texture environments, indoor scenes are often omitted. Generally, it appears that the application to indoor scenarios, despite the practical importance, is not so much in the focus of SfM researchers as it is for the SLAM community.

Feature detection and matching is one of the most important stages in any SfM pipeline, for which reason renewed interest has been shown in the problems persistent in indoor applications. Some works explicitly avoid problematic interest point-based features by building on line [16] or curve segments [17], which are prevalent indoors. The combination of line segments aligned with dominant vanishing points and general point-based features was shown in [19]. The work of [26] builds on the assumption that there are planes in the scene. This is the same assumption as ours, but the authors impose this constraint already in the SfM stage. We discuss the difference in more detail later. Another approach is more hardware related: Employing a very wide angle or omnidirectional camera, see, *e.g.* Google Street View and [23], helps to overcome the problem with a few number of matches and in addition to avoid critical cases when only one plane is visible in the image. One way or the other, often an initial SfM estimate can be obtained, which is sufficiently well behaved for further refinement.

In this work, we show that the raw estimates can be vastly improved by using planarity constraints in a final optimization stage. Imposing planarity directly in the early frame-to-frame association stage, as Zhou *et al.* proposed, is not always reliable in practice and may accumulate drift in case of not perfectly flat surfaces, resulting in instable estimates. In [26] the planarity constraint is locally enforced per a flat patch, as a result of an automatic labeling of feature tracks. If two patches on the same plane appear in the sequence one after the other, they would be considered as two different planes, and the global constraint would be lost. We overcome this issue and show results and benefits when spatially distant planar patches, *e.g.* posters on opposite sides of a long corridor, are still constrained to be on one plane. This introduces a potent global planar constraint which greatly improves the final structure and motion estimate. As a difference to [3], we use a parametrization that relies on the unconstrained SfM for the initialization. Also, we confirm the feasibility of the method on by a magnitude larger datasets.

3   Concept

Let us assume a multiple camera setup with J internally calibrated cameras. The extrinsic parameters of the cameras are represented by rotation matrices

R_j and translation vectors t_j, where $j = 1 : J$. Furthermore, we assume that image point correspondences across cameras are established. These correspondences serve as the basic primitives for utilizing the underlying geometry and to estimate structure and motion parameters simultaneously. The structure is captured by 3D points X_i, $i = 1 : I$, initialized by the triangulation of the image correspondences.

Ignoring camera and point subscripts for simplicity, we may write the projection of a 3D point X to the unit length vector p (the projective ray) as

$$p = \frac{RX + t}{\|RX + t\|_2}. \tag{1}$$

By working with unit vectors lying on the unit sphere, we can cope with perspective as well as with omnidirectional images in the same manner, see [9]. In order to keep the generality of the sensing device, our error functions are accordingly defined in 3D instead of the commonly used image space re-projection error [10].

Having initial values for all R, t, and X, a standard step of a SfM pipeline is the non-linear sparse BA, aiming at minimizing the re-projection or the angular error in our case,

$$\mathcal{S}_1 = \arg \min_{\mathcal{X},\mathcal{R},\mathcal{T}} \sum_{ij} \left\| \tilde{p}_{ji} - \frac{R_j X_i + t_j}{\|R_j X_i + t_j\|_2} \right\|_2^2, \tag{2}$$

where $\tilde{p}_{ji}$ stands for a unit vector computed from the measured image feature point. The attained local minimum of the objective function gives the final estimate of the SfM problem, *i.e.* the set of all 3D points $\mathcal{X} = \{X_i : i = 1 : I\}$, the set of all camera rotations $\mathcal{R} = \{R_j : j = 1 : J\}$, and the set of all translations $\mathcal{T} = \{t_j : j = 1 : J\}$.

In our work, we suppose to be given additional information about some of the points. Specifically, we assume that a number of points lie in the same planes in the scene, see Fig. 3 for an example. Note, however, that the planes, or to be more specific their parameters, are *unknown*. Planarity is one of the common and unique properties of indoor spaces. Unfortunately, this additional constraint is hard to enforce in the triangulation step in case of unknown plane parameters.

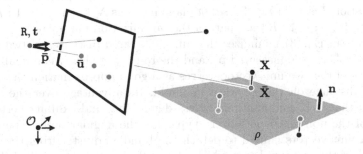

Fig. 3. Concept. A camera observing a scene with general and planar 3D points.

It only can be solved algebraically if either camera or plane parameters are kept fixed [6,12,13]. To get around the problem, we bootstrap the planes by using the output of standard SfM. That is, we initialize the plane parameters using a set of 3D points and subsequently refine them simultaneously with the remaining structure and motion parameters in a non-linear optimization step.

Let the *unknown* plane ρ in 3D be defined by the following equation

$$\rho: \quad \mathbf{n}\bar{\mathbf{X}} + d = 0,$$

where $\mathbf{n}$ denotes the unit plane normal and d the negative distance from the plane to the origin of the coordinate frame. Furthermore, let us assume that some regions in certain images are labeled to be planar patches, see Fig. 2. Then, the 3D points whose image projections fall into the annotated patches are utilized for initialization of the planes in 3D. Specifically, we fit a plane to all the points assigned to one plane in a robust manner, utilizing MLESAC [24] in order to avoid the need to of an explicit inlier-outlier threshold for a point on the plane. This is important because typically the SfM output is up to scale and in classic RANSAC, such a threshold would need to vary for each dataset. To add the planarity constraint into the objective function of Eq. (2) we proceed as follows. The plane labeled 3D points $\mathbf{X}$ is perpendicularly projected into $\bar{\mathbf{X}}$ as

$$\bar{\mathbf{X}} = \mathbf{X} - (\mathbf{n}\mathbf{X} + d)\,\mathbf{n}.$$

Then the corresponding re-projected unit vector $\bar{\mathbf{p}}$ is used instead of the vector $\mathbf{p}$. The new geometric objective function reads as

$$\mathcal{S}_2 = \arg\min_{\mathcal{X},\mathcal{R},\mathcal{T},\mathcal{N},\mathcal{D}} \sum_{ij} w_i \left\| \tilde{\mathbf{p}}_{ji} - \frac{\mathsf{R}_j\bar{\mathbf{X}}_i + \mathbf{t}_j}{\|\mathsf{R}_j\bar{\mathbf{X}}_i + \mathbf{t}_j\|_2} \right\|_2^2 \tag{3}$$

where

$$\bar{\mathbf{X}}_i = \begin{cases} \mathbf{X}_i - (\mathbf{n}_k\mathbf{X}_i + d_k)\,\mathbf{n}_k & \text{if } \mathbf{X}_i \text{ on plane } \rho_k \\ \\ \mathbf{X}_i & \text{if } \mathbf{X}_i \text{ not on plane} \end{cases}$$

is the 3D point, which is either projected onto the plane, or not. The weight w_i reflects the importance of the 3D point. In our experiments, we set it to 2 for the points on the planes, and to 1 otherwise. In addition to the point $\mathcal{X}$, rotation $\mathcal{R}$ and translation $\mathcal{T}$ sets in Eq. (3), set of plane normals $\mathcal{N} = \{\mathbf{n}_k : i = 1 : K\}$ and depths $\mathcal{D} = \{d_k : i = 1 : K\}$ are part of the optimization parameters.

The error in Eq. (3) minimizes the sum of squared distances of two vectors on the unit sphere, the measured $\tilde{\mathbf{p}}_{ji}$ and the re-projected $\bar{\mathbf{p}}_{ji}$. For small errors, the distance of the two unit vectors serves as a good approximation the angular (geodesic) distance on the spherical manifold. Its advantage over the angular measure is that it is computationally less demanding and exhibits better convergence of the non-linear optimizer. Typically, the angular object space error defined on unit vectors is used to detach the bundle adjuster from the projection model of the sensing device which allows thus easier handling of different omnidirectional cameras.

Note that the non-planarity penalty, the distance between $\mathbf{X}$ and $\bar{\mathbf{X}}$, is transformed into the (approximated) angular error in order to naturally combine the plane enforced points with the general ones. In comparison to [26] our structure and motion recovery geometric objective function looks different. It directly minimizes over 3D points like the standard bundle adjustment which allows to combine planar and non-planar points more feasibly. In [26] the non-planar points add a constant penalty which is difficult to tune.

3.1   Solution

To solve the minimization problem in Eq. (3) we use the CERES solver [1]. Before the optimization, the usual trick is applied to increase numerical stability, *i.e.* the point cloud is centered on the origin of the coordinate frame and its standard deviation is scaled to 1. The rotation matrix is parametrized by the axis-angle representation, the translation vector and the 3D points both as 3 vectors, the plane depth as a scalar. The unit plane normal is parametrized with two scalars as suggested in [10] (A6.9.3). First the normal $\mathbf{n}$ is transformed via the Householder matrix $H_{\mathbf{v}}$ as $\mathbf{a} = H_{\mathbf{v}}\mathbf{n}$ to $(0, 0, 1)^{\top}$, where the minimal parametrization behaves locally stable. The Householder matrix is computed as

$$H_{\mathbf{v}} = I - 2\mathbf{v}\mathbf{v}^{\top}/\mathbf{v}^{\top}\mathbf{v}$$

with

$$\mathbf{v} = \mathbf{n} - (0, 0, 1)^{\top}.$$

Then, the way from the unit vector to the minimal representation $\mathbf{a} \rightarrow \mathbf{m}$ is

$$\mathbf{m}_{2\times 1} = \frac{2\mathrm{asin}(\|(a_x, a_y)^{\top}\|)}{\|(a_x, a_y)^{\top}\|} \begin{pmatrix} a_x \\ a_y \end{pmatrix}$$

and the way from the minimal representation to the unit vector $\mathbf{m} \rightarrow \mathbf{a}$ is

$$\mathbf{a}_{3\times 1} = \left(\frac{1}{2}\mathrm{sinc}(\frac{\|\mathbf{m}\|}{2})\mathbf{m}^{\top} \;,\; \cos(\frac{\|\mathbf{m}\|}{2}) \right)^{\top}$$

and $\mathbf{n} = H_{\mathbf{v}}\mathbf{a}$ since $H_{\mathbf{v}}^{-1} = H_{\mathbf{v}}$.

In preliminary experiments, this minimal parametrization was to be advantageous for the convergence of the minimization problem in Eq. (3), where it outperformed the naïve (also minimal) two angle parametrization of a unit 3D vector.

4   Experiments

4.1   Synthetic Data Experiment

In order to investigate the impact of the plane constraint under noisy conditions, we performed the following simulations. We generated 20 cameras at every 30

36   B. Micusik and H. Wildenauer

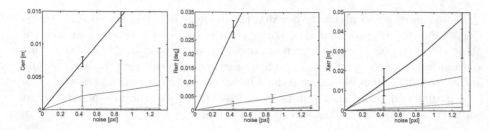

Fig. 4. Error distribution on camera centers, rotations, and 3D points. In **black** is the error without BA, in red the classic BA, in green is the proposed plane constrained BA, and in **blue** is the proposed plane constrained BA with fixed ground truth plane parameters. The contribution of the planarity constraint is evident on all three errors. (Color figure online)

cm along a line, parallel to the ground plane and three sets of 3D points. Each set contains 40 points, where one set contains general points, one set points on the ground plane and one set points on the vertical wall. We added noise to the camera centers (0, 5, 10, 15 mm), to camera rotations (0, 0.02, 0.04, 0.06 deg), and to image measurements (0, 0.45, 0.90, 1.35 pxl). At each noise level, the experiment was repeated several hundred times. At each value set, we linearly triangulated the 3D points using noisy image measurements and noisy camera poses. The resulting 3D points were then used to fit the two planes. The plane parameters along with the camera poses, 3D points, and image measurements were fed into the BA in Eq. (3).

Figure 4 shows the result. As expected, the case without BA is the worst. Classic BA worked better, but was outperformed significantly by the proposed plane constrained bundler with mutable plane parameters. We also considered the case when the equations of the plane are known. This corresponds to a very practical setup where planar markers are placed on the ground plane and where the camera is calibrated *w.r.t.* the ground plane and undergoes a planar motion. This setup can be found typical in mobile robotics.

Fig. 5. CASTLE-P30 dataset [20]. Two rightmost images show the annotation of a common plane enforced as a constraint.

4.2 Real Data Experiments

Indoor spaces are difficult for classic SfM pipelines due to poor or very repetitive texture, wiry structures, and reflections. To handle such environments we

Table 1. Accuracy of the pose estimation on the benchmark datasets of our line based to two point based SfM methods.

	CASTLE-P30	
	R_{err} [°]	C_{err} [mm]
Sweeney *et al.* [21]	–	32.4
VSFM [25]	0.097	29.9
VSFM [25] with plane	0.088	23.9
Our SfM	0.090	26.3
Our SfM with plane	0.089	25.4

therefore use an omnidirectional camera with very wide (more than 180°) FOV fisheye lens. We use our own sequential SfM to cope with such a lens as *e.g.* the standard VisualSFM pipeline of [25] did not deliver plausible result in the real indoor sequences. In order to guarantee to be at the state-of-the-art level and such that the results are not biased by using an out-of-date pipeline we first compared our sequential SfM pipeline which we use for all real experiments to the classic VisualSFM [25]. For that purpose we used the benchmarking CASTLE-P30 dataset of Strecha *et al.* [20].

CASTLE-P30. The dataset of 30 images inside a castle courtyard, shown in Fig. 5, provides the ground truth necessary for the quantitative comparison. Table 1 summarizes the result and Fig. 5 shows a plane which was annotated in two images in order to enforce the proposed planar constraint. The result shows that a simple plane constraint enforced only in two images already helps to estimate camera poses more accurately. This dataset is a gold standard dataset used to evaluate SfM pipelines where state-of-the-art SfM pipelines perform already very well. Despite the fact that the scene is not an indoor space which we target at, the impact of the plane constraint is evident.

UNIVERSITY. This dataset represents a typical difficult indoor environment, see some representative pictures in Figs. 2 and 6. The camera mounted on a mobile robot moves along a loop like trajectory, starting and ending in the same room, making a loop in 12 m long corridor. It contains the classic difficult maneuver for sequential SfM and SLAM pipelines, passing an open door. This transition

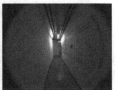

Fig. 6. UNIVERSITY dataset.

from one room to another often breaks down state-of-the-art approaches. The floor, ceiling and the walls exhibit little texture with the exception of a small number of posters on the wall. Under these hard conditions, using a wide angle lens was a necessity. One plane constraint was enforced on the small marker in the starting room, Fig. 2, and one constraint on the two posters on the walls. Figure 1 demonstrates the qualitative results of the raw and the refined SfM. The plane constraints significantly helped to polish the 3D structure and the camera poses. It can be seen that planar structures which were not forced to be planar became so. Also, the camera underwent a planar motion. This was, however, not enforced in the estimation process, but rather served as a criterion to judge the quality of the SfM output. It is evident from Fig. 1 that the plane constraint reduced the bias in the pose estimate significantly, and that the poses became planar after the refinement stage.

TRAIN STATION. This dataset represents a large indoor environment of about 70 m length, showing a train station. The environment contains many repetitions, as well as glass surfaces resulting in reflections and wiry structures. Here, classic SfM trying to match all the images together in order to build the viewing graph [21, 25] failed, most likely because of the repetitiveness of the environment. The train

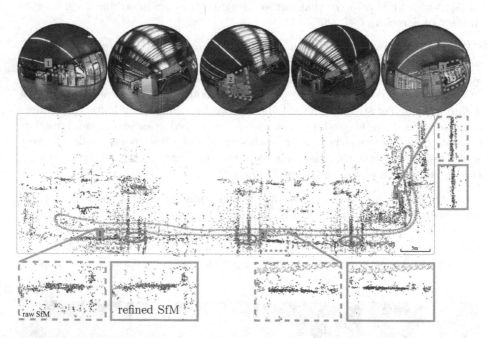

Fig. 7. Result on TRAIN STATION. Top row shows some of the images of the sequence with annotated planar patches, shown in green. Some of the regions, marked by blue dashed quadrilaterals are shown in detail in the 3D model. In red is the camera trajectory. Note that the points in the blue rectangles were not constrained or forced to be planar. They became planar as the consequence of the planarity constraint on the points in the green regions. (Color figure online)

station consists of almost identical subparts which misled the matching process. Our sequential SfM can cope with such a problem as the viewing graph is given by the ordering of images in the sequence. Figure 7 shows qualitative results before and after applying the plane constraint. The plane constraint was enforced on three small platform signs only. This constraint not only corrected planarity of the signs in the 3D model, but did so for other patches shown in blue in Fig. 7. This is a strong indication of the strength of the constraint: a small planar patch can correct the camera pose such that other patches are correctly reconstructed as well.

4.3 Discussion

Manual image annotation for enforcing planarity constraints might be seen as a limiting factor. However, the annotation is sufficient to be done in one image, is only region wise, and is much easier than establishing *e.g.* point correspondences across multiple images. To automate this process, one of the ways to enforce homography between two images in order to detect features on the same plane, *e.g.* [5], can be used. Another practical, more controlled, way would be to place well detectable and recognizable planar markers in he scene, *e.g.* those in ARToolKit [2]. Those could be distributed in the scene on a chosen set of planes and be planarity constrained accordingly in the final refinement step.

5   Conclusion

We presented an effective way to enhance the quality of the output of SfM pipelines for difficult indoor scenery. To this end, we proposed to enforce the planarity constraint of some of the 3D points in the final bundle adjustment refinement stage. Here, plane parameters are assumed unknown, and only the association of some of the 3D points to a plane is required. The association to planes represents somewhat weaker information than point correspondences, but is less laborious to be annotated than *e.g.* the latter. Also, enforcing planarity can complement missing loop closure constraints in hard indoor environments.

The explicit utilization of structural constraints demonstrates that planar structures can leverage poorly performing, general SfM pipelines to better levels. This is especially important when SfM output is utilized by visual navigation methods, where accurate 3D maps are a prerequisite.

References

1. Agarwal, S., Mierle, K., et al.: Ceres solver (2010). https://code.google.com/p/ceres-solver/
2. ARToolKit: Open source augmented reality SDK (2000). http://artoolkit.org
3. Bartoli, A., Sturm, P.: Constrained structure and motion from multiple uncalibrated views of a piecewise planar scene. IJCV **52**(1), 45–64 (2003)
4. Camposeco, F., Pollefeys, M.: Using vanishing points to improve visual-inertial odometry. In: Proceeding of ICRA, pp. 5219–5225 (2015)

5. Chum, O., Werner, T., Matas, J.: Two-view geometry estimation unaffected by a dominant plane. In: Proceedings of CVPR (2005)
6. Chum, O., Pajdla, T., Sturm, P.F.: The geometric error for homographies. CVIU **97**(1), 86–102 (2005)
7. Coughlan, J.M., Yuille, A.L.: Manhattan world: orientation and outlier detection by Bayesian inference. Neural Comput. **15**(5), 1063–1088 (2003)
8. Elqursh, A., Elgammal, A.M.: Line-based relative pose estimation. In: Proceedings of CVPR, pp. 3049–3056 (2011)
9. Geyer, C., Daniilidis, K.: Catadioptric projective geometry. Int. J. Comput. Vis. (IJCV) **45**(3), 223–243 (2001)
10. Hartley, R., Zisserman, A.: Multiple View Geometry in Computer Vision. Cambridge University Press, Cambridge (2000)
11. Ikehata, S., Yan, H., Furukawa, Y.: Structure indoor modeling. In: Proceedings of ICCV (2015)
12. Kanatani, K.: Statistical Optimization for Geometric Computation, Theory and Practice. Dover Publications, New York (1996)
13. Kanatani, K., Niitsuma, H.: Optimal two-view planar scene triangulation. In: Kimmel, R., Klette, R., Sugimoto, A. (eds.) ACCV 2010. LNCS, vol. 6493, pp. 242–253. Springer, Heidelberg (2011). doi:10.1007/978-3-642-19309-5_19
14. Ke, Q., Kanade, T.: Transforming camera geometry to a virtual downward-looking camera: robust ego-motion estimation and ground-layer detection. In: Proceedings of CVPR, pp. 390–397. IEEE Computer Society (2003)
15. Micusik, B., Kosecka, J.: Multi-view superpixel stereo in urban environments. IJCV **89**(1), 106–119 (2010)
16. Micusik, B., Wildenauer, H.: Descriptor free visual indoor localization with line segments. In: Proceedings of CVPR (2015)
17. Nurutdinova, I., Fitzgibbon, A.: Towards pointless structure from motion: 3D reconstruction and camera parameters from general 3D curves. In: Proceedings of ICCV (2015)
18. Schindler, G., Dellaert, F.: Atlanta world: an expectation maximization framework for simultaneous low-level edge grouping and camera calibration in complex man-made environments. In: Proceedings of CVP, pp. 203–209 (2004)
19. Sinha, S.N., Steedly, D., Szeliski, R.: A multi-stage linear approach to structure from motion. In: Kutulakos, K.N. (ed.) ECCV 2010. LNCS, vol. 6554, pp. 267–281. Springer, Heidelberg (2012). doi:10.1007/978-3-642-35740-4_21
20. Strecha, C., von Hansen, W., Van Gool, L., Fua, P., Thoennessen, U.: On benchmarking camera calibration and multi-view stereo for high resolution imagery. In: Proceedings of CVPR (2008)
21. Sweeney, C., Sattler, T., Höllerer, T., Turk, M., Pollefeys, M.: Optimizing the viewing graph for structure-from-motion. In: Proceedings of ICCV (2015)
22. Teller, S., Antone, M., Bodnar, Z., Bosse, M., Coorg, S., Jethwa, M., Master, N.: Calibrated, registered images of an extended urban area. IJCV **53**(1), 93–107 (2003)
23. Torii, A., Havlena, M., Pajdla, T.: From Google street view to 3D city models. In: Proceedings of ICCV Workshop on Omnidirectional Vision, Camera Networks and Non-classical Cameras (2009)
24. Torr, P., Zisserman, A.: MLESAC: a new robust estimator with application to estimating image geometry. CVIU **78**(1), 138–156 (2000)
25. Wu, C.: VisualSFM: a visual structure from motion system (2011)
26. Zhou, Z., Jin, H., Ma, Y.: Robust plane-based structure from motion. In: Proceedings of CVPR (2012)

A Time-Efficient Optimisation Framework for Parameters of Optical Flow Methods

Michael Stoll[✉], Sebastian Volz, Daniel Maurer, and Andrés Bruhn

Institute for Visualization and Interactive Systems,
University of Stuttgart, Stuttgart, Germany
{stoll,volz,maurer,bruhn}@vis.uni-stuttgart.de

Abstract. Due to the increase of optical flow benchmark data, concerning both amount and resolution, learning parameters from training sequences with ground truth has become significantly more challenging in recent years. Moreover, optical flow methods are much more complex than a few years ago resulting in a larger amount of model parameters and a noticeably increased runtime. As a consequence, even optimising a small set of suitable parameters may take hours or even days which makes hand tuning infeasible. Hence, time-efficient strategies for automatic parameter optimisation become more and more important. In this context, our work addresses three important aspects. First, we provide an overview of different optimisation strategies and juxtapose them in the context of different optical flow methods and different evaluation benchmarks. Second, we focus on choosing a suitable subset of the training data to speed up the computation while still obtaining meaningful results. Finally, we also consider different strategies for distributing the evaluation on hardware infrastructures which allows to further reduce the run time. Experiments show that the proposed methodology allows to obtain good results while keeping the overall effort reasonably low.

Keywords: Performance evaluation · Parameter optimisation · Distributed optimisation · Adaptive scheduling · Optical flow

1 Introduction

Among the variety of computer vision tasks, optical flow estimation sticks out in some important aspects. First of all, it is applied to sequences of images, in contrast to single images. Secondly, it covers a 2D search space (in contrast to e.g. 1D stereo problems). Additionally, there is already a variety of benchmarks which allow for a comparison of different methods.

Evaluating an optical flow method is usually done by comparing its performance quantitatively to other methods. Recent benchmarks provide training datasets containing image sequences *with* ground truth data and testing datasets containing only image sequences. The estimation results are evaluated against the given ground truth by computing error measures. Depending on the benchmark, different error measures such as the *average angular error* (AAE), the

© Springer International Publishing AG 2017
P. Sharma and F.M. Bianchi (Eds.): SCIA 2017, Part I, LNCS 10269, pp. 41–53, 2017.
DOI: 10.1007/978-3-319-59126-1_4

average endpoint error (AEE) or the *average bad pixel* (BP3, using a 3 pixel threshold) are chosen for comparison. Hence, the typical optimisation task comes down to choosing a benchmark and the corresponding error measure and finding the parameters that minimise the overall error value.

In the early days of quantitative evaluation, there was a single artificial image sequence with ground truth, the *Yosemite* sequence [2] having a resolution of 316×252 pixels. In 2007, the Middlebury benchmark was presented providing 8 training and 8 testing datasets with resolutions from 316×252 to 640×480 pixels [1]. Most of these image sequences have been created synthetically, not providing realworld challenges such as significant illumination changes, considerable out-of-plane motion or disturbances in data such as lens flares, under- and oversaturations or noise. These types of challenges are provided by the KITTI Vision benchmark which was proposed in 2012 [10]. It contains 194 training and 195 testing datasets covering resolutions from 1226×370 to 1238×374 (training data) and 1242×375 pixels (testing data), respectively. In the same year, 2012, also the MPI Sintel Flow dataset was proposed, which contains artificial data on the one hand, but provides tough challenges by creating specular reflections, motion blur, defocus blur and atmospheric effects [6]. There are 1064 training and 564 testing image datasets at a resolution of 1024×436 pixels. The Middlebury benchmark provides two important rankings, one w.r.t. the *AAE* and another w.r.t. the *AEE*. The MPI Sintel benchmark makes use of the *AEE* while the KITTI benchmark ranks methods according to their average BP3 error.

There is an increasing amount of data – provided by publicly available benchmarks – and there are optical flow methods with increasing complexity (e.g. [23–26]) – developed by researchers. But to the best of our knowledge there is no easy-to-use publicly available framework that connects data and methods by finding suitable parameters in a reasonable amount of time. An open-source variant that provides a basis of common parameter optimisation algorithms which can be extended through contributions by experienced researchers could be valuable for the whole community.

Contributions. We present such a framework which does not require any specific properties of the underlying objective function (such as e.g. its derivatives) and by design can be generalised to other computer vision tasks comprising a scalar error measure. In this context our contributions are fourfold: First, we briefly introduce different classes of derivative-free algorithms for parameter optimisation which are juxtaposed later in the evaluation. Second, we reduce the computational burden by applying a selection strategy in order to determine suitable subsets of the datasets for evaluation. Third, we argue how all the individual sampling steps can be distributed efficiently among different computers within the typical hardware infrastructure of a research institute. Finally, we provide an open-source framework for both, single-machine- as well as distributed parameter optimisation with easy adaptability and extendibility regarding both the interfaces to the optical flow methods and the implementation of derivative-free algorithms for parameter optimisation. The framework is publicly available at http://go.visus.uni-stuttgart.de/cvis-optimizer/.

Related Work. While there are a few methods that estimate their parameters jointly with the optical flow [12,17], most of the current algorithms rely on some kind of a-priori parameter selection based on training data; see e.g. [5,9,15,22,26]. Surprisingly, most of these approaches do not comment on the underlying optimisation strategy. The few exceptions can essentially be divided into two classes: On the one hand, there are *derivative-based* approaches that rely on gradient descent or Newton-like techniques to estimate the optimal parameters as minimiser of a cost functional that relates the obtained results to ground truth data [15,22,26]. However, since the underlying optical flow models are rather complex – in contrast to parameter estimation for image restoration [14,21] – the functional gradient is typically not computed analytically but approximated stochastically [15,22,26]. On the other hand, there are *derivative-free* techniques that are solely based on multiple evaluation runs for different parameter settings. Although some of those methods implicitly look for the downhill direction, they can also be applied in those cases where no analytic or stochastic gradient is available. Typical representatives for this class of method are the multidimensional sampling [8] as well as the Downhill Simplex method [9,16]. Moreover, also evolutionary algorithms such as the Covariance Matrix Adaptation Evolution Strategy [11,20] or nature inspired algorithms [18] belong to this class.

The only related work regarding the comparison of parameter optimisation methods for optical flow estimation is given by [18]. However, the current paper improves upon this work by addressing significantly more aspects. In particular, it considers three different classes of derivative free methods (fixed sampling, geometric sampling, stochastic sampling), it evaluates the performance for two optical flow algorithms, it considers all major benchmarks, it investigates the usage of suitable subsets of image sequences and it elaborates on the distribution of evaluation runs on homogeneous and heterogeneous hardware infrastructures.

There is also a lot of work in the context of auto-tuning of image processing and computer vision algorithms like e.g. [7,13,19] which are helpful to split up an image processing algorithm into parallelisable pieces and schedule them on a multicore system. Our goal, however, is not to split up an algorithm and schedule its parts in the context of parallelization. We rather want to distribute multiple, independent executions of an algorithm (i.e. optical flow estimations) among the available hardware, measured by the number of CPU cores in one or more computing systems. Regarding distributed optimisation, [3] provides domain decomposition schemes for general optimisation problems, discusses such examples in statistics and machine learning and gives hints on the implementation of general distributed optimisation. It does, however, not focus on the scheduling of the distributed tasks. [4] gives an overview of different scheduling heuristics to map independent tasks on heterogeneous distributed computing systems.

Organisation. Section 2 reviews different derivative-free strategies for parameter optimisation. Moreover, it discusses how to speed up the computation by selecting a suitable subset of the training sequences. Section 3 then sketches the distributed computation, while Sect. 4 presents an evaluation of the different strategies. Section 5 concludes with a summary.

2   The Optimisation Process

For an optimisation problem with objective function $f(\mathbf{x})$, the goal is to find the parameters $\mathbf{x}_{\min} \in \mathbb{R}^P$ that minimise f where P is the number of parameters. In case of optical flow, this objective function depends on the optical flow method, its parameters $\mathbf{x}$, the training data with ground truth and the respective error measure d, e.g. AEE, AAE or BP(3). Please note that the objective function is often nonconvex w.r.t. the parameters such that one typically relaxes the task to finding a suitable local minimum. In the following, the whole optimisation process consists of many *optimisation runs*, each evaluating f on the whole dataset with a different parameter set $\mathbf{x}$. Each optimisation run, in turn, consists of several *evaluation tasks*, each evaluating d on a single element i of the dataset yielding the deviation $d_i(\mathbf{x})$. The value of the objective function is then given by $f(\mathbf{x}) = \text{avg}_i\, d_i(\mathbf{x})$, which we call the *error value* of the optimisation run.

2.1   Continuous Parameters

In order to find optimal parameters that can be sampled from a continuous space, different derivative-free parameter optimisation methods are known that solely rely on the values of the objective function $f(\mathbf{x})$ at specific sampling points $\mathbf{x}_i$. These methods can be further classified according to the heuristics they use. Examples for methods without a heuristic are equidistant (cascadic) sampling or logarithmic cascadic sampling [8]. Given a sampling interval, they compute sample points and determine the best result. In the cascadic case, they repeat this procedure with a smaller interval around the best sampling point so far. An example for a method with a geometric heuristic is the so-called Downhill Simplex method, a.k.a. Nelder-Mead method [16], which constructs an initial simplex of sampling points in $\mathbb{R}^P$ and moves this simplex in a downhill manner towards a minimum by different geometrically inspired steps which are called *reflection, expansion, contraction* and *reduction*. The Covariance Matrix Adaptation Evolution Strategy (CMA-ES) [11] is an example for a method with a stochastic heuristic. It belongs to the class of evolutionary algorithms, i.e. is based on populations of parameter sets and consists of two iterated main steps: *mutation* and *recombination*.

We will use these methods exemplarily in the evaluation. However, any other derivative-free method is also implementable in our system.

2.2   Discrete Parameters

Besides the continuous parameters like e.g. the weights of different assumptions in a model, there can also be discrete parameters like booleans or enumerations which can only take a finite, a priori known set of values. These discrete parameters are separated from the continuous ones, all possible combinations are computed and for each set of discrete parameters the remaining continuous parameters are optimised using one of the strategies from above.

2.3 Reducing the Amount of Evaluations

Instead of using all datasets from provided image and ground truth data, it may be sufficient to use a subset of them for the parameter optimisation. The question then arises which datasets to keep.

As the total error value is the average error of each dataset w.r.t. the chosen error measure, a promising approach is to take a look at those datasets that have the highest variance among the error values for an initial set of optimisation runs, thus being sensitive to parameter variations and influencing the variations of the average error the most. Hence, we evaluate an initial set of optimisation runs (i.e. the first level of a cascade, the first simplex or the first population) on all given datasets, determine the range r_i of error values for each dataset i and - given a percentage value q as a parameter - we select those q percent of datasets with the highest r_i and re-compute the average error values using only the selected datasets. After the parameter optimisation strategy has terminated, we compute the errors for the ignored datasets using the estimated parameters and re-compute the average error using all datasets in order to obtain an average error for the complete given dataset.

3 Distributed Computation of Evaluation Tasks

After each step of the chosen parameter optimisation algorithm there is a set of evaluation tasks to be performed. Each of these tasks can be computed independently from the others and thus may be distributed to a separate thread. These threads can either run on the same computer or in a distributed fashion on different computers. Taking into account only one computer means that each thread of a thread pool can simply fetch a task, evaluate it and fetch the next one as long as there are evaluation tasks left.

Using more computers, however, makes things more complicated. On the one hand, one needs a central instance managing the optimisation process and clients that take care of the evaluations as well as a communication layer between them. On the other hand, different computers may show different performance. Hence, the question arises how to distribute the tasks such that they are computed as fast as possible. In case each client is allowed to evaluate tasks as long as there are any left - named as *Opportunistic Load Balancing (OLB)* [4] - it may happen that the slowest client fetches the last task while faster clients stay unemployed. For an optimisation run, this leads to a longer computation time than necessary.

3.1 Minimal Completion Time Algorithm

The *Minimal Completion Time (MCT)* algorithm [4] is a heuristic that uses an estimation of the run time for a task in order to assign it to the client which is assumed to have the smallest time until it completes all computations. As the highest completion time (CT) among all clients coincides with the completion of the current bunch of evaluation tasks, achieving a minimal CT avoids unnecessary waiting for the tasks to be finished. To this end, we need to keep track of the performance of all clients, measured by the average run time t_{avg}.

The algorithm is a greedy algorithm. It iterates through all available tasks and assigns each task to the client whose estimated CT - including the task it is currently computing - would be the smallest one. The CT of the client is the maximum CT among all its threads. For the computation of the CT, the current task is virtually assigned to the thread with the smallest CT as we expect this thread to first acquire the next task. This way, a small amount of (remaining) tasks is assigned to fast clients instead of slow clients.

Timing Statistics. The central quantity is the average evaluation time t_{avg} of each client. Upon the first acquisition of a task by the client, it is initialised with a small value bigger than zero in order to get a reasonable first distribution of the tasks. Otherwise, tasks would probably be assigned to only the first client or the one with $t_{avg} = 0$. On the first submission of an evaluation result, t_{avg} is set to the evaluation time t_{curr} of the submitted task. In subsequent submissions, we update t_{avg} with a weighted averaging:

$$t_{avg} = \left(1 - \frac{1}{q}\right) \cdot t_{avg} + \frac{1}{q} \cdot t_{curr} \tag{1}$$

We set $q = 6$ in order to make the estimation of t_{avg} robust against outliers.

If the run times are not expected to be rather equal for all datasets (e.g. due to a different resolution of the data), a so-called run time factor can be assigned to those datasets that lead to a significantly different evaluation effort. The evaluation time t_{curr} is then normalised before the computation of t_{avg}.

The performance of a client may undergo variations, e.g. due to workload, thermal issues etc. In particular, it might become significantly slower than estimated. In order to be robust against this, we also keep track of the time since the last task acquisition or submission, respectively, which we call $t_{silence}$. In the worst case, a client evaluates only tasks with the maximal run time factor and may thus be expected to be busy for $t_{busy} = \max(R) \cdot t_{avg}$ (if R denotes the set of all run time factors). In case, it is not interacting for a longer time (i.e. $t_{silence} > t_{busy}$) we assume a decrease in performance, thus replacing t_{avg} by an effective average evaluation time $t_{eff} = \frac{t_{silence}}{\max(R)}$.

If there are multiple optimisations at the same time, they can involve different methods and/or different data leading to different average run times. Computing only a single, joint average run time t_{avg} could lead to strong over- or underestimations of the run times of different evaluation tasks, thus leading to inappropriate task distributions to the available clients. Hence, the average run time t_{avg} is measured for each optimisation separately.

4   Evaluation

In order to perform an evaluation of our optimiser, we applied it to the publicly available methods of Brox and Malik (LDOF) [5] and Weinzaepfel et al. (DF)

[26]. The datasets are the eight sequences of the Middlebury training dataset (MB) [1], a subset of 20 sequences of the KITTI training dataset (seq. 0–19) [10] and a subset of 69 sequences of the Sintel training dataset [6] (final version, using the middle image pair as well as the pairs five frames before and five frames after the middle of each scene). The restriction to these subsets of KITTI and Sintel, which we expect to be representative, had to be performed in order to make the evaluation, which incorporates a large amount of optimisation processes, feasible. Otherwise, each single optimisation process on KITTI or Sintel would have already taken several days of run time. Please note, that we only evaluate on these subsets and they thus serve as starting point for the proposed selection of even smaller subsets in Sect. 4.2. Furthermore, we restrict the optimisation to those parameters that define weights for different terms in the functional (3 parameters for LDOF, 4 parameters for DF). The initial search interval for all parameters was $[0, 200]$ (or they were initialised with 100 for CMA-ES which constitutes the centre of the interval $[0, 200]$) except for the matching term weights which take the discrete values 200, 300 or 400. Both equidistant cascadic sampling and logarithmic cascadic sampling use five samples per parameter and four cascade levels.

4.1 Comparison of the Parameter Optimisation Methods

In our first experiment, we compare the different combinations of optical flow methods, optimisation methods and benchmarks in terms of both the achieved error and the required number of optimisation runs. Table 1 shows both values where the number of optimisation runs can be found in brackets. As one can see, CMA-ES performs best for DeepFlow (DF) but it gets trapped in a local minimum for LDOF. In contrast, the logarithmic cascadic approach avoids this minimum and provides the best result for this optical flow method. Regarding the number of optimisation runs, all parameter optimisation strategies stay in the same order of magnitude for LDOF (3 parameters) while things change for DeepFlow (4 parameters). Here, the equidistant and logarithmic sampling strategies need one order of magnitude more runs than Downhill Simplex (DS). In all cases CMA-ES requires a medium amount of runs. Regarding both error and number of optimisation runs DS is the favourable method: It does not achieve the absolute best results but stays within a range of less than 1% from the top results while only requiring at most 25% of the number of runs compared to CMA-ES. Moreover, its results are superior to the baseline that has been computed using the specified parameters in the paper. The only slight exception is DeepFlow on the Middlebury dataset but the baseline parameters ignore the - in this setting - inappropriate matching term which we however keep active.

For the DeepFlow algorithm on the KITTI training subset, Fig. 1 displays how the individual algorithms converge to their final error value while the number of runs is growing. Please note, that we only show a single instance of each algorithm, i.e. for the best discrete choice of the matching weight (being either 200, 300 or 400). We can observe that logarithmic cascadic sampling passes a

Table 1. Error values and number of optimisation runs for different methods and benchmarks.

	Equid. (C)	Log. (C)	DS	CMA-ES	Baseline
LDOF (MB, AAE)	4.349 (240)	**3.976** (300)	3.993 (106)	4.318 (763)	4.100
DF (MB, AEE)	0.252 (1344)	0.251 (1500)	0.252 (138)	0.251 (552)	**0.250**
DF (KITTI, BP)	8.340 (1341)	8.335 (1500)	8.296 (201)	**8.292** (880)	9.303
DF (Sintel, AEE)	8.732 (1341)	8.743 (1500)	8.739 (144)	**8.720** (848)	11.026

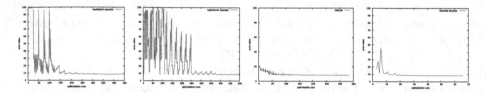

Fig. 1. Convergence of the parameter optimisation strategies. **From left to right:** equidistant cascadic sampling, logarithmic cascadic sampling, CMA-ES, DS.

large variety of error values indicating that it tests many different orders of magnitude of ratios of the different weighting parameters. In contrast, equidistant cascadic sampling only covers a small range of error values despite some single peaks. CMA-ES already starts at a low error level and successively oszillates to the minimum while DS already converges after a few optimisation runs.

4.2   Influence of Using Subsets

Our second experiment evaluates the effect of reducing the amount of evaluation tasks per optimisation run by selecting a suitable subset of the dataset for evaluation. To this end, we automatically choose this subset according to the highest variance (cf. Sect. 2.3) and consider both, the resulting errors and the corresponding number of evaluations. Table 2 shows that using around 10% of all datasets leads to only a moderate degradation. In this context, the loss in quality is considerably higher if either the baseline dataset is already small (only 20 KITTI sequences) or the method does not perform many optimisation runs like DS, i.e. if the total number of evaluations has already been small. In the other cases, the loss in quality is not that significant. When using around 20% of the baseline dataset for evaluation, there is hardly any degradation - in particular for CMA-ES or Sintel. It is worth noting that the amount of optimisation runs can change depending on the subset due to different adaptions of the respective algorithm to the intermediate errors. In some cases, using less sequences may even lead to more evaluations (cf. Table 2, KITTI (DS), last two rows).

Table 2. Using the sequences with the **highest** variances of the error values for optimisation. The entries have the format *error (#runs, #evaluations)*.

	KITTI (CMA-ES)	KITTI (DS)	Sintel (CMA-ES)	Sintel (DS)
100%	8.292 (880, 17600)	8.296 (201, 4020)	8.720 (848, 58512)	8.739 (144, 9936)
50%	8.311 (960, 9690)	8.429 (198, 2030)	8.722 (864, 29691)	8.737 (159, 5581)
20%	8.322 (920, 3824)	8.920 (138, 632)	8.760 (832, 11320)	8.737 (129, 1957)
10%	8.734 (936, 2034)	9.031 (189, 652)	8.777 (840, 5607)	8.931 (165, 1305)

4.3 Task Distribution Strategies

In our final experiment, we compare the average optimisation times for DeepFlow with the Downhill Simplex method for different levels of parallelism. To this end, we compute the times of an optimisation process on a typical office computer (4 cores), a workstation (24 cores) and a combination of a workstation and some office PCs (9 additional cores). The latter is evaluated using both, opportunistic load balancing (OLB) and load balancing using the minimal completion time algorithm (MCT). We investigate the two cases: First, having less evaluation tasks than CPU cores which are evaluated in parallel (using the 20 sequences of the KITTI subset) and second, having more evaluation tasks than CPU cores (using the 69 sequences of the Sintel subset). The DS algorithm is - besides its fast convergence - an interesting algorithm for investigation because it is mostly sequential in terms of optimisation runs. Hence, a bad distribution of evaluation tasks leads to unnecessary waiting times at almost each optimisation run which sum up in the total optimisation time.

Table 3. Optimisation times (given in hours) for DS on DeepFlow w.r.t. different distribution strategies in a heterogeneous infrastructure. 1 and 2 indicate different sets of additional computers (each giving 9 additional cores, i.e. 33 cores in total).

	4 cores	12 cores	24 cores	OLB[1]	MCT[1]	OLB[2]	MCT[2]
KITTI (20 seq.)	08:43	03:54	02:55	02:56	02:43	02:33	**02:32**
Sintel (69 seq.)	-	-	06:38	05:04	05:04	05:01	**04:49**

From Table 3, we can see that using more cores on a single machine is of course beneficial, but the gain in performance might not always be as high as expected. One reason is the sequential nature of DS regarding the optimisation runs. But there is another important reason: the behaviour of some modern many-core CPUs. When using only a small subset of all cores, these cores run at a higher frequency compared to the case when all cores are busy. In the latter case, an individual task is evaluated more slowly. Hence, even when the single machine with 24 cores is not working to capacity, using more cores on different machines and distributing the workload among those is beneficial despite of the overhead due to communication and synchronisation.

When comparing the different distribution strategies, the benefit of the MCT heuristic becomes evident. Either it is better than OLB or at least they show similar performance. The latter can be explained by the already mentioned behaviour of some modern CPUs. For them, the estimation of a task's expected run time - which is the basis for the MCT heuristic - is nontrivial. If these CPUs are assigned many tasks, they become slow. As a consequence, they are assigned only a few tasks afterwards, making them faster again. Using an estimation that respects this behaviour is future work and may improve the outcome of the MCT heuristic. Nevertheless, MCT can already reduce the run time by up to 8%.

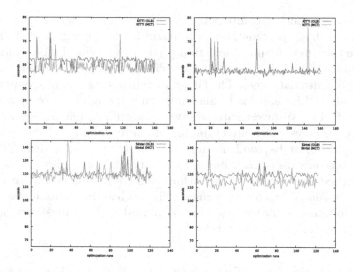

Fig. 2. Run times for subsequent optimisation runs. Red stands for the OLB heuristic, green for the MCT heuristic. **From left to right:** The sets 1 and 2 of additional computers according to Table 3. **From top to bottom:** KITTI, Sintel. (Color figure online)

In this context, we also compared the run times of the individual, subsequent optimisation runs by the example of DS. To this end, we considered only those steps of DS that led to an evaluation of only a single optimisation run (i.e. reflection, expansion and contraction steps) in order to have comparable time values. The graphs in Fig. 2 confirm the tendency of the total optimisation times in Table 3. These graphs show that the findings that hold for the total run time are also valid for the individual run times of the optimisation runs. When comparing the graphs for OLB (red) and MCT (green), one can see that the MCT heuristic leads to smaller run times for configuration 1 of additional computers on the KITTI dataset and for configuration 2 on the Sintel dataset. In the other two cases, both graphs indicate similar run times. The sparse peaks in all graphs are related to peripheral effects which do not depend on the heuristic.

5 Conclusion

In this work, we investigated three important aspects of time-efficient parameter optimisation for optical flow. To this end, we first compared different optimisation methods w.r.t. quality and workload. In this context, the Downhill Simplex method showed up to be a favourable compromise as it achieves good results using only a small amount of optimisation runs. Furthermore, we proposed a variance based strategy to reduce the number of training datasets (thus reducing the number of evaluation tasks per optimisation run) while retaining a good level of quality - in particular for large baseline datasets. Finally, we adopted a promising heuristic for distributing the evaluation tasks among different computers which allowed us to further reduce the overall optimisation time. To encourage the use of our entire optimisation framework, we provide a publicly available implementation as open source.

Acknowledgements. We thank the German Research Foundation (DFG) for financial support within project B04 of SFB/Transregio 161.

References

1. Baker, S., Roth, S., Scharstein, D., Black, M.J., Lewis, J.P., Szeliski, R.: A database and evaluation methodology for optical flow. In: Proceedings of IEEE International Conference on Computer Vision (ICCV). IEEE Computer Society Press (2007)
2. Barron, J.L., Fleet, D.J., Beauchemin, S.S.: Performance of optical flow techniques. Int. J. Comput. Vis. **12**(1), 43–77 (1994)
3. Boyd, S., Parikh, N., Chu, E., Peleato, B., Eckstein, J.: Distributed optimization and statistical learning via the alternating direction method of multipliers. Found. Trends Mach. Learn. **3**(1), 1–122 (2011)
4. Braun, T.D., Siegel, H.J., Beck, N., Bölöni, L.L., Maheswaran, M., Reuther, A.I., Robertson, J.P., Theys, M.D., Yao, B., Hensgen, D., Freund, R.F.: A comparison of eleven static heuristics for mapping a class of independent tasks onto heterogeneous distributed computing systems. J. Parallel Distrib. Comput. **61**(6), 810–837 (2001)
5. Brox, T., Malik, J.: Large displacement optical flow: descriptor matching in variational motion estimation. IEEE Trans. Pattern Anal. Mach. Intell. **33**(3), 500–513 (2011)
6. Butler, D.J., Wulff, J., Stanley, G.B., Black, M.J.: A naturalistic open source movie for optical flow evaluation. In: Fitzgibbon, A., Lazebnik, S., Perona, P., Sato, Y., Schmid, C. (eds.) ECCV 2012. LNCS, vol. 7577, pp. 611–625. Springer, Heidelberg (2012). doi:10.1007/978-3-642-33783-3_44
7. Datta, K., Murphy, M., Volkov, V., Williams, S., Carter, J., Oliker, L., Patterson, D., Shalf, J., Yelick, K.: Stencil computation optimization and auto-tuning on state-of-the-art multicore architectures. In: Proceedings of the 2008 ACM/IEEE Conference on Supercomputing, pp. 4:1–4:12. IEEE Press (2008)
8. Demetz, O.: Feature Invariance versus Change Estimation in Variational Motion Estimation. Ph.D. Thesis, Faculty of Mathematics and Computer Science, Saarland University (2015)
9. Drayer, B., Brox,T.: Combinatorial regularization of descriptor matching for optical flow estimation. In: British Machine Vision Conference (BMVC). BMVA Press (2015)

10. Geiger, A., Lenz, P., Urtasun, R.: Are we ready for autonomous driving? The KITTI vision benchmark suite. In: Proceedings of IEEE Conference on Computer Vision and Pattern Recognition (CVPR), pp. 3354–3361. IEEE Computer Society Press (2012)
11. Hansen, N., Ostermeier, A.: Completely derandomized self-adaptation in evolution strategies. Evol. Comput. **9**(2), 159–195 (2001)
12. Krajsek, K., Mester, R.: A maximum likelihood estimator for choosing the regularization parameters in global optical flow methods. In: Proceedings of IEEE International Conference on Image Processing (ICIP), pp. 1081–1084. IEEE Computer Society (2006)
13. Kulkarni, T., Kohli, P., Tenenbaum, J.B., Mansinghka, V.: Picture: a probabilistic programming language for scene perception. In: Proceedings of IEEE Conference on Computer Vision and Pattern Recognition (CVPR), pp. 4390–4399. IEEE Computer Society Press (2015)
14. Kunisch, K., Pock, T.: A bilevel optimization approach for parameter learning in variational models. SIAM J. Imaging Sci. **6**(2), 938–983 (2013)
15. Li, Y., Huttenlocher, D.P.: Learning for optical flow using stochastic optimization. In: Forsyth, D., Torr, P., Zisserman, A. (eds.) ECCV 2008. LNCS, vol. 5303, pp. 379–391. Springer, Heidelberg (2008). doi:10.1007/978-3-540-88688-4_28
16. Nelder, J.A., Mead, R.: A simplex method for function minimization. Comput. J. **7**(4), 308–313 (1965)
17. Memin, E., Heas, P., Herzet, C.: Bayesian inference of models and hyperparameters for robust optic-flow estimation. IEEE Trans. Image Process. **21**(4), 1437–1451 (2012)
18. Perreira, D.R., Delpiano, J., Papa, J.P.: On the optical flow model selection through metaheuristics. EURASIP J. Image Video Process. **2015**, 11 (2015)
19. Ragan-Kelley, J., Barnes, C., Adams, A., Paris, S., Durand, F., Amarasinghe, S.: Halide: a language and compiler for optimizing parallelism, locality, and recomputation in image processing pipelines. In: Proceedings of the 34th ACM SIGPLAN Conference on Programming Language Design and Implementation, pp. 519–530. ACM (2013)
20. Salmen, J., Caup, L., Igel, C.: Real-time estimation of optical flow based on optimized haar wavelet features. In: Takahashi, R.H.C., Deb, K., Wanner, E.F., Greco, S. (eds.) EMO 2011. LNCS, vol. 6576, pp. 448–461. Springer, Heidelberg (2011). doi:10.1007/978-3-642-19893-9_31
21. Samuel, K.G.G., Tappen, M.F.: Learning optimized map estimates in continuously-valued MRF models. In: Proceedings of IEEE Conference on Computer Vision and Pattern Recognition (CVPR), pp. 477–484. IEEE Computer Society Press (2009)
22. Sun, D., Roth, S., Lewis, J.P., Black, M.J.: Learning optical flow. In: Forsyth, D., Torr, P., Zisserman, A. (eds.) ECCV 2008. LNCS, vol. 5304, pp. 83–97. Springer, Heidelberg (2008). doi:10.1007/978-3-540-88690-7_7
23. Sun, D., Sudderth, E.B., Black, M.J.: Layered segmentation and optical flow estimation over time. In: Proceedings of IEEE Conference on Computer Vision and Pattern Recognition (CVPR), pp. 1768–1775. IEEE Computer Society Press (2012)
24. Sun, D., Roth, S., Black, M.J.: A quantitative analysis of current practices in optical flow estimation and the principles behind them. Int. J. Comput. Vis. **106**(2), 115–137 (2013)

25. Volz, S., Bruhn, A., Valgaerts, L., Zimmer, H.: Modeling temporal coherence for optical flow. In: Proceedings of IEEE International Conference on Computer Vision (ICCV), pp. 1116–1123. IEEE Computer Society Press (2011)
26. Weinzaepfel, P., Revaud, J., Harchaoui, Z., Schmid, C.: DeepFlow: large displacement optical flow with deep matching. In: Proceedings of IEEE International Conference on Computer Vision (ICCV), pp. 1385–1392. IEEE Computer Society Press (2013)

Subpixel-Precise Tracking of Rigid Objects in Real-Time

Tobias Böttger$^{(\boxtimes)}$, Markus Ulrich, and Carsten Steger

MVTec Software GmbH, Munich, Germany
{boettger,ulrich,steger}@mvtec.com
http://www.mvtec.com

Abstract. We present a novel object tracking scheme that can track rigid objects in real time. The approach uses subpixel-precise image edges to track objects with high accuracy. It can determine the object position, scale, and rotation with subpixel-precision at around 80 fps. The tracker returns a reliable score for each frame and is capable of self diagnosing a tracking failure. Furthermore, the choice of the similarity measure makes the approach inherently robust against occlusion, clutter, and nonlinear illumination changes. We evaluate the method on sequences from rigid objects from the OTB-2015 and VOT2016 dataset and discuss its performance. The evaluation shows that the tracker is more accurate than state-of-the-art real-time trackers while being equally robust.

Keywords: Visual object tracking · Real-time tracking · Template matching

1 Introduction

Visual object tracking is a fundamental problem in computer vision that is concerned with estimating the 2D pose of an object in a video sequence. It has a wide range of applications, such as robotics, human-computer interaction, and visual surveillance [3,6,9].

To cover the large amount of applications, the performance of trackers is usually evaluated on very diverse, publicly available, benchmarks, such as VOT2016 [7], OTB-2015 [17], or MOT16 [10]. Although the videos in the benchmarks are very diverse, they do not necessarily cover specific applications, but rather try to be as general as possible. This leads to the fact that, in general, trackers are optimized towards their generalization capabilities and not to a specific application [3,4,11]. Furthermore, the objects in these datasets are manually labeled with either axis aligned or oriented bounding boxes and the accuracy of a tracker is measured by its bounding box overlap [8]. Hence, trackers with subpixel precise localization or ones not restricted to oriented or axis-aligned rectangles do not necessarily have higher overlap scores in the benchmarks. Nevertheless, many industrial applications such as autonomous driving or the visual monitoring of industrial production processes require a good localization accuracy

© Springer International Publishing AG 2017
P. Sharma and F.M. Bianchi (Eds.): SCIA 2017, Part I, LNCS 10269, pp. 54–65, 2017.
DOI: 10.1007/978-3-319-59126-1_5

in real-time. For example, when picking an object from a conveyor belt with a robot, the bounding box of the object is not sufficient.

We present a real-time capable tracker that is able to determine the similarity transformation of a rigid object between frames. We leverage the fact that image edges can be determined with subpixel-precision to obtain a subpixel-precise object localization and do not restrict the object to a bounding box. The tracker returns a reliable score for each frame and is capable of self diagnosing a tracking failure. The two examples sequences in Fig. 1 show the localization quality of our approach and the fact that it is virtually drift-free for a rigid object in a sequence over 3000 frames. In the experiments section, we evaluate the method on further sequences from the OTB-2015 and VOT2016 dataset and comment on the performance characteristics of the presented approach.

Correlation-based tracker [2] ——— STAPLE [1] ——— Our approach

Fig. 1. Vase and Car24 from the OTB-2015 [17] benchmark. We compare our approach to two equally fast, state-of-the-art trackers: STAPLE [1] and a scale adaptive correlation tracker [2]. Our approach is able to accurately determine the position, scale, *and* rotation of the objects, as opposed to just the axis aligned bounding box. The tracker is virtually drift-free in the second sequence (which has over 3000 frames). All three trackers run in real-time at 100 fps on both sequences.

2   Related Work

Within visual object tracking, immense progress has been made in recent years. For example, the best performing tracker in the VOT-2014 challenge was only ranked 35th in the 2016 challenge, with around half of the expected average overlap (AEO) of the VOT2016 winner [7]. The great gain in performance is mostly due to the widespread adoption of discriminative learning methods such as discriminative correlation filters with complex features [3,5,6,18] and deep convolutional neural networks (CNNs) for tracking [4,11,16] (2015 VOT winner and the 2016 baseline for runner up). Although CNN-based trackers show impressive generalization properties and can cope with diverse sequences, most approaches are restricted to axis-aligned bounding boxes [16] and their accuracy is not on par with their robustness [7]. Furthermore, the computational complexity of many approaches is infeasible and only few approaches are real-time capable with a high performance GPU [4], which is not an option in many industrial applications.

In terms of speed, very robust trackers have emerged in the last few years which are real-time capable. Most of them build on discriminative correlation filters and extensions thereof [1,3,5,6,18]. For example, the STAPLE tracker [1] combines a HOG-based correlation filter with a model based on color statistics and achieved the best real-time performance at the VOT-2016 challenge [7]. Danelljan *et al.* [3] go beyond the ordinary discriminative correlation framework and train continuous convolution filters. Their approach performs on par with trackers based on CNNs and can be extended to subpixel-precise feature point tracking. Unfortunately, all of the mentioned approaches are restricted to axis aligned bounding boxes.

Similar to our approach, Lepetit and Fua [9] present a tracker that estimates the pose of rigid objects. Their keypoint-based recognition system is robust to occlusion and clutter, but requires an extensive offline training phase to generate the tracking model. In contrast to our approach, their tracking is restricted to textured objects that exhibit sufficient keypoints for reliable tracking.

As opposed to the above mentioned approaches, our approach is capable of estimating the position, scale, *and* rotation of an arbitrarily shaped object in real-time and does not require an extensive offline training nor is it restricted to textured objects.

3   Shape-Based Tracking

Our tracking approach builds on the efficient shape-based object recognition technique of Steger [13]. In the first frame, a shape-based model is generated from the arbitrarily shaped ROI of the detected or marked object. The model is used to determine the optimum object pose in the subsequent frames. After each successful tracking step, the model is updated and unstable points are filtered out and new points are added. The three steps (1) model generation, (2) model localization and, (3) model update are explained in more detail in the following section. Furthermore, we describe how the approach is made efficient and able to track most objects in the VOT2016 [7] and OTB-2015 [17] datasets at around 80 fps without using the GPU on an IntelCore i7-4810 CPU @2.8 GHz with 16 GB of RAM with Windows 7 (x64).

3.1   Model Generation

In the first frame, the tracking model $\mathcal{M}$ is constructed from the ROI of the automatically detected or manually marked object. The model consists of a set of n points $p_i = (x_i, y_i)^T$ and their corresponding direction vectors $d_i = (t_i, u_i)^T$:

$$\mathcal{M} = \{(p_i, d_i) \in \mathbb{R}^2 \times \mathbb{R}^2, \text{for } i = 1, \ldots, n\}. \tag{1}$$

In a first step, point candidates are extracted by applying a threshold on the Sobel filter edge amplitude of the input ROI. To thin out the number of points, non-maximum suppression is applied with automatically estimated thresholds, see the patent [14] for details. The remaining points can then be refined to subpixel-precision which is described in more detail in Chapt. 3.3 of [12]. The coordinates of the model points are all expressed relative to an arbitrary reference point. We use the center of the bounding box for simplicity. An exemplary model is displayed in Fig. 2.

Please note, that we do not have a training phase of our model. The model consists of a single set of points and their directions. The model transformation to different poses is done on the fly in the localization step.

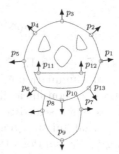

(a) Example Image from the OTB-2015 benchmark [17]

(b) Model Points p_i and their direction vectors

Fig. 2. The model in (b) is generated from the ROI in the first input frame displayed in (a). The model point are generated from non-maximum suppressed edges with a high enough edge amplitude (see [14]). The model in (b) is simplified for better visualization. The real model has over 400 points.

3.2   Model Localization

The model localization essentially amounts to finding the best matching candidate within the target image in a template matching framework. Hence, we compare a transformed model to the target image at a particular location by a similarity measure. By setting a minimal required similarity, we are able to avoid a very large number of computations.

In a first step, we calculate a direction vector for each pixel within the current frame and identify them as $e_{x,y} = (v_{x,y}, w_{x,y})$. We can then evaluate the similarity of the tracking model $\mathcal{M}$ to the current frame at various image locations and for different transformations of the tracking model. The location and transformation parameters with the highest similarities are the most probable object locations. The similarity transformation of a model point is given by:

$$p_i' = \underbrace{\begin{pmatrix} \sigma\cos\theta & -\sigma\sin\theta \\ \sigma\sin\theta & \sigma\cos\theta \end{pmatrix}}_{T_{\theta,\sigma}} p_i + \begin{pmatrix} x_t \\ y_t \end{pmatrix},$$
(2)

where θ and σ are the rotation and scale parameters, respectively. Similarly, the transformed direction vectors are obtained by

$$d_i' = (T_{\theta,\sigma}^{-1})^T d_i.$$
(3)

As similarity measure we use the normalized sum of dot products of the normalized direction vectors of the transformed model and the target image:

$$s(x_t, y_t, \theta, \sigma)_{\mathcal{M}} = \frac{1}{n} \left| \sum_{i=1}^{n} \frac{\langle d_i', e_{p_i'} \rangle}{\|d_i'\| \cdot \|e_{p_i'}\|} \right|,$$
(4)

with $s : \mathbb{R}^4 \to [0,1]$. The similarity measure is robust to occlusion, clutter, non-linear illumination changes, and a moderate amount of defocusing [13]. The robustness to non-linear illumination change comes from the fact that all direction vectors are scaled to unit-length. The robustness to occlusion comes from the fact that missing points in the target image will, on average, contribute nothing to the sum of (4). Similarly, clutter lines or points in the target image do not only need to coincide with the sparse set of model points, but also need to have a similar direction vectors to contribute to the similarity.

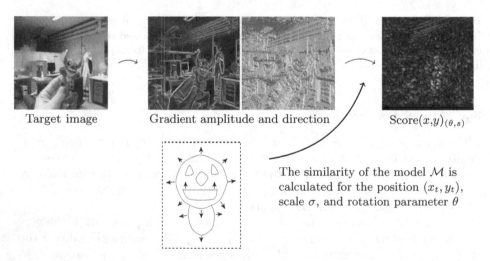

Target image   Gradient amplitude and direction   Score$(x,y)_{(\theta,s)}$

The similarity of the model $\mathcal{M}$ is calculated for the position (x_t, y_t), scale σ, and rotation parameter θ

Fig. 3. In a first step, the gradient amplitudes and direction of the target image are calculated. The amplitudes are required for the subpixel-precise refinement of the object position. The maximum similarity of the model $\mathcal{M}$ from Fig. 2 is calculated for the position, scale and angle within the discretized 4d search space.

The localization process is visualized in the flowchart of Fig. 3. At this point the optimal position, angle, and scale are determined with pixel accuracy. In the following subsection, we display how the 4d optima $(\tilde{x}_t, \tilde{y}_t, \tilde{\theta}, \tilde{\sigma})$ is refined to subpixel accuracy.

3.3 Subpixel-Precise Refinement

The accuracy of the localization step depends on the chosen discretization of σ and θ as well as the pixel resolution. To refine the match, we apply the concept of the 4d facet model. We approximate the 4d parameter space by calculating a second order Taylor polynomial around the $3 \times 3 \times 3 \times 3$ best match and extracting the maximum of this polynomial [13].

To further improve the localization accuracy and the robustness of the tracking to small model deformations and transformations that cannot be captured by a similarity transformation, we use a modified version of the least squares refinement described in [15]. The least-squares refinement assumes a good initial approximation of the current transformation and improves the global similarity transformation for all points. For each model point p_i, the best point match in the direction of $\pm d_i'$ is determined. The concept is displayed in Fig. 4 and explained in more detail in [15]. In contrast to [15], we do not restrict the length of the search line to 1, but rather allow an arbitrarily long search line. We found that a single least squares iteration was sufficient for most tracking sequences.

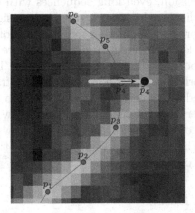

Fig. 4. The model is adapted to transformations that cannot be captured by a similarity transformation of the complete model. After the optimal model position, scale, and rotation have been determined, each point searches for its best match along a 1d search line perpendicular to its image tangent. The length of the search line is variable, but has a significant impact on the runtime. (Color figure online)

3.4 Model Update

After we have successfully refined the object pose to subpixel-precision, we conduct one final search for the best corresponding target image point for each model point in the direction of $\pm d_i'$. This time we do not update the global similarity transformation of the model, but rather update the relative positions of the points themselves. This improves how well the model will fit to the target image at future timesteps.

In the example shown in Fig. 4, we shift p_4 towards the best match $\tilde{p}_4$ that is found along the yellow line. We regularize the model update with a parameter λ to be more robust to noisy object deformations. At frame t we update each point p_i^t with its best match $\tilde{p}_i^t$ such that:

$$p_4^{t+1} = p_4^t + \lambda \tilde{p}_4^t. \tag{5}$$

The update step does not only allow our approach to capture small model deformations, but also weakens the restriction of our approach to similarity transformations of the model, consequently projective transformations that increment over time may be captured by locally deforming the model points.

Please note that in tracking settings the model update step always needs to find the balance between keeping the localization accuracy high and generalizing well to new representations of the model. In our approach, too large parameter values of λ may add drift and can lead to a degeneration of the tracking model if no extra care is taken. Nevertheless, since the model transformation is determined with subpixel-precision, even long sequences with over 3000 frames, like the one in the example displayed in Fig. 1, do not drift significantly.

During the tracking process, we further monitor how often every model point is found. This step enables us to identify points that are not significantly contributing to the model localization and to remove them. These points may have either emerged from poorly initialized points in the first frame or by parts of the object becoming occluded or changing in later frames. To prevent us from deleting all points and degenerating the model, we sample new points in sparse areas of the model after a successful tracking step. This allows us to capture newly emerging object edges. The example in Fig. 5 shows how the model update helps to capture deformations of the object.

3.5 Implementation Details

In tracking we do not need to search for the model in the target image exhaustively in each frame. To reduce the workload, we restrict the possible parameter values of $(\theta, \sigma) \in [\theta_c - 0.1, \theta_c + 0.1] \times [\sigma_c - 0.2, \sigma_c + 0.2]$, where θ_c and σ_c refer to the current object rotation and scale. Furthermore, for the translation, we define a circular search region with a radius of $1/2$ of the object diagonal. Although the search space was adequate for all of the test sequences, the size of the search region may be increased for fast moving objects.

If no parameter set of $(x_t, y_t, \theta, \sigma)$ that has a score $> s_{\min}$ is found in a frame, we increase the search region and the parameter ranges of (θ, σ) for the

Fig. 5. Sequence Coupon from OTB-2015. If no model update is performed ($\lambda = 0.0$) the tracker jumps to the wrong dollar note when the folded top note is moved. If the model update parameter λ is set high enough, the model is updated when the note is folded and the tracker succeeds.

next frames successively. As soon as we find the object again, we reset the search parameters to their initial value. To prevent the workload from becoming too large, we decrease the discretization of the search space when it becomes bigger. Although this decreases the accuracy, it gives us the chance to re-detect lost objects and improve the accuracy in the subsequent frames.

To achieve further speed-ups, we stop calculating a score for a model transformation as soon as it cannot reach a predefined minimal score $s_{\min}$ anymore. To obtain an even larger speed-up it is possible to be even stricter, please refer to [15] for further details.

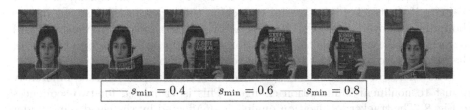

Fig. 6. The results for 3 different parameters of $s_{\min}$ are displayed for the FaceOcc1 sequence from OTB-2015. The value of $s_{\min}$ is an indicator of how much the object is allowed to be occluded. Lower values improve the robustness to occlusion but also require more time, since more score values need to be computed. For $s_{\min} = 0.8$, the object is not detected for the third to fifth image, while for $s_{\min} = 0.6$ it is lost for the fourth and fifth image. However, all approaches recover when the occlusion ends.

The parameter $s_{\min}$ gives a good estimation of the allowed object occlusion. If half of the model points are occluded in the target image, the maximum score that can be obtained is 0.5. This is displayed in Fig. 6, where the face is essentially occluded by over 50% and hence, only values of $s_{\min} < 0.5$ are able to track the object through all frames. Please note that a low value of $s_{\min}$ increases the number of points for which the score needs to be calculated and hence has a negative impact on the runtime.

Further speed-ups can be obtained by only using a fixed number of points for tracking the object. Before each localization step, a random subset of points is selected from the model $\mathcal{M}$ and used for tracking. Although some accuracy may be lost, the execution time can be reduced.

4   Experiments

The restriction to rigid objects and subpixel-precise localization makes it difficult to compare the approach to the vast majority of existing schemes in general. First of all, the data of existing benchmarks, such as VOT2016 [7] and OTB-2015 [17], are not annotated with sufficient accuracy and focus on robust, generalizable tracking. Hence, we focus our evaluation on selected sequences of rigid objects from both datasets and point out the strengths and weaknesses of the proposed approach.

To get a fair comparison, we compare our method to the state of the art STAPLE tracker [1], which was the best real-time tracker at the VOT-2016 challenge [7], and a very fast correlation filter tracker with scale adaption, based on [2]. We evaluate the average bounding box overlap [17] on a selection of rigid objects from both the OTB-2015 and VOT-2016 datasets in Fig. 7. Please note that the ground truth data is mostly only labeled as axis aligned bounding boxes which puts a heavy bias on the obtained overlaps. For example, for the sequence Vase we visually clearly outperform both approaches, as is seen nicely in Fig. 1. Nevertheless, the bounding box abstraction lets the overlap drop very low.

To measure the robustness, we do not use the VOT-2016 measures, but rather evaluate if the tracker is successfully tracking the target in the last few frames (bounding box overlap > 50%). Here STAPLE (22/26) and our approach (21/26) perform equally well, while the correlation based approach drops off (13/26). In the following we will discuss individual sequences in more detail.

The choice of the similarity measure in (4) makes the tracker inherently robust to nonlinear illumination changes. This is visualized for two sequences in Fig. 8. The tracker localization quality is unaffected by the car driving under

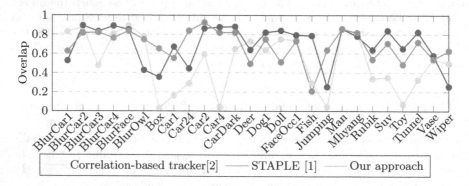

Fig. 7. The average bounding box overlap for rigid sequences within the VOT2016 and OTB-2015 datasets. We compare our overlap scores to STAPLE [1] and a scale adaptive correlation tracker based on [2].

| Correlation-based tracker[2] —— STAPLE[1] —— Our approach |

Fig. 8. Our tracking scheme is inherently robust to illumination changes because of the similarity measure we use in the localization step (4). The method performs comparable to STAPLE and outperforms the scale adaptive correlation tracker, which fails in the first sequence, in terms of accuracy.

the bridge or the light being turned on and off in the second sequence. However, very strong changes of the illumination can make it difficult to segment the edge orientation of the target image robustly and lead to tracking failure, which is what happens in the sequence `Fish`.

The fact that our approach is essentially a local object detector with a meaningful score allows our framework to detect tracking failure reliably and recover from complete object occlusion. In Fig. 9, the object is completely occluded in the middle of the sequence and hence the STAPLE and correlation-based tracker start adapting their filters to the foreground. Neither of the approaches is able to recover from the complete occlusion. Our approach, on the other hand, detects the tracking failure and is able to re-detect the object when it reemerges.

The fact that our approach searches for the best similarity transformation between the frames leads to the fact that our approach is weak in sequences where the object has strong local deformations. Furthermore, strong camera motion and image blur can make it fail. In the sequence in Fig. 10, the object is lost whenever the camera motion is too strong. Fortunately, the object is always

| Correlation-based tracker[2] —— STAPLE[1] —— Our approach |

Fig. 9. A further advantage of our approach is the possibility of self-diagnosing when the object is lost. The score is a reliable indicator of how much of the model is visible. In the sequence `Box` from OTB-2015, all three trackers fail when the object is strongly occluded. Nevertheless, our approach recovers when the occluded object reappears.

re-detected when the camera motion stops and the edges become clear enough. In the respective frame, the STAPLE tracker fails near the end when the camera motion is extremely high.

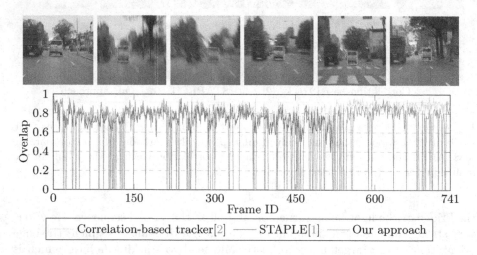

Fig. 10. Edge-based tracking has difficulties when the image blur becomes too large. Neighboring edges may merge into each other or disappear completely. This becomes evident in the sequences `BlurCar1` sequence from OTB-2015 [17]. Our tracker loses the object when the camera motion is too large. Nevertheless, in both sequences the tracker is always able to recover and finishes the sequence with a very good localization. The STAPLE tracker loses the track at frame 543 and does not recover.

5   Conclusion

In this paper, we have proposed an efficient object tracker that is able to determine the position, scale *and* rotation of a rigid objects in various different sequences with high accuracy. We validated our framework on a rigid subset of the VOT-2016 [7] and OTB-2015 [17] datasets and were able to perform on par with real-time state-of-the art approaches in terms of robustness. As opposed to the existing schemes, our approach is more accurate in terms of localization. On the one hand this is due to the subpixel-precise refinement of the object pose and, on the other hand, due to estimating the object rotation.

Unfortunately, the label data of the existing benchmarks is restricted to axis-aligned and oriented bounding boxes, which makes it difficult to quantize the localization gains in the established performance measures. A subpixel-precise tracking dataset and evaluation framework that is not restricted to bounding boxes would be very helpful for future evaluations.

References

1. Bertinetto, L., Valmadre, J., Golodetz, S., Miksik, O., Torr, P.H.S.: Staple: complementary learners for real-time tracking. In: IEEE CVPR, pp. 1401–1409 (2016)
2. Danelljan, M., Häger, G., Khan, F.S., Felsberg, M.: Accurate scale estimation for robust visual tracking. In: BMVC (2014)
3. Danelljan, M., Robinson, A., Shahbaz Khan, F., Felsberg, M.: Beyond correlation filters: learning continuous convolution operators for visual tracking. In: Leibe, B., Matas, J., Sebe, N., Welling, M. (eds.) ECCV 2016. LNCS, vol. 9909, pp. 472–488. Springer, Cham (2016). doi:10.1007/978-3-319-46454-1_29
4. Held, D., Thrun, S., Savarese, S.: Learning to Track at 100 FPS with deep regression networks. In: Leibe, B., Matas, J., Sebe, N., Welling, M. (eds.) ECCV 2016. LNCS, vol. 9905, pp. 749–765. Springer, Cham (2016). doi:10.1007/978-3-319-46448-0_45
5. Henriques, J.F., Caseiro, R., Martins, P., Batista, J.: Exploiting the circulant structure of tracking-by-detection with kernels. In: Fitzgibbon, A., Lazebnik, S., Perona, P., Sato, Y., Schmid, C. (eds.) ECCV 2012. LNCS, vol. 7575, pp. 702–715. Springer, Heidelberg (2012). doi:10.1007/978-3-642-33765-9_50
6. Henriques, J.F., Caseiro, R., Martins, P., Batista, J.: High-speed tracking with kernelized correlation filters. IEEE Trans. Pattern Anal. Mach. Intell. **37**(3), 583–596 (2015)
7. Kristan, M., et al.: The visual object tracking VOT2016 challenge results. In: Hua, G., Jégou, H. (eds.) ECCV 2016. LNCS, vol. 9914, pp. 777–823. Springer, Cham (2016). doi:10.1007/978-3-319-48881-3_54
8. Kristan, M., Matas, J., Leonardis, A., Vojír, T., Pflugfelder, R.P., Fernández, G., Nebehay, G., Porikli, F., Cehovin, L.: A novel performance evaluation methodology for single-target trackers. IEEE Trans. Pattern Anal. Mach. Intell. **38**(11), 2137–2155 (2016)
9. Lepetit, V., Fua, P.: Keypoint recognition using randomized trees. IEEE Trans. Pattern Anal. Mach. Intell. **28**(9), 1465–1479 (2006)
10. Milan, A., Leal-Taixé, L., Reid, I.D., Roth, S., Schindler, K.: MOT16: a benchmark for multi-object tracking. CoRR, abs/1603.00831 (2016)
11. Nam, H., Han, B.: Learning multi-domain convolutional neural networks for visual tracking. In: IEEE CVPR, pp. 4293–4302 (2016)
12. Steger, C.: An unbiased detector of curvilinear structures. IEEE Trans. Pattern Anal. Mach. Intell. **20**(2), 113–125 (1998)
13. Steger, C.: Similarity measures for occlusion, clutter, and illumination invariant object recognition. In: Radig, B., Florczyk, S. (eds.) DAGM 2001. LNCS, vol. 2191, pp. 148–154. Springer, Heidelberg (2001). doi:10.1007/3-540-45404-7_20
14. Ulrich, M., Steger, C.: System and methods for automatic parameter determination in machine vision. US Patent 7,953,290 (2011)
15. Ulrich, M., Steger, C.: Performance evaluation of 2d object recognition techniques. Technical report PF-2002-01, Lehrstuhl für Photogrammetrie und Fernerkundung, Technische Universität München (2002)
16. Wang, L., Ouyang, W., Wang, X., Lu, H.: Visual tracking with fully convolutional networks. In: IEEE ICCV, pp. 3119–3127 (2015)
17. Yi, W., Lim, J., Yang, M.-H.: Object tracking benchmark. IEEE Trans. Pattern Anal. Mach. Intell. **37**(9), 1834–1848 (2015)
18. Zhang, K., Zhang, L., Liu, Q., Zhang, D., Yang, M.-H.: Fast visual tracking via dense spatio-temporal context learning. In: Fleet, D., Pajdla, T., Schiele, B., Tuytelaars, T. (eds.) ECCV 2014. LNCS, vol. 8693, pp. 127–141. Springer, Cham (2014). doi:10.1007/978-3-319-10602-1_9

Wearable Gaze Trackers: Mapping Visual Attention in 3D

Rasmus R. Jensen[1], Jonathan D. Stets[1(✉)], Seidi Suurmets[2], Jesper Clement[2], and Henrik Aanæs[1]

[1] Technical University of Denmark, Kongens Lyngby, Denmark
stet@dtu.dk
[2] Copenhagen Business School, Frederiksberg, Denmark

Abstract. The study of visual attention in humans relates to a wide range of areas such as: psychology, cognition, usability, and marketing. These studies have been limited to fixed setups with respondents sitting in front of a monitor mounted with a gaze tracking device. The introduction of wearable mobile gaze trackers allows respondents to move freely in any real world 3D environment, removing the previous restrictions.

In this paper we propose a novel approach for processing visual attention of respondents using mobile wearable gaze trackers in a 3D environment. The pipeline consists of 3 steps: modeling the 3D area-of-interest, positioning the gaze tracker in 3D space, and 3D mapping of visual attention.

The approach is general, but as a case study we created 3D heat maps of respondents visiting supermarket shelves as well as finding their in-store movement relative to these shelves. The method allows for analysis across multiple respondents and to distinguish between phases of in-store orientation (far away) and product recognition/selection (up close) based on distance to shelves.

1 Introduction

The study of human visual attention relates to a wide range of areas such as: psychology, cognition, useability, and marketing. In order to directly study this in various settings, eye tracking has become a standard method. A common way of visualizing and analysing gaze data is using Areas Of Interest (AOI) and attentional heat maps [13]. The heat maps represent the spatial distribution of eye movement throughout the AOI and can often be used for quantitative analysis. The most common method of visualizing heat maps is using a Gaussian based solution. Here, four parameters are used to determine the appearance of the heat map: the width of the basic construct, the use of fixations vs. raw data, whether accounting for duration of fixation and the mapping color altitude form [3]. For many years, mapping visual attention as heat maps has been limited to static setups with respondents sitting in front of a screen mounted with a stationary calibrated gaze tracker. Such a setup can accurately map the visual attention as a heat map of what is projected on the screen, but obviously limits

© Springer International Publishing AG 2017
P. Sharma and F.M. Bianchi (Eds.): SCIA 2017, Part I, LNCS 10269, pp. 66–76, 2017.
DOI: 10.1007/978-3-319-59126-1_6

Fig. 1. Supermarket vegetables shown as a 3D model with heatmap and respondent viewing points.

the visual attention to a 2D surface. The recent introduction of mobile wearable gaze trackers (Fig. 2) enables data collection in about any real-world environment. On mobile wearable eye-trackers, the scene is recorded using a front facing camera, and gaze data collected from eye tracking cameras can be projected onto this video. Despite the potential of introducing recordings of three dimensional scenes, common for both the stationary and mobile wearable eye-tracker is that ultimately the data is still recorded and analysed in 2D.

Mapping visual attention data recorded in a 3D space to a 2D heatmap is not straightforward. A simple approach is to find the best homographic correspondence between a reference image and a given frame from the eye-tracker, and then map the gaze according to this homography [4,12]. Figure 3 shows common errors in mapping using a homography relative to the actual mapping onto a 3D AOI. We argue that gaze collected in 3D mapped onto a 2D reference image using a homography will always be limited as a result of incorrect mappings.

Fig. 2. Tobii Pro Glasses 2 [12]. A wearable gaze tracker that tracks a respondents eye movements using IR cameras, while also recording the environment with a front facing video camera.

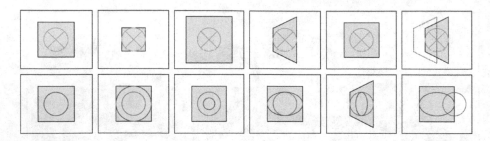

Fig. 3. This figure shows the common errors relating to mapping gaze. The top row shows a respondents viewpoint of an AOI with the gaze point in green, while the second row shows a reference view of an AOI with gaze mapped as homography mapping in red and mapping according to 3D structure in blue. First column shows mapping, when the viewpoint of the respondent and reference coincide. In this case, homography mapping and mapping according to 3D structure will be identical and perfectly overlapping. Column 2 and 3 show the mapping, when respondent is closer to or further from the AOI. Using homography mapping, the gaze point does not change along with the movement of the respondent. Row 4 and 5 show how the mapped gaze changes shape according to changes in viewpoint for the 3D mapping, while homography mapping does not change accordingly. The final column shows the error, when the homography is offset from the plane of the actual viewpoint, which introduces parallax error. (Color figure online)

We propose a solution to these problems and limitations by modelling an AOI in 3D as a reference for mapping gaze data. The reference model is reconstructed from photographs of the AOI to establish a good base for image feature matching and a high quality model mesh. We demonstrate a fully automatic pipeline for generating a 3D attention heat map, and furthermore the possibility of calculating the respondent viewing points as shown in Fig. 1. Our pipeline enables spatial filtering, positioning and orientation relative to the selected AOI, as well as correlation of multi-respondent data. We use supermarket shelves as a case study, but our pipeline is not limited to this setup. Our method requires a standard digital camera to capture images of the reference model, and a wearable gaze tracker with a front facing camera, such as the one shown in Fig. 2, for recording the scene and gaze data.

There are a number of recent studies that addresses the need to move mapping of visual attention to 3D. [11] introduces the potential of measuring 3D gaze coordinates from head-mounted gaze trackers, and [9] proposes visualisation of 3D gaze data on to virtual computer generated models. A method similar to our pipeline is described in [10], which demonstrates the use of a Microsoft Kinect to create a 3D reference model. Our method differs by using images to create a more dense point cloud, which also enables us to backproject the heat map to a traditional 2D visualization for comparison.

2   Data

We have collected data in both a real world supermarket and using a mock-up supermarket shelf in our lab. Reference data of the AOIs have been captured using a digital mirrorless camera: a Panasonic GH4 with a 12 mm lens (24 mm in 35 mm equivalent). To collect respondent data we have used the Tobii Pro Glasses 2 wearable gaze tracker [12] (Fig. 2), which collects the respondents view using a front facing video camera, while also recording the respondent gaze direction using 4 infrared cameras facing the eyes. Both cameras were calibrated using a standard checkerboard approach [16]. Data was collected of four in-store product sections in a supermarket: wine, vegetables, flour and cereal, as well as a mock-up of the cereal section in our lab. We used the digital camera to capture sets of reference images to cover the desired AOIs (12–20 images of each AOI). Gaze and video data were collected of respondents visiting the given sections (16 sets), visiting the store but acquired to get cereal (4 sets), and finally, presented for a mock-up of the cereal section in the lab (6 sets). All gaze data samples are raw, so no fixation filtering has been applied [3].

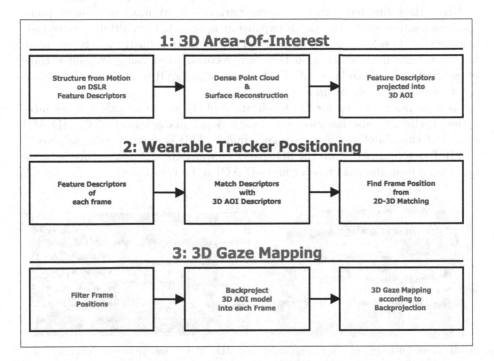

Fig. 4. The 3 steps in our proposed pipeline to construct 3D gaze mapping: Modelling of an Area-Of-Interest, Eye-tracker frame positioning, and finally the gaze mapping.

3   Method

In order to map gaze data onto a 3D AOI, we propose a pipeline consisting of three parts (Fig. 4): construction of the 3D AOI reference model, localization of the wearable gaze tracker frames relative to the reference model, and finally gaze mapping onto the AOI as a heat map.

3.1   Modelling a 3D Area-Of-Interest

The 3D AOI reference model is built using a series of images of the AOI. This task is divided further into three steps (Fig. 4). First, we use structure from motion to find the spatial camera positions and a sparse point cloud representation. We have opted for a structure from motion (SfM) [6] implementation, which requires a sequence of images followed by an image rectification based on the parameters obtained from the camera calibration. SIFT descriptors [7,15] are found in each image and sequentially matched across the sequence of images in an iterative fashion. Images with sufficient feature matches are included, while the extrinsic camera parameters are estimated and refined using bundle adjustment [14].

Given the estimated extrinsic camera parameters, we move onto dense point cloud estimation using the patch-expansion approach to multiview stereopsis proposed by Furukawa and Ponce [2]. This method robustly produces dense representations from which a surfaces are reconstructed using Poisson surface reconstruction by Kazhdan et al. [5]. A 3D modelled AOI from the cereal section in a supermarket is shown in Fig. 5(a).

As a preparation step for the localization of the wearable gaze tracker later in the pipeline, we use backprojection with depth management of the 3D AOI to project the model into each reference image. This is done in order to project 2D SIFT descriptors [7] into the 3D space, allowing the 2D descriptors between each frame from the gaze tracker and 3D AOI to be compared.

(a) 3D Area-Of-Interrest reference model. (b) 3D AOI backprojected onto a gaze tracker frame.

Fig. 5. The 3D AOI in (a) is backprojected onto an undistorted gaze tracker frame and the gaze point with trace from previous frames (b). The frame is shown in black and white, while the projection is shown in color.

3.2 Wearable Gaze Tracker Frame Localization

In order to correctly map gaze data on the 3D AOI, each frame from the gaze tracker has to be positioned relative to the AOI (if visible). The SIFT descriptors in each frame are matched with the reference 2D descriptors projected into 3D space, when constructing the 3D AOI. These correspondences are sent to a 2D to 3D pose solver, which finds the best fit camera pose using a RANSAC approach discarding outliers [1]. Given sufficient corresponding points, the solver will return the correct camera pose relative to the AOI. Without sufficient good matches the solver either fails or returns a false camera pose. Since a given frame might not cover any part of the AOI, the resulting matching consists of either a lot of true positive or a few false positive matches. Figure 5(b) shows the 3D AOI backprojected into a frame from the wearable tracker using the estimated camera pose. This backprojection is an immediate sanity check, showing the correctness of the pose estimation. Incorrect pose estimates tends to be very inconsistent from one frame to the next. To speed things up we have used the above approach to find the pose in keyframes, which are followed by frames, where the correspondence points are tracked using optical flow [8] (1 keyframe followed by 5 optical flow frames). This is substantially faster than finding and matching features in each frame. The pose solver is initialized with the pose from the previous frame, which along with the optical flow gives timewise consistency in the pose estimation.

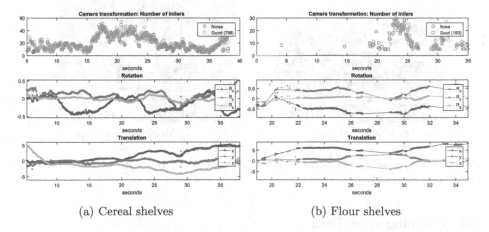

(a) Cereal shelves   (b) Flour shelves

Fig. 6. Estimated respondent poses from visits to the cereal and flour shelves with timestamp. First row of plot: the framewise number of inliers (y-axis) in the camera positioning solver. Green points are included as reliable and red points are considered noise. Noise points are filtered out based on spatial and rotational inconsitency. Second and third row of plots are the rotation and translation with inliers shown as a connected graph and the few outliers as single points. (Color figure online)

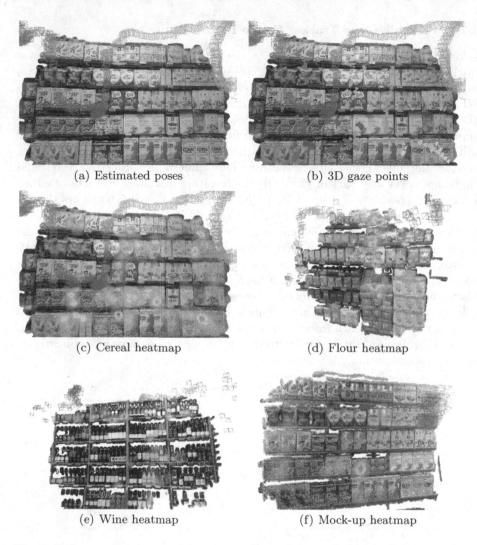

(a) Estimated poses   (b) 3D gaze points

(c) Cereal heatmap   (d) Flour heatmap

(e) Wine heatmap   (f) Mock-up heatmap

Fig. 7. Poses, gaze points and heatmaps obtained from the data of the 5 sections included in our study.

3.3 Mapping Gaze Data

The pose estimation is unbiased and may results in a few faulty poses. We consider these noise and use the following approach to filter them from the good poses. Correct pose estimates between consecutive frames is assumed to have small variation, while incorrect poses are very inconsistent. This inconsistency is used to identify and discard faulty poses. In Fig. 6 the good pose estimates is shown as a connected graphs with discarded poses as outlying points. The number of inliers returned from the 2D to 3D pose solver is a good estimator of correctness, but thresholding this number is not as robust as filtering the pose.

A respondent moving in front of the AOI is shown in Fig. 7(a). Using the good poses, the gaze can be mapped onto the 3D AOI model creating a 3D heat map as seen in Fig. 7(b). The gaze intersection with the 3D model is found using backprojection with depth management into the current frame, which is significantly faster than calculating the intersection between the line of sight and the 3D model.

A similar approach is taken, when creating the heatmap. Here a predetermined symmetric 2D Gaussian function with center at the gaze coordinate is added to a sum map of Gaussians in 3D space. Using a Gaussian serves both the fact that sight is not an infinitely small point, while also incorporating some uncertainty in the gaze estimates. Discussions about the size of the Gaussian, and whether the raw gaze data or fixation filtered data should be used is beyond the scope of this work. The resulting heatmaps visualized on the 3D AOIs can be seen in Figs. 1 and 7(c) to (e).

One benefit worth noticing is, that the approach of mapping Gaussian to the backprojection of the AOI allows for a normalization of the contribution from each gaze point. It also addresses the problems shown in Fig. 3. When a respondent is close, the covering of the Gaussian gaze point of the 3D AOI will be small with a locally high intensity. Respondents far away will cover a larger area in the 3D AOI, which will result in less locally intensive mapping. It also handles change in perspective, while effectively shaping the Gaussian according to the viewpoint without introducing parallax error. Since the sum of Gaussian gaze points is done on a 3D model, the heatmap can be projected into any frame or reference image. The backprojection of the a heatmap is shown in Fig. 8(a) as an overlay to the original image.

(a) 3D heatmap backprojected into reference image   (b) 2D heatmap from iMotions software

Fig. 8. Heatmaps based on 3D gaze mapping and 2D gaze mapping. For the 3D mapping the heatmap has been backprojected into the DSLR frame used for the 2D mapping.

4   Results

The core of our presented pipeline is the ability to correctly find the pose of the wearable gaze tracker relative to the 3D AOI in a given frame. Validating this after filtering puts each frame in one of four categories:

True positive correctly detecting the AOI.
True negative correctly not detecting the AOI.
False positive incorrect detection of the AOI.
False negative incorrectly not detecting the AOI.

Reviewing the output videos with 3D AOI overlay backprojected as presented in Fig. 5(b) is an easy way to quickly asses the quality of the AOI detection. Such a review shows non or only a very few false positives, but some false negatives. Since the gaze tracker has a very small sensor, the sensor struggles with low indoor light, which results in both frames with motion blur from head movement and rolling shutter. In the supermarket setting, these frames provide the vast majority of false negatives and one could debate, whether they are actually false negatives. Occlusion from people or other shelves can also cause false negatives. Reviewing both the frame positions as a graph in Fig. 6 or the resulting spatial positions in Fig. 7(a) are can also provide quick qualitative verifications in addition to reviewing a video with backprojected 3D AOI.

We have reprojected the heatmap into a reference image, which has also been applied homography gaze mapping using the iMotions 6.2 software [4] and the results are shown in Fig. 8. The heatmaps are both based on raw data samples but using different techniques: 3D mapping and homography mapping respectively. This means they cannot be compared directly, however there have clear similarities of the path pattern and duration of attention.

5   Conclusion

We have successfully created 3D AOIs and heat maps for respondents visiting the five sections in our data set: vegetables, cereal, flour, wine, and cereal mock-up. Our proposed pipeline does away with the problems relating to mapping gaze using a homography. The proposed pipeline is fully automatic and runs at ~2 fps using a combination of Matlab, mex, vlfeat and OpenCV. A full C++ implementation will provide further speed up, but the scope was to demonstrate a feasible pipeline, which allows researchers to spend time only on the results, once processed. Concerns such as visual attention mapped based on fixation vs. raw data, size of the Gaussian gaze point, and cross respondent analysis has not been evaluated. We found that our pipeline works well in in-store settings, since store products tend to have very distinct image features. However, settings with only repeating image features, such as frames with only the same product present, complicates the feature matching. This is often the case, when the respondent is very near a product shelf. On the other hand, detection works well in the case where the respondent is inspecting the shelves at an arm length distance, which in many cases is the important frames for generating heat maps. Our approach provides a fully automatic method of mapping gaze data and positioning of the respondent relative to the AOI, thus adding another dimension to the resulting data.

Acknowledgements. This work has been funded by the Innovation Fund (Denmark) and carried out in collaboration with iMotions. We would like to thank both for the support and collaboration.

References

1. Bradski, G.: The OpenCV Library. Dr. Dobb's J. Softw. Tools, Article id 2236121 (2000). http://code.opencv.org/projects/opencv/wiki/CiteOpenCV
2. Furukawa, Y., Ponce, J.: Accurate, dense, and robust multi-view stereopsis. IEEE Trans. Pattern Anal. Mach. Intell. **1**(1), 1–14 (2007). ISSN 01628828. http://doi.ieeecomputersociety.org/10.1109/TPAMI.2009.161
3. Holmqvist, K., Nyström, M., Andersson, R., Dewhurst, R., Jarodzka, H., Van de Weijer, J.: Eye Tracking: A Comprehensive Guide to Methods and Measures. OUP, Oxford (2011)
4. iMotions. iMotions biometric research platform (2017). https://imotions.com/
5. Kazhdan, M., Bolitho, M., Hoppe, H.: Poisson surface reconstruction. In: Proceedings of the Fourth Eurographics Son Geometry Processing (SGP 2006), pp. 61–70, Aire-la-Ville, Switzerland. Eurographics Association (2006). ISBN 3-905673-36-3
6. Koenderink, J., Van Doorn, A.: Affine structure from motion. Optical Soc. Am. A **8**(2), 337–385 (1991). doi:10.1364/JOSAA.8.000377, ISSN 1084–7529
7. Lowe, D.G.: Distinctive image features from scale-invariant keypoints. Int. J. Comput. Vis. **60**(2), 91–110 (2004)
8. Lucas, B.D., Kanade, T.: An iterative image registration technique with an application to stereo vision (1981). ISSN 17486815, http://www.ic.unicamp.br/rocha/teaching/2013s1/mc851/aulas/additional-material-lucas-kanade-tracker.pdf
9. Maurus, M., Hammer, J.H., Beyerer, J.: Realistic heatmap visualization for interactive analysis of 3d gaze data. In: Proceedings of the Symposium on Eye Tracking Research and Applications, pp. 295–298. ACM (2014)
10. Paletta, L., Santner, K., Fritz, G., Mayer, H., Schrammel, J.: 3d attention: measurement of visual saliency using eye tracking glasses. In: CHI 2013 Extended Abstracts on Human Factors in Computing Systems, pp. 199–204. ACM (2013)
11. Pfeiffer, T.: Measuring and visualizing attention in space with 3d attention volumes. In: Proceedings of the Symposium on Eye Tracking Research and Applications, pp. 29–36. ACM (2012)
12. Tobii Pro. Tobii pro glasses 2 (2017). http://www.tobiipro.com/product-listing/tobii-pro-glasses-2/
13. Purucker, C., Landwehr, J.R., Sprott, D.E., Herrmann, A.: Clustered insights: improving eye tracking data analysis using scan statistics. In: Psychological Considerations on Car Designs-An Investigation of Behavioral and Perceptual Aspects Using Eye Tracking and Cross-Cultural Studies (2012)
14. Triggs, B., McLauchlan, P.F., Hartley, R.I., Fitzgibbon, A.W.: Bundle adjustment — a modern synthesis. In: Triggs, B., Zisserman, A., Szeliski, R. (eds.) IWVA 1999. LNCS, vol. 1883, pp. 298–372. Springer, Heidelberg (2000). doi:10.1007/3-540-44480-7_21, http://www.springerlink.com/content/plvcrq5bx753a2tn

15. Vedaldi, A., Fulkerson, B.: Vlfeat. In: Proceedings of the International Conference on Multimedia (MM 2010), vol. 3, no. 1, p. 1469 (2010). doi:10.1145/1873951. 1874249, http://dl.acm.org/citation.cfm?doid=1873951.1874249
16. Zhang, Z.: Flexible camera calibration by viewing a plane from unknown orientations. In: Proceedings of the Seventh IEEE International Conference on Computer Vision, vol. 1, no. c, p. 7 (1999). doi:10.1109/ICCV.1999.791289, ISSN 01628828

Image Processing of Leaf Movements in *Mimosa pudica*

Vegard Brattland[1], Ivar Austvoll[1], Peter Ruoff[2], and Tormod Drengstig[1(✉)]

[1] Department of Electrical Engineering and Computer Science,
University of Stavanger, Stavanger, Norway
v.brattla@hotmail.com, {ivar.austvoll,tormod.drengstig}@uis.no
[2] Centre for Organelle Research, University of Stavanger, Stavanger, Norway
peter.ruoff@uis.no
http://www.uis.no

Abstract. In this paper the plant *Mimosa pudica*'s response to changed illumination conditions is being examined. An image processing routine, using the HSV color model and triangle intensity threshold segmentation, is developed to segment time-lapse image series of *Mimosa pudica*, quantifying the plant's image pixel count as a measure of movement. Furthermore, the method of Farnebäck is used to estimate dense optical flow (both magnitude and direction), describing the plants movement orientation in the image plane. The pixel count results indicate that the plant exhibits an anticipatory behavior in that it starts to close its leaves prior to the light-to-dark transition. Furthermore, the optical flow results indicate that each compound leaf show different behavior depending on the whereabouts in the circadian rhythm cycle. This suggests that a complex regulating structure lies behind the plant's response to different illumination regimes.

1 Introduction

Observations of the plant *Mimosa pudica* under natural/wild conditions reveal that the plant opens its leaves prior to sun rise and closes its leaves prior to sun set, i.e. anticipating the shift in illuminative conditions. These responses are described as being related to the plants circadian rhythm, which is a periodic, self-sustaining and temperature compensated biological process [2,7,9–11].

We wanted to investigate in controlled conditions how this behaviour is related to the day length, to the dark/light ratio and to phase shifts in the entraining dark/light cycle. Our overall goal in the project is to describe the observed behaviour in mathematical terms using non-linear differential equations, and relating these to mathematical models of circadian rhythms. By investigating the influence of different light/dark cycles and constant illumination we seek to get valuable insights into the regulatory mechanisms behind circadian rhythms. The work presented in this paper, however, is related to the image processing tool used to analyse the time lapse images of *Mimosa pudica* in order to get a quantitative measure of leaf movements, and is based on the master's thesis of the first author [1].

© Springer International Publishing AG 2017
P. Sharma and F.M. Bianchi (Eds.): SCIA 2017, Part I, LNCS 10269, pp. 77–87, 2017.
DOI: 10.1007/978-3-319-59126-1_7

Mimosa pudica originates from the tropical parts of America. It is a perennial plant (can live several years) and it is usually between 15–45 cm tall. It is also known as the *sensitive plant* as it closes its leaves upon touching. The young plants will usually consist of two to four *compound leaves* which again contain between 7 and 10 pairs of *leaflets* arranged in a bi-pinnate (feather like) pattern, see Fig. 1(a). These leaflets are observed to be in the color range of dark greenish to yellowish with an oblong shape. One of the favorable properties of *Mimosa pudica* from an experimental viewpoint is its rapid response to a stimulus. These responses originate from the plant's different pulvini which can be described as motor organs that cause movement by increasing or decreasing the turgor pressure within the pulvini cells [14]. The location of the three different pulvini are shown in Fig. 1, i.e. (*i*) the primary pulvini at the joint of the petiole and stem, (*ii*) the secondary pulvini located in the joints between the petiole and rachis, and (*iii*) the tertiary pulvini found at the base of each leaflet.

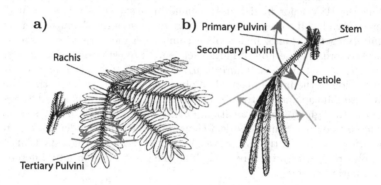

Fig. 1. *Mimosa pudica* stem and leaf description [13]. The drawings correspond to light (panel a) and dark (panel b) conditions. We have added colored lines to indicate the typical movements caused by the different pulvini, red lines correspond to primary pulvini, green to secondary pulvini and blue to tertiary pulvini. (Color figure online)

The different pulvini have also been proven to respond differently [4], where, as an example, the primary and tertiary pulvini responds to touch or vibration, while the secondary pulvini do not. The most apparent difference between the leaves at light and dark conditions is caused by the tertiary pulvini which cause the folding/unfolding of each leaflet. Our aim is therefore to capture the motion caused by these cells.

2   Experimental Setup

Each experiment is run for several weeks where the light/dark stimulus is either a combination of several different periodic patterns, or is kept constant for several days. The plant movement will in this period be recorded as a 5 min interval

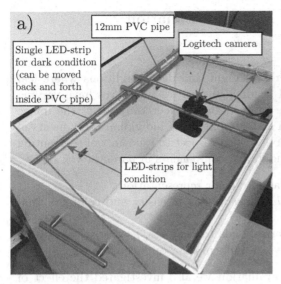

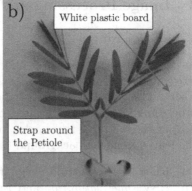

Fig. 2. (Panel a): The experimental setup where the plant is placed at the bottom of a container with white walls and white lid resulting in good light distribution conditions. There are 4 LED strips to provide light, and one single LED strip for dark conditions to allow monitoring of leaf movements. Each LED strip consists of 9 LEDs with a total power consumption of 0.7 W. (Panel b): Picture taken by the Logitech camera mounted above the plant. The leaf is resting on the surafe of a plastic board, and the petiole is loosely strapped to avoid the vertical movement caused by primary pulvini.

time-lapse series by a Logitech C615 full HD 1080 p web-camera. The camera is installed vertically above the plant in a sealed container with light emitting diode (LED) strips mounted on each of the four walls, see Fig. 2(a).

The reason for using LED as a light source is its low heat dissipation, resulting in neglectable temperature increase. In order to adjust the amount of light at dark condition, a single LED strip can be moved back and forth inside a 12 mm gray PVC pipe. Usually dark conditions are made up of 3–4 singel LEDs.

As shown in Fig. 1, the primary and secondary pulvini create motions in the petiole and rachis, respectively. For this reason it is therefore more or less impossible to isolate the folding/unfolding motions caused by the tertiary pulvini cells using a single camera mounted above the plant. However, by strapping a single leaf on a horizontally mounted plastic board enables us to isolate motions caused by the secondary and tertiary pulvini cells, see Fig. 2(b). The light/dark conditions used in this paper is 8 h of darkness and 16 h of light, giving a circadian rhythm of 24 h.

3 Image Processing and Motion Estimation

3.1 Color Model

As the images are captured under both light and dark conditions where leaf movements cause changes in local shading and where the leaves have varying color and shape, the choice of the color model will certainly affect the image processing. We therefore evaluated and compared the individual components of each of the following three color models with respect to the effect of segmenting the leaves from the background (individual components in parenthesis):

(i) RGB (*red, green* and *blue*), standard color representation.
(ii) HSV (*hue, saturation* and *value*), decoupling intensity and color information.
(iii) CIELAB (L^*, a^*, b^*, which corresponds to lightness, green-red colors, and blue-yellow colors, respectively), a uniform color opponent space.

Furthermore, as a part of the evaluation we also investigated the effect of using a colored background. From the initial results with a white background, we observed that most of the *hue* component of the plant was gathered in the yellow/green area. Hence, by using a blue colored background, the HSV and CIELAB color models gave a large distance between the background and foreground colors. Therefore, the evaluation is based on using a blue colored background, and the best individual component of each color model was

(i) the *blue* component from RGB
(ii) the *hue* component from HSV, and
(iii) the b^* component from CIELAB.

A comparison of these three representations are shown Fig. 3, where two pictures (one at dark and one at light conditions) are used as test pictures. As we see, the blue RGB component is most affected by changes in the illumination and is therefore discarded. On the other hand, both *hue* from HSV and b^* from CIELAB showed promising results, and there is no reason for choosing one over the other. However, based on the more intuitive representation of the *hue* component, the HSV color model was chosen.

3.2 Image Segmentation

One of the challenges with image segmentation is the lack of a universal method [8], and the method must therefore be chosen based on the problem to be solved. Based on the observation that the blue and green colors are well distinguished using the HSV color model, we found that an intensity-based segmentation method localizing a threshold value from the *hue* component histogram was suitable. The hue component of the image pixels $\in [0°\ 360°]$ is assigned into 100 histogram bins, and the almost uniform blueish background will be seen as a narrow symmetric uni-modal distribution in the histogram.

Fig. 3. The *dark* regime (top row) and the *light* regime (bottom row) used in the color space comparison (Color figure online)

This suggests that a suitable intensity threshold can be found by implementing a triangle based method [15]. This method is based on locating (i) the first non-zero bin and (ii) the largest bin by iterating through the histogram data. Their location is then used to parameter fit a straight line between the two bins, and the most suitable pixel intensity threshold is assumed to be the histogram bin with the largest distance to this line [15]. Compared to the manually segmented pictures, the triangle threshold segmentation performs pretty well as shown in Fig. 4. Some background pixels are though included in the foreground where the spacing between leaflets is small. An objective evaluation of the segmentation can be done using the Dice similarity coefficient as a measure, where the quotient of similarity (QS) is found as

$$QS = \frac{2|X \cap Y|}{|X| + |Y|}. \tag{1}$$

The Dice similarity coefficient is a metric that describes the similarity of two images, X and Y, in the range [0 1], where the coefficients 0 and 1 is the points of non- and full spatial overlap respectively [12]. The quotient of similarity (QS) is found by first multiplying the intersecting samples of binary value 1 in image X and Y by two, before the product is divided by the total amount of samples with binary value of 1 in both images. The calculation of the Dice similarity coefficients is executed with the *DiceScore* function developed by Dr. Hanno Scharr [12]. Compared to the ground truth image, the image at dark conditions in Fig. 4 has $QS = 0.9556$, whereas the image at light condition has $QS = 0.9366$.

3.3 Motion Estimation

The idea of estimating the movement of the leaves from the time-lapse series has its root in a recent published article [5] where two frame motion estimation is used to locate vertical movements of plants. They have used their own implementation based on the method by Lucas and Kanade [6]. The algorithm that we have chosen for the estimation of motion is the method of Farnebäck [3].

Fig. 4. Comparing manually segmentet images against the triangle threshold method. (Color figure online)

This method has the benefit of being able to estimate motion at coarser scale due to its pyramid layer implementation, and thus making it more suitable to estimate the larger displacements associated with e.g. leaf contraction. It also has the benefit of being tensor based algorithm, which is ideal for Matlab implementation in terms of running speeds. We used the implementation found in the Mathworks Matlab 2015b *Computer Vision System Toolbox*. This algorithm computes a dense *optical flow* field, i.e. for each pixel in the image a 2D velocity vector is estimated.

The method is based on approximating an image neighborhood by a quadratic polynomial. The velocity vector is modeled as a displacement of the neighborhood and it is assumed that the displacement field is slowly varying in the spatial coordinates of the image frame. The result is that the optical flow field is smeared (in the spatial image coordinates). The intensity (pixel value) on the leaves has a low variation. Therefore it is difficult to estimate reliable optical flow vectors for some parts of the leaves where the intensity is nearly constant. An example of the optical flow is shown in Fig. 5.

4   Results

Our purpose is to analyse plant behavior as a function of time when a plant is exposed to different light-dark regimes, and where the overall goal is to identify properties that will support the development of a dynamic mathematical model of the plant. An example of such a property is the total leaf motion represented by the pixel count performed on the segmented images, and a typical result is shown in Fig. 6. The curve represents the total white pixel count as a function

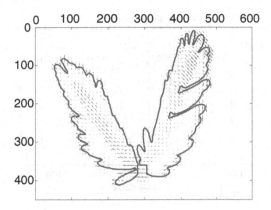

Fig. 5. Optical flow field, blue arrows. The border of the two compound leaves are shown in red. (Color figure online)

of an entrained circadian light rhythm of 8 h of darkness and 16 h of light during a 7 days experiment. The plant was prior to the experiment kept in constant light conditions for 10 days, and the result displays a number of interesting observations. First of all, the plant uses three days to be entrained by the new light/dark cycle. This can e.g. be seen at the end of each light period where the leaf first opens up *more* (indicated with three red arrows), and from the fourth day displays the anticipatory behaviour of initiating leaf closure prior to the light-to-dark transition (indicated with green arrows).

Secondly, we saw from inspection of the time lapse movie that the *leaflets* behaved differently hours before the light-to-dark transition compared to the hours before the dark-to-light transition. The behavior of the leaflets consisted of rapid motions, similar to shivering, which is recorded in the pixel count as noise, see e.g. in the range from frame number 1100 to 1200. To capture and quantify this behavior, we estimated this type of motion by optical flow. However, the complete optical flow for each frame is difficult to use, and in order to find a more simple description we computed the distribution of velocity magnitudes and velocity directions. As a measure we used the mean flow magnitude and the main flow direction of motion computed for each image frame. The main direction is computed as the maximum of the histogram for the directions. If the directions are too diverse, i.e. several peaks in the histogram, no value is computed.

As the leaf used in the experiment consists of *two* compound leaves (see Fig. 5), the motion estimation will therefore result in averaged values based on often opposite individual values, and is thereby of limited relevance. Hence, in order to capture the individual motion of each compound leaf, we separated the leaf by introducing the stem indicated in red in Fig. 7.

These images correspond to the frames 880, 881, 892 and 893 from Fig. 6, and represents the situation prior to and right at the light-to-dark transition. For the two lower images, the contraction of the compound leaves results in a

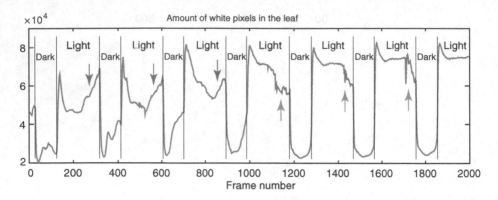

Fig. 6. Amount of white pixels in the leaf during a 7 days experiment as a function of frame number. Time distance between each frame is 5 min. The dark period starts at 12 p.m. and lasts for 8 h, whereas the light period starts at 8 a.m. and lasts 16 h. The dark/light period is therefore 24 h. (Color figure online)

Fig. 7. Separation of the leaf into the two compound leaves. In the situation when the compound leaves interfere, the result is summed and divided by two. The two upper images correspond to the frames 880 and 881 prior to the light-to-dark transition shown in Fig. 6. The two lower images correspond to the frames 892 and 893 coinciding with the same transition. (Color figure online)

situation where the leaf is considered as one, and the motion estimation for the two parts are calculated from the entire leaf.

An example of (*a*) white pixel count, (*b*) mean optical flow and (*c*) flow direction is shown in Fig. 8, which corresponds to the frames 850 to 1050 from Fig. 6 and where the light-to-dark transition occurring at frame 892 is indicated.

In order to relate the upper two images of Fig. 7 to the results in Fig. 8, we have indicated frame 880 and we see that there are individual results (one black and one red curve) for the white pixel count, the mean optical flow and the flow direction. We also see that around the light-to-dark transition at frame 892, the calculated values are similar for the black and red curves (corresponds to the leaf being considered as one in the lower two images of Fig. 7).

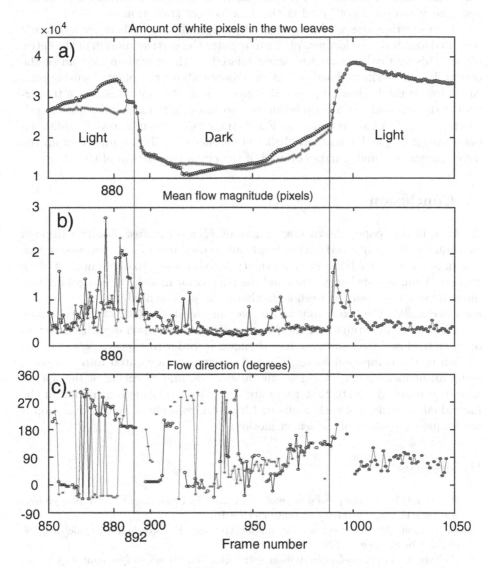

Fig. 8. Detailed results for the frames 850 to 1050 in Fig. 6. The black curve represents one of the two compound leaves and the red curve the other. There is a light-to-dark transition at frame number 892 and a dark-to-light transition at frame number 989, both marked by a vertical line. Frame 880 is indicated to relate the result to the two upper images of Fig. 7. (Panel a): White pixel count. (Panel b): Optical flow magnitude. (Panel c): Optical flow direction. (Color figure online)

An interesting result is that the shivering we manually observed in the time lapse movie prior to the light-to-dark transition is found in the flow magnitude and direction in Fig. 8 (panels b) and (c). We see that the optical flow has large variations between frames around the periode of light-to-dark transition, though a similar variation is not found at the dark-to-light transition.

To summarize, our work was originally inspired by [5] where cameras were mounted alongside *Arabidopsis* plants to register the vertical motion of the entire plant. This vertical motion was again related to the circadian rhythm of the plants. In our work, we have placed the camera above the plant, focusing on a single leaf in the horizontal plane. We apply (*i*) pixel count to estimate the circadian rhythm and (*ii*) motion estimation to investigate motion details for each compound leaf. Thus, as shown in Fig. 8, these two measures reveal additional details regarding leaf motion. We believe that these details are important for the development of a mathematical model of the circadian rhythm of the plant.

5   Conclusion

We have in this paper shown how images of *Mimosa pudica* obtained through experiments in a controlled environment can be used in an image processing routine in order to reveal behavioral information with respect to different light/dark regimes. Using a blue background and the HSV color model, the images are segmented using the triangle threshold method. These segmented images are further used to calculate the pixel count which gives an overall estimate of the leaf movement and also an estimate of the plants anticipatory behavior with respect to open or to close leaves prior to light changes. In order to capture details in the motion of the compound leaves, the leaf images are separated into two parts using an artificial stem. Based on the magnitude and direction of the optical flow, the results indicate that there are behavioral information hidden in the individual movement of each compound leaf. An example of such information is the frequency content of the leaflet motion.

References

1. Brattland, V.: Image processing and analysis of leaf movement in *Mimosa pudica*. Master's thesis, University of Stavanger (2016)
2. De Mairan, J.J.: Observation botanique. Histoire de l'Academie Royale des Sciences. Des Sciences (1729)
3. Farnebäck, G.: Two-frame motion estimation based on polynomial expansion. In: Bigun, J., Gustavsson, T. (eds.) SCIA 2003. LNCS, vol. 2749, pp. 363–370. Springer, Heidelberg (2003). doi:10.1007/3-540-45103-X_50
4. Fromm, J., Eschrich, W.: Transport processes in stimulated and non-stimulated leaves of *Mimosa pudica*. Trees **2**(1), 18–24 (1988)
5. Greenham, K., Lou, P., Remsen, S.E., Farid, H., McClung, C.R.: TRiP: Tracking Rhythms in Plants, an automated leaf movement analysis program for circadian period estimation. Plant Methods **11**(33), 1 (2015). doi:10.1186/s13007-015-0075-5

6. Lucas, B.D., Kanade, T.: An iterative image registration technique with an application to stereo vision. In: Proceedings of DARPA Image Understanding Workshop, pp. 121–130 (1981)
7. Mancuso, S., Shabala, S.: Rhythms in Plants. Springer, Heidelberg (2007)
8. Marques, O.: Practical Image and Video Processing using MATLAB. Wiley, Hoboken (2011)
9. McClung, C.R.: Plant circadian rhythms. Plant Cell **18**(4), 792–803 (2006)
10. Ruoff, P., Rensing, L.: Temperature effects on circadian clocks. J. Thermal Biol. **29**, 445–456 (2004)
11. Ruoff, P., Vinsjevik, M., Monnerjahn, C., Rensing, L.: The Goodwin Model: simulating the effect of light pulses on the circadian sporulation rhythm of Neurospora crassa. J. Theoret. Biol. **209**, 29–42 (2001)
12. Scharr, H., Minervini, M., French, A.P., Klukas, C., Kramer, D.M., Liu, X., Luengo, I., Pape, J.M., Polder, G., Vukadinovic, D., et al.: Leaf segmentation in plant phenotyping: a collation study. Mach. Vis. Appl. **27**(4), 585–606 (2015)
13. Taubert, P.H.W.: Leguminosae. Natürliche Pflanzenfamilien III (1891)
14. Taya, M.: Bio-inspired design of intelligent materials. In: Smart structures and materials, pp. 54–65. International Society for Optics and Photonics (2003)
15. Zack, G.W., Rogers, W.E., Latt, S.A.: Automatic measurement of sister chromatid exchange frequency. J. Histochem. Cytochem. **25**(7), 741–753 (1977)

Evaluation of Visual Tracking Algorithms
for Embedded Devices

Ville Lehtola, Heikki Huttunen, Francois Christophe$^{(\boxtimes)}$, and Tommi Mikkonen

Tampere University of Technology, Tampere, Finland
{ville.lehtola,heikki.huttunen,
francois.christophe,tommi.mikkonen}@tut.fi

Abstract. Today's embedded platforms enable executing difficult tasks such as visual tracking. However, such resource-constrained systems are still facing challenges regarding the performance and accuracy in executing these tasks. This paper presents the evaluation of 5 open-source visual tracking implementations available from the contributions branch of the Open Computer Vision (OpenCV) library. This evaluation is performed based on the performance and accuracy of these implementations when embedded in a Raspberry Pi. The algorithms evaluated are On-Line Boosting, Multiple Instance Learning (MIL), Median Flow, Tracking-Learning-Detection (TLD), and Kernelized Correlation Filters (KCF). Even if commercial implementations of these algorithms perform better than their open-source version, the popularity of OpenCV motivates this evaluation. Tests are based on a benchmark of 100 video streams from which the tracking implementations should follow moving objects. The algorithms are evaluated for accuracy using averaged Jaccard indices and for performance by measuring their frame rate. We want to find an open-source implementation that performs well on these two criteria when tested on an embedded platform. Results show Median Flow being the fastest but its accuracy is the lowest. We therefore recommend KCF as it is the second fastest and the most accurate.

Keywords: Tracker algorithms · OpenCV

1 Introduction

Object detection from still images and object tracking in video streams are important problems in computer vision. During the recent years, there have been many breakthroughs in object detection and localization exploiting the advances in deep learning—often surpassing human level accuracy [6,11]. While it is possible to track objects by re-detecting them in each frame (tracking by detection), there are several drawbacks to this approach: tracking is limited to categories used at training time; the detector may lose track of the object in poor illumination, pose changes, etc.; and most importantly, object detection based on modern deep learning requires heavy computation usually done on a GPU.

© Springer International Publishing AG 2017
P. Sharma and F.M. Bianchi (Eds.): SCIA 2017, Part I, LNCS 10269, pp. 88–97, 2017.
DOI: 10.1007/978-3-319-59126-1_8

Therefore, it is often advantageous to approach the problem from the traditional angle of generic object tracking.

Tracking algorithms typically differ from detection in that they learn the changes in object appearance over time. Trackers typically implement two functions: add and update, with the first initialized every time a new object appears in the scene, and the latter applied subsequently for each frame. The variations between the algorithms reside in the update step, where each method essentially learns the most recent appearance of each target. Moreover, some algorithms may exploit the historical information about motion trajectories of objects, for example in order to limit the search region. Tracking algorithms tend to have a lower computational complexity than a full-scale detection approach would have. Although the accuracy of different tracking algorithms has been widely studied (see, *e.g.,* [12,13]), the computational load has gained less attention. The execution speed varies across algorithms, which easily becomes a limitation when implementing real-time detection in low-resource platforms without any hardware acceleration (such as a GPU).

From the point of view of embedded systems and the increasing interest in the Internet of Things (IoT), the accuracy is often secondary to the use of resources: the burden of both computation and integration tends to dictate the choice of algorithms. Therefore, we consider five tracking algorithms readily implemented in OpenCV library, which is used as a component in many open source projects today. Moreover, we will get an insight into the actual implementations of this widely used library, which are not exactly the same as used in the original papers where the algorithms were first proposed.

This paper is structured as follows. Section 2 introduces the trackers available in the OpenCV library and their original papers. Section 3 introduces the dataset and equipment used in the experiments and defines the methods used to determine the quality of the tested trackers. Section 4 presents the results. Section 5 concludes.

2   Tracking Algorithms

This section presents a brief description of the trackers evaluated in this study. The algorithms evaluated are available from the contributions branch of the Open Computer Vision (OpenCV) library [2].

On-Line Boosting. On-line boosting is a tracking algorithm that considers tracking as a binary classification task [4,5]. At each tracking update step, the algorithm updates the object model by using AdaBoost for training a collection of weak classifiers. The training step uses the target region as a positive example and samples patches from the vicinity of the currently tracked object region. The object location in the next frame is then estimated by applying the binary classifier in the next frame, and choosing the most likely location as predicted by the AdaBoost classifier.

Multiple Instance Learning (MIL). Similarly to the on-line boosting app-roach, Multiple Instance Learning algorithm (MILBoost) poses the tracking problem as a classification task [14]. For classification, the method uses the multiple instance learning approach, which considers *bags* of objects by group-ing similar samples into bags, where each bag is considered an overall positive sample if at least one of the individual samples it contains is positive, and nega-tive otherwise. This attempts to avoid confusing the classifier with sub-optimal samples being labeled as positive, instead giving the classifier a more vague understanding of what positive samples should be like. More recently, an on-line modification of the learning method was proposed [1], which is more suitable for tracking due to lower computational load.

Median Flow. Median Flow is presented in [9]. The authors present a tracking error measure, where points of each object are tracked both forward and back-ward in time and the resulting trajectories are compared. By the assumption that a correct tracking method would yield the same but opposite trajectory when running on time reversed input, any divergence of the forward and back-ward tracking trajectories indicates a tracking error. The authors use this error measure to propose a tracking method where points inside a bounding box are tracked and measured for error, then classified to inliers and outliers by the result. Outliers are filtered out and the bounding box motion is estimated based on the inliers.

Tracking-Learning-Detection (TLD). Tracking-Learning-Detection is a method for long-term tracking tasks [10]. The authors describe it as a frame-work rather than a tracking method as the different stages, tracking, learning and detection, are performed by separate components of the overall system. The goal of the method is to improve the robustness of tracking by disabling the on-line learning if the object is out of frame or completely occluded by other objects, thus avoiding learning from misinformation. The detection component also allows the method to re-detect the object, should it reappear in the video later.

Kernelized Correlation Filter (KCF). The Kernelized Correlation Filter method [8] employs the shift invariance property of Fourier transform for design-ing a fast algorithm for correlation filter based matching. The Fourier transform simplifies the correlation computation to make it extremely fast. Further, the method generalizes by applying the kernel trick to allow nonlinear correlation measures. The authors note that one of the main challenges in tracking is the inability of using a large enough number of training data available from each frame of input due to high computational load. This problem is avoided in KCF due to its lightweight implementation.

3 Evaluation Method

The primary interest of evaluation was not only to find out how accurate the trackers are, but also to measure the viability of the OpenCV implementations as real-time tracking solutions on hardware with severe memory and performance constraints. Each algorithm was initialized with the ground-truth bounding box from the first frame of a sequence and only the default parameters for the tracker were used.

Dataset Used for Evaluation. The dataset of 100 video sequences used in the evaluation is from the visual tracking benchmark by Wu and Lim, who originally ran the benchmark on 29 trackers in [13]. From those trackers, three—MIL, TLD and Boosting (OAB in original benchmark)—are implemented in OpenCV and the other two provide completely new benchmarks on the full data, although the authors of KCF have run their implementation of the algorithm on a 50 sequence subset of the dataset [8].

Description of Material and Experiment. A Raspberry Pi 3 B v1.2 was used as the hardware. The board has a 64 bit CPU with 4 cores clocked at 1.2 GHz, and 1 GB RAM memory, with a default of 100 MB of virtual memory. The RAM is shared with the GPU, leaving 862 MB for general use by the operating system. The board was installed with a Raspbian GUI-based OS. This was necessary as the board was intended to be used in various tasks and it also helped with debugging the tracking programs by allowing the user to see the input and output of the software.

OpenCV was compiled without OpenMP [3] installed on the Raspberry Pi and as such the algorithms were tested on single core executions.

Evaluation of Performance. The performance of the algorithms was determined by measuring the processor time spent in the methods that update the bounding box of the tracker. The time spent initializing the tracker was not measured. The number of frames in a sequence was divided by the total time spent on all frames, producing a measure for frames per second, FPS.

Evaluation of Accuracy with the Jaccard Index of Similarity. The Jaccard index is a common measure of similarity between two sets A and B. This index is expressed as follows:

$$J(A, B) = \frac{A \cap B}{A \cup B} \tag{1}$$

When following an object, the Jaccard index presents a good measure of accuracy when comparing bounded boxes of the object followed drawn by the tested algorithm with bounded boxes of reference. When considering n frames of a video, R as the set of ground-truth bounding boxes in each frame and A the set of bounding boxes drawn by the tested algorithm, we derive a global similarity between A and R from Eq. 1 as follows:

$$J_R(A) = \frac{\sum\limits_{i=1}^{n} J(A_i, R_i)}{n} \tag{2}$$

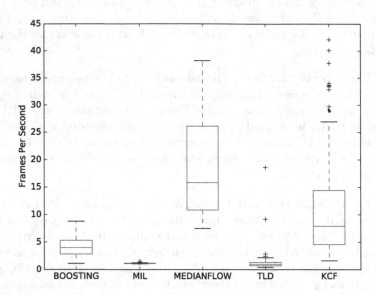

Fig. 1. Box plot illustrating averages and distributions of frame rates. Four (4) outliers above 50 FPS between Median Flow and KCF are missing to make the y axis scale better.

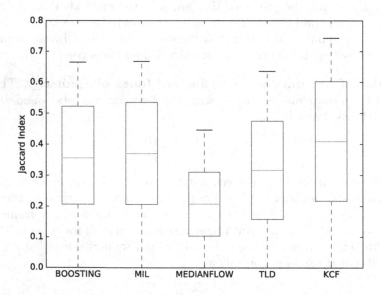

Fig. 2. Box plot illustrating averages and distributions of Jaccard indices.

4   Results

This section presents a visualization of results obtained from the benchmark. The code for the benchmark and the table of results are available from github[1].

The tests showed that from the implemented trackers only Median Flow and TLD were able to adapt to scale changes in the tracked object, that is to the change of the size and shape of the detected bounding box.

Figure 1 illustrates the typical performance of the trackers in terms of processed frames per second. Median Flow is typically the fastest and KCF the second fastest, although KCF reaches very high performance on a few sequences. The Boosting algorithm sits between the fastest and slowest, and MIL and TLD are the slowest ones.

Only Median Flow and KCF trackers are able to perform at speeds that could be considered as real-time. The performance of any of these trackers in a practical application can be increased by reducing the resolution of the video feed, so the result can also be interpreted such that Median Flow and KCF allow for the largest resolution while maintaining real-time performance.

Figure 2 compares the statistics of the trackers with respect to the measured Jaccard indices. The Median Flow algorithm has a notably low accuracy. The rest share similar average accuracies, but KCF has the best average and can reach better accuracy than the others in best cases.

Figures 3(a) to (e) show the Jaccard index as a function of video frame for the sequence 'Football1'. The sequence shows players clashing in a game of American football, with the tracking target as one of the players faces. The images above the charts show selected frames from the sequence with the same frames shown for every tracker. The green rectangle in the images is the ground truth and the yellow rectangle is the tracking result.

Most of the trackers yield good results on this sequence and surpass their average performance. The boosting tracker (Fig. 3a) comes close to losing the target around the middle of the sequence, but is able to recover before finally losing the target at the end. The MIL tracker (Fig. 3b) shows a gradual degradation of tracking accuracy with some sudden jumps and finally loses the target. Despite being the lowest accuracy tracker, the Median Flow tracker (Fig. 3c) performs exceptionally well on this sequence; in fact better than any of the others. The low overall accuracy of this tracker seems to be caused by its tendency to completely lose its target and when this occurs, it moves its bounding box to the upper left corner of the frame and resizes it to zero, giving it no chance to recover. It seems to be a powerful tracker, if only it could recover from losing its target by e.g. re-detection. The TLD tracker (Fig. 3d) performs somewhat poorly on the sequence. The tracking bounding box jumps around erratically in the middle part of the sequence implying that this implementation of the tracker is relying mostly on tracking by detection rather than moving the bounding box in small increments. The KCF tracker (Fig. 3e) shows a mostly reliable tracking

---

[1] https://github.com/lehtolav/tracker-benchmark.

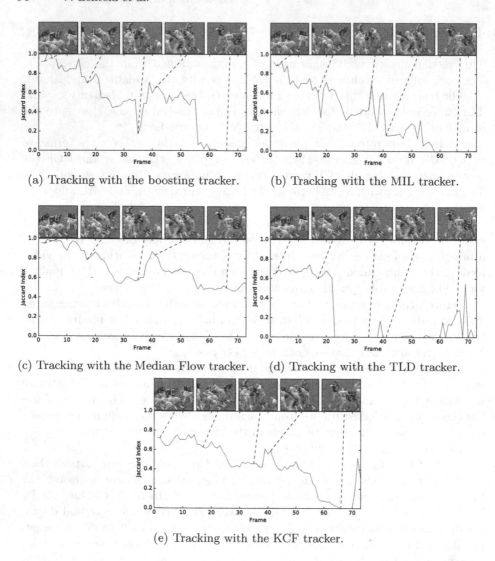

(a) Tracking with the boosting tracker.   (b) Tracking with the MIL tracker.

(c) Tracking with the Median Flow tracker.   (d) Tracking with the TLD tracker.

(e) Tracking with the KCF tracker.

Fig. 3. Examples of the evolution of Jaccard index with the 5 implementations on the *football*1 video stream

accuracy with some degradation over time until in the end it finally loses the target and suddenly finds it again.

The success rate is represented as a function of overlap threshold in Fig. 4 similarly to the benchmark in [13]. The overlap threshold is the minimum Jaccard index required per frame. The success rate is the amount of frames that reached more than the threshold for all sequences. For example, 51% of the frames tracked with KCF have a Jaccard index higher than 0.4. The number in square brackets is the area under curve (AUC) of the tracker. We see that the MIL and TLD

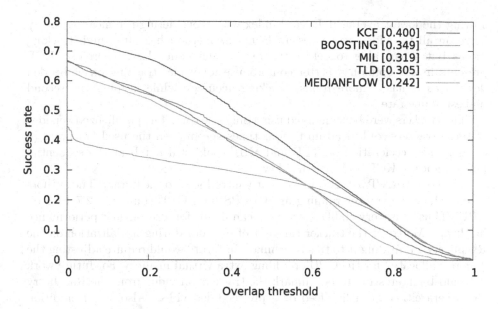

Fig. 4. Success plots of the five tested trackers showing the fraction of frames successfully tracked as a function of required Jaccard index.

implementations in OpenCV are inferior to those used in the benchmark, with MIL scoring 0.319 versus 0.362 and TLD scoring 0.305 versus 0.437. Notably, even the best of the OpenCV trackers, KCF, falls short of the performance of the TLD tracker of the benchmark, although TLD was only the third best scoring method there, with Struck scoring 0.473 and SCM scoring 0.499. While running the benchmarks, it became clear that the TLD implementation does not allow the object detection to fail, but forced it to detect the most likely candidate from the frame. This led to many false positives that confused the method.

5 Conclusion

In this paper, we presented an evaluation of 5 tracker implementations. These 5 trackers are Median Flow, MIL, On-line Boosting, TLD, and KCF all of which are available from the contributions branch of th OpenCV library. This evaluation is based on two performance criteria. The first criterion is related to the rapidity of execution of implementations when embedded on a Raspberry Pi. The second criterion evaluates the accuracy in tracking with an averaged Jaccard index. According to this criterion, KCF scores highest. This quick evaluation method by averaged Jaccard indices (Fig. 2) gives the same ranking as the area under curve (Fig. 4) so it can be used as a fast approximation for selecting a tracking implementation.

It appears that Median Flow presents a high frame rate and is the fastest of the 5 implementations. Despite this rapidity of execution, it shows a low accuracy scoring the lowest averaged Jaccard index when considering the mean accuracy score on all the video sequences. In fact, this method has often high tracking scores but is not able to detect the target again once lost. Therefore, KCF presents in our opinion a better compromise for visual tracking on embedded devices as it has a more robust tracking accuracy while keeping the second highest frame rate.

The trackers were benchmarked using only one core. The parallelized versions of these trackers could lead up to four times speedup on the used hardware (4 cores). This acceleration on TLD and MIL would still not beat the sequential performance of KCF and Median Flow.

The GOTURN Tracker [7] was recently introduced to the library. The authors results show the tracker running at 165 FPS on a GPU, but only 2.7 FPS on CPU. Thus it is very unlikely to be a candidate for use on low performance hardware. Moreover the tracker runs out of memory during initialization on the Raspberry Pi. Adding it to this benchmark in future would require adjusting the memory allocated for the GPU or adding extra virtual memory. For future work it would be interesting to benchmark the trackers on video from the Raspberry Pi camera either on a live feed or a pre-recorded video. Also, as an addition to the results in this paper it would be possible to measure the accuracy and performance of the trackers as a function of video resolution by using several downsized versions of the sequences.

Acknowledgement. This research is funded by the Academy of Finland under project named "Bio-integrated Software Development for Adaptive Sensor Networks", project number 278882. The work presented in this article is part of Ville Lehtola's Master thesis. The authors also thank Jani Boutellier for his valuable comments to this article.

References

1. Babenko, B., Belongie, S.: Visual tracking with online Multiple Instance Learning. In: 2009 IEEE Conference on Computer Vision and Pattern Recognition, pp. 983–990 (2009)
2. Bradski, G.: The OpenCV library. Dr. Dobb's J. Softw. Tools **25**(11), 122–125 (2000). http://www.drdobbs.com/open-source/the-opencv-library/184404319
3. Dagum, L., Menon, R.: OpenMP: an industry standard API for shared-memory programming. IEEE Comput. Sci. Eng. **5**(1), 46–55 (1998)
4. Grabner, H., Bischof, H.: On-line boosting and vision. In: Proceedings of the IEEE Computer Society Conference on Computer Vision and Pattern Recognition, vol. 1, pp. 260–267 (2006)
5. Grabner, H., Grabner, M., Bischof, H.: Real-time tracking via on-line boosting. In: Proceedings of the British Machine Vision Conference, vol. 1, pp. 1–10 (2006)
6. He, K., Zhang, X., Ren, S., Sun, J.: Deep residual learning for image recognition. In: 2016 IEEE Conference on Computer Vision and Pattern Recognition (CVPR). IEEE, Las Vegas (2016). http://ieeexplore.ieee.org/document/7780459/
7. Held, D., Thrun, S., Savarese, S.: Learning to track at 100 FPS with deep regression networks. In: Leibe, B., Matas, J., Sebe, N., Welling, M. (eds.) ECCV 2016. LNCS, vol. 9905, pp. 749–765. Springer, Cham (2016). doi:10.1007/978-3-319-46448-0_45

8. Henriques, J.F., Caseiro, R., Martins, P., Batista, J.: High-speed tracking with kernelized correlation filters. IEEE Trans. Pattern Anal. Mach. Intell. **37**(3), 583–596 (2015)
9. Kalal, Z., Mikolajczyk, K., Matas, J.: Forward-backward error: automatic detection of tracking failures. In: Proceedings - International Conference on Pattern Recognition, pp. 2756–2759 (2010)
10. Kalal, Z., Mikolajczyk, K., Matas, J.: Tracking-learning-detection. IEEE Trans. Pattern Anal. Mach. Intell. **34**(7), 1409–1422 (2012)
11. Redmon, J., Divvala, S., Girshick, R., Farhadi, A.: You only look once: unified real-time object detection. In: CVPR 2016, pp. 779–788 (2016)
12. Smeulders, A.W.M., Chu, D.M., Cucchiara, R., Calderara, S., Dehghan, A., Shah, M.: Visual tracking: an experimental survey. IEEE Trans. Pattern Anal. Mach. Intell. **36**(7), 1442–1468 (2014)
13. Wu, Y., Lim, J., Yang, M.H.: Online object tracking: a benchmark. In: Proceedings of the IEEE Computer Society Conference on Computer Vision and Pattern Recognition, pp. 2411–2418 (2013)
14. Zhang, C., Platt, J.C., Viola, P.A.: Multiple instance boosting for object detection. Neural Inf. Process. Syst. **74**, 1769–1775 (2005). http://papers.nips.cc/paper/2926-multiple-instance-boosting-for-object-detection.pdf

Multimodal Neural Networks: RGB-D for Semantic Segmentation and Object Detection

Lukas Schneider[1,2]([✉]), Manuel Jasch[3], Björn Fröhlich[1], Thomas Weber[3], Uwe Franke[1], Marc Pollefeys[2,4], and Matthias Rätsch[3]

[1] Daimler AG, Stuttgart, Germany
lukas.schneider@daimler.com
[2] ETH Zurich, Zurich, Switzerland
[3] Reutlingen University, Reutlingen, Germany
[4] Microsoft Corporation, Seattle, USA

Abstract. This paper presents a novel multi-modal CNN architecture that exploits complementary input cues in addition to sole color information. The joint model implements a mid-level fusion that allows the network to exploit cross-modal interdependencies already on a medium feature-level. The benefit of the presented architecture is shown for the RGB-D image understanding task. So far, state-of-the-art RGB-D CNNs have used network weights trained on color data. In contrast, a superior initialization scheme is proposed to pre-train the depth branch of the multi-modal CNN independently. In an end-to-end training the network parameters are optimized jointly using the challenging Cityscapes dataset. In thorough experiments, the effectiveness of the proposed model is shown. Both, the RGB GoogLeNet and further RGB-D baselines are outperformed with a significant margin on two different task: semantic segmentation and object detection. For the latter, this paper shows how to extract object-level groundtruth from the instance level annotations in Cityscapes in order to train a powerful object detector.

1 Introduction

Semantic interpretation of image content is one of the most fundamental problems in computer vision and is of highest importance in various applications. The availability of extremely large datasets has pushed the development of strongly data-driven machine learning methods. In particular, convolutional neural networks (CNNs) have pushed the state of the art in image understanding in various different tasks and applications. Simultaneously, the costs for cameras with increasing resolution have decreased substantially in the last years. We expect this trend to continue and thus focus on methods that can deal with such high resolution images. At the same time, we are interested in efficient methods that can meet high real-time requirements as in e.g. robotics or autonomous driving. Naturally, the main focus in the computer vision community has been in the interpretation of color images which neglects the availability of complementary

© Springer International Publishing AG 2017
P. Sharma and F.M. Bianchi (Eds.): SCIA 2017, Part I, LNCS 10269, pp. 98–109, 2017.
DOI: 10.1007/978-3-319-59126-1_9

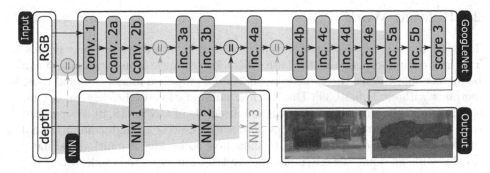

Fig. 1. Structure of our mid-level fusion approach. A GoogLeNet is used on top of the RGB input image and a NiN for the depth information. Alternative processing paths to fuse both networks are shown grayed.

inputs from other domains, e.g. depth, infrared, or motion. In this work we focus on depth data as additional input to CNNs. However, the presented approach is easily adaptable to other modalities.

Only using state-of-the-art CNN approaches for the multi-modal data is not optimal, since huge datasets such as ImageNet [33], MS COCO [28] or places [40] only provide color images and do thus not allow the training of large multi modal CNNs. Two main different approaches have emerged to deal with this problem. Either, only the small amount of data is used for training while accepting the resulting degraded performance. Or existing RGB networks are simply applied to the new domain and fused with those fork responsible for the color domain.

This paper proposes a novel network architecture, c.f. Fig. 1, that implements a mid-level fusion of features from the individual input domains. This combines both advantages of the previous approaches: first, the network can exploit highly complex intra-domain dependencies through the joint feature processing in order to maximize the semantic accuracy of the network Secondly, it allows the reuse of the existing initialization on large datasets. Furthermore, we demonstrate that using a network designed and trained for color inputs is suboptimal in the depth domain and propose a superior adapted architecture together with an initialization scheme yielding significant improvements in terms of semantic accuracy. The experiments show that filters learned on depth data with this approach differ substantially to those obtained by a training on RGB data.

Overall, this paper presents a simple yet effective novel network architecture together with an initialization scheme that exploits depth data in addition to sole color information. This approach leads to a significant improvement on two different common tasks in computer vision: semantic segmentation and object detection. It is based on an standard state-of-the-art network architecture and is easily adaptable to different modalities as well as tasks.

2 Related Work

The vast amount of relevant literature can be split into three different groups. The first comprises methods that use CNNs for semantic segmentation and can be further split into methods using an additional graphical model [2,3] and those methods without [1,8,19,39]. Due to their computational efficiency we opt for a purely CNN-based approach. Using an ensemble of CNNs [18,19,38] can lead to significant performance gains, however, with the cost of high computational burden. The scope of this paper is to show how to benefit of multi-modal data. To this end, we don't use ensembles and restrict ourselves to standard network architectures and training schemes.

The second line of work is formed by CNN's for object detection. Current literature differs between two basic approaches. First, methods such as RCNN [14], Fast RCNN [13], or R-FCN [23] require a previous hypotheses generation step and finally classify each hypothesis. As another group of methods there are e.g. Overfeat [34], YOLO [32] or SSD [29] without extra hypotheses generation. For a more detailed overview and comparison of multiple state of the art object detection methods we refer to [22]. Due to the excellent trade off between computational time and performance, we focus in this work on SSD.

Finally, we identify those methods basing on CNNs that exploit depth data, as the most related line of work. Apparently, most works use additional input features such as height, depth or angle of gradient [4,15,21,26]. Instead, we simply rely on inverse depth as input in addition to the color image. Some methods use graphical models to increase the semantic accuracy with the cost of more computational demands [9,24]. The depth input has been used to select the scale in a scale pyramid [25]. This way, however, no depth features such as depth discontinuities can be exploited. This method serves as baseline in our experimental section, c.f. Section 4. A further distinction between methods in this group is the level of fusion. Fusing color and depth data at an early level, i.e. concatenating the inputs directly, has been studied by [4,7,37]. But [4] report better results with a late fusion. We address this observation to the little availability of labeled multi-modal data. Most existing works, on the other hand, opt for a late fusion, i.e. separate network streams for depth and color data. Either a classifier is applied on the independently trained networks [4,15,16] or the networks are fused in one of the last layers and a joint training is carried out [10,17,21,26]. In the spirit of end-to-end learning, we also perform joint training. In contrast to these methods, this work shows the benefit of a mid-level fusion of learned features from the depth and color domain.

Most related to this paper is recent work that also implements a mid-level fusion with RGB-D data [17]. However, some significant differences exist: first, they use a decoder architecture with unpooling layers based on the SegNet architecture [1]. Due to the poor reported results on the Cityscapes dataset [5], we opt for a learned transpose convolutional (also referred to as deconvolutional) decoder instead. It seems that the SegNet architecture suffers from high-resolution input images. Also [17] use a small resolution of 224×224 px as input although the dataset Sun RGBD [36] provides varying resolutions around

640×480 px. This paper focuses on high resolution images - we use input resolutions up to 2048×1024 px during both: training and testing. Finally, an initialization of the depth branch with ImageNet pre-trained weights is needed in that work. In this paper, we show that this is non-optimal: the parameters trained for depth data lead to different filters in the CNN and to superior results. We hope that this paper will stimulate more works that exploit depth data on the challenging high-resolution dataset Cityscapes [6].

We consider the following as main contributions of this paper: (1) A novel generic mid-level fusion network architecture is proposed together with an experimentally grounded initialization scheme for the additional modality. This network is simple yet effective and can be easily adapted to different modalities and tasks. (2) In thorough experiments on the Cityscapes dataset, the effectiveness of the proposed approach as well as the influence of the important design choices is demonstrated. Both, the RGB as well as an RGB-D baseline are outperformed with a significant margin on two different challenging tasks: semantic segmentation and object detection. (3) Finally, we show how to use the pixel-level annotations of the Cityscapes dataset to train an powerful neural network for object detection. To this end, the well-known SSD approach is adapted to the GoogLeNet and extended to the proposed multi-modal CNN.

3 Method

In this work, we propose a novel deep neural network architecture that can exploit other modalities such as depth images in addition to sole color information. Since in many cases no large datasets like ImageNet exist for the new modalities, c.f. Sect. 2, simply using an existing state-of-the-art CNN architecture and performing a training for multi-modal data is unfortunately not possible. Instead, we adapt the frequently used GoogLeNet [38] and fuse it with a network branch optimized for depth data. Note that the modifications described in this work are easily adaptable to other modalities, e.g. optical flow or infrared, as well as other network architectures, e.g. Network-in-Network (NiN) [27], VGG [35] or ResNet [18].

Depth Network. For the depth branch, we train and adapt a NiN [27] variant for sole depth data and use the large semi-supervised part of the Cityscapes dataset [6] for initialization. A NiN consists of multiple modules, each being further composed of one convolutional layer with a kernel size larger than one that captures spatial information and multiple 1×1 convolutional kernels. Such a module is equivalent to a multi-layer perceptron (MLP). For classification, a global average pooling layer yields one score per class. We follow [31] and discard the global average pooling resulting in a FCN [30] that predicts one score per pixel and class.

We argue that depth data requires filters that differ significantly from those obtained via training on RGB data. For instance, we expect edge and blob filters to be wider in order to be robust to the noisy depth estimates. For this reason we use random initialization for training and reduce the channel count in each layer to $\frac{1}{3}$ considering that depth yields only one instead of the three color input channels.

RGBD Network. The GoogLeNet consists of a first part with convolutional and max pooling layers that quickly reduce the spatial resolution. This part is followed by nine inception modules including further pooling layers each halving the spatial dimension as illustrated in Fig. 1. We identify different points for joining the depth and RGB network. First, the RGB and depth input can be concatenated directly resulting in a new first convolutional layer. We will refer to this model as *early fusion*. Second, the scores of the RGB network and depth branch can be concatenated at the end, followed by a 1×1 convolutional as classifier. We will refer to this as *late fusion*. Finally, scores of the depth branch can be merged in the RGB network before one of it's max-pooling layer, again followed by a 1×1 convolutional layer. The number of NiN modules used in this mid-level fusion approach is determined by the required spatial dimensional in the RGB network. Thus, we call these models according to the number of NiN modules, e.g. *NiN 1*.

In theory, a multi-modal CNN with an early fusion as described above can develop independent network streams by learning features that only take one input modality into account. Thus, an early fusion is generally more expressive then a mid-level fusion, it can exploit correlations between the modalities already on a low-level of CNN computation. However, the higher expressivity comes with the price that larger amounts of data might be required for training. The benefit of a later fusion is that most of the network initialization can be reused directly without the necessity to adapt the network weights to the additional input cue. Unfortunately, it does not allow the network to learn this high-level interdependencies between the individual input modalities, since only the resulting scores on classification level are fused.

4   Experiments

We evaluate our proposed model on two different tasks: semantic segmentation, the task to assign a semantic label to each pixel in an image c.f. Sect. 4.2, and object detection, c.f. Sect. 4.3. The initialization of the depth network branch is described and evaluated in Sect. 4.1.

Dataset. Throughout our experiments, we use the Cityscapes dataset that provides a high number of pixel-level semantic annotations with 19 classes, e.g. person, car, road etc., in challenging inner city traffic scenarios. In addition to this fine annotations, 20 000 coarsely annotated images are provided. The coarse labels are more quickly and thus more cheaply annotated images where objects are labeled via polygons. Although this way many pixels remain unlabeled, each annotated pixel is defined to be correct.

4.1   Depth Network

Evaluation and Training Details. We use the 20 000 coarsely annotated images to train a NiN for scene labeling c.f. Sect. 3, consisting of three NiN modules with two 1×1 convolutional layers each. We follow [30] and add two skip layers in order to exploit low-level image features for the expanding part of the network. We opt for a batch size of ten and use random crops during training to account for the GPU memory limitations. As depth input, the publicly available stereo data obtained via semi-global matching [20] is used. More precisely, we follow [31] and encode depth as disparities, i.e. inverse depth. Missing measurements are encoded as -1, the mean value is subtracted. After this initialization phase, the network is fine-tuned on the 2975 finely annotated training image of Cityscapes. For evaluation, we use the 500 validation images. As evaluation metric, we use Intersection-over-Union (IoU) [6] defined as $IoU = \frac{TP}{TP+FP+FN}$, where TP, FP, and FN are the numbers of true positive, false positive and false negative pixels determined over the whole dataset.

The 19 Cityscapes classes are grouped to seven categories: flat, construction, nature, vehicle, sky, object, and human. In addition to the IoU on the 19 classes (*IoU class*), we measure the performance in terms of IoU on this seven categories.

Table 1. Impact of the initialization scheme for the disparity network on the semantic segmentation performance. The upper part of the table shows results with random weight initialization. The high amount of channels in the original NiN prevents a successful training. The influence of different initialization schemes is shown in the lower part. The proposed initialization on the coarsely annotated images yields significantly improved results compared to a variant initialized on ImageNet [33].

	Initialization	IoU class [%]	IoU category[%]
Original NiN [27]	Random	< 5.0	< 10.0
NiN (ours)	Random	30.5	61.3
Original NiN [27]	ImageNet	35.0	64.0
Original NiN [27]	Cityscapes coarse	< 5.0	< 10.0
NiN (ours)	Cityscapes coarse	**37.3**	**66.5**

Initialization Method. In Sect. 3, we argued that a CNN for depth data should significantly differ from a CNN on RGB data. First, we train a CNN solely on depth data discarding the available RGB input. The results of this proposed model in comparison to the original NiN are given in Table 1. The first observation from the upper half of the table is that we were not able to train the original NiN on the cityscapes dataset only. The proposed variant with $\frac{1}{3}$ of the channels, however, yields surprisingly good results. Second, initialization with the weights trained on the RGB ImageNet data guides the learning process and yields an improvement of 4.5%. Nevertheless, an initialization on actual depth data leads to significant improvements. Overall, the number of parameters of

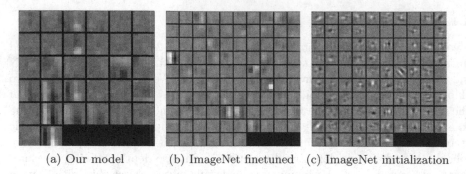

(a) Our model   (b) ImageNet finetuned   (c) ImageNet initialization

Fig. 2. Filters of the first convolutional layer in the NiN architecture. The filters initialized randomly and trained on Cityscapes coarse labels (left) differ significantly of those trained on color data (right). Mainly gradient, mean and blob filters are developed during training. Fine-tuning of the color filters on depth data (middle) yields smaller amount of sharp filters.

the network was reduced to $\frac{1}{3}$ leading to $\frac{1}{3}$ of the computational costs. On the other hand the results were improved significantly. The resulting filters in the first convolutional layer differ substantially between the depth and color input respectively, c.f. Fig. 2. Observably, the amount of meaningful filters is higher in our model which we address to the reduced number of filters in the network.

4.2 RGBD Semantic Segmentation

Evaluation and Training Details. We leverage the 5000 finely annotated images, i.e. 2975 for training, the 500 validation images for testing, of the Cityscapes dataset and IoU as evaluation metric as before, c.f. Sect. 4.1. We do not use the images in the validation set for training. A batch-size of two and the maximal learning rate were used. After convergence, we decrease the learning rate step by step with a factor of $\frac{1}{10}$ until no further improvements on the validation set are observed. For each method, we report the best results according to the IoU on the validation set. The results of best model according to the following experiments is submitted to the Cityscapes benchmark server for evaluation on the remaining ~1500 test images.

Level of Fusion. First, the optimal level for fusing the color and depth branch is determined. To this end, we train and evaluate the early-fusion, late-fusion and all five mid-level fusion models NiN-1 to NiN-7 and compare it to the RGB baseline. The results in Fig. 3 first show that the additional depth input helps in all fusion variants significantly. The RGB baseline achieves 63.9% IoU (class wise) compared to the 69.1% of the NiN-2 model. This is a considerable relative improvement of about 10%. Furthermore, it becomes apparent that a mid-level fusion after 2 NiN modules leads to the best results, the most frequently used late fusion only yields 67.1% IoU. Surprisingly, the NiN-1 and NiN-7 variants perform worst. The feature concatenation in the NiN-1 model takes place directly

Table 2. Comparison to baselines on the Cityscapes dataset [5]. Both: the RGB baseline as well as an external RGB-D baseline (with and without additional CRF) are outperformed by the proposed model without the need for a CRF in terms of semantic accuracy on all 19 classes respectively the seven categories.

Method	Input	IoU class [%]	IoU category[%]
[25] w/o CRF	RGB-D	62.5	N/A
[25] with CRF	RGB-D	66.3	85.0
GoogleNet	RGB	63.0	85.8
Ours	RGB-D	**67.4**	**87.5**

after a local-response-normalization that might harm the interplay with the non-normalized features of the depth branch (Fig. 3).

Comparison to Baselines. So far, all experiments have been carried out on the validation set, for the comparison with external baselines, the test set is used. On the Cityscapes dataset, the results of only one work has been reported that exploits depth information: [25], c.f. Sect. 2. We report the results according to the Cityscapes benchmark server [5]. Secondly, the GoogLeNet trained with the same scheme naturally serves as additional baseline.

Visually, the proposed RGB-D model outperforms the RGB baseline particularly at objects in farther distance, e.g. the car in the left image as well as the pedestrian and traffic sign in the right image of Fig. 5. Although the detection of these objects can be of highest important for e.g. autonomous vehicles, the influence on the pixel-level IoU score is rather low.

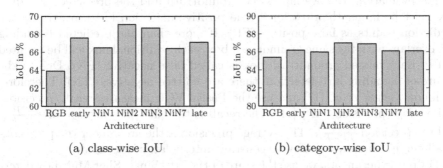

(a) class-wise IoU   (b) category-wise IoU

Fig. 3. Where is the optimal level for modal fusion? The semantic accuracy for different levels of fusion of RGB and depth data with the proposed CNN.

4.3 RGBD Object Detection

Dataset, Evaluation, and Training Details. For object detection, we also use the Cityscapes dataset [6]. Due to the highly accurately labeled instances of all object types, bounding boxes can simply be extracted from the pixel-wise

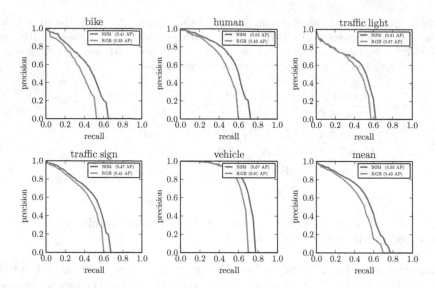

Fig. 4. Precision-recall curves for all object classes as well as the mean precision-recall curve for both: the RGB only baseline and our approach.

annotation. For training, we use the publicly available training data set with 2975 fully annotated images. Since the groundtruth for the test data is not publicly available, we test on the 500 images of the validation data set. Since not all classes are "object-like", we only use a subset of Cityscapes: *vehicle* (in Cityscapes: *car, truck, bus*), *bike* (in Cityscapes: *motorcycle, bicycle*), *traffic sign, traffic light, human* (in Cityscapes: *person, rider*).

For evaluation, the overlap of the groundtruth and the predicted bounding box must be larger than 0.5 for a true positive detection (TP), otherwise the prediction counts as false positive (FP). If more than one predicted bounding box overlaps with the same groundtruth box, each additional box will be counted as FP. Each missed groundtruth box is called false negative (FN). Due to the nature of bounding box detection, there are no true negatives (TN). Therefore, we follow Geiger *et al.* [12] and use the Pascal VOC measures recall, precision, and average precision (AP) [11]. The recall is the class-wise average of $\frac{TP}{TP+FN}$ and the precision $\frac{TP}{TP+FP}$. The average precision is the area under the precision-recall curve, whereby a piecewise constant interpolation was used.

For our experiments, we use the state of the art "Single Shot Multibox detector" frame work (SSD) [29]. Following our experiments from Sect. 4.2, we use a fully convolutional approach based on GoogLeNet and extend the RGB framework with the proposed and pre-trained NiN architecture for depth images.

Results. First, we adapt SSD to the GoogLeNet architecture with RGB input only and second, we add a depth branch as proposed in Sect. 3. Both the class-wise and the mean precision-recall curves are shown in Fig. 4. The classification of all classes benefits in similar fashion from depth data. Especially the performance

input

RGB

RGB-D

RGB

RGB-D

sidewalk	building	vegetation	traffic light	traffic sign	bicycle	motorcycle

terrain	road	wall	pole	rider	truck	bus	train	fence	person	sky	car

Fig. 5. Incorporating depth information (RGB-D, third line for semantic segmentation and fifth line for detection) leads to better segmentation of small objects, more details in the classification results, and tighter bounding boxes.

for the classes human and bike increases significantly. As shown in Fig. 5 objects in far distances are detected more robustly and more accurately by using our approach in comparison to the traditional RGB only approach.

5 Conclusion

This paper presented a novel generic CNN architecture that exploits input cues from other modalities in addition to sole color information. To this end, the GoogLeNet was extended with a branch specifically adapted to depth as complementary input. Together, the joint network implemented a mid-level fusion that allowed the network to exploit cross-modal interdependencies already on a medium feature-level. So far, state-of-the-art RGB-D CNNs have used network weights pre-trained on color data. In contrast, a superior initialization scheme was proposed to pre-train the depth branch of the multi-modal CNN independently. In an end-to-end training the network parameters were optimized jointly using the challenging Cityscapes dataset. The evaluation is carried on two different common computer vision tasks namely semantic segmentation and object detection. For the latter this paper furthermore showed how to extract object-level groundtruth from the instance level annotations in Cityscapes in order to train a powerful SSD object detector. In thorough experiments, the effectiveness of the proposed multi-modal CNN was shown. Both, the RGB GoogLeNet and further RGB-D baselines were outperformed significantly.

References

1. Badrinarayanan, V., Kendall, A., Cipolla, R.: SegNet: a deep convolutional encoder-decoder architecture for image segmentation. In: CVPR (2015)
2. Chen, L.C., Papandreou, G., Kokkinos, I., Murphy, K., Yuille, A.L.: Semantic image segmentation with deep convolutional nets and fully connected CRFs. In: ICLR (2014)
3. Chen, L.C., Yuille, A.L., Urtasun, R.: Learning deep structured models. In: ICML (2015)
4. Chen, X., Kundu, K., Zhu, Y., Berneshawi, A., Ma, H., Fidler, S., Urtasun, R.: 3D object proposals for accurate object class detection. In: NIPS (2015)
5. Cordts, M., Omran, M., Ramos, S., Rehfeld, T., Enzweiler, M., Benenson, R., Franke, U., Roth, S., Schiele, B.: Benchmark of Cityscapes dataset. www.cityscapes-dataset.com/benchmarks/, Accessed 27 Aug 2016
6. Cordts, M., Omran, M., Ramos, S., Rehfeld, T., Enzweiler, M., Benenson, R., Franke, U., Roth, S., Schiele, B.: The Cityscapes dataset for semantic urban scene understanding. In: CVPR (2016)
7. Couprie, C., Farabet, C., Najman, L., LeCun, Y.: Indoor semantic segmentation using depth information. In: ICLR (2013)
8. Couprie, C., Najman, L., Lecun, Y.: Learning Hierarchical features for scene labeling. Trans. PAMI $35(8)$, 1915–1929 (2013)
9. Deng, Z., Todorovic, S., Jan Latecki, L.: Semantic segmentation of RGBD images with mutex constraints. In: ICCV (2015)
10. Eitel, A., Springenberg, J.T., Spinello, L., Riedmiller, M., Burgard, W.: Multimodal deep learning for robust RGB-D object recognition. In: IROS (2015)
11. Everingham, M., Eslami, S.M.A., Van Gool, L., Williams, C.K.I., Winn, J., Zisserman, A.: The pascal visual object classes challenge: a retrospective. IJCV $111(1)$, 98–136 (2015)
12. Geiger, A., Lenz, P., Urtasun, R.: Are we ready for autonomous driving? The KITTI vision benchmark suite. In: CVPR (2012)
13. Girshick, R.: Fast R-CNN. In: ICCV (2015)
14. Girshick, R., Donahue, J., Darrell, T., Malik, J.: Rich feature hierarchies for accurate object detection and semantic segmentation. In: CVPR (2014)
15. Gupta, S., Girshick, R., Arbeláez, P., Malik, J.: Learning rich features from RGB-D images for object detection and segmentation. In: Fleet, D., Pajdla, T., Schiele, B., Tuytelaars, T. (eds.) ECCV 2014. LNCS, vol. 8695, pp. 345–360. Springer, Cham (2014). doi:10.1007/978-3-319-10584-0_23
16. Gupta, S., Hoffman, J., Malik, J.: Cross modal distillation for supervision transfer. In: CVPR (2015)
17. Hazirbas, C., Ma, L., Domokos, C., Cremers, D.: Fusenet: incorporating depth into semantic segmentation via fusion-based CNN architecture. In: ACCV (2016)
18. He, K., Zhang, X., Ren, S., Sun, J.: Deep residual learning for image recognition. In: CVPR (2015)
19. Zhao, H., Jianping Shi, X.Q., Wang, X., Jia, J.: Pyramid scene parsing network. ArXiv (2016)
20. Hirschmüller, H.: Stereo processing by semiglobal matching and mutual information. Trans. PAMI $30(2)$, 328–341 (2008)

21. Hou, S., Wang, Z., Wu, F.: Deeply exploit depth information for object detection. In: CVPRW (2016)
22. Huang, J., Rathod, V., Sun, C., Zhu, M., Korattikara, A., Fathi, A., Fischer, I., Wojna, Z., Song, Y., Guadarrama, S., Murphy, K.: Speed/accuracy trade-offs for modern convolutional object detectors. ArXiv (2016)
23. Jifeng Dai, Y.L., He, K., Sun, J.: R-FCN: object detection via region-based fully convolutional networks. In: NIPS (2016)
24. Khan, S.H., Bennamoun, M., Sohel, F., Togneri, R., Naseem, I.: Integrating geometrical context for semantic labeling of indoor scenes using RGBD images. IJCV **117**(1), 1–20 (2016)
25. Krešo, I., Čaušević, D., Krapac, J., Šegvić, S.: Convolutional scale invariance for semantic segmentation. In: Rosenhahn, B., Andres, B. (eds.) GCPR 2016. LNCS, vol. 9796, pp. 64–75. Springer, Cham (2016). doi:10.1007/978-3-319-45886-1_6
26. Li, Z., Gan, Y., Liang, X., Yu, Y., Cheng, H., Lin, L.: LSTM-CF: unifying context modeling and fusion with LSTMs for RGB-D scene labeling. In: Leibe, B., Matas, J., Sebe, N., Welling, M. (eds.) ECCV 2016. LNCS, vol. 9906, pp. 541–557. Springer, Cham (2016). doi:10.1007/978-3-319-46475-6_34
27. Lin, M., Chen, Q., Yan, S.: Network in network. In: ICLR (2013)
28. Lin, T.-Y., Maire, M., Belongie, S., Hays, J., Perona, P., Ramanan, D., Dollár, P., Zitnick, C.L.: Microsoft COCO: common objects in context. In: Fleet, D., Pajdla, T., Schiele, B., Tuytelaars, T. (eds.) ECCV 2014. LNCS, vol. 8693, pp. 740–755. Springer, Cham (2014). doi:10.1007/978-3-319-10602-1_48
29. Liu, W., Anguelov, D., Erhan, D., Szegedy, C., Reed, S., Fu, C.-Y., Berg, A.C.: SSD: single shot multibox detector. In: Leibe, B., Matas, J., Sebe, N., Welling, M. (eds.) ECCV 2016. LNCS, vol. 9905, pp. 21–37. Springer, Cham (2016). doi:10.1007/978-3-319-46448-0_2
30. Long, J., Shelhamer, E., Darrell, T.: Fully convolutional networks for semantic segmentation. In: CVPR (2014)
31. M. Jasch, T. Weber, M.R.: Fast and robust RGB-D scene labeling for autonomous driving. In: ICSCC, JCP (2016, to appear)
32. Redmon, J., Divvala, S., Girshick, R., Farhadi, A.: You only look once: unified, real-time object detection. In: CVPR (2016)
33. Russakovsky, O., Deng, J., Su, H., Krause, J., Satheesh, S., Ma, S., Huang, Z., Karpathy, A., Khosla, A., Bernstein, M., Berg, A.C., Fei-Fei, L.: ImageNet large scale visual recognition challenge. IJCV **15**(3), 211–252 (2015)
34. Sermanet, P., Eigen, D., Zhang, X., Mathieu, M., Fergus, R., Lecun, Y.: Overfeat: integrated recognition, localization and detection using convolutional networks. In: ICLR (2014)
35. Simonyan, K., Zisserman, A.: Very deep convolutional networks for large-scale image recognition. Ecology (2015)
36. Song, S., Lichtenberg, S.P., Xiao, J.: SUN RGB-D: a RGB-D scene understanding benchmark suite. In: CVPR (2015)
37. Song, S., Yu, F., Zeng, A., Chang, A.X., Savva, M., Funkhouser, T.A.: Semantic scene completion from a single depth image. In: CVPR (2017, to appear)
38. Szegedy, C., Liu, W., Jia, Y., Sermanet, P.: Going deeper with convolutions. In: CVPR (2014)
39. Yu, F., Koltun, V.: Multi-scale context aggregation by dilated convolutions. In: ICLR (2015)
40. Zhou, B., Lapedriza, A., Xiao, J., Torralba, A., Oliva, A.: Learning deep features for scene recognition using places database. In: NIPS (2014)

Uncertainty Computation in Large 3D Reconstruction

Michal Polic[(⊠)] and Tomas Pajdla

Department of Cybernetics, Faculty of Electrical Engineering,
Center for Machine Perception, Czech Technical University in Prague,
Prague, Czech Republic
policmic@fel.cvut.cz
http://cmp.felk.cvut.cz/~policmic

Abstract. Many automatic methods for reconstructing camera motion and scene geometry from a large number of images evaluate the quality of the reconstruction by error propagation from measurements to the estimated parameters. Unfortunately, uncertainty propagation is computationally challenging for large scenes and hence cannot be used for large scenes in practice. We present a new algorithm for efficient uncertainty propagation which works with millions of feature points, thousands of cameras and millions of 3D points on a single computer and achieves about twenty times speedup.

Keywords: Uncertainty propagation · 3D reconstruction · Schur complement · Information matrix · Taylor expansion

1 Introduction

Recent work in Structure from Motion (SfM) has demonstrated the possibility of reconstructing geometry from large photo collections [1,16]. Efficient non-linear refinement [3,34] of camera and point parameters was developed to produce optimal reconstruction. However, the important information about the reconstruction quality [12] is missing in state of the art pipelines [28,33] and optimization solvers [2] because of the computational and memory requirements. If the quality evaluation is fast enough with reasonable memory requirements, it can be used for filtering cameras and 3D points before Bundle Adjustment which usually run in each step of SfM and therefore remove the bottleneck of a reconstruction pipelines [28,29]. The second usage may be sophisticated smoothing in the most unconstrained directions in dense reconstructions [19,27].

The mathematics of uncertainty propagation is well understood [15], and specific equations have been derived for many camera setups, e.g. for stereo rigs [22,25], laser scans [17,26], lines [4] and edges [6]. Except for few a works, e.g. [20], general error propagation has not received significant attention in the computer vision community. We believe that it is due to the high computational complexity of uncertainty propagation, which is, as a function of the number

© Springer International Publishing AG 2017
P. Sharma and F.M. Bianchi (Eds.): SCIA 2017, Part I, LNCS 10269, pp. 110–121, 2017.
DOI: 10.1007/978-3-319-59126-1_10

of camera parameters, cubic in time and quadratic in memory. To evaluate the quality, classical photogrammetric methods propagate the uncertainty of image measurements to the uncertainty of estimated parameters.

General propagation requires transport of covariances of measurements to parameters. The standard non-linear projection function is usually simplified by replacing it by its first order approximation, and the covariance of the estimated parameters is then computed by inverting its information matrix. Common objective function [15] of reconstruction from images is the sum of squared reprojection errors.

Camera rotation angle and radial distortion parameters are usually much smaller than coordinates of 3D points and, at the same time, have a large impact on the objective function. Hence, the Jacobian of the objective function often contains a large range of values which are further squared to information matrix. Thus, the information matrix becomes numerically rank deficient for larger collections of images. The information matrix can be decomposed into blocks which simplify its inversion. The block related to 3D points parameters is, up to special artificial configurations, full rank block diagonal matrix which can be easily inverted. The block related to camera parameters appears as over-parametrised. We have to find the closest one-one mapping from measurements to parameters with respect the Mahalanobis norm to be able to inverse submatrix of camera parameters [15]. The inversion should be computed as Moore-Penrose (MP) inversion [2]. The state of the art algorithm use Singular Value Decomposition (SVD) [20]. This algorithm works well for problems with a few hundreds of photos. As the number of images in reconstructions grows up to several hundred thousand, the efficiency and scalability of the quality evaluation become a critical issue.

We present an iterative algorithm based on the Taylor expansion of Schur complement matrix Z in a regularization point λ, which computes an approximation of the MP inverse. This algorithm is approximately three times faster and reduces approximately two times the memory requirements in comparison with SVD. The second improvement is that we use the structure of matrices for uncertainty propagation from camera parameters to the point parameters, which increases the speed about two orders of magnitude for this particular step for real datasets. The output of our work is publicly available source code which can be used as an external library in nonlinear optimization pipelines like Ceres Solver [2]. The code, test problems, and detailed experiments are available online at cmp.felk.cvut.cz/~policmic.

The rest of the paper is organized as follow. Section 2 overviews previous works. Then, Sect. 3 provides our formulation of the problem of uncertainty propagation and Sect. 4 introduced our Taylor expansion algorithm. Section 5 describe how to use the structure of information matrix to compute covariances of point parameters. Finally, Sect. 6 reports the results of the experiments.

2   Previous Work

Many methods for 3D reconstruction of camera motion and scene geometry from a large number of images have been developed [1,13,16]. To evaluate the quality of the reconstruction, classical photogrammetric methods propagate the uncertainty in image measurements to the uncertainty of estimated parameters. The propagation is often based on the first order approximations [12,20,21]. This classical approach becomes computationally challenging with a large number of cameras and points [30,32] and hence is often not fully implemented in computer vision reconstruction pipelines [2].

Let us next review previous work related to computing uncertainties in 3D reconstruction from images. Blostein and Huang [8] provided closed form expressions for evaluating parameters of error distributions of the disparity for a stereo rig. The confidence intervals for stereo setup were presented in [18]. Work [25] minimizes the Mean Square Error of measurement errors to compute optimal camera poses and to compute covariance matrices representing their uncertainties. The influence of errors in calibration on the errors in 3D reconstruction for stereo setup was investigated in [22]. In the recent paper [5], uncertainties of estimated parameters are studied for each step of a stereo-vision based 3D reconstruction (e.g., for calibration, contour segmentation, matching, and reconstruction), and used to estimate the error distributions of the output geometry. Inverse perspective equations for describing 3D line error were investigated in [4] and later generalized for reconstruction from edges in images [6]. In [23] the uncertainties are expressed as likelihood map of 3D points position created by aggregating local linear extrapolations computed from weighted least squares fits of 2D curves between 3D points.

The most relevant previous work. The most relevant work is from Lhuillier and Perriollat [20] where forming information matrix of estimated parameters and its inverse has been investigated. Lhuillier has computed MP inverse of Schur complement [35] of the submatrix of point parameters using the SVD algorithm. We extend Lhuillier work using faster MP inverse and improve 3D points covariance computation speed using the structure of Jacobian. A general algorithm for covariance propagation forward/backward for system linear/nonlinear which is or isn't overparameterized was also described by Hartley in the book [15].

3   Problem Formulation

We next review the standard approach for computing covariances of reconstructed parameters. Our contributions are described in Sects. 4 and 5.

We consider a setup with n cameras $C = \{C_1, C_2, \ldots, C_n\}$ where $C_i \in \mathbb{R}^p$ is p-dimensional vector of camera parameters, m points $X = \{X_1, X_2, \ldots, X_m\}$ in 3D and k image observations represented by vector $u \in \mathbb{R}^{2k}$. Each observation $u_{i,j} \in u$, i.e. an image point, is a projection of 3D point X_j by camera C_i, using projection function $p(C_i, X_j)$. Noise $\epsilon_{i,j}$ has zero mean, $E(\epsilon_{ij}) = 0$, and standard deviation $V(\epsilon_{i,j})$. All pairs of indices (i,j) are in an index set S that determines which point X_j is visible in which camera view C_i.

$$u_{i,j} = p(C_i, X_j) + \epsilon_{i,j} \qquad \forall (i,j) \in S \tag{1}$$

Further, we define the vector θ which represents all 3D reconstruction parameters $[C_1, \ldots, C_n, X_1, \ldots, X_m]$, the vector ϵ which gathers all $\epsilon_{i,j}$ where $(i,j) \in S$ and the function f which is composed from projection functions p and project vector θ into the image observations u.

$$u = f(\theta) + \epsilon \tag{2}$$

The function (2) leads to a non-linear least squares optimization

$$\theta^* = \arg\min_\theta \|u - f(\theta)\|_2^2 \tag{3}$$

minimizing the sum of squares of the differences $r(\theta) = u - f(\theta)$ between observations u and reprojections $f(\theta)$. The standard way to solve the nonlinear differentiable system is to use its first order approximation

$$r(\theta + \delta\theta) \approx r(\theta) + J(\theta)\,\delta\theta \tag{4}$$

One step $\delta\theta$ towards the solution θ^* is obtained by solving

$$J^\top J\,\delta\theta = -J^\top r \tag{5}$$

with $J(\theta)$ and $r(\theta)$ replaced by simpler J and r. Using Gauss-Markoff theorem, the covariance $\hat{V}(\theta)$ can be computed as

$$\hat{V}(\theta) = \sigma^2 (J^\top J)^{-1} \tag{6}$$

Parameter σ can be estimated from residuals as $\sigma^2 \approx \|r\|^2/(p\,n)$, which is described in Hartley [15]. If the content of Jacobian is permuted to have cameras followed by points, i.e. $J = [J_C\ J_X]$, the information matrix

$$(J^\top J)^{-1} = \begin{bmatrix} U & W \\ W^\top & V \end{bmatrix}^{-1} \tag{7}$$

will be sparse and block diagonal [11]. See the information matrix for *cube* dataset in Fig. 1(a).

To compute the inverse of $J^\top J$, we introduce $Y = -V^{-1}W^\top$. We note, first, that V is composed of 3×3 blocks on the diagonal and its inverse can be done separately for each block, and then also that forming Y is fast thanks to the sparsity of V and W. The Upper triangular–Diagonal–Lower triangular (UDL) decomposition of the block matrix $\hat{V}(\theta) = \sigma^2 (J^\top J)^{-1}$ leads to

$$\hat{V}(\theta) = \sigma^2 \left(\begin{bmatrix} I & -Y^\top \\ 0 & I \end{bmatrix} \begin{bmatrix} Z & 0 \\ 0 & V \end{bmatrix} \begin{bmatrix} I & 0 \\ -Y & I \end{bmatrix} \right)^{-1} \tag{8}$$

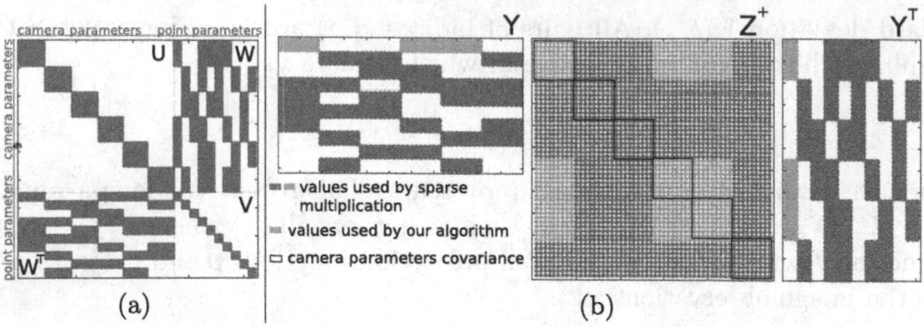

Fig. 1. (a) Structure of information matrix $J^T J$ for *cube* dataset. (b) The structure of matrix Y, Z^+ and Y^T for *cube* dataset, where sparse multiplication use red and green values, our approach use green values only and the black boxes highlight final covariances of camera parameters (Color figure online)

where the matrix Z is the Schur complement of the block V of the matrix $J^T J$

$$Z = U + WY \tag{9}$$

Notice that we are not interested in off-diagonal blocks. All covariances of parameters are in the blocks on the diagonal of the dense matrix $(J^T J)^{-1}$. Hence, Lhuillier and Perriollat [20] computed the interesting sub-matrices as

$$(J^T J)^{-1} \simeq \begin{bmatrix} Z^{-1} & - \\ - & YZ^{-1}Y^\top + V^{-1} \end{bmatrix} \tag{10}$$

The blocks of size $\mathbb{R}^{p \times p}$ on the diagonal Z^{-1} are covariances of camera parameters. Matrix Z is rank deficient and its inversion has to be solved by MP inversion. The blocks of size $\mathbb{R}^{3 \times 3}$ on the diagonal of sub-matrix $YZ^{-1}Y^\top + V^{-1}$ are covariances of point parameters.

4   The Taylor Expansion Approach

We next denote the MP inverse of matrix Z as Z^+. Assume general vectors x and b such that

$$Zx = b \tag{11}$$

holds true. Our approach is to redefine x as a function where

$$Zx(0) = b \tag{12}$$

holds true. Unfortunately, the pseudoinverse can't be computed as $(Z^T Z)^{-1}Z$ since $Z^T Z$ is also rank deficient. Hence, we compute the inversion for $x(\lambda)$ as

$$x(\lambda) = (Z^\top Z + \lambda I)^{-1} Z^\top b \tag{13}$$

where λ is close to zero. Finally, we estimate $x(0)$ and Z^+ from Taylor series of function $x(\lambda)$ at the point 0.

We need a general formula for the k-th derivative with respect λ

$$x^k(\lambda) = -k(ZZ + \lambda I)^{-1} x^{k-1}(\lambda) \tag{14}$$

for building Taylor series. The matrix Z is symmetric and $Z^\top Z = ZZ$. Using the derivatives, we can approximate $x(\lambda)$ around λ using the t-th order Taylor series

$$x(0) = x(\lambda) + \sum_{k=1}^{t} \frac{(-1)^{k-1}\lambda^k}{k!} x^k(\lambda). \tag{15}$$

Next we substitute $B = (ZZ + \lambda I)^{-1}$, which is computed using Cholesky decomposition [10]. $x(\lambda) = BZ$ can be factored out and we get the formula for Z^+ as

$$x(0) = Z^+ b \qquad Z^+ = \left(I + \sum_{k=1}^{t} (-1)^{k-1} \lambda^k B^k \right) BZ \tag{16}$$

In our experiments, the algorithm converges for $t \in [2, 5]$. Each iteration performs one symmetric matrix multiplication and one addition. If we scale the matrix ZZ by scalar $c = 1/mean(ZZ)$, we get $\hat{\lambda} = c\lambda$, $\hat{B} = (c\hat{Z}\hat{Z} + \hat{\lambda}I)^{-1}$ and

$$Z^+ = \left(I + \sum_{k=1}^{t} (-1)^{k-1} c^k \hat{\lambda}^k \hat{B}^k \right) \hat{c}\hat{B}\hat{Z} \tag{17}$$

This scaling decrease the range of values and allows computation in double precision. Numerical stability of the algorithm depends on the value of $\hat{\lambda}$. We use $\hat{\lambda} = trace(ZZ)\, c/10^{16}$ which is usually $[10^{-8}, 10^{-10}]$. The parameter $\hat{\lambda}$ depends on the size of Schur complement matrix which depends on the number of parameters of cameras. Reconstructions, which contains more than 10^6 cameras, may require higher parameter $\hat{\lambda}$, i.e. above 10^{-4}.

5   Point Covariances

We compute the covariance matrices of point parameters from Eq. 10 using the structure of the matrix Y. The matrix Y contains values in blocks where 3D points (triples of rows) are seen by cameras (p columns). We compute only blocks $\bar{H} = Z^+ Y^\top$ which are required for multiplication $Y\bar{H}$, Fig. 1(b). The speed up in comparison with sparse multiplication is $n\,m/k$ times.

6   Experimental Evaluation

6.1   Datasets

We experimented with realistic synthetic scenes, as well as a number of medium to large-scale Internet datasets: Flat, Tower of London, Notre Dame, Trench, Seychelles. The Cube dataset was used for visualization of the matrices. Flat is

Table 1. The table summarize number of cameras N_{Cams}, number of points N_{Pts}, number of observations N_{Obs} and one side length N_{Schur} of rectangular Schur complement matrix Z for compared datasets

Dataset	SfM	N_{Cams}	N_{Pts}	N_{Obs}	N_{Schur}
Flat	Synthetic	361	251	10 517	3249
Tower of London	COLORMAP	530	65 768	508 579	4770
Trench	YaSfM	698	434 049	1 017 909	6282
Notre Dame	Bundler	715	127 431	748 003	6435
Seychelles	COLORMAP	1400	407 193	2 098 201	12600

the only dataset that is synthetic. Tower of London and Seychelles are reconstructed with COLORMAP [28], Notre Dame is reconstructed by Bundler [29] and the remaining one by our internal reconstruction pipeline YaSfM. The images for Tower of London and Notre Dame were collected in projects [29,32] from Flickr by using geotag. The remaining datasets are available on web page cmp.felk.cvut.cz/~policmic. The parameters of datasets are summarized in Table 1.

6.2 Algorithms

We compared the performance, memory requirements and the sensitivity on input parameters for three algorithms: qr-iteration, divide-and-conquer and taylor-expansion. The state of the art method is done using SVD [20]. The algorithms qr-iteration, divide-and-conquer are described in [7]. We do not compare our algorithm with recent [24] paper because the Cholesky decomposition on Schur complement matrix is performed there. However, the Schur complement matrix is numerically rank deficient for middle and large reconstructions (i.e. Cholesky decomposition can't be used). This subsection has following structure: we describe how the Schur complement matrix was formed, how its MP inversion was computed and lastly focus on the precision of Taylor expansion approach. Let us pinpoint the main figures and tables: Fig. 2 shows the structure of C++ code and labels used for comparison of the speed. The comparisons of the speeds are in Tables 2, 3. The Fig. 3 visualize the uncertainty ellipsoids for different values of parameters γ, λ.

Forming Schur Complement. Given Jacobian $J(\theta) = [J_C(\theta) \, J_X(\theta)]$ from Ceres [2], we form an information matrix in Eq. 7 and Schur complement in Eq. 9 using spare matrix multiplication for all algorithms. For each experiment, $\theta = [\theta_C \theta_X]$ were $\theta_C = [\theta_{C_1}, \dots, \theta_{C_n}]$ and $\theta_X = [\theta_{X_1}, \dots, \theta_{X_m}]$. One camera $\theta_{C_i} \in \mathbb{R}^9$ is represented by 3 angle-axis, 3 camera center, 1 focal length, 2 radial distortion parameters and point $\theta_{X_i} \in \mathbb{R}^3$ by its 3D coordinates.

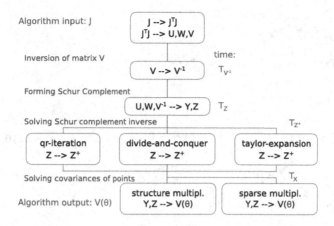

Fig. 2. The diagram of algorithm implementation

Solving Schur Complement Inverse. The first two methods qr-iteration, divide-and-conquer are SVD algorithms described in [7]. Both algorithms decompose Schur complement into $Z = \bar{U}\bar{S}\bar{V}^\top$ and invert the diagonal values $\bar{S}'_{i,i} = 1/\bar{S}_{i,i}$ for which holds $\bar{S}_{i,i} > \gamma$ where $i \in 1, 2, \ldots, N_{Schur}$. The remaining values on the diagonal $\bar{S}_{i,i} <= \gamma$ are set to zero. Note that no exact estimate of γ exists. The parameter γ, i.e. the number of inverted values, is crucial for the output values. Figure 3 visualize the influence of γ (SVD algorithms) and $\hat{\lambda}$ (Taylor expansion algorithm) on uncertainty ellipsoids where the ellipsoid show the position of a 3D point or a 3D camera center. Taylor expansion algorithm compute inverse $(cZZ + \hat{\lambda}I)^{-1}$ using Cholesky factorization. The parameter $\hat{\lambda} = trace(ZZ)c/10^{16}$ appears as reasonable trade-off between regularization cZZ and distance from zero which we try to approximate. The algorithm performs several iterations. The i-th iteration computes the i-th term T_i of Taylor series. We compute next T_i until $l_\infty(T_i) > 10^{-5}$. The usual number of iterations is [2,5].

Table 2. Additional processing times, especially the time $T_{V^{-1}}$ required for inversion of the matrix V, the time T_Z required for forming the Schur complement matrix, values of parameter $\hat{\lambda}$, the time T_{X_sparse} required for *sparse* multiplication and the time $T_{X\_structure}$ for dense multiplication with using the *structure* of matrix Y

Dataset	$\hat{\lambda}$	T_{V-1}	T_Z	T_{X_sparse}	$T_{X\_structure}$	$T_{X\_speedup}$
Flat	5.20^{-9}	0.001 s	4.12 s	0.90 s	0.10 s	8.62
Tower of London	2.28^{-9}	0.09 s	5.02 s	244.68 s	3.57 s	68.54
Trench	3.95^{-9}	0.62 s	3.31 s	720.28 s	2.43 s	297.64
Notre Dame	2.42^{-8}	0.17 s	7.51 s	1131.6 s	9.29 s	121.81
Seychelles	5.08^{-4}	0.52 s	15.32 s	2982.94 s	10.98 s	271.69

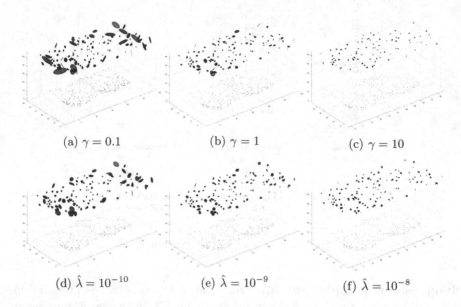

(a) $\gamma = 0.1$ (b) $\gamma = 1$ (c) $\gamma = 10$

(d) $\hat{\lambda} = 10^{-10}$ (e) $\hat{\lambda} = 10^{-9}$ (f) $\hat{\lambda} = 10^{-8}$

Fig. 3. Parameter sensitivity of the SVD based algorithms (parameter γ) and Taylor expansion algorithm (parameter $\hat{\lambda}$) for *flat* dataset

Assuming that all algorithms store Z and Z^+ in the same format, the difference between their memory usages depend on how they work with matrix Z. Taylor expansion algorithm requires one additional symmetric matrix of size $\mathbb{R}^{pn \times pn}$. Both remaining algorithms require matrices $\bar{U}, \bar{V} \in \mathbb{R}^{pn \times pn}$ and vector $diag(\bar{S}) \in \mathbb{R}^{pn}$. Further qr-iteration require additional $(pn)^2 + 67(pn)$ and divide-and-conquer additional $4(pn)^2 + 7(pn)$ values for SVD.

Precision of the MP Inverse Approximation. We computed the ground truth using SVD with 100 significant digits and compare it with Taylor expansion and SVD algorithm computed in double (≈ 15 significant digits). The mean difference of camera parameters (angle axis and camera position) between the ground truth covariance matrix and the covariance matrix computed by Taylor expansion approach for $\lambda = 10^{-9}$ was 3.910^{-3} while the difference for SVD for $\gamma = 10^{-7}$ was 1.410^{-4} for Cube dataset. We are not able to test the precision of our approximations for larger dataset because of computational complexity. The covariance matrices computed by Taylor expansion approach may be useful because we get large ellipsoids for the most unconstrained cameras and points and vice versa, Fig. 3. So, we are able to determine which cameras are more and which less uncertain.

All algorithms were implemented as part of a single C++ code base. They share inversion of matrix V and forming Schur complement matrix. We use Intel MKL 11.3.1 [9] implementation of Blas and Lapack for dense linear algebra, Eigen 3.3 [14] for sparse linear algebra and matrix representation. Blas and

Table 3. Memory and time requirements. The columns M_{Z+} and T_{Z+} represent memory and time requirements for computing PM inverse of Schur complement. $\sum T$ expresses the complete run-time of the algorithm. The parameter γ was equal to 10^{-10} in all evaluations.

Flat	M_{Z+}	T_{Z+}	$\sum T$
qr-iteration	424 MB	12.57 s	13.14 s
divide-and-conquer	676 MB	10.61 s	11.18 s
taylor-expansion	**253 MB**	**3.82 s**	**4.40 s**
Tower of London	M_{Z+}	T_{Z+}	$\sum T$
qr-iteration	913 MB	46.79 s	52.13 s
divide-and-conquer	1 456 MB	31.02 s	36.36 s
taylor-expansion	**546 MB**	**8.05 s**	**13.39 s**
Trench	M_{Z+}	T_{Z+}	$\sum T$
qr-iteration	1 582 MB	90.91 s	102.04 s
divide-and-conquer	2 526 MB	69.15 s	80.28 s
taylor-expansion	**947 MB**	**19.7 s**	**30.83 s**
Notre Dame	M_{Z+}	T_{Z+}	$\sum T$
qr-iteration	1 660 MB	76.20 s	96.72 s
divide-and-conquer	2 651 MB	65.77 s	86.29 s
taylor-expansion	**994 MB**	**21.24 s**	**41.76 s**
Seychelles	M_{Z+}	T_{Z+}	$\sum T$
qr-iteration	6 357 MB	464.89 s	502.07 s
divide-and-conquer	10 161 MB	379.13 s	416.81 s
taylor-expansion	**3 810 MB**	**139.59 s**	**176.47 s**

Lapack functions are called from MAGMA 2.2.0 [31]. All experiments were performed on a single computer with one 2.6 GHz Intel Core i7-6700HQ with 32 GB RAM running a 64-bit Windows 10 operating system.

7   Conclusion

The state of the art method for finding covariances is based on SVD [20] which is computationally challenging for a large number of cameras and points. We propose a new method which works for scenes with 10^3 cameras and 10^6 points in reasonable time 60 s on a single computer. The algorithm output changes smoothly with input parameter which is beneficial for large reconstructions. We scale Schur complement to sufficiently reduce numerical errors when using double precision. We hope that more sophisticated pre-conditioners used in Bundle Adjustment solvers [3, 34] can scale values to the range where single precision will be sufficient. That representation will allow GPU usage and order of magnitude speedup.

Acknowledgment. This work has been supported by the EU-H2020 project LADIO (number 731970) and Grant Agency of the CTU Prague project SGS16/230/OHK3/3T/13.

References

1. Agarwal, S., Furukawa, Y., Snavely, N., Simon, I., Curless, B., Seitz, S.M., Szeliski, R.: Building rome in a day. Commun. ACM **54**(10), 105–112 (2011)
2. Agarwal, S., Mierle, K., et al.: Ceres solver. http://ceres-solver.org
3. Agarwal, S., Snavely, N., Seitz, S.M., Szeliski, R.: Bundle adjustment in the large. In: Daniilidis, K., Maragos, P., Paragios, N. (eds.) ECCV 2010. LNCS, vol. 6312, pp. 29–42. Springer, Heidelberg (2010). doi:10.1007/978-3-642-15552-9_3
4. Balasubramanian, R., Das, S., Swaminathan, K.: Error analysis in reconstruction of a line in 3-D from two arbitrary perspective views. Int. J. Comput. Math. **78**(2), 191–212 (2001)
5. Belhaoua, A., Kohler, S., Hirsch, E.: Estimation of 3D reconstruction errors in a stereo-vision system. In: SPIE Europe Optical Metrology, p. 73900X. International Society for Optics and Photonics (2009)
6. Belhaoua, A., Kohler, S., Hirsch, E.: Error evaluation in a stereovision-based 3D reconstruction system. EURASIP J. Image Video Process. **2010**(1), 1 (2010)
7. Ake, B.: Numerical Methods for Least Squares Problems. SIAM, Philadelphia (1996)
8. Blostein, S.D., Huang, T.S.: Error analysis in stereo determination of 3-D point positions. IEEE Trans. Pattern Anal. Mach. Intell. PAMI **9**(6), 752–765 (1987)
9. Intel Corporation. Intel math kernel library. https://software.intel.com/en-us/intel-mkl
10. Dong, T., Haidar, A., Tomov, S., Dongarra, J.: A fast batched Cholesky factorization on a GPU. In: 2014 43rd International Conference on Parallel Processing, pp. 432–440. IEEE (2014)
11. Faugeras, O., Luong, Q.-T., Papadopoulou, T.: The Geometry of Multiple Images: The Laws That Govern the Formation of Images of a Scene and Some of Their Applications. MIT Press, Cambridge (2001)
12. Förstner, W.: Uncertainty and projective geometry. Handbook of Geometric Computing, pp. 493–534. Springer, Heidelberg (2005)
13. Frahm, J.-M., et al.: Building Rome on a cloudless day. In: Daniilidis, K., Maragos, P., Paragios, N. (eds.) ECCV 2010. LNCS, vol. 6314, pp. 368–381. Springer, Heidelberg (2010). doi:10.1007/978-3-642-15561-1_27
14. Guennebaud, G., Jacob, B., et al.: Eigen v3.3 (2010). http://eigen.tuxfamily.org
15. Hartley, R., Zisserman, A.: Multiple View Geometry in Computer Vision. Cambridge University Press, New York (2003)
16. Heinly, J., Schönberger, J.L., Dunn, E., Frahm, J.-M.: Reconstructing the world* in six days *(as captured by the Yahoo 100 million image dataset). In: Computer Vision and Pattern Recognition (CVPR) (2015)
17. Höhle, J., Höhle, M.: Accuracy assessment of digital elevation models by means of robust statistical methods. ISPRS J. Photogrammetry Remote Sens. **64**(4), 398–406 (2009)
18. Kamberova, G., Bajcsy, R.: Precision in 3-D points reconstructed from stereo. Technical reports (CIS), p. 201 (1997)

19. Langguth, F., Sunkavalli, K., Hadap, S., Goesele, M.: Shading-aware multi-view stereo. In: Leibe, B., Matas, J., Sebe, N., Welling, M. (eds.) ECCV 2016. LNCS, vol. 9907, pp. 469–485. Springer, Cham (2016). doi:10.1007/978-3-319-46487-9_29
20. Lhuillier, M., Perriollat, M.: Uncertainty ellipsoids calculations for complex 3D reconstructions. In: Proceedings 2006 IEEE International Conference on Robotics and Automation, ICRA 2006, pp. 3062–3069. IEEE (2006)
21. Min, S., Rixin, H., Daojun, W.: Precision analysis to 3D reconstruction from image sequences. In: The 5th ISPRS Workshop on DMGISs. CiteSeer (2007)
22. Park, S.-Y., Subbarao, M.: A multiview 3D modeling system based on stereo vision techniques. Mach. Vis. Appl. **16**(3), 148–156 (2005)
23. Pauly, M., Mitra, N.J., Guibas, L.J.: Uncertainty and variability in point cloud surface data. In: Symposium on point-based graphics, vol. 9 (2004)
24. Polok, L., Ila, V., Smrz, P.: 3D reconstruction quality analysis and its acceleration on GPU clusters (2016)
25. Rivera-Rios, A.H., Shih, F.-L., Marefat, M.: Stereo camera pose determination with error reduction and tolerance satisfaction for dimensional measurements. In: Proceedings of the 2005 IEEE International Conference on Robotics and Automation, pp. 423–428. IEEE (2005)
26. Schaer, P., Skaloud, J., Landtwing, S., Legat, K.: Accuracy estimation for laser point cloud including scanning geometry. In: Mobile Mapping Symposium 2007, Padova, number TOPO-CONF-2008-015 (2007)
27. Schönberger, J.L., Zheng, E., Frahm, J.-M., Pollefeys, M.: Pixelwise view selection for unstructured multi-view stereo. In: Leibe, B., Matas, J., Sebe, N., Welling, M. (eds.) ECCV 2016. LNCS, vol. 9907, pp. 501–518. Springer, Cham (2016). doi:10.1007/978-3-319-46487-9_31
28. Schönberger, J.L., Frahm, J.-M.: Structure-from-motion revisited. In: IEEE Conference on Computer Vision and Pattern Recognition (CVPR) (2016)
29. Snavely, N., Seitz, S.M., Szeliski, R.: Photo tourism: exploring photo collections in 3D. In: ACM transactions on graphics (TOG), vol. 25, pp. 835–846. ACM (2006)
30. Snavely, N., Seitz, S.M., Szeliski, R.: Skeletal graphs for efficient structure from motion. In: CVPR, vol. 1, p. 2 (2008)
31. Tomov, S., Dongarra, J., Baboulin, M.: Towards dense linear algebra for hybrid GPU accelerated manycore systems. Parallel Comput. **36**(5–6), 232–240 (2010)
32. Wilson, K., Snavely, N.: Robust global translations with 1DSfM. In: Fleet, D., Pajdla, T., Schiele, B., Tuytelaars, T. (eds.) ECCV 2014. LNCS, vol. 8691, pp. 61–75. Springer, Cham (2014). doi:10.1007/978-3-319-10578-9_5
33. Wu, C.: Towards linear-time incremental structure from motion. In: 2013 International Conference on 3D Vision-3DV 2013, pp. 127–134. IEEE (2013)
34. C. Wu, Agarwal, S., Curless, B., Seitz, S.M.: Multicore bundle adjustment. In: IEEE Conference on Computer Vision and Pattern Recognition (CVPR), pp. 3057–3064. IEEE (2011)
35. Zhang, F.: The Schur Complement and Its Applications. Springer, New York (2005)

Robust and Practical Depth Map Fusion for Time-of-Flight Cameras

Markus Ylimäki[1]([✉]), Juho Kannala[2], and Janne Heikkilä[1]

[1] Center for Machine Vision Research, University of Oulu, Oulu, Finland
{markus.ylimaki,janne.heikkila}@ee.oulu.fi
[2] Department of Computer Science, Aalto University, Espoo, Finland
juho.kannala@aalto.fi

Abstract. Fusion of overlapping depth maps is an important part in many 3D reconstruction pipelines. Ideally fusion produces an accurate and nonredundant point cloud robustly even from noisy and partially poorly registered depth maps. In this paper, we improve an existing fusion algorithm towards a more ideal solution. Our method builds a nonredundant point cloud from a sequence of depth maps so that the new measurements are either added to the existing point cloud if they are in an area which is not yet covered or used to refine the existing points. The method is robust to outliers and erroneous depth measurements as well as small depth map registration errors due to inaccurate camera poses. The results show that the method overcomes its predecessor both in accuracy and robustness.

Keywords: Depth map merging · RGB-D reconstruction

1 Introduction

Merging partially overlapping depth maps into a single point cloud is an essential part of every depth map based 3-dimensional (3D) reconstruction software. A simple registration of depth maps may lead to a huge number of redundant points even with relatively small objects. That will make the further processing very slow.

The amount of points could be reduced afterwards by simplifying the cloud but it is more reasonable to aim directly at a nonredundant point cloud. This will save both time and needed memory capacity.

In this paper, we further develop a method which merges a sequence of depth maps into a single nonredundant point cloud [7]. The method takes the measurement accuracy of obtained depths into account and merges nearby depth measurements into a single point in 3D space by giving more weight to the more certain measurement. Thus, only those points that do not have other neighbouring points are added to the cloud. The proposed method significantly reduces the amount of outliers in the depth maps and rejects incorrectly measured or badly registered points.

P. Sharma and F.M. Bianchi (Eds.): SCIA 2017, Part I, LNCS 10269, pp. 122–134, 2017.
DOI: 10.1007/978-3-319-59126-1_11

Fig. 1. Illustration of the multipath interference error in time-of-flight cameras. Left: A Poisson reconstructed surface [6] created from a point cloud which was back projected from a single depth map. Right: the same surface part but now created from the output point cloud of the proposed method.

One major issue in time-of-flight cameras, such as Kinect V2, is the problem called multipath interference [13]. It occurs when the depth sensor receives multiple scattered or reflected signals from the same direction and causes a positive bias to the depth measurements. As illustrated in the left part of Fig. 1, the problem especially occurs in concave corners, which in this case are formed by the table and the backrests of the chairs. Our method, proposed in this paper, is able to correct those errors as shown in the right part of Fig. 1.

2 Related Work and Our Contributions

Fusion of depth maps from the aspect of 3D reconstruction has been studied widely during recent years [4,9,11,18,21]. The most relevant work regarding to our work is the one presented in [11]. There, the authors proposed a depth map fusion method which is capable of building 3D reconstructions from live video in real time. The method is designed for passive stereo depth maps, and thus, does not use uncertainties for depth measurements.

Since the release of Microsoft Kinect, the interest in the real-time reconstruction has increased widely. These methods mostly represent the models as voxels [1,14,16,17,19] which means that their resolution is limited by the available memory. However, this restriction is successfully avoided especially in [14] but this method is designed for live video, and therefore, it may not work that well with wide baseline depth maps. Choi et al. [1] have also achieved impressive results recently. In their method, the loop closures play a significant role which have to be taken into account when capturing the data. The voxel based approach is also used in [3] in the merging of depth maps with multiple scales but the depth maps were acquired with a range scanner or with a multi-view stereo system.

Kyöstilä et al. proposed a method where the point cloud is created iteratively from a sequence of depth maps so that the added depth maps do not increase the redundancy of the cloud [7]. That is, starting with a point cloud, back projected from a single depth map, the method either creates new points to the cloud from other depth maps if they are in an area which has not yet been covered by other points or uses the new measurements to refine the existing points. The refinement merges nearby points by giving more weight to measurements that have lower empirical, depth dependent variances.

However, Kyöstilä's method is mainly designed for merging redundant depth maps and it cannot handle outliers. In addition, the method was designed for the first generation Kinect device (Kinect V1), and thus, it does not take all the characteristics of the newer Kinect device (Kinect V2) into account. These differences and our solutions are discussed in more detail in Sect. 2.1.

2.1 Our Contributions

As described in Sect. 2, the method in [7] cannot handle outliers and does not work properly with Kinect V2. Regarding to our method, the most essential difference between the Kinect devices is the depth measuring technique. Kinect V1 calculates the depths using an infrared dot pattern projected into the space, whereas Kinect V2 is based on time-of-flight (ToF) technique and predicts the depths from the phase shift between an emitted and received infrared signals [15]. Generally, the measurements acquired with Kinect V2 are more accurate, but in certain cases the sensor might receive multiple reflected or scattered signals from the same direction which might cause significant measurement errors as presented in Fig. 1. This multipath interference problem [13] is not taken into account in [7].

Thus, in this paper we propose three extensions to the method in [7] to overcome its weaknesses. The extensions provide three different ways to measure the errors which occur in ToF measurements and our method tries to replace and refine the erroneous points with more accurate measurements from other redundant depth maps. That, is the contributions of this paper are

1. pre-filtering of depth maps to reduce the amount of outliers,
2. improved uncertainty covariance to compensate for the measurement variances and make the method more accurate and
3. filtering of the final point cloud based on a simple visibility violation rule to reduce the amount of erroneous and badly registered measurements due to the multipath interference [13] and incorrect camera poses, respectively.

The experiments show that the extensions significantly improve the results when compared with [7] which make the proposed method a potential postprocessing step for methods like ORB-SLAM [12] or [2]. In addition, the nonredundant point clouds produced with the proposed method can be further transformed into a mesh, like e.q. in [1,11], using [6] or [8] for example.

3 Method

As presented in Fig. 2, the proposed method takes a set of depth maps and calibrated RGB images with known camera poses as input and outputs a point cloud. The method improves the algorithm described in [7] with three extensions

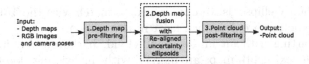

Fig. 2. An overview of the proposed fusion pipeline. In this paper, we propose three extensions (rectangles with a grey background) to the fusion algorithm in [7].

which are marked with darker boxes in Fig. 2. Similarly to [7], our method can be used as a pipeline to process one depth map at a time and therefore the only thing that limits the size of the reconstruction is the available memory for storing the created point cloud.

The pipeline consists of three steps: (1) depth map pre-filtering, (2) actual depth map fusion with re-aligned uncertainty ellipsoids and (3) post-filtering of the final point cloud. The steps are described in more detail in the following sections. Section 3.1, describes the pre-filtering step. The re-aligned uncertainty ellipsoids extension is described together with the fusion step in Sect. 3.2 and the filtering of the final point cloud is presented in Sect. 3.3.

3.1 Pre-filtering of Depth Maps

Typically, backprojected Kinect depth maps (both V1 and V2) have outliers or inaccurate measurements near depth edges and near the corners of the depth image. Usually, their distances to the nearest neighbouring points are much above the average. To remove such measurements from the depth maps, we first calculate a reference curve which describes the average distance from a point to its nth nearest neighbour (NN) ($n = 4$ in all our experiments) in the 3D space at a certain depth. The left part of Fig. 3 presents the calculation of a reference distance at depth d_z for one pixel. The final reference distance at depth d_z is the average of such distances of all pixels. The average distances are calculated for depths from 0.5 m to 4.5 m with 0.1 m interval and the reference curve (blue solid line in the right sub figure) is then acquired by fitting a line to these values.

Now in the pre-filtering, the distance d_m to the 4th nearest neighbour is calculated for every pixel in the input depth map and compared with the reference value d_{ref} at the same depth. The measurement is removed as an outlier if

$$d_m > \frac{d_{ref}}{\sqrt{0.3}}. \tag{1}$$

The red dashed line in the right part of Fig. 3 illustrates the equation. That is, the measurements whose distance value is above the line are removed.

3.2 Improved Depth Map Fusion

The actual depth map fusion is based on [7] with two exceptions: (1) the device dependent parameter values were calibrated for Kinect V2 and (2) the orientations of uncertainty ellipsoids were improved to match with the ToF measuring technique. The details are described later in the section.

That is, starting with an initial point cloud, backprojected from a single depth map, the next depth maps are merged with the existing cloud so that the new measurements are either added to cloud if there is no other points nearby or used to refine the existing measurements without increasing the point count. As described in Sect. 2, the refinement gives more weight to the measurement with lower empirical variance, i.e. uncertainty. The uncertainty of a measurement is described as a covariance $\mathbf{C}$ which determines the location uncertainty of the measurement in x, y and z directions as depth dependent variances

$$\mathbf{C} = \begin{bmatrix} \lambda_1(\beta_x z/\sqrt{12})^2 & 0 & 0 \\ 0 & \lambda_1(\beta_y z/\sqrt{12})^2 & 0 \\ 0 & 0 & \lambda_2(\alpha_2 z^2 + \alpha_1 z + \alpha_0)^2 \end{bmatrix}, \qquad (2)$$

where z is the measured depth and λ_1, λ_2, β_x, β_y, α_2, α_1 and α_0 are parameters which were calibrated for Kinect V2 using the approach presented in [7].

The covariance matrix corresponds to an ellipsoid in the 3D space and in [7] it is aligned so that the z-axis of the ellipsoid is parallel to the optical axis of the camera. However, as described in Sect. 2.1, Kinect V2 measures the depth by comparing the phase shift between the emitted and received signals which travel to the sensor along the line of sight. Therefore, in the proposed method, the covariance ellipsoids are aligned parallel to the line of sights, which means that their orientations depend on the locations of the measurements in the original depth maps. That is, given the rotations $\mathbf{R}$ between the world frame and the camera coordinate frame and $\mathbf{R}_{los}$ between the optical axis of the camera and the line of sight, the covariance $\mathbf{C}$ can be expressed in the world frame with

$$\mathbf{C}_{world} = \mathbf{R}^T \mathbf{R}_{los}{}^T \mathbf{C} \mathbf{R}_{los} \mathbf{R} \qquad (3)$$

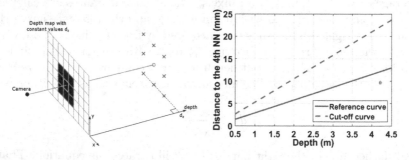

Fig. 3. Illustration of the calculation of the reference distances (left) and the reference and cut-off curves (right). The reference distance for the backprojected pixel (green circle) is the distance to its 4th nearest neighbour. The value of reference curve (blue solid line) at the depth d_z is the average of such distances of all backprojected pixels at the depth d_z. The pre-filtering removes points whose distance to the 4th nearest neighbour is above the cut-off curve (red dashed line). (Color figure online)

As in [7], an existing measurement is refined by the new measurement nearby. First, the refined location is calculated using the best linear unbiased estimator (BLUE) [10], which gives

$$\mathbf{p}'_e = \mathbf{p}_e + \mathbf{C}'_e \mathbf{C}_n{}^{-1}(\mathbf{p}_n - \mathbf{p}_e), \tag{4}$$

where $\mathbf{p}_e$ is the location estimation of the existing point which has been added to the cloud earlier, $\mathbf{p}_n$ is the new measurement with the covariance $\mathbf{C}_n$ and $\mathbf{C}'_e$ is the covariance of the refined point defined by

$$\mathbf{C}'_e = (\mathbf{C}_e{}^{-1} + \mathbf{C}_n{}^{-1})^{-1}, \tag{5}$$

where $\mathbf{C}_e$ is the covariance of the existing measurement estimation.

Now, the Mahalanobis distances d_1 and d_2 between $\mathbf{p}'_e$ and $\mathbf{p}_e$ and $\mathbf{p}'_e$ and $\mathbf{p}_n$, respectively, are calculated using the corresponding covariances

$$d_1 = \sqrt{(\mathbf{p}'_e - \mathbf{p}_e)\mathbf{C}_e{}^{-1}(\mathbf{p}'_e - \mathbf{p}_e)} \tag{6}$$

$$d_2 = \sqrt{(\mathbf{p}'_e - \mathbf{p}_n)\mathbf{C}_n{}^{-1}(\mathbf{p}'_e - \mathbf{p}_n)} \tag{7}$$

If both distances are below the threshold τ, the existing estimate is updated with

$$\mathbf{p}_e \leftarrow \mathbf{p}'_e \quad \text{and} \quad \mathbf{C}_e \leftarrow \mathbf{C}'_e. \tag{8}$$

3.3 Post-filtering of the Final Point Cloud

If in the refinement part of the fusion, at least one of the distances d_1 and d_2 (Eqs. (6) and (7), respectively) is bigger than the threshold τ, the existing measurement is not updated but the measurements might violate the visibility of each other depending on their locations. To solve possible visibility violations, we need normals for every point. The normals are estimated by a plane fitted to the k nearest neighbours of the point ($k = 50$ in all our experiments) in the original back projected depth map.

In this paper, we consider three alternatives, illustrated in Fig. 4, how the measurements may locate with respect to each other. In the first case, point $\mathbf{A}$ occludes point $\mathbf{B}$ but they are far away from each other so that is not a visibility violation. Next, the point $\mathbf{C}$ is occluding point $\mathbf{D}$ nearby but this time the normal of measurement $\mathbf{D}$ is not pointing towards the half space where the camera under consideration is located, and therefore, this is not a visibility violation either. In the third case, the point $\mathbf{E}$ occludes the nearby point $\mathbf{F}$ whose normal is towards the camera. In this case, there is a visibility violation because it is very unlikely that both of these measurements really exist in the scene. In practice, the points are near enough when the distance between them is less than 10% of the depth of the new measurement. This kind of violation may occur due to the inaccuracy of the camera poses or calibration, noise or the multipath interference.

The post-filtering consists of two parts. The first part is built-in to the depth map fusion and it collects some point-wise statistics which are utilized in the

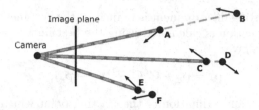

Fig. 4. Three alternatives considered in this paper how the points that project onto the same pixel may locate in the 3D space. The lowest case is the only one causing a visibility violation between points because the points are nearby and their normals point towards the same half space where the camera under consideration is located.

second part that does the actual filtering after the fusion. The statistics are two values which record the number of merges and the number of visibility violations.

That is, if two points that project onto the same pixel are not close enough to be merged together but still violate the visibility of each other in the 3D space, either the existing measurement or the new one is probably an outlier or too inaccurate to be added to the final cloud of points. If the existing measurement has already been merged with another point more than once, it can be considered more reliable and the visibility violation value of the new measurement is incremented by one. Otherwise, the reliability is based on an unreliability weight $w = (1/cos(\alpha))^2$, where α is the angle between the line of sight and the normal of the point, i.e. the bigger the angle the more unreliable the point is and the violation value of the more unreliable measurement is incremented.

Finally in the second part when the fusion has stopped, the points whose visibility violation count is bigger than the value which measures the count of merges, are removed from the cloud.

4   Experiments

The experiments were carried out using three data sets captured with Kinect V2: CCorner, Office1 and Office2. The last two are complicated office environments whereas the first one is a simple concave corner bounded by floor and two walls. Figure 5 presents a sample image of each data set. The checker boards on CCorner data set were used to acquire the poses of the cameras as well as to create a ground truth for quantitative evaluation. The data sets consist of RGB images and depth maps and they were captured with Kinect by moving the device around the room and holding it still while capturing. The sets were captured so that the depth maps had redundant measurements and sequential RGB images had common areas with rich texture in order to gain as good camera pose estimations as possible as described below.

Kinect device was calibrated using the method in [5] which was slightly modified in order to use it with Kinect V2. In the office data sets, the camera poses

Fig. 5. Sample images of the data sets used in the experiments. From left to right: CCorner, Office1 and Office2.

were obtained via structure from motion using VisualSFM[1]. The calibration parameters, the depth maps and the sparse point cloud, produced by SfM, were used to set the scale of the obtained poses to match with the metric system used by the depth sensor of Kinect V2. The poses, the depth maps and the RGB images were then fed to the algorithm pipeline.

The method in [7] was used as a baseline in the evaluations. The results presented in the following sections show step by step how each extension iteratively enhances the results made with the baseline algorithm. In Sect. 4.1, we present the enhancement achieved with pre-filtering. Then, Sect. 4.2 compares the results produced by our method and the baseline extended with the pre-filtering, and finally, in Sect. 4.3 the influence of every extension, including the pre-filtering (PRF), re-aligned covariances (RAC) and post-filtering (POF), is shown by three quantitative analyses.

4.1 Depth Map Pre-filtering

Figure 6 illustrates the pre-filtering result on a single depth map. As expected, the filtering removes measurements near depth edges and image corners where the outliers typically exist or the measurement accuracy is worse due to the lens distortion. In addition, the filtering removes points in darker areas where measurement noise is bigger and on surfaces whose normal create too big angle with the optical axis of the camera and thus are unreliable. The redundancy in the input depth maps guarantees that the removed points does not make holes in the final point cloud.

In Fig. 7, we present the results of Office2 data set after the depth map fusion made with the baseline algorithm and the baseline extended with the pre-filtering. In the figure, the improvement is clearly visible. The filtering has removed a vast majority of the outliers around the laptop as well as from the air beyond the wall (the green solid ellipses) for example. The filtering also improves the details in the view like the area in front of the two computer screens (the red dashed ellipse).

---

[1] http://ccwu.me/vsfm/.

Fig. 6. An illustration of the depth map pre-filtering. Left: the original depth map, right: the filtered depth map. The filter removes incorrect or inaccurate measurements near depth edges and image corners. In the fusion, the holes are filled with more accurate points from other depth maps.

Fig. 7. Comparison between Office2 results made with the baseline algorithm without (left) and with the depth map pre-filtering. The filtering removes a great number of outliers around the laptop and from the air beyond the wall (the green solid ellipses). Pre-filtering also enhances the visibility of the details (the red dashed ellipse). (Color figure online)

4.2   Re-Aligned Covariances and Post-filtering

Figure 8 shows the comparison of the results made with the baseline method extended with the pre-filtering and the proposed method. Our method is able to remove the outliers between the backrests of the chairs and the table as shown in the top part of the figure (green rectangles), but as the bottom part of the figure illustrates, the method is also able to remove the incorrect measurements under the table (red dashed ellipses) and the misplaced measurements above (green solid ellipses). The incorrect measurements below the table have suffered from the multipath interference via backrest of the chair and Kinect had obtained too long distances for those measurements (cf. Figure 1). The misplaced measurements above the table exists due to an inaccurate pose of the camera where the measurements originate from.

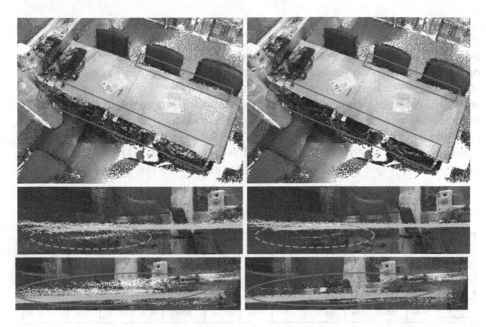

Fig. 8. Comparison between Office1 results made with the baseline algorithm with pre-filtering extension (left) and with the proposed method (right). The proposed method is able to significantly decrease the amount of outliers between the table and the backrests of the chairs (green rectangles) as well as incorrect measurements below the table (red dashed ellipses) and misplaced measurements above (green solid ellipses). (Color figure online)

The noise on the measured surfaces is usually parallel to the line of sight and therefore, by re-aligning the covariance also parallel to that line helps the refinement to move the point in the right direction. The post-filtering instead handles the overlapping points by preserving the one which seems to be more reliable. For example, if the same surface is captured in two ways; first so that the surface in perpendicular to the camera and second so that the surface is slanted and may more probably suffer from multipath interference, the perpendicularly measured points remain in the cloud and others are removed.

4.3   Quantitative Analyses

In the last experiment, the methods and extensions were tested against each other with three quantitative analyses. First, Table 1 illustrates an overview of the sizes of the used data sets and the sizes of the final results. The abbreviations PRF, RAC and POF refer to the proposed extensions to the baseline method, i.e. pre-filtering, re-aligned covariances and post-filtering, respectively. As the table shows, every extension increases the ratio of reduction of the point count.

Then the results from CCorner data set were compared against the ground truth consisting three planes defined by the backprojected checker board corners. The extrinsic parameters of the cameras were obtained by the non-linear optimization where the errors between the detected and projected checker board corners were minimized while the intrinsic parameters of the camera were kept constant. Now for each fusion result, the distances from the points to the nearest plane (floor, right wall or left wall) were calculated and presented as cumulative error curves shown on the left in Fig. 9. The value on the y-axis is the percentage of points whose error is below the value on the x-axis. 100% contains all the points in each fused point cloud (absolute point counts are listed in Table 1).

Table 1. An overview of the sizes of used data sets and achieved point reduction ratios.

Dataset	CCorner				Office1				Office2			
Method	[7]	PRF	PRF+RAC	PRF+RAC+POF	[7]	PRF	PRF+RAC	PRF+RAC+POF	[7]	PRF	PRF+RAC	PRF+RAC+POF
View count	59				98				114			
Original point count	9 307 296				16 690 662				20 400 588			
Final point count	1 299 555	1 123 701	955 161	939 730	5 930 663	4 533 418	4 408 269	4 352 962	6 777 222	5 578 252	5 382 525	5 221 117
Ratio of reduction	86.0%	87.9%	89.7%	89.9%	64.5%	72.8%	73.6%	73.9%	66.7%	72.7%	73.6%	74.4%

As shown in the left sub figure, each extension enhances the accuracy of the fusion. Especially the re-aligned covariance extension significantly improves the result (the red square curve versus the green diamond curve). Pre-filtering and post-filtering bring only moderate improvement in this data set because, due to the simplicity of the data set, the amount of outliers is moderate and practically there are no badly misplaced measurements because the camera poses were acquired relatively accurately as described earlier.

The right part of Fig. 9 was produced with the voxel based evaluation method presented in [20]. The figure illustrates the coverage and the compactness of the reconstructions. The coverage is presented as Jaccard index indicating the proportion of the ground truth which is covered by the reconstruction within a certain threshold. The coverage value is calculated between the voxel representations of the ground truth and the reconstruction so the above-mentioned threshold is the width of a voxel edge. The compactness is presented as a compression ratio which is the ratio of the number of points in the ground truth and the reconstruction. Now, one can see from Fig. 9 that the completeness of the result made with the proposed method is at least equal to that of the baseline method depending on the width of a voxel while the compression ratio is clearly better. That is although the pre-filtering may also have removed some possible correct points on slanted surfaces, that did not make any holes in the reconstruction.

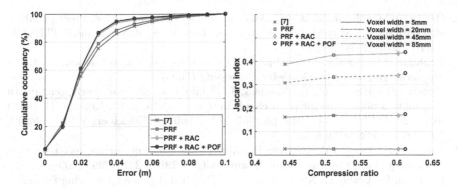

Fig. 9. Evaluation of the leftover errors after the fusion pipeline (left) and evaluation of the coverage (Jaccard index) and compactness (compression ratio) of the reconstructions (right) [20]. PRF, RAC and POF refer to the proposed extensions, i.e. prefiltering, re-aligned covariances and post-filtering, respectively. Jaccard index indicates the proportion of the ground truth which is covered by the reconstruction within the certain threshold represented by the width of a voxel. Compression ratio is the ratio of the number of points in the ground truth and the reconstruction.

5 Conclusion

In this paper, we proposed a method for merging a sequence of overlapping depth maps into a single non-redundant point cloud. Starting with a point cloud back projected from a single depth map, the method iteratively adds points from other depth maps so that the new measurements refine the existing points in overlapping areas. The refinement is based on an uncertainty covariance calculated for every measurement. The proposed method improves the algorithm [7] with three extensions: (1) depth map pre-filtering, (2) depth map fusion with directed uncertainty covariances and (3) post-filtering of the final point cloud. The performance of each extension was demonstrated with several experiments. The proposed method outperformed the baseline algorithm both in robustness and accuracy.

References

1. Choi, S., Zhou, Q.Y., Koltun, V.: Robust reconstruction of indoor scenes. In: IEEE Conference on Computer Vision and Pattern Recognition (CVPR), pp. 5556–5565 (2015)
2. Córdova-Esparza, D.M., Terven, J.R., Jiménez-Hernández, H., Herrera-Navarro, A.M.: A multiple camera calibration and point cloud fusion tool for kinect v2. In: Science of Computer Programming (2017, inpress)
3. Fuhrmann, S., Goesele, M.: Fusion of depth maps with multiple scales. In: Proceedings of the 2011 SIGGRAPH Asia Conference, pp. 148:1–148:8. ACM (2011)
4. Goesele, M., Curless, B., Seitz, S.M.: Multi-view stereo revisited. In: IEEE Conference on Computer Vision and Pattern Recognition (CVPR) (2006)

5. Herrera, C.D., Kannala, J., Heikkilä, J.: Joint depth and color camera calibration with distortion correction. IEEE Trans. Pattern Anal. Mach. Intell. (TPAMI) **34**(10), 2058–2064 (2012)
6. Kazhdan, M., Bolitho, M., Hoppe, H.: Poisson surface reconstruction. In: Eurographics Symposium on Geometry Processing (2006)
7. Kyöstilä, T., Herrera C., D., Kannala, J., Heikkilä, J.: Merging overlapping depth maps into a nonredundant point cloud. In: Kämäräinen, J.-K., Koskela, M. (eds.) SCIA 2013. LNCS, vol. 7944, pp. 567–578. Springer, Heidelberg (2013). doi:10. 1007/978-3-642-38886-6_53
8. Labatut, P., Pons, J.P., Keriven, R.: Robust and efficient surface reconstruction from range data. Comput. Graph. Forum (CGF) **28**(8), 2275–2290 (2009)
9. Li, J., Li, E., Chen, Y., Xu, L., Zhang, Y.: Bundled depth-map merging for multiview stereo. In: IEEE Conference on Computer Vision and Pattern Recognition (2010)
10. Mendel, J.: Lessons in Estimation Theory for Signal Processing, Communications and Control. Prentice Hall, Englewood Cliffs (1995)
11. Merrell, P., et al.: Real-time visibility-based fusion of depth maps. In: IEEE International Conference on Computer Vision (ICCV) (2007)
12. Mur-Artal, R., Montiel, J.M.M., Tardós, J.D.: ORB-SLAM: a versatile and accurate monocular SLAM system. IEEE Trans. Robot. **31**(5), 1147–1163 (2015)
13. Naik, N., Kadambi, A., Rhemann, C., Izadi, S., Raskar, R., Kang, S.B.: A light transport model for mitigating multipath interference in time-of-flight sensors. In: IEEE Conference on Computer Vision and Pattern Recognition, pp. 73–81 (2015)
14. Nießner, M., Zollhöfer, M., Izadi, S., Stamminger, M.: Real-time 3D reconstruction at scale using voxel hashing. ACM Trans. Graph. (TOG) **32**(6), 169 (2013)
15. Pagliari, D., Pinto, L.: Calibration of kinect for xbox one and comparison between the two generations of microsoft sensors. Sensors **15**(11), 27569–27589 (2015)
16. Richard A., N., Shahram, I., Otmar, H., David, M., David, K., Andrew J., D., Pushmeet, K., Jamie, S., Steve, H., Andrew, F.: KinectFusion: real-time dense surface mapping and tracking. In: IEEE International Symposium on Mixed and Augmented Reality (ISMAR), pp. 127–136, October 2011
17. Roth, H., Vona, M.: Moving volume KinectFusion. In: British Machine Vision Conference (2012)
18. Tola, E., Strecha, C., Fua, P.: Efficient large-scale multi-view stereo for ultra high-resolution image sets. Mach. Vis. Appl. **23**(5), 903–920 (2012)
19. Whelan, T., Kaess, M., Maurice, F., Johannsson, H., Leonard, J., McDonald, J.: Kintinuous: spatially extended KinectFusion. Technical report (2012)
20. Ylimäki, M., Kannala, J., Heikkilä, J.: Optimizing the Accuracy and Compactness of Multi-view Reconstructions, pp. 171–183, September 2015
21. Zach, C., Pock, T., Bischof, H.: A globally optimal algorithm for robust TV-L^1 range image integration. In: IEEE International Conference on Computer Vision (ICCV) (2007)

An Error Analysis of Structured Light Scanning of Biological Tissue

Sebastian Nesgaard Jensen[(✉)], Jakob Wilm, and Henrik Aanæs

Department of Applied Mathematics and Computer Science,
Technical University of Denmark,
Richard Petersens Plads B321, Kongens Lyngby, Denmark
{snje,jakw,aanes}@dtu.dk

Abstract. This paper presents an error analysis and correction model for four structured light methods applied to three common types of biological tissue; skin, fat and muscle. Despite its many advantages, structured light is based on the assumption of direct reflection at the object surface only. This assumption is violated by most biological material e.g. human skin, which exhibits subsurface scattering. In this study, we find that in general, structured light scans of biological tissue deviate significantly from the ground truth. We show that a large portion of this error can be predicted with a simple, statistical linear model based on the scan geometry. As such, scans can be corrected without introducing any specially designed pattern strategy or hardware. We can effectively reduce the error in a structured light scanner applied to biological tissue by as much as factor of two or three.

Keywords: 3D reconstruction · Error modeling · Structured light

1 Introduction

Structured light has proven to be very useful for 3D scene acquisition. This is due to its high speed, precision and versatility. As such a wide array of related techniques have been developed in the past decades, facilitating everything from high precision metrology to real-time guidance of automation [8].

Structured light uses a calibrated camera-projector pair as shown in Fig. 1. A series of time multiplexed patterns is projected onto the scene, which can be used for matching and triangulation with the camera. This active approach makes correspondence searching much simpler than passive stereo approaches, and is applicable to scenes with poor texturing. A very important application for structured light is 3D scanning of biological materials, especially human tissue. Examples include head tracking for medical motion correction [22], vision guided surgery [18,23], medical diagnostics [1,4,28] and automation in agriculture and farming [7,21,25]. While the progress in the field has been impressive, one must understand that many target materials are quite problematic. Indeed, they violate the inherent assumption of direct, diffuse surface reflection that most

© Springer International Publishing AG 2017
P. Sharma and F.M. Bianchi (Eds.): SCIA 2017, Part I, LNCS 10269, pp. 135–145, 2017.
DOI: 10.1007/978-3-319-59126-1_12

structured light methods are built on. The Fresnel equations predict that when light transitions from one media to another a portion is directly reflected and another is transmitted into the media itself. In the media the light is scattered one or multiple times until it is absorbed or retransmitted into the environment. The proportion between reflected and refracted light is determined by the specific media's optical properties. For example only 5–7% of human skin reflectance is direct, the remainder is emitted via subsurface scattering [14]. It is therefore of paramount importance that the effect of this violation on structured light is understood and quantified.

In this study, we show that in general, a structured light scan of biological tissue deviates significantly from reference measurements, even with patterns designed specifically to reduce these effects. A large portion of the error can be predicted with a simple, stochastic linear model based on the incident ray geometry. Scans can then be corrected without the need for advanced pattern strategies or special hardware. We can effectively reduce the error in any structured light scanner applied to biological tissue by as much as factor of two or three.

Our study focuses on three types of biological tissue (fat, muscle and skin) with an emphasis on human applications. However we are using porcine materials as a substitute due to its availability and optical similarity to human tissue [26, 27]. Through empirical study we quantify the error induced in structured light by the biological material's optical properties. This results in a linear error model based on the view geometry fitted to each method, material combination that can be used to predict and correct the scan error.

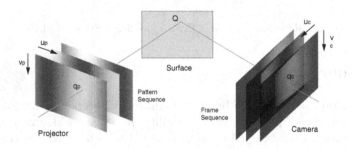

Fig. 1. The structured light principle: a number of patterns are projected onto the scene, and images are captured by a camera. Correspondences are determined by different encoding algorithms, and used to triangulate points on the object surface. In this example, 3-step Phase Shifting patterns are shown.

2   Related Work

The issue of global lighting effects in the context of structured light has been recognized by many authors, e.g. in the acquisition of a human face reflectance field [6].

In order to reduce these effects, hardware modifications such as polarization have been used [2]. Recent attempts have been to design structured light encoding patterns such that they are less susceptible to global lighting effects. The underlying observation is, that with high-frequent patterns, global lighting effects can be considered constant, and invariant to a spatial shift of the pattern. This allows for efficient separation of the observed light intensities into direct and global light [20]. In Modulated phase Shifting [3], structured light patterns are modulated by means of carrier patterns, such that they become high-frequent in both spatial dimensions, thereby improving their separation power. Micro Phase Shifting [10] makes use of sinusoidal patterns in a narrow high-frequency band, promising robustness to global lighting effects and stable phase unwrapping with an optimal number of patterns. It should be noted, that the decoding process in conventional Phase Shifting methods (e.g. [13]) also implicitly performs direct/global light separation. This is true in particular for high frequency scene coding patterns. Since lower frequency phase unwrapping patterns are affected differently by global lighting effects, this can lead to gross outliers. Hence, the advantage of Micro Phase Shifting is not in higher accuracy, but rather in improved robustness (fewer outliers), and more efficient use of information in the encoding patterns.

A newer approach is unstructured light [5], in which the pattern frequency can be high in both dimensions. However the number of patterns is not ideal, and the matching procedure rather inefficient. For binary encoding methods, exclusively high or low-frequency pattern schemes can be considered robust against different global illumination effects [9].

An approach to compensate for the measurement error in isotropic semi-transparent material caused by subsurface scattering was presented in [16]. Similarly to our approach, this work empirically determines the measurement error and explains it by means of a single variable (the projected light angle), albeit only with a single verification object and structured light method. In [15], a Monte-Carlo simulation of the measurement situation was presented, which gives some insight into the error forming process.

In [11], an analytical derivation of the measurement error is given for the Phase Shifting method. This error model predicts the error to decrease with increased spatial frequency of the pattern. The model does not however take into account the loss of amplitude at higher frequency patterns, which increases noise in the measurement data. Furthermore it requires precise knowledge about the scanned material's optical properties (extinction coefficient, phase function and index of refraction), all of which can be difficult to find or estimate.

Computer simulations of structured light scans were performed in [19] to benchmark encoding methods with respect to various parameters, and were found to have similar robustness with respect to subsurface effects.

To our knowledge, no study has thus far quantified the amount of error in scans of biological tissue, or provided a means of correcting for it.

3   Statistical Error Model

Our principle assumption is that the error is composed of a deterministic part, which once determined can be subtracted from future scans, in order to improve the accuracy. Previous work gives some hints as to which parameters to include in a statistical error model [11,16].

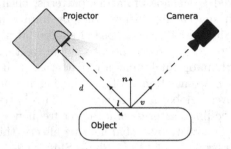

Fig. 2. The structured light scan geometry with the parameters of our error model. The surface normal is n, view direction v, light direction l and the projector-surface distance is d.

Considering the scan geometry, as shown in Fig. 2, we include three variables in our error model: the view angle (given by $n \cdot v$), the light angle (given by $n \cdot l$) and the distance from projector to object, d. We then formulate the following error model:

$$y = \begin{bmatrix} 1 & n \cdot v & n \cdot l & d \end{bmatrix} \begin{bmatrix} \beta_0 \\ \beta_1 \\ \beta_2 \\ \beta_3 \end{bmatrix}, \tag{1}$$

where

    y is the predicted error in mm,
    β_i is a weight,
    n, v, l and d are shown in Fig. 2.

We also tried including many other variables, including reflected light to view angle and coding direction to normal vector angles. These variables are inspired by the analytical error model of Holroyd [11], but did not explain sufficient variance to justify their inclusion in our model. We also fitted Holroyd's error model directly, but our linear model provided more explanatory power.

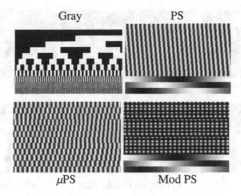

Fig. 3. Structured light patterns used in our experiments. In each case, 12 patterns were used.

4 Experiments

In order to gather data for the error quantification we scanned surfaces made of one of three porcine tissue types; fat, muscle or skin. All samples were raw and unprocessed with 8 samples of each type. The samples were placed individually in the scan volume and spanned many view and light angles. Their distance to the projector also varied from approximately 200 mm to 400 mm. Each scan produced approximately $5 \cdot 10^5$ data points resulting in millions for each tissue type.

In optical metrology it is common practice to prepare optically challenging surface with a spray [12]. This makes the surface optically diffuse while preserving the geometry. The method was used to acquire a ground truth surface to which each scan was compared. Specifically, after each scan the object was sprayed and covered with a thin layer of chalk. Then the reference scan was obtained. While we cannot assume that the chalk coated surfaces to be perfect, we consider them ground truth as they provide very clear contrast with virtually no global illumination. In order to verify that this procedure does not alter surface geometry, we applied two separate layers of chalk to a sample object, and compared the scan result after each layer. The mean signed distance was 0.037 mm, indicating that chalk spraying the surfaces does not significantly bias the result. As can be seen in Fig. 4 the effect of chalk spraying is relatively pronounced, increasing reflectance and counteracting the pattern blurring caused by subsurface scattering.

In our experiments, we used four different structured light methods:

- Binary Gray coding [24]: one completely lit and one completely dark image were used to define the binary threshold individually in each camera pixel. The remaining patterns were used to encode $2^{10} = 1024$ individual lines on the object surface.

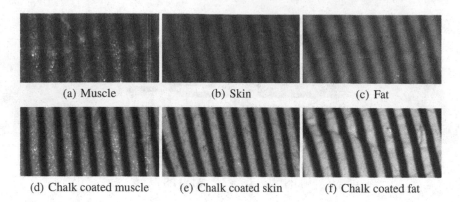

(a) Muscle   (b) Skin   (c) Fat

(d) Chalk coated muscle   (e) Chalk coated skin   (f) Chalk coated fat

Fig. 4. Fine grained binary structured light pattern projected onto various types of tissues. The effect of subsurface scattering is clearly seen the pattern becomes blurred without chalk coating.

- N-step Phase Shifting was used with 9 shifts of a high-frequency sinusoid of frequency $1/76\,\mathrm{px}^{-1}$, corresponding to approximately $1/10\,\mathrm{mm}$ on the object surface. Three additional patterns were used for phase-unwrapping [13].
- Micro Phase Shifting [10] using frequencies in the band $[1/80.00 - 1/70.00]\,\mathrm{px}^{-1}$. These frequencies corresponds to a spatial frequency on the object surface of approximately $1/10\,\mathrm{mm}$. Slightly different from [10], the specific values were determined using a derivative free non-linear pattern search.
- Modulated Phase Shifting [3] with three shifts of a sinusoid of frequency $1/76\,\mathrm{px}^{-1}$ ($1/10\,\mathrm{mm}$ on the object surface). Each of these sinusoids was modulated in the orthogonal direction using a sinusoidal carrier with the same frequency. Three additional patterns were used for phase-unwrapping.

For the sake of brevity these will henceforth be referred respectively to as; Gray, PS, Micro PS and Mod PS. The former two are standard methods of structured light and can be expected to perform very similar to many derived methods. The latter two are state-of-the-art and have been specifically designed to mitigate the effects of global illumination, as described in Sect. 2. As such we asses and compare the progress in the state-of-the-art in terms of counteracting the influence of global illumination. A pattern budget[1] of 12 was settled on for each method as it provided a reasonable balance in acquisition time and accuracy. For all Phase Shifting methods, pattern frequency was set so that each period would be approximately 10 mm on the object surface. The remaining frequencies needed in micro Phase Shifting were determined using simplex optimization as suggested in the original paper [10]. Figure 3 shows the pattern sequences used in our experiments.

For every sample, we defined a binary mask within which all possible surface points were reconstructed. This ensured that the exact same surface region was used in the evaluation of each method.

---

[1] Pattern budget is the number of projected patterns allowed in a single scan.

The error of each surface point was quantified by determining its signed distance to the corresponding point in the chalk sprayed reference. For Gray code scans we define the corresponding points as being the pair with the smallest absolute normal distance. With the other methods, we compared points using their position in the pixel grid.

5   Results and Discussion

The parameters obtained after fitting the error model to our data are seen in Tables 1, 2 and 3. These shows the estimated parameters as well as the RMS of data compared to the chalk coated reference before and after correction (respectively RMS_{raw} and RMS_{cor}) in units of mm. The two latter were evaluated through a process of leave-one-out k-fold cross validation with 5 partitions. In addition we have also estimated the degree of variance explained, R^2, as well as the P-values for the statistical significance of our model against a constant model. All model dependencies were subject to an analysis of variance (ANOVA) [17].

In general the model provides a significant reduction in RMS for all methods with the greatest effect for muscle and skin. It is interesting to note that R^2 is

Table 1. Muscle model estimate and regression quality

	β_0	β_1	β_2	β_3	RMS_{raw}	RMS_{cor}	R^2	P
Gray	0.13	0.15	−0.026	2.3×10^{-4}	**0.42**	0.27	0.0082	0
Phase Shifting	0.25	0.47	−0.18	-2.5×10^{-5}	0.5	**0.21**	0.06	0
Micro PS	0.21	0.36	−0.12	-4.1×10^{-6}	0.45	0.23	0.034	0
Modulated PS	0.27	0.077	0.053	-9.7×10^{-5}	0.42	0.26	0.0037	0

Table 2. Skin model estimate and regression quality

	β_0	β_1	β_2	β_3	RMS_{raw}	RMS_{cor}	R^2	P
Gray	−0.48	0.018	0.43	1.3×10^{-3}	**0.4**	0.19	0.069	0
Phase Shifting	0.27	0.28	0.26	-5.9×10^{-4}	0.54	**0.17**	0.13	0
Micro PS	0.45	0.27	0.21	-1.0×10^{-3}	0.52	0.19	0.13	0
Modulated PS	0.34	0.1	0.27	-6.7×10^{-4}	0.46	0.22	0.054	0

Table 3. Fat model estimate and regression quality

	β_0	β_1	β_2	β_3	RMS_{raw}	RMS_{cor}	R^2	P
Gray	−0.12	0.13	0.039	2.0×10^{-4}	0.26	0.24	0.016	0
Phase Shifting	−0.18	0.31	−0.11	3.9×10^{-4}	0.22	**0.16**	0.084	0
Micro PS	−0.13	0.2	−0.043	3.0×10^{-4}	**0.2**	0.16	0.043	0
Modulated PS	−0.06	0.15	−0.029	1.6×10^{-4}	0.2	0.17	0.018	0

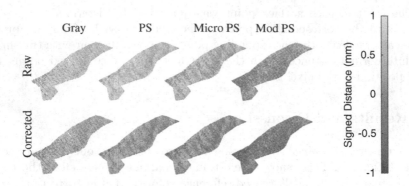

Fig. 5. Signed distance (sd) between scan and reference on a single sample of muscle. Top row: before applying the linear correction model. Bottom row: after correction.

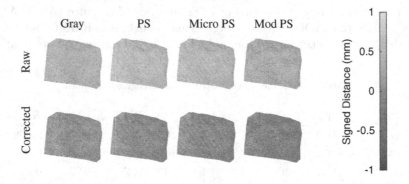

Fig. 6. Signed distance (sd) between scan and reference on a single sample of skin. Top row: before applying the linear correction model. Bottom row: after correction.

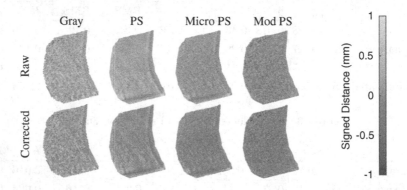

Fig. 7. Signed distance (sd) between scan and reference on a single sample of fat. Top row: before applying the linear correction model. Bottom row: after correction.

in general relatively low; at best 13% and at worst 0.8%. Such measure might dispute model's validity, but the statistical test versus a constant model proves otherwise. In all cases we can conclude that our model is statistical significant within almost a 100% confidence interval, as indicated by the P-values tested against a constant model. While this might seem improbably low, bear in mind that the models was estimated using millions of points which assists in obtaining a statistical significant results. The model estimate itself is rather stable, yielding almost the same error measure for every iteration in the cross validation. This is to be expected due to the high number of training samples and the low dimensionality of the model.

It is seen that most methods have a positive intercept, meaning that regardless of measurement conditions the surface seems to be further away from the camera. The Phase Shifting methods are especially affected by this bias. This effect is further amplified under ideal scanning conditions, where view and light angle are approximately perpendicular to the measured surface. Since β_1 and β_2 are in most cases positive it will further add to positive surface bias. It is also interesting to note that for Phase Shifting methods distance adds a negative weight. This means that distant measurement will effectively have less of a positive bias than close ones. The worst bias can be observed in standard Phase Shifting applied to skin were error can climb to approximately 0.75 mm.

This positive trend can be illustrated by visualizing the per point error as a heat map upon an obtained point cloud, an interesting trend can be observed. Figures 5, 6 and 7 shows the signed error on a single sample visually before and after applying the correction model. All have a positive bias which is very strong for muscle and skin. This alludes to a general trend, subsurface scattering causes the estimated surface to lie further away from the scanner. This is intuitively correct as subsurface scattering is caused by light entering the material for a bit before it is reflected.

In all cases the application of the proper linear model reduces the error's RMS significantly. With a relatively low reduction for fat and a high reduction for skin and muscle. Skin seems to be especially interesting for application as it has the highest error RMS and also receives the largest reduction from error prediction. The remaining unmodeled variance can probably be attributed to variance in chalk thickness, material inhomogeneity and slight vibrations in the recording environment.

6   Conclusion

Structured light is greatly affected by the optical properties of biological materials such as subsurface scattering. By comparing structured light scans of a biological object with scans of the same object covered with a thin chalk layer, we have successfully quantified the resulting error. Our study shows a general positive bias resulting in a surface that lies further away from the scanner than an identical diffuse surface. Due to this positive bias, the RMS of the error can be as high as 0.54 mm. We described the error by fitting a stochastic linear model based on view geometry to the obtained data. Using it, a large portion of the error can be predicted and compensated for. For instance, applying this model to Phase Shifting scans of skin reduces error RMS from 0.54 mm to 0.17 mm.

As opposed to the solutions to global illumination proposed in [3, 10] our approach requires no specially designed pattern strategy or hardware. It can simply be applied directly to the obtained geometry. Additionally our methodology can be applied to any given structured light method and subsurface scattering material. From a pragmatic view, one must conclude that standard Phase Shifting is the superior choice for scanning biological tissue. Not because it shows the lowest error, but rather because the error can be predicted well and compensated using our method.

References

1. Ackerman, J.D., Keller, K., Fuchs, H.: Surface reconstruction of abdominal organs using laparoscopic structured light for augmented reality. In: Proceedings of SPIE, vol. 4661, pp. 39–46 (2002)
2. Chen, T., Lensch, H.P.A., Fuchs, C., Seidel, H.P.: Polarization and phase-shifting for 3D scanning of translucent objects. In: Proceedings of IEEE CVPR (2007)
3. Chen, T., Seidel, H.-P., and Lensch, H.P.: Modulated phase-shifting for 3D scanning. In: Proceedings of IEEE CVPR, pp. 1–8 (2008)
4. Clancy, N.T., Lin, J., Arya, S., Hanna, G.B., Elson, D.S.: Dual multispectral and 3D structured light laparoscope. In: Proceedings of SPIE, vol. 9316, p. 93160C (2015)
5. Couture, V., Martin, N., Roy, S.: Unstructured light scanning robust to indirect illumination and depth discontinuities. Int. J. Comput. Vis. 108(3), 204–221 (2014)
6. Debevec, P., Hawkins, T., Tchou, C., Duiker, H.-P., Sarokin, W., Sagar, M.: Acquiring the reflectance field of a human face. In: Proceeding of SIGGRAPH, pp. 145–156 (2000)
7. Feng, Q.C., Cheng, W., Zhou, J.J., Wang, X.: Design of structured-light vision system for tomato harvesting robot. Int. J. Agric. Biol. Eng. 7(2), 19–26 (2014)
8. Geng, J.: Structured-light 3D surface imaging: a tutorial. Adv. Optics Photonics 160(2), 128–160 (2011)
9. Gupta, M., Agrawal, A., Veeraraghavan, A., Narasimhan, S.G.: A practical approach to 3D scanning in the presence of interreflections, subsurface scattering and defocus. Int. J. Comput. Vis. 102(1–3), 33–55 (2012)
10. Gupta, M., Nayar, S.K.: Micro phase shifting. In: Proceeding of IEEE CVPR, pp. 813–820 (2012)
11. Holroyd, M., Lawrence, J.: An analysis of using high-frequency sinusoidal illumination to measure the 3D shape of translucent objects. In: Proceedings of IEEE CVPR, pp. 2985–2991 (2011)
12. Huang, Z., Ni, J., Shih, A.J.: Quantitative evaluation of powder spray effects on stereovision measurements. Meas. Sci. Technol. 19(2), 025502 (2008)
13. Huntley, J.M., Saldner, H.: Temporal phase-unwrapping algorithm for automated interferogram analysis. Appl. Optics 32(17), 3047–3052 (1993)
14. Krishnaswamy, A., Baranoski, G.: A biophysically-based spectral model of light interaction with human skin. Comput. Graph. Forum 23(3), 331–340 (2004)

15. Lutzke, P., Heist, S., Kühmstedt, P., Kowarschik, R., Notni, G.: Monte Carlo simulation of three-dimensional measurements of translucent objects. Optical Eng. **54**(8), 084111 (2015)
16. Lutzke, P., Kühmstedt, P., Notni, G.: Measuring error compensation on three-dimensional scans of translucent objects. Optical Eng. **50**(6), 063601 (2011)
17. Madsen, H., Thyregod, P.: Introduction to General and Generalized Linear Models. CRC Press, Taylor & Francis Group (2011)
18. Maurice, X., Albitar, C., Doignon, C., De Mathelin, M.: A structured light-based laparoscope with real-time organs' surface reconstruction for minimally invasive surgery. In: Proceedings of International Conference of IEEE EMBS 2012, pp. 5769–5772 (2012)
19. Medeiros, E., Doraiswamy, H., Berger, M., Silva, C.T.: Using physically based rendering to benchmark structured light scanners. Pacific Graph. **33**(7), 71–80 (2014)
20. Nayar, S.K., Krishnan, G., Grossberg, M.D., Raskar, R.: Fast separation of direct and global components of a scene using high frequency illumination. ACM Trans. Graph. **25**(3), 935 (2006)
21. Nguyen, T., Slaughter, D.C., Max, N., Maloof, J.N., Sinha, N.: Structured light-based 3D reconstruction system for plants. Sensors **15**(8), 18587–18612 (2015)
22. Olesen, O.V., Paulsen, R.R., Højgaard, L., Roed, B., Larsen, R.: Motion tracking for medical imaging: a nonvisible structured light tracking approach. IEEE Trans. Med. Imaging **31**(1), 79–87 (2012)
23. Paquit, V., Price, J.R., Seulin, R., Meriaudeau, F., Farahi, R.H., Tobin, K.W., Ferrell, T.L.: Near-infrared imaging and structured light ranging for automatic catheter insertion. In: Proceedings of SPIE, vol. 6141 (2006)
24. Posdamer, J., Altschuler, M.: Surface measurement by space-encoded projected beam systems. Comput. Graph. Image Process. **18**, 1–17 (1982)
25. Rosell-Polo, J.R., Cheein, F.A., Gregorio, E., Andjar, D., Puigdom-nech, L., Masip, J., Escol, A.: Chapter three - advances in structured light sensors applications in precision agriculture and livestock farming. In: Advances in Agronomy, vol. 133, pp. 71–112. Academic Press (2015)
26. Tfaili, S., Gobinet, C., Josse, G., Angiboust, J.-F., Manfait, M., Piot, O.: Confocal raman microspectroscopy for skin characterization: a comparative study between human skin and pig skin. Analyst **137**(16), 3673–3682 (2012)
27. Weigmann, H.-J., Schanzer, S., Patzelt, A., Bahaban, V., Durat, F., Sterry, W., Lademann, J.: Comparison of human and porcine skin for characterization of sunscreens. J. Biomed. Optics **14**(2) (2009). Article No. 024027
28. Wissel, T., Stüber, P., Wagner, B., Bruder, R., Schweikard, A., Ernst, F.: Enriching 3D optical surface scans with prior knowledge: tissue thickness computation by exploiting local neighborhoods. Int. J. Comput. Assist. Radiol. Surg. (2015)

Structure from Motion by Artificial Neural Networks

Julius Schöning$^{(\boxtimes)}$, Thea Behrens, Patrick Faion, Peyman Kheiri,
Gunther Heidemann, and Ulf Krumnack

Institute of Cognitive Science, Osnabrück University, Osnabrück, Germany
{juschoening,tbehrens,pfaion,pkheiri,gheidema,krumnack}@uos.de

Abstract. Retrieving the 3D shape of an object from a collection of
images or a video is currently realized with multiple view geometry algorithms, most commonly Structure from Motion (SfM) methods. With the
aim of introducing artificial neuronal networks (ANN) into the domain of
image-based 3D reconstruction of *unknown* object categories, we developed a scalable voxel-based dataset in which one can choose different
training and testing subsets. We show that image-based 3D shape reconstruction by ANNs is possible, and we evaluate the aspect of scalability by examining the correlation between the complexity of the reconstructed object and the required amount of training samples. Along with
our dataset, we are introducing, in this paper, a first baseline achieved
by an only five-layer ANN. For capturing life's complexity, the ANNs
trained on our dataset can be used a as pre-trained starting point and
adapted for further investigation. Finally, we conclude with a discussion
of open issues and further work empowering 3D reconstruction on real
world images or video sequences by a CAD-model based ANN training
data set.

1   Introduction

Nowadays, image-based 3D reconstruction, also known as multiview stereo
approaches, are usually realized as an analytic solution, based on multiple view
geometry. In general, almost all approaches use the principle of *Structure* from
Motion (SfM)—or rephrased *real wold object shapes* from *different 2D images
captured from different locations*. The reconstructed 3D objects are usually represented by a set of s points in a 3D space calculated from $w \geq 2$ 2D images
captured from different camera positions. This kind of representation leads to
a nonlinear least squares problem with $2 \cdot w \cdot s$ constraints and $6 \cdot w + 3 \cdot s$
unknowns. Benchmarks of state of the art commercial 3D reconstruction soft-
ware like *123D Catch* [2], or *PhotoScan* [1], as well as academic approaches e.g.
VisualSfM [29], or *ARC 3D* [26], show that the accuracy of the reconstructed
3D point cloud with respect to the ground truth is mostly sufficient [23] for
the specific use case, but exponentially impacts the processing time. For solving the nonlinear least squares problem in these SfM approaches, methods such
as gradient descent, conjugate gradient, Gauss-Newton, Levenberg Marquardt,

© Springer International Publishing AG 2017
P. Sharma and F.M. Bianchi (Eds.): SCIA 2017, Part I, LNCS 10269, pp. 146–158, 2017.
DOI: 10.1007/978-3-319-59126-1_13

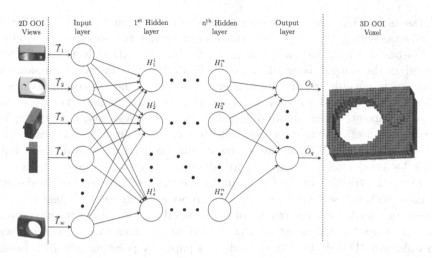

Fig. 1. Simplified scheme of an ANN system architecture for image-based 3D reconstruction. As input vectors $(\vec{I}_w)$ for the ANN, a number of w 2D gray scale projections of a 3D object, taken from particular viewpoints, are used. The kind of ANN (feedforward, convolutional, recurrent, etc.) and setup (numbers of layers [n], numbers of neurons within each layer, kind of meshing, etc.) has to be varied to find the best possible network for this task. The output layer of the ANN consists of v^3 voxels, depending on the resolution of the voxel space. Our current training dataset consists of several $10,0000$ objects with different voxel resolutions. Each training item consists of $w = 12$ images from various viewpoints of each object that serve as input—seen on the left hand side of the figure—as well as v^3 binary voxel values as output—seen on the right hand side. Currently, we train and test our ANNs only on our virtual computer generated dataset. In the future, these pre-trained ANNs shall be applied for the training on "real-world" images and objects.

and singular value decomposition are commonly used. The complete processing pipeline of SfM approaches consists of a concatenation of algorithms, some of which have already been shown to be realizable by ANNs [4,10,20,27]. Hence, we introducing a scalable dataset, which provides 3D objects and 2D images of different complexity. Such a dataset, we argue, would allow for a systematic investigation of ANNs for multiview reconstruction, hopefully leading to a better understanding of which architectures are suitable for this task. Highlighting these opportunities, we include a first baseline showing that a single simple feedforward ANN can replace the whole 3D reconstruction pipeline. It already allows us to analyze some of its weaknesses, like a low reconstruction accuracy of occluded parts.

While recent deep neural network models have achieved promising results on related tasks, e.g. 3D object recognition, depth prediction, and single-view 3D reconstruction, which are introduced in detail in Sect. 2, the task of image-based 3D reconstruction of unknown object categories using ANNs is, to our knowledge, not yet systematically examined. This fact is partly due to the lack of datasets for this task, containing images and 3D ground truth of different 3D objects. Filling this gap, we introduce in Sect. 3 the general architecture of ANN-based 3D reconstruction and discuss how the 3D objects can be represented in the output layer of an ANN. Following this initial discussion, we introduce our voxel-based dataset in Sect. 4 and release the less complex cases of $3 \times 3 \times 3$, $4 \times 4 \times 4$ and $8 \times 8 \times 8$ voxel spaces as a static dataset. For providing a scalable dataset, which allows different degrees of complexity, a generator for $n \times n \times n$ datasets is introduced as well. In Sect. 5, we describe the design of a simple feedforward ANN and show that it can reconstruct 3D objects in both the $3 \times 3 \times 3$ and $4 \times 4 \times 4$ setup with an adequate accuracy. Based on the results we achieved by training our ANN, we introduce the first baseline on this dataset in Sect. 6. Section 7 sets our approach in a broader context and discusses opportunities for transferring the results of ANN-based multiview 3D reconstruction from our computer generated dataset to real world images or video sequences with unknown 3D objects. We conclude this paper by pointing out open issues, proposing further working directions, and describing ongoing work such as our CAD model-based ANN training data set.

Table 1. Overview of the cube dataset showing randomly picked objects in different setups as example.

setup	images of the object seen from 12 different viewpoints
$3 \times 3 \times 3$	cube *012497* with its pattern $(1,0,1,0,0,0,0,0,1,0,1,0,1,1,1,0,1,0,1,1,0,0,1,1,1,1,0)$
$4 \times 4 \times 4$	cube *030290* with its pattern $(0,1,0,0,0,0,0,0,1,1,1,\ldots,0,0,1,0,0,0,1,0,0,0,1,1)$
$8 \times 8 \times 8$	cube *029598* with its pattern $(1,0,0,0,0,0,0,0,1,0,0,\ldots,1,1,0,0,1,1,1,0,1,1,1)$
$16 \times 16 \times 16$	cube *000776* with its pattern $(1,1,1,0,1,0,0,1,0,1,0,1,1\ldots,0,1,1,0,1,0,1,0,0,0,1)$

2   State of the Art

The reconstruction of 3D objects with ANNs is still in its infancy. To our knowledge, there is no approach that does 3D reconstruction of *unknown* 3D objects using only ANNs. For known 3D object categories, recently few approaches [7,30] were introduced. However, our approach worked with unknown categories just like any SfM approach. We summarize in this section work on related topics that seem relevant in this context, for example 3D object recognition, depth map generation as well as prediction, stereo image generation, and single-view 3D reconstruction.

For depth prediction, Eigen et al. [8] use series of convolutional neural network (CNN) stacks applied at increasing resolution for surface normal estimation and semantic labeling. The first scale predicts a set of coarse features for the entire image, the second scale produces mid-level resolution predictions, and the third scale refines the predictions to higher resolution. This work is based on a multi-scale deep network for depth estimation [9]. A global coarse-scaled network consisting of convolutional and max-pooling layers estimates the depth at a global level. A fully convolutional fine-scaled network aligns these coarse predictions with local details. For surface normal estimation, Wang et al. [28] implement two CNNs. A top-down CNN takes the whole image as input and captures the coarse structures which cannot be decoded by local evidence alone. A bottom-up CNN acts on local patches extracted from the image and captures local evidence at a higher resolution. The output of these two networks is combined with a fusion network that learns how to incorporate their predictions.

In 2016, Liu et al. [18] propose a single-view 3D reconstruction method using a CNN to estimate per-pixel depth, normal, and symmetry correspondence. Rectifying the depth information, they set up the symmetry correspondences as an optimization problem in their network. Another work [17] introduces a CNN to learn unary and binary potentials for the continuous conditional random field layer that estimates depth on single images which are over-segmented into superpixels. Li et al. [16] use a method that extracts multi-scale image patches around the superpixel centers, and a CNN learns the relationship between these patches. Thus, the estimation of depth can be formulated as a regression problem.

Roy et al. [22] present a convolutional regression forest where each node in the forest is associated with a CNN which makes a depth estimation for a window around each pixel with a corresponding probability. Without requiring labeled data by using only pairs of images with a small camera motion between themselves, Garg et al. [11] propose an unsupervised framework to train a deep neural network for single-view depth prediction.

Already back in 2004, Peng et al. [20] showed that object reconstruction by an ANN is more accurate than object generation by 3rd order polynomials. The ANNs they used in their experiments were completely based on multilayer feedforward network designs.

3 Multiview 3D Reconstruction by ANNs

Images as input for ANNs have become quite common for, e.g., object recognition tasks [12,14,24]. For these tasks, the output layer of an ANN is usually defined in a way that each output neuron represents a single object category. However, in computer graphics and computer vision, the shape of 3D objects is usually described by vertices, edges, and faces which vary depending on the complexity of the object. A simplified representation of 3D objects for computers was introduced by the first computer games, which use volumetric pixels—so called voxels. Thus, when designing an ANN to output 3D information one has to choose a suitable representation for 3D objects, i.e., the output layer of the ANN must encode either (i) vertices, (ii) edges, (iii) faces, or (iv) voxels of the object.

In connection with the performance of ANNs in binary classification tasks, we chose a voxel-based representation of the 3D objects for the output layer. This form of representation has the additional advantage that the resolution and therefore its accuracy, as well as its computational complexity, can be scaled. The input consists of images captured from the object from different viewpoints, as is common for any SfM approach. These considerations lead to the scheme of a general architecture for ANN-based 3D reconstruction, illustrated in Fig. 1. Like in digital imaging, a general working architecture with a low resolution should establish the basis for higher resolution results. Hence, we designed our dataset with low resolutions, but with an option to scale up.

4 Dataset

As a first idea for creating a dataset which provides the necessary amount of training and test samples to train larger ANNs, we considered a database generator based on a geometry definition file format like Wavefront *obj*. Since a dataset of a variety of 3D *obj* objects will consist of simple and complex objects at the same time, and because the complexity of an object is quite difficult to quantize, we turned towards designing a more conservative dataset. Thus, we brought up a cube-based dataset, cf. exemplary objects of it in Table 1. The generation of this dataset is done in two steps. First, a *cube generator* computes random cubes in a $r \times r \times r$-space and stores them as 3D *obj* objects. Second, the *images and voxel generator* creates w images showing an object from different viewpoints and its corresponding voxel cloud in a defined resolution.

4.1 Cube Generator

This generator[1], written in Matlab, randomly generates n 3D objects. Each such object is created by taking a unit cube in $\mathbb{R}^3$ and subdividing it into $r \times r \times r$ subcubes. The parameter r can be defined by the user. By ensuring the uniqueness of the cube distribution in the $r \times r \times r$ grid, this generator

---

[1] Cf. supplementary material on https://ikw.uos.de/%7Ecv/publications/SCIA17.

is able to generate 2^{r^3} different 3D objects and export them as 3D *obj* object files. Each 3D object is generated by filling the $r \times r \times r$ grid with random binary values, where 1 is interpreted as a filled subcube, while 0 signifies an empty space. To generate the object description, each subcube in the $r \times r \times r$ grid is described by 8 vertices connected by 6 rectangular faces. Before exporting them as 3D *obj* objects, duplicate vertices are combined and inner faces, where two subcubes are connected with each other, are deleted.

4.2 Images and Voxel Generator

For the generation of the w input images and a voxel cloud with the resolution v, a Matlab script is provided[1] which accepts 3D *obj* files as input data. For this generator, the user can define the number of images w taken from the 3D object, its pixel resolution $x \times x$, and the resolution of the voxel cloud $v \times v \times v$.

When generating w images showing the object from w different viewpoints, the generator will choose viewpoints that are uniformly distributed around the object to provide as different perspectives as possible. This is achieved by considering a sphere enclosing the object and evenly distributing the w viewpoints on the sphere using the Fibonacci lattice [13]. From these viewpoints, gray scale images with a resolution of $x \times x$ are rendered. For each scene, a light source is added next to the viewpoint. The images, as well as the voxel cloud of the

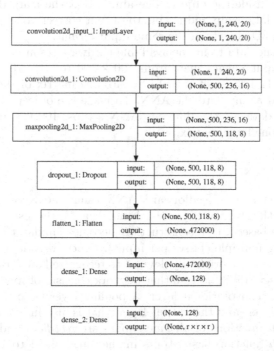

Fig. 2. The activation functions for all inner nodes are *rectifiers*, while the *tanh* is used for the output layer. The loss function used for training is the mean squared error.

objects, are created after scaling the object to the $v \times v \times v$ grid, so that all voxels laying inside the object are defined as cubes.

4.3   Cube Datasets

In addition to the generators, we have released three datasets—a $3 \times 3 \times 3$, a $4 \times 4 \times 4$ and a $8 \times 8 \times 8$ setup, cf. Table 1—as static benchmarks. This ensures reproducibility for the proposed baseline. These static snapshots consist of $100,000$ different objects for the $3 \times 3 \times 3$ setup, $300,000$ different objects for the $4 \times 4 \times 4$ setup and $430,000$ different objects for the $8 \times 8 \times 8$ setup. The objects in both datasets are shown from $w = 12$ different viewpoints with a resolution of 100×100 pixel. This number of objects is, in our opinion, a sufficient amount for training an ANN. Further, we chose $w = 12$ different viewpoints since this is the minimum amount allowing us to solve this task by hand.

5   3D Reconstruction with ANNs

To show the usability of our dataset in a 3D reconstruction task, we trained an ANN to predict the 3D voxel representation of the object from its images. For each of the objects, there were 12 images, taken from different viewpoints (cf. Table 1), and the values in the binary voxel grid are either 0 when outside the object or 1 when inside the object. The aim is that the trained ANN acquires a model of this relationship and is able to output correct voxel grid labels, also for new unknown 3D objects in the same input format.

We prepared the data to fit the available hardware resources by downscaling the size of the input images, here from 100×100 to 20×20 pixels. We then concatenated all 12 images of one object into one matrix of the size 240×20, which then served as input to the ANN. To make the objects easier to load at once as training data for the different ANN setups, $100,000$ different objects were stored into one large matrix.

5.1   Baseline ANN

We created a first simple feedforward ANN using the Keras [6] neural networks library, with Theano [25] as backend. Due to the use of binary voxels as output, image-based 3D reconstruction becomes similar to a classification task (but allowing multiple classes per input). Hence, we adapt a simple example ANN [5,24], that was originally devised to recognize handwritten digits from the MNIST dataset [15]. The resulting network consists of five main layers i.e. the input layer, a convolutional layer, a pooling layer (max pooling), a fully connected dense layer, and the output layer, that is also fully connected.

The activation functions of all inner nodes are *rectifiers* and for the output *tanh*. As the *tanh* function has outputs in the range of -1 to 1, the voxel grid labels are adapted accordingly. Because tuning the ANN is not the primary objective in this paper, the architecture of the layers are chosen without much

Table 2. Baseline of the image-based cube 3D reconstruction dataset. All results are generated by using the feedforward ANN seen in Fig. 2. The object accuracy is a measure of how many objects were reconstructed completely correct or more than 80% of their voxels.

	Training cubes	Epochs	Voxel accuracy	Object 100% accuracy	Object 80% accuracy
$3 \times 3 \times 3$	1–10,000	4	78.01%	0.52%	40.80%
	1–10,000	8	82.03%	2.39%	63.05%
	1–30,000	2	84.66%	1.38%	60.91%
	1–30,000	4	**88.86%**	4.84%	81.72%
$4 \times 4 \times 4$	1–10,000	8	66.01%	0.00%	1.53%
	1–30,000	2	67.09%	0.00%	0.90%
	1–30,000	4	**70.80%**	0.00%	4.47%

consideration. Therefore, it is possible that other sizes would be more suitable for the problem—the values used can be found in Fig. 2. For the objective of providing a dataset for a sytematical development of multiview reconstruction by ANN, the generated results provide an initial step (cf. Sect. 7) as a starting point for further research.

For creating a more robust network, a dropout rate of 20% during training in the connections between the pooling layer and the dense layer is implemented in addition to the main layers cf. Fig. 2. For training we used, the *training sample size* and *epochs* hyperparameters mentioned in Table 2, a batch size of 200, the mean squared error as loss function and the *Adam* optimizer.

6   Baseline and Results

The best performance we achieved when training the ANN shown in Fig. 2 was 88.68% overall voxel accuracy on the $3 \times 3 \times 3$ setup, using objects 1 to 30,000 as the training set during 4 epochs of learning. In the $4 \times 4 \times 4$ setup, a maximum accuracy over all voxels of 70.80% could be reached, using the objects 1 to 30,000 as the training set over 4 epochs. In Table 2, a selected overview of trained ANNs is given as baseline, where we also provide the percentages of 3D objects which could be reconstructed 100% and 80% correct, respectively. The footprint on the hard disk of all trained ANNs was about 725 MB per network. The overall run time varied from 2 to 6 hours using a non GPU accelerated version of Theano.

To investigatie how well a voxel at a certain position was learned, the accuracy for each voxel is calculated. Figure 3 shows the average accuracy for (a) the $3 \times 3 \times 3$ setup and (b) the $4 \times 4 \times 4$ setup, where an accuracy of random chance—50%— is marked blue and an accuuracy of 100% is marked dark red.

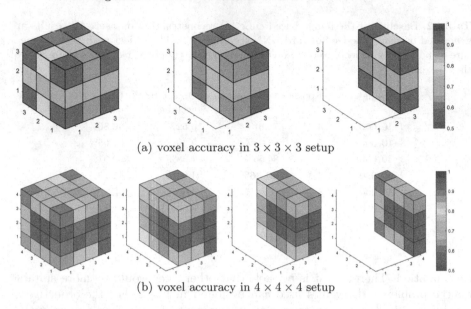

(a) voxel accuracy in $3 \times 3 \times 3$ setup

(b) voxel accuracy in $4 \times 4 \times 4$ setup

Fig. 3. Average accuracy of each voxel in two different setup (a) and (b); from the left to the right hand side the slice by slice walk through; accuracy is color coded from 0.5 random chance–blue to 1.0 every time correct–dark red (Color figure online)

7 Discussion

There is an ongoing discussion on what reconstruction accuracy can be reached with an ANN approach for image-based 3D reconstruction. During the preparation of this dataset, its generator, and while designing the first ANN architecture, our estimations on the expected accuracy varied from about 60% up to 98%. As the standard processing pipeline used for image-based 3D reconstruction is a concatenation of algorithms, most of them applying only linear operations, one would expect that ANNs should be able to perform quite well on this kind of problem. However, the first observations, presented in Table 2, suggest that the problem is not as simple as one might expect. One possible reason is that classical SfM approaches [1,19,21,29] reconstruct vertices and faces instead of reconstructing the object as voxel space. This issue might increase the complexity, due to the *hidden* inner object voxels. These inner voxels may be occluded and thus not detectable on any image. Therefore, the additional interrelationship must be learned, that, e.g., inner object voxels are always to be filled.

Based on these considerations, we were interested if there are positions that are harder to predict within the voxel space. Thus, we expanded the results, with Fig. 3 showing the average accuracy for each voxel. It can be seen that the accuracy of the outer voxels of the cube in both setups is very high, while the accuracy drops towards the center of the voxel cube. This can be explained by the fact that these voxels are often blocked from view by other parts of the object.

Considering the issues that occur even on this simple cube dataset, one can expect further insights from a more thorough analysis of ANNs trained on this dataset. It may turn out that 2D convolution is not the appropriate architecture for this 3D scenario and that other architectures are more suitable. The recent technique of layer-wise relevance propagation [3] may help to identify problematic regions in the input images and suggest changes to the design of the ANN. Furthermore, one may consider using the classification, in a non binary manner. Instead, one can create *half* cubes by incorporating the correspondences to the surrounding blocks. In this way, more accurate objects can be reconstructed.

8 Conclusion

In this paper, we introduced a cube-based dataset to be used as a reference dataset for evaluating multiview 3D reconstruction algorithms. Also, we devised a first baseline ANN architecture to be trained on that dataset. Our initial results suggest that multiview 3D reconstruction of unknown objects by a simple five layer feedforward ANN is possible. Based on the results observed on our simple ANN, we expect that a significant improvement in the performance can be achieved. These enhancements can be done, e.g. through another design of the layers, as well as its connections, more training examples or epochs, another size of the training set or even through another non-voxel-based representation of the 3D object. As already stated in the introduction, the main goal of this paper is not to introduce an ANN with the best possible accuracy, but rather to introduce a dataset for a systematic investigation of ANNs for multiview reconstruction.

Consequently, the dataset presented here is the basis on top of which we will develop, design, train, test, and benchmark various ANNs. To get a deeper understanding, we plan to visualize the features learned on each layer and in each node of the network. We hope that the release of a standard scalable set of 3D training data will empower the community to head in the same or similar direction. In addition, such a set allows evaluating the best performing ANN against *classical* SfM approaches, which are nowadays used for exactly these tasks—the image- and video-based 3D reconstruction of objects. Such an evaluation will only lead to meaningful results if the SfM, as well as the ANN approaches, use the same input data for the 3D object and a comparable output format for the reconstruction process.

Achieving this ambitious goal, further work must be done, which also includes the identification of open issues and the discussion of ANN architectures whether they are promising or misleading. For this purpose, our roadmap of further work consists of five packages which not necessarily have to be processed sequentially:

- Proving a widely diversified set of ANNs which outperform the current baseline on our dataset by a significant value of at least 5%.
- Visualization of learned features by each layer and node to understand the general working principle of each designed ANN.

- Developing, generating, and introducing a virtual 3D CAD dataset with a voxel resolution of at least $100 \times 100 \times 100$.
- Developing, generating, and introducing a real 3D object dataset with the corresponding images captured from different viewpoints and including the 3D ground truth data of the object.
- Analyzing if voxel-based multiview 3D reconstruction by ANN can be scaled up, e.g. by the use of pre-trained computer-generated training data, to reach a suitable voxel resolution with available hardware resources.

One forthcoming work package is the generation of an extensive dataset based on 3D CAD objects. Although those objects will introduce an additional level of complexity and hence may be less suited for initial experiments, they definitely deserve attention in the future due to their practical relevance. Thus, we will expand our *images and voxel generator* to allow for any volumetric *obj* 3D object to be converted into a set of w images and a voxel cloud with a certain resolution. Furthermore, we plan to design, train, and test a diversified set of ANNs on our dataset to support our claim that ANN-based 3D reconstruction from images or video sequences is feasible. We encourage everyone to outperform our baseline and to improve image-based 3D reconstruction using ANNs.

References

1. Agisoft: Agisoft PhotoScan, January 2017. http://www.agisoft.ru/
2. Autodesk Inc.: Autodesk 123D Catch — 3D model from photos. http://www.123dapp.com/catch
3. Bach, S., Binder, A., Montavon, G., Klauschen, F., Müller, K.R., Samek, W.: On pixel-wise explanations for non-linear classifier decisions by layer-wise relevance propagation. PLoS ONE **10**(7), 1–46 (2015)
4. Brabandere, B.D., Jia, X., Tuytelaars, T., Gool, L.V.: Dynamic filter networks. In: Advances in Neural Information Processing Systems (NIPS). Curran Associates, Inc. (2016)
5. Brownlee, J.: Machine learning mastery. http://machinelearningmastery.com/handwritten-digit-recognition-using-convolutional-neural
6. Chollet, F.: Keras. https://github.com/fchollet/keras
7. Choy, C.B., Xu, D., Gwak, J.Y., Chen, K., Savarese, S.: 3D-R2N2: a unified approach for single and multi-view 3D object reconstruction. In: Leibe, B., Matas, J., Sebe, N., Welling, M. (eds.) ECCV 2016. LNCS, vol. 9912, pp. 628–644. Springer, Cham (2016). doi:10.1007/978-3-319-46484-8_38
8. Eigen, D., Fergus, R.: Predicting depth, surface normals and semantic labels with a common multi-scale convolutional architecture. In: IEEE International Conference on Computer Vision (ICCV), pp. 2650–2658 (2015)
9. Eigen, D., Puhrsch, C., Fergus, R.: Depth map prediction from a single image using a multi-scale deep network. In: Advances in Neural Information Processing Systems (NIPS), pp. 2366–2374. Curran Associates, Inc. (2014)
10. Elizondo, D., Zhou, S.-M., Chrysostomou, C.: Surface reconstruction techniques using neural networks to recover noisy 3D scenes. In: Kůrková, V., Neruda, R., Koutník, J. (eds.) ICANN 2008. LNCS, vol. 5163, pp. 857–866. Springer, Heidelberg (2008). doi:10.1007/978-3-540-87536-9_88

11. Garg, R., B.G., V.K., Carneiro, G., Reid, I.: Unsupervised CNN for single view depth estimation: geometry to the rescue. In: Leibe, B., Matas, J., Sebe, N., Welling, M. (eds.) ECCV 2016. LNCS, vol. 9912, pp. 740–756. Springer, Cham (2016). doi:10.1007/978-3-319-46484-8_45

12. Girshick, R., Donahue, J., Darrell, T., Malik, J.: Rich feature hierarchies for accurate object detection and semantic segmentation. In: IEEE Conference on Computer Vision and Pattern Recognition (CVPR) (2014)

13. González, À.: Measurement of areas on a sphere using fibonacci and latitude-longitude lattices. Math. Geosci. **42**(1), 49–64 (2009)

14. Krizhevsky, A., Sutskever, I., Hinton, G.E.: ImageNet classification with deep convolutional neural networks. In: Pereira, F., Burges, C.J.C., Bottou, L., Weinberger, K.Q. (eds.) Advances in Neural Information Processing Systems (NIPS), pp. 1097–1105. Curran Associates, Inc. (2012)

15. Lecun, Y., Bottou, L., Bengio, Y., Haffner, P.: Gradient-based learning applied to document recognition. Proc. IEEE **86**(11), 2278–2324 (1998)

16. Li, B., Shen, C., Dai, Y., van den Hengel, A., He, M.: Depth and surface normal estimation from monocular images using regression on deep features and hierarchical crfs, pp. 1119–1127. Institute of Electrical and Electronics Engineers (IEEE) (2015)

17. Liu, F., Shen, C., Lin, G.: Deep convolutional neural fields for depth estimation from a single image. In: IEEE Conference on Computer Vision and Pattern Recognition (CVPR), pp. 5162–5170 (2015)

18. Liu, G., Yang, C., Li, Z., Ceylan, D., Huang, Q.: Symmetry-aware depth estimation using deep neural networks. arXiv preprint arXiv:1604.06079 (2016)

19. Pan, Q., Reitmayr, G., Drummond, T.: ProFORMA: probabilistic feature-based on-line rapid model acquisition. In: British Machine Vision Conference (BMVC), pp. 112.1–112.11 (2009)

20. Peng, L.W., Shamsuddin, S.M.: 3D object reconstruction and representation using neural networks. In: Proceedings of the International Conference on Computer Graphics and iIteractive Techniques in Austalasia and Southe East Asia (GRAPHITE), pp. 139–147. Association for Computing Machinery (ACM) (2004)

21. Pollefeys, M., Nistér, D., Frahm, J.M., Akbarzadeh, A., Mordohai, P., Clipp, B., Engels, C., Gallup, D., Kim, S.J., Merrell, P., et al.: Detailed real-time urban 3D reconstruction from video. Int. J. Comput. Vis. **78**(2–3), 143–167 (2008)

22. Roy, A., Todorovic, S.: Monocular depth estimation using neural regression forest. In: IEEE Conference on Computer Vision and Pattern Recognition (CVPR) (2016)

23. Schöning, J., Heidemann, G.: Evaluation of multi-view 3D reconstruction software. In: Azzopardi, G., Petkov, N. (eds.) CAIP 2015. LNCS, vol. 9257, pp. 450–461. Springer, Cham (2015). doi:10.1007/978-3-319-23117-4_39

24. Simard, P., Steinkraus, D., Platt, J.: Best practices for convolutional neural networks applied to visual document analysis. In: Seventh International Conference on Document Analysis and Recognition (ICDAR), vol. 3, pp. 958–962 (2003)

25. Theano Development Team: Theano: A Python framework for fast computation of mathematical expressions. arXiv e-prints arXiv.abs/1605.02688 (2016)

26. Vergauwen, M., Van Gool, L.: Web-based 3D reconstruction service. Mach. Vis. Appl. (MVA) **17**(6), 411–426 (2006)

27. Waller, L., Tian, L.: Computational imaging: machine learning for 3D microscopy. Nature **523**(7561), 416–417 (2015)

28. Wang, X., Fouhey, D.F., Gupta, A.: Designing deep networks for surface normal estimation. In: IEEE Conference on Computer Vision and Pattern Recognition (CVPR), pp. 539–547, June 2015

29. Wu, C.: VisualSFM: a visual structure from motion system. http://ccwu.me/vsfm/
30. Wu, Z., Song, S., Khosla, A., Yu, F., Zhang, L., Tang, X., Xiao, J.: 3D ShapeNets:
    a deep representation for volumetric shapes. In: IEEE Conference on Computer
    Vision and Pattern Recognition (CVPR). Institute of Electrical and Electronics
    Engineers (IEEE) (2015)

Pattern Detection and Recognition

Computer Aided Detection
of Prostate Cancer on Biparametric MRI
Using a Quadratic Discriminant Model

Carina Jensen[1](✉), Anne Sofie Korsager[2], Lars Boesen[3],
Lasse Riis Østergaard[2], and Jesper Carl[1,4]

[1] Department of Medical Physics, Oncology,
Aalborg Hospital, 9000 Aalborg, Denmark
carje@rn.dk, JesperCarl@outlook.dk
[2] Department of Health Science and Technology,
Aalborg University, 9000 Aalborg, Denmark
{asko,lasse}@hst.aau.dk
[3] Department of Urology, Herlev University Hospital, 2730 Herlev, Denmark
Lars.Ploug.Boesen@regionh.dk
[4] Department of Clinical Medicine, Aalborg University, 9000 Aalborg, Denmark

Abstract. This paper presents a computer-aided detection (CAD) algorithm for detection of prostate cancer (PCa) in biparametric magnetic resonance imaging (bpMRI). Using image intensity, gradient and gradient direction from T2-weighted (T2 W), diffusion weighted imaging (DWI) and apparent diffusion coefficient (ADC) MRI series, together with a distance feature, a quadratic discriminant analysis (QDA) model was evaluated in 18 patients. A 3D probability map was created for each patient and the number of true- and false positive tumors was determined. Visual assessment showed that for the majority of patients, highest tumor probability was found within the expert annotated volume. The algorithm successfully located 21 of 22 tumors with 0 to 4 false positive per patient. However, the algorithm had a tendency of under-estimating the tumor volume compared to the expert. The study suggests that features extracted from bpMRI can be used for automatic detection of PCa with performance comparable to existing CAD algorithms.

Keywords: Prostate cancer · Magnetic resonance imaging · CAD · Multiparametric · mpMRI

1 Introduction

In 2012 over 1 million men were diagnosed with prostate cancer (PCa) worldwide and around 300.000 men died from the disease in 2012 [1]. The current diagnostic tool for PCa diagnosis is systematic transrectal ultrasound guided biopsies (TRUS+B) due to suspicious elevated prostate specific antigen (PSA) and/or an abnormal digital rectal examination (DRE) [2]. The biopsies are used to grade the PCa according to the Gleason system, which describes the microscopic appearance of PCa. In practice, the Gleason score ranges from 6–10, with 6 being the lowest tumor aggressiveness and 10

© Springer International Publishing AG 2017
P. Sharma and F.M. Bianchi (Eds.): SCIA 2017, Part I, LNCS 10269, pp. 161–171, 2017.
DOI: 10.1007/978-3-319-59126-1_14

being the most aggressive [3]. DRE is not effective in detecting small tumors and tumors located in the anterior or central part of the gland [4]. TRUS+B has a risk of missing tumors that are not palpable by DRE and not visible on ultrasound [5]. Using standard TRUS+B only 0.05–0.5% of the prostate volume is sampled [6]. Thus, TRUS +B entails a risk of missing significant tumors, under-grading cancer burden, and conversely detecting small insignificant tumors that might lead to over detection and possible overtreatment [7, 8]. In patients with persistent suspicion of PCa, despite previous negative, or inconclusive TRUS+B, repeated biopsy procedures are performed in around 31% of the patients [9, 10]. The detection rates of second to fifth set of biopsies range from 12.5 to 16.9% [10].

Magnetic resonance imaging (MRI) provides excellent contrast between soft tissues, which makes it suitable for PCa examination [11, 12]. Recent studies suggest that multiparametric MRI (mpMRI) guided biopsies improve the detection of clinically significant tumors compared to TRUS+B [13]. Furthermore, it can help reduce the number of unnecessary biopsies and allows better assessment of the cancer aggressiveness [14, 15].

Using mpMRI data for PCa screening is a labor intensive task; it requires a high level of expertise, which is not widely available, and is affected by inter-observer variation [7, 16, 17]. This motivates the need for semi- or fully automatic methods, such as computer-aided detection (CAD) algorithms that holds the potential of reducing reading time and inter-observer variation, and may improve the detection rate of clinically significant PCa [18].

Different semi- or fully automatic CAD algorithms have been designed, but it still is a novel technique that remains a challenging issue to improve [7, 19]. The first CAD system to identify cancerous regions in the peripheral zone (PZ) was proposed in 2003 by Chan et al. [20]. Since then, a substantial number of papers have been published on the subject along with detailed overviews of the current literature on prostate CAD algorithms [7, 21, 22]. The methodology behind the published algorithms vary greatly regarding region of interest (peripheral zone (PZ) or whole prostate), MRI sequences, definition of ground truth, features and classifiers used [7, 21]. The best combination of these parameters used for the CAD algorithm remains unsolved and might be scanner and dataset dependent.

The most commonly used mpMRI sequences for prostate CAD algorithms are T2W, DWI (ADC) and DCE. The first two sequences, T2W and DWI (ADC), take less than 20 min to acquire, while adding the DCE sequence prolongs scanning time by up to 45 min. Furthermore, the DCE sequence requires administration of an expensive contrast agent [23]. The long scanning time and contrast costs could pose a limitation on a more widespread distribution of mpMRI diagnostics. Limiting the number of sequences used may partly resolve those limitations [24].

The aim of the present study was to establish a new algorithm for detection of PCa suspicious foci using biparametric MRI (bpMRI) based on T2W and DWI (ADC) MRI sequences and compare it to expert annotations.

2 Materials and Methods

2.1 Patient Data

Eighteen patients were scanned at Herlev Hospital, Denmark using a 3.0T MRI scanner (Ingenia, Philips Healthcare) with an anterior pelvic phased-array coil. One mg intramuscular Glucagon combined with 1 mg hyoscine butylbromid (Buscopan) intravenous injection was administered to the patient to reduce peristaltic movement during the MR examination. MR series were axial T2W and DWI including four b-values (0, 100, 800, and 1400 s/mm^2)). An ADC map (b-values 100 and 800 s/mm^2) was calculated for each patient using the MR-scanner software. For details about the MRI protocol, see Table 1.

Table 1. Sequence parameters for 3 Tesla Ingenia MRI with pelvic phased-array coil

Sequence	Pulse sequence	TR (ms)	TE (ms)	FA (°)	FOV (cm)	ACQ Matrix	Number of slices	Slice thickness (mm)
Axial DWI*	SE-EPI	4916	76	90	18 × 18	116 × 118	25	4
Axial T2 W	SE-TSE	4228	90	90	18 × 18	248 × 239	31	3

SE = spin echo, EPI = echo planar imaging, TSE = turbo spin echo, TR = repetition time, TE = echo time, FA = flip angle, ACQ matrix = acquisition matrix. * = b = 0, b = 100, b = 800, b = 1400 s/mm^2.

All patients had at least one negative or inconclusive TRUS+B prior to the MRI examination. Patients underwent a new TRUS + B with either 10 standard biopsies and 1–3 biopsies from MR positive areas, or only biopsies from MR positive areas (3–4 biopsies). All patients were diagnosed with local or locally advanced PCa. Patient and tumor characteristics are listed in Table 2. Fusion of MRI and real-time ultrasound was done using a Hitachi Medical Systems, Real-time Virtual Sonography (RVS) setup.

Table 2. Patient and tumor characteristics. Prostate and tumor volume is based on expert delineation on T2 W. Gleason scores were obtained from prostate biopsies.

	Median (range, STD)
Patient age [years]	66 (53–78, 6.57)
Prostate volume [cc]	46 (34–85, 14)
Tumor volume [cc]	2.09 (1.02–10.38, 2.19)
All Tumors (22)	
Gleason score 6 (n = 4)	
Gleason score 7 (n = 13)	
Gleason score 8 (n = 3)	
Gleason score 9 (n = 2)	

2.2   Image Pre-processing

T2 W series were manually cropped to exclude some of the normal tissues surrounding the prostate gland. To correct for non-uniformity in MRI intensities the images were normalized using the N3 algorithm, and made isotropic using tri-linear interpolation ($1 \times 1 \times 1$ mm^3 voxels) [25]. DWI (b = 1400 s/mm^2) and ADC series were resampled to match the world coordinate system of the T2 W series, using coordinate information from the image headers. For one patient the DWI and ADC images were manually co-registered to T2 W images using 3dSlicer since there was clear displacement between the image series [26, 27]. The remaining 17 patients were visually inspected for any displacement and no co-registration was done.

2.3   Expert Delineation

The prostate contour was delineated on T2 W images for all patients by an expert (>5 year experience in prostate MRI) to focus the analysis on prostate tissue only. Furthermore, expert tumor contours on biopsy confirmed areas were annotated on T2 W images using the combined MRI series, see Fig. 1. All contours were made in Eclipse™ Treatment Planning System (Varian Medical System).

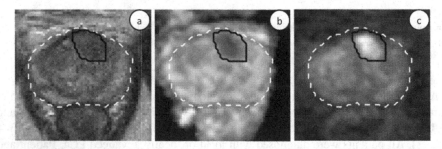

Fig. 1. Example of expert delineation of prostate boundary (dashed white) and tumor boundary (solid black) on T2 W (a), ADC (b) and DWI (c) for patient 1.

2.4   Voxel Feature Extraction

Intensity features from T2W, high b-value DWI (b = 1400 s/mm^2) and ADC, together with 3D image gradient magnitude and gradient direction for T2 W, ADC and DWI images were used as features. The gradient magnitude is the square root of the sum of squares of the individual gradients in x, y and z direction. Gradient direction feature indicates in which direction the image intensity changes most rapidly using the Azimuth angle (measured in the xy-plane from the x-axis). Furthermore, a Euclidean distance feature, measuring the shortest distance from each voxel within the prostate to the prostate boundary, was used.

The 10 features used for this study are listed in Table 3.

Table 3. The features used for the classifier

Feature type	Images
Intensity	T2W, DWI, ADC
Gradient magnitude	T2W, DWI, ADC
Gradient direction	T2W, DWI, ADC
Distance	Prostate contour on T2 W

2.5 Voxel Classification

The intensity features (T2W, DWI and ADC) for each patient were normalized to zero mean and unit variance to account for interpatient-variation. Afterwards, all feature vectors were normalized to zero mean and unit variance.

The classifier used for this study was a quadratic discriminant analysis (QDA) model. The 10 features listed in Table 3 were used in the final model using leave-one-out cross-validation. In leave-one-out cross-validation one patient is kept outside the training set and used for subsequent testing of the model. This is repeated until all patients have been used for testing. The result of the classifier was a probability map per-voxel-basis for each 3D prostate volume with values between 0 and 1, where 1 is indicating highest suspicion of PCa.

2.6 Evaluation

A true positive (TP) was defined as a model detected volume (connected voxels with >0.5 probability) of >0.2 cc within the expert tumor contour. False positives (FP) were defined as model detected volumes outside the expert tumor contour of volumes >0.2 cc. The number of TP and FP, and percentage of TP and FP voxels were evaluated for >0.5 probability obtained from the probability map. Furthermore, the receiver operating characteristics area under curve (ROC-AUC) was calculated (voxel-wise) for each patient and overall for the algorithm.

3 Results

Figure 2 shows the probability maps for image slices at tumor location (approx. center) for each patient.

Visual assessment of the probability maps show that the highest tumor probability corresponds well with the expert annotated area for many patients (e.g. Fig. 2a, i–j, n–o and p–s)

In several patients (Fig. 2d, g–h and m) the tumor region has been identified, although the detected area is smaller than the expert annotation. In some of the patients a high tumor probability is found near the expert annotation (Fig. 2e, k–l). One tumor

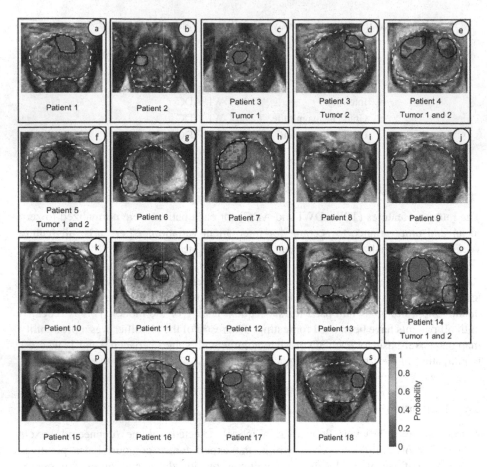

Fig. 2. Probability maps (0 probability being transparent) overlaid T2 W images presented for all patients.

shows no area with high tumor probability within the expert annotation (Fig. 2o) and some tumors only show a small area with high tumor probability (e.g. Fig. 2b and f).

Table 4 shows the quantitative performance of the algorithm with the number of TP and FP for detected volumes >0.2 cc for >0.5 probability.

Of the 22 tumors 21 were detected by the algorithm with a median number of FP per patient of 1. The number of FP ranged from 0 to 4 per patient with a total of 28. The detected TP volumes ranged from 10.71% to 97.31% (median 38.58%) with actual volumes of 0.24 to 5.45 cc (median: 1.14 cc). Three FP volumes were >1.50 cc, the remaining were < 1.00 cc (median: 0.52 cc). The ROC-AUC ranged from 0.69–0.98 with a mean of 0.83.

Table 4. Overview of output from the algorithm showing percentage true positive (for each tumor) and false positive voxels. Furthermore, the number of TP and FP for each patient for lesion volumes >0.2 cc at >0.5 probability is shown along with overall and per-patient ROC-AUC.

Patient	True positive voxels [%]		False positive voxels [%]	No. true positives	No. false positives	ROC-AUC
	Tumor 1	Tumor 2				
1	56.49	–	2.96	1/1	1	0.88
2	20.67	–	5.13	1/1	4	0.76
3	60.71	34.9	3.56	2/2	2	0.82
4	34.34	10.71	2.45	2/2	1	0.71
5	16.94	25.47	3.83	2/2	2	0.74
6	31.53	–	2.14	1/1	1	0.69
7	38.24	–	1.95	1/1	1	0.80
8	38.92	–	3.19	1/1	1	0.86
9	41.44	–	2.07	1/1	0	0.69
10	15.71	–	3.25	1/1	3	0.81
11	27.75	–	1.85	1/1	1	0.84
12	54.57	–	3.57	1/1	2	0.84
13	73.84	–	2.61	1/1	1	0.91
14	90.13	0.00	0.94	1/2	1	0.80
15	83.18	–	3.26	1/1	1	0.95
16	74.27	–	4.21	1/1	2	0.93
17	97.31	–	5.09	1/1	2	0.98
18	77.54	–	3.89	1/1	2	0.91
				21/22	28	0.83

4   Discussion

In this study we presented a CAD algorithm based on bpMRI that can locate the majority of PCa annotated by an expert. A probability map was calculated for each patient using intensity features from T2W, DWI and ADC along with gradient magnitude and direction, and a distance feature. Both visual and quantitative evaluation showed good performance of the algorithm with only one missed tumor. Thus, the algorithm can potentially aid physicians in detecting PCa on MRI for biopsy guidance.

The most used evaluation metric for CAD algorithms is ROC-AUC [21]. We found a ROC-AUC of 0.83, which is in line with the ROC-AUC (0.80–0.89) reported by others [7, 21, 22]. However, some studies report higher values >0.89. Ehrenberg et al. [28] obtained a ROC-AUC of 0.92, also detecting 21 out of 22 tumors with a low number of FP.

We found a low number of FP ranging from 0–4 per patient [15]. Giannini et al. [29] found a per lesion sensitivity of 96% with a median of 3 FP per patient when considering only PZ tumors. Their results are comparable to our per-lesion sensitivity of 95% (21/22 detected tumors) and a median number of 1 FP per patient. A FP in a healthy patient will lead to unnecessary biopsy and healthcare cost, whereas in repeat biopsy patients a high sensitivity is more important than a high specificity [30].

Support Vector Machines (SVM) are the most studied classifier for PCa CAD algorithms. However, other classifiers such as Random Forest, Naïve Bayes and Linear Discriminant Analysis, have been used [21]. S. E. Viswanath [31] compared 12 different classifiers, including QDA, for PCa detection and found that QDA was the best forming classifier in terms of accuracy, execution time and overall evaluation.

Our algorithm has a tendency of under-estimating tumor volume compared to expert annotation (e.g. Fig. 2d). MRI series have been shown to generally under-estimate tumor volume compared to histopathological estimated volumes, although more prominent on ADC than T2W [24, 32]. However, the intent of the algorithm was not to segment the tumor volume but to determine the location of the tumor in order to target biopsies.

Even though tumor volumes >0.5 cc usually are deemed clinically significant, a threshold of 0.2 cc was used for detecting TP and FP in this study [33]. However, tumor volume alone does not determine the PCa risk as some small tumors (0.2–0.5 cc) have high Gleason grade components (Gleason grade 4) and are therefore clinically significant tumors [34, 35]. Thus, a volume threshold <0.5 cc on MRI seems appropriate [24].

We acknowledge certain limitations to this study; Firstly, the prostate was not automatically segmented but annotated by an expert. For a clinical useful CAD algorithm, prostate segmentation should be done automatically as well, however, much research has already been done within this subject and was not within the aim of this study [36]. Another limitation is the use of biopsy results with expert annotation as ground truth. The optimal ground truth would have been the pathological results from radical prostatectomy specimens. Since PCa often is a multifocal disease it is possible that some of the FP lesions found actually are TP not detected by the expert on bpMRI [37]. Furthermore, standard biopsies detected additional small, low grade cancers missed by the expert, which corresponds well with the fact that MRI often overlook clinically insignificant PCa [38]. Finally, no co-registration was done between T2W images and DWI and ADC images (except manual registration in one patient). Thus, geometric image mismatch could have affected the results of the CAD algorithm. A volume with high tumor probability was found in some patients (e.g. Fig. 2e and k–i) at the same location as the expert annotation although with a slight displacement. This might be the results of deformation and/or movement of the prostate during the MRI examination. Automatic registration methods have been explored in the literature, however, this is not a trivial problem to solve [7, 39]. According to Wang et al. [7] registration using the coordinate information in the image header is often sufficient when there is limited patient motion. In this study, prostate movement was visually accessed, and found to be minimal.

The PIRADS v2 guidelines recommend the use of DCE series for expert assessment of PCa, however, it is not clear whether DCE is necessary for CAD algorithms in order to obtain good performance [40]. The combination of T2W, DWI (ADC) and DCE are the most commonly used sequences for PCa diagnostic. However other imaging modalities, such as proton density, diffusion tensor and MR spectroscopy, have been applied for CAD algorithms as well [7, 39].

CAD algorithms are intended to assist radiologists in their workflow by selecting key images and highlight suspicious areas for further evaluation. This might decrease the workload and inter-observer variance among radiologists [7]. Hambrock et al. [16] showed that their CAD algorithm could assist less-experienced radiologists in evaluating PCa on mpMRI reaching performance levels similar to experienced radiologists.

In this study all patients had at least one prior negative biopsy and all were PCa positive. A future study is needed to assess the algorithms performance in PCa negative patients to test the algorithms ability to exclude PCa negative patients from further diagnosis.

5 Conclusion

This study demonstrates that a new algorithm based on bpMRI can be used for PCa detecting with only one missed tumor and a low number of false positives. The quantitative results are within the range of existing CAD algorithms using MRI data for PCa detection.

References

1. WHO Cancer Registry. Prostate cancer estimated incidence, mortality and prevalence worldwide in 2012, http://globocan.iarc.fr/Pages/fact_sheets_cancer.aspx. Accessed 27 Mar 2017
2. Shariat, S.F., Roehrborn, C.G.: Using biopsy to detect prostate cancer. Rev. Urol. **10**, 262–280 (2008)
3. Arasi, E., Kausar, Z., Lakkarasu, S.K.: Prognostic role of new contemporary grading system in prostate cancer. Eur. J. Pharm. Med. Res. **3**, 243–250 (2016)
4. Sumura, M., Shigeno, K., Hyuga, T., Yoneda, T., Shiina, H., Igawa, M.: Initial evaluation of prostate cancer with real-time elastography based on step-section pathologic analysis after radical prostatectomy: a preliminary study. Int. J. Urol. **14**, 811–816 (2007)
5. Hwang, S.I., Lee, H.J.: The future perspectives in transrectal prostate ultrasound guided biopsy. Prostate Int. **2**, 153–160 (2014)
6. Hoogland, A.M., Kweldam, C.F., van Leenders, G.J.L.H.: Prognostic histopathological and molecular markers on prostate cancer needle-biopsies: a review. Biomed Res. Int. **2014**, 1–12 (2014)
7. Wang, S., Burtt, K., Turkbey, B., Choyke, P., Summers, R.M., Wang, S., Burtt, K., Turkbey, B., Choyke, P., Summers, R.M.: Computer aided-diagnosis of prostate cancer on multiparametric MRI: a technical review of current research. Biomed. Res. Int. **2014**, 1–11 (2014)
8. Hoeks, C.C.M. a, Barentsz, J.J.O., Hambrock, T., Yakar, D., Somford, D.M., Heijmink, S. W.T.P.J., Scheenen, T.W.J., Vos, P.C., Huisman, H., van Oort, I.M., Witjes, J.A., Heerschap, A., Fütterer, J.J.: Prostate cancer: multiparametric MR imaging for detection, localization, and staging. Radiology **261**, 46–66 (2011)
9. Boesen, L., Noergaard, N., Chabanova, E., Logager, V., Balslev, I., Mikines, K., Thomsen, H.S.: Early experience with multiparametric magnetic resonance imaging-targeted biopsies under visual transrectal ultrasound guidance in patients suspicious for prostate cancer undergoing repeated biopsy. Scand. J. Urol. **49**, 25–34 (2015)

10. Ploussard, G., Nicolaiew, N., Marchand, C., Terry, S., Allory, Y., Vacherot, F., Abbou, C. C., Salomon, L., De La Taille, A.: Risk of repeat biopsy and prostate cancer detection after an initial extended negative biopsy: longitudinal follow-up from a prospective trial. BJU Int. **111**, 988–996 (2013)

11. Low, R.N., Fuller, D.B., Muradyan, N.: Dynamic gadolinium-enhanced perfusion MRI of prostate cancer: assessment of response to hypofractionated robotic stereotactic body radiation therapy. Am. J. Roentgenol. **197**, 907–915 (2011)

12. Ozer, S., Langer, D.L., Liu, X., Haider, M. a., van der Kwast, T.H., Evans, A.J., Yang, Y., Wernick, M.N., Yetik, I.S.: Supervised and unsupervised methods for prostate cancer segmentation with multispectral MRI. Med. Phys. **37**, 1873 (2010)

13. Lee, D.J., Ahmed, H.U., Moore, C.M., Emberton, M., Ehdaie, B.: Multiparametric magnetic resonance imaging in the management and diagnosis of prostate cancer: current applications and strategies. Curr. Urol. Rep. **15**, 390 (2014)

14. Kitajima, K., Kaji, Y., Fukabori, Y., Yoshida, K.I., Suganuma, N., Sugimura, K.: Prostate cancer detection with 3 T MRI: comparison of diffusion-weighted imaging and dynamic contrast-enhanced MRI in combination with T2-weighted imaging. J. Magn. Reson. Imaging **31**, 625–631 (2010)

15. Litjens, G.J.S., Vos, P.C., Barentsz, J.O., Karssemeijer, N., Huisman, H.J.: Automatic computer aided detection of abnormalities in multi-parametric prostate MRI. SPIE Med. Imaging. **7963**, 79630T (2011)

16. Hambrock, T., Vos, P.C., de Kaa, C.A.H., Barentsz, J.O., Huisman, H.J.: Prostate cancer: computer-aided diagnosis with multiparametric 3-T MR imaging - effect on observer performance. Radiology **266**, 521–530 (2013)

17. Litjens, G.J.S., Barentsz, J.O., Karssemeijer, N., Huisman, H.J.: Automated computer-aided detection of prostate cancer in MR images: From a whole-organ to a zone-based approach. In: Proceedings of the SPIE 8315, Medical Imaging 2012: Computer-Aided Diagnosis, p. 83150G (2012)

18. Vos, P.C., Barentsz, J.O., Karssemeijer, N., Huisman, H.J.: Automatic computer-aided detection of prostate cancer based on multiparametric magnetic resonance image analysis. Phys. Med. Biol. **57**, 1527 (2012)

19. Rampun, A., Zheng, L., Malcolm, P., Tiddeman, B., Zwiggelaar, R.: Computer-aided detection of prostate cancer in T2-weighted MRI within the peripheral zone. Phys. Med. Biol. **61**, 4796–4825 (2016)

20. Chan, I., Wells, W., Mulkern, R.V., Haker, S., Zhang, J., Zou, K.H., Maier, S.E., Tempany, C.M.C.: Detection of prostate cancer by integration of line-scan diffusion, T2-mapping and T2-weighted magnetic resonance imaging; a multichannel statistical classifier. Med. Phys. **30**, 2390–2398 (2003)

21. Lemaître, G., Martí, R., Freixenet, J., Vilanova, J.C., Walker, P.M., Meriaudeau, F.: Computer-Aided Detection and diagnosis for prostate cancer based on mono and multi-parametric MRI: a review. Comput. Biol. Med. **60**, 8–31 (2015)

22. Liu, L., Tian, Z., Zhang, Z., Fei, B.: Computer-aided detection of prostate cancer with MRI: technology and applications. Acad. Radiol. **23**, 1024–1046 (2016)

23. Puech, P., Sufana-Iancu, A., Renard, B., Lemaitre, L.: Prostate MRI: can we do without DCE sequences in 2013? Diagn. Interv. Imaging. **94**, 1299–1311 (2013)

24. Radtke, J., Boxler, S., Kuru, T., Wolf, M., Alt, C., Popeneciu, I., Steinemann, S., Huettenbrink, C., Bergstraesser-Gasch, C., Klein, T., Kesch, C., Roethke, M., Becker, N., Roth, W., Schlemmer, H.-P., Hohenfellner, M., Hadaschik, B.: Improved detection of anterior fibromuscular stroma and transition zone prostate cancer using biparametric and multiparametric MRI with MRI-targeted biopsy and MRI-US fusion guidance. Prostate Cancer Prostatic Dis. **18**, 288–296 (2015)

25. Montreal Neurological Institute (MNI): MINC ToolKit. https://en.wikibooks.org/wiki/MINC
26. Fedorov, A., Beichel, R., Kalpathy-Cramer, J., Finet, J., Fillion-Robin, J.-C., Pujol, S., Bauer, C., Jennings, D., Fennessy, F., Sonka, M., Buatti, J., Aylward, S., Miller, J.V., Pieper, S., Kikinis, R.: 3D Slicer as an image computing platform for the Quantitative Imaging Network. Magn. Reson. Imaging 30, 1323–1341 (2012)
27. 3DSlicer. https://www.slicer.org/
28. Ehrenberg, H.R., Cornfeld, D., Nawaf, C.B., Sprenkle, P.C., Duncan, J.S.: Decision forests for learning prostate cancer probability maps from multiparametric MRI. In: Tourassi, G.D., Armato, S.G. (eds.) SPIE Medical Imaging, p. 97851 J. International Society for Optics and Photonics (2016)
29. Giannini, V., Mazzetti, S., Vignati, A., Russo, F., Bollito, E., Porpiglia, F., Stasi, M., Regge, D.: A fully automatic computer aided diagnosis system for peripheral zone prostate cancer detection using multi-parametric magnetic resonance imaging. Comput. Med. Imaging Graph. 46, 219–226 (2015)
30. Cheikh, A.: Ben, Girouin, N., Colombel, M., Maréchal, J.-M., Gelet, A., Bissery, A., Rabilloud, M., Lyonnet, D., Rouvière, O.: Evaluation of T2-weighted and dynamic contrast-enhanced MRI in localizing prostate cancer before repeat biopsy. Eur. Radiol. 19, 770–778 (2009)
31. Viswanath, S.E.: A Quantitative Data Representation Framework for Structural and Functional MR Imaging with Application to Prostate Cancer Detection (2012)
32. Le Nobin, J., Orczyk, C., Deng, F.-M., Melamed, J., Rusinek, H., Taneja, S.S.S., Rosenkrantz, A.B.B.: Prostate tumour volumes: evaluation of the agreement between magnetic resonance imaging and histology using novel co-registration software. BJU Int. 114, E105–E112 (2014)
33. Mottet, N., Bellmunt, J., Briers, E., Bolla, M., Cornford, P., De Santis, M., Henry, A., Joniau, S., Lam, T., Mason, M.D., Matveev, V., van der Poel, H., Van Der Kwast, T.H., Rouvière, O., Wiegel, T.: EAU-ESTRO-SIOG Guidelines on Prostate Cancer (2016)
34. Cheng, L., Jones, T.D., Pan, C.-X., Barbarin, A., Eble, J.N., Koch, M.O.: Anatomic distribution and pathologic characterization of small-volume prostate cancer (<0.5 ml) in whole-mount prostatectomy specimens. Mod. Pathol. 18, 1022–1026 (2005)
35. Van Der Kwast, T.H., Roobol, M.J.: Defining the threshold for significant versus insignificant prostate cancer. Nat. Publ. Gr. 10, 473–482 (2013)
36. Ghose, S., Oliver, A., Marti, R., Llado, X., Vilanova, J., Freixenet, J., Mitra, J., Fabrice, M., Survey, A.: A survey of prostate segmentation methodologies in ultrasound, magnetic resonance and computed tomography images. Comput. Methods Programs Biomed. 108, 262–287 (2012)
37. Le, J.D., Tan, N., Shkolyar, E., Lu, D.Y., Kwan, L., Marks, L.S., Huang, J., Margolis, D.J. A., Raman, S.S., Reiter, R.E.: Multifocality and prostate cancer detection by multiparametric magnetic resonance imaging: correlation with whole-mount histopathology. Eur. Urol. 67, 569–576 (2015)
38. Renard-Penna, R., Roupret, M., Compérat, E., Rozet, F., Granger, B., Barkatz, J., Bitker, M. O.O., Lucidarme, O., Cussenot, O., Mozer, P.: Relationship between non-suspicious MRI and insignificant prostate cancer: results from a monocentric study. World J. Urol. 34, 673–678 (2016)
39. Litjens, G., Debats, O., Barentsz, J., Karssemeijer, N., Huisman, H.: Computer-aided detection of prostate cancer in MRI. IEEE Trans. Med. Imaging 33, 1083–1092 (2014)
40. Weinreb, J.C., Barentsz, J.O., Choyke, P.L., Cornud, F., Haider, M.A., Macura, K.J., Margolis, D., Schnall, M.D., Shtern, F., Tempany, C.M., Thoeny, H.C., Verma, S.: PI-RADS prostate imaging - reporting and data system: 2015, version 2. Eur. Urol. 69, 16–40 (2015)

Pipette Hunter: Patch-Clamp Pipette Detection

Krisztian Koos[1], József Molnár[1], and Peter Horvath[1,2(✉)]

[1] Synthetic and Systems Biology Unit,
Hungarian Academy of Sciences, BRC, Szeged, Hungary
{koos.krisztian,molnar.jozsef,horvath.peter}@brc.mta.hu
[2] Institute for Molecular Medicine Finland,
University of Helsinki, Helsinki, Finland

Abstract. Segmentation of objects with known geometries in an image is a wide research area. In this paper we show an energy minimization model to detect the tip of glass pipettes in microscopy images. The described model fits two rectangles with a common reference point to dark image regions, which are the sides of a pipette. The model is minimized using gradient descent. The low number of parameters result in a fast evolution and noise insensitivity. The algorithm is tested on label-free and fluorescent microscopy images. The error of the tip detection is only a few micrometers. Automatic pipette tip detection is a step forward to automate the patch-clamping process. The described method can be extended to 3 dimensions or other applications.

Keywords: Shape detection · Patch-clamp · Pipette detection · Autopatching · Energy minimization · Label-free

1 Introduction

Segmentation of objects with well defined geometries is a vital problem in image analysis. Several methods were proposed to detect lines, ellipses or rectangles to identify roads [6], trees [11], or houses [7], respectively using marked point processes. Another way is to compromise strict geometries and use variational methods. For example higher order active contours (HOAC) that can describe various objects with defined shape allowing slight variations of the boundaries. HOACs were successfully used to model circular objects [9] or complex road structures [12]. Recently a family of hybrid variational models was proposed [13,14] that is capable of capturing circular and elliptical objects by minimizing only a few parameters. Here we present a variational method that extends the latter model to detect elongated straight object pairs that have a common reference point. We use this model to segment pipette tips under a microscope and automatically navigate these tips with micrometer precision for patch-clamping and measure properties of neuron cells.

Patch-clamping is a technique to study ion channels in cells. The technique was invented by Erwin Neher and Bert Sakmann in the early 1980s who received the Nobel Prize in Physiology or Medicine in 1991 for their work. Although the

© Springer International Publishing AG 2017
P. Sharma and F.M. Bianchi (Eds.): SCIA 2017, Part I, LNCS 10269, pp. 172–183, 2017.
DOI: 10.1007/978-3-319-59126-1_15

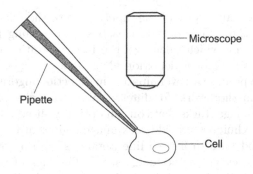

Fig. 1. Schematic whole-cell patch-clamping.

technique can be applied to a wide variety of cells, it is especially useful for measuring the electrophysiological properties of nerve cells (neurons).

The schematic process of patch-clamping a single cell is the following. A glass pipette is pulled onto an electrode. The tip of the glass pipette is open, thus the measured signal originates only from the pipette tip, because the glass does not transfer electricity. The pipette is then pushed next to a cell. When a tight connection, called 'gigaseal' is formed between the cell and the pipette, the cell membrane is broken by vacuum or relatively high voltage pulses. This way the whole-cell patch-clamping configuration is established as illustrated in Fig. 1, the electric signal is passed to an amplifier and then it is ready to be recorded.

The patch-clamping process has to be repeated manually for every target cell. Experienced biologists can usually do only 10–30 successful patch-clamping a day. The process is repetitive and monotonous, thus error prone as the researchers get fatigued. Recently, efforts have been made to automate the technique. In [3] authors used their automatic patch-clamp setup for *in vivo* applications. In [10] the authors extended the technique to a multi-electrode system using up to 12 pipettes. A MATLAB implementation of an automatic patch-clamp software is publicly available [2]. A detailed description of building an automatic patch-clamp setup can be found in [4]. In [1] automatic patch-clamping has been successfully used for cardiomyocytes. An issue of automatic patch-clamping is that glass pipettes has to be changed after every patch-clamping, limiting the throughput. In [5] the authors show a way to clean the pipettes which allows them to be used about 10 times. Patch-clamping is often used in tissue slices when there is an imaging modality to see the target cells and the pipette, unlike to *in vivo* applications. However, changing the pipettes introduces another problem. The pipettes are not perfectly identical and the tip can be slightly translated after the change. In [8] a method is proposed for automatic pipette tip detection using fluorescent channels. However, fluorescent materials can damage cell functions and are not always applicable. Recently, a method for pipette detection in label-free microscopy was proposed in combination with fluorescent cell detection in tissues [15]. The method is used in low magnification

(4x) which provides sharp image of the pipette due to the relatively thick focus plane, even if it is tilted. The detection is based on finding intersecting lines using Hough transform, which calculates the lateral position of the tip. The z position is refined using a focus detection algorithm.

In this paper we propose a novel pipette tip detection algorithm using energy minimization. The method works for differential interference contrast (DIC) and oblique microscopy image stacks that contain optically sliced images of a pipette. These microscopy techniques provide dark image regions at the edge of the glass pipette. The method tries to fit two line segments with a common end point on the different projections of the image stack. The idea of fitting a primitive shape to the image is inspired by the Snakuscule [14] algorithm that segments circular objects. Besides the exact location of the tip's endpoint in 3D, our detection algorithm determines the orientation and tilt angle as well. The algorithm can be extended to fluorescent pipette tip detection, after an edge detection preprocessing step.

2   Methods

Our tip detection method relies on the image formation of DIC (and similarly of oblique) microscopy. DIC is able to show very small optical path length differences of the objects. Diffraction, refraction, reflection and too high optical path length differences can cause effects that are not shown correctly by DIC. The light rays hit the sides of the pipette in a very flat angle, which results in strong effects of the mentioned principles. Due to these side-effects, the regions of pipette edges in the image will be dark, which is information about the position of the pipette. Our method relies on this observation. The pipette is usually not completely in focus in a single image, which is illustrated in Fig. 2a–b. We have developed a 2D model that works on a minimum intensity projection (MIP) of an image stack. The MIP image contains dark stripes in the position of the pipette edges as shown in Fig. 2c. We apply the proposed algorithm for all three possible projection directions to determine the exact position of the pipette tip.

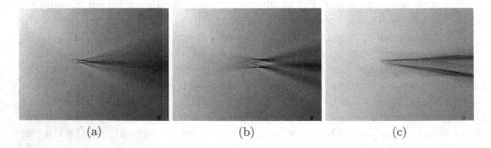

(a)   (b)   (c)

Fig. 2. Example images of a pipette. (a) The pipette tip is nearly in focus. (b) The z level is moved 20 micrometers up compared to (a). (c) Minimum intensity projection along the third dimension of the stack image.

2.1   The Pipette Hunter

The pipette capturing configuration is illustrated in Fig. 3. Let $\mathbf{i}$ and $\mathbf{j}$ define our standard basis at the image origin. The coordinates in the standard basis are denoted by x^1, x^2 and $I(x^1, x^2)$ is the image data. The main idea is to cover dark image regions (the edges of the pipette) with two wide rectangles given some constraints. The rectangles are aligned with two line segments that have a common end point, also called as pivot point. The pivot point is the reference point of the two rectangles. The line segments are called legs. The pivot point is given by its position vector $\mathbf{r} = x^1 \mathbf{i} + x^2 \mathbf{j}$. The rotation of the legs around the pivot point φ^1, φ^2 are measured from the $\mathbf{i}$ axis in the positive direction. The unit direction vectors of the legs are noted as $\mathbf{e}_1, \mathbf{e}_2$ and their unit normals as $\mathbf{n}_1, \mathbf{n}_2$ respectively. ξ_{i1} and ξ_{i2} are distances from the pivot point in the direction of $\mathbf{e}_i$ that define the placement along the leg and the length of a rectangle ($i \in \{1, 2\}$, $\xi_{i2} > \xi_{i1} \geq 0$). η_{i1} and η_{i2} are distances in the normal directions that define the perpendicular placement and the thickness of a rectangle ($i \in \{1, 2\}$, $\eta_{i2} > \eta_{i1}$). Note, that η values can be given such that the rectangles are not symmetric to the legs, which will allow fine tuning of the pivot point later in the algorithm. The model has 4 degrees-of-freedom (DOF), 2 for the coordinates of $\mathbf{r}$ and 1-1 for φ^1 and φ^2.

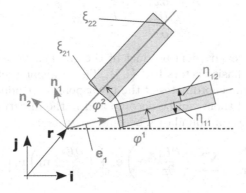

Fig. 3. The pipette hunter.

2.2   The Associated Extreme-Value Problem

The points of the rectangles $\mathbf{p}_i$, $i \in \{1, 2\}$ can be decomposed either in the directions of the standard basis vectors or in the directions defined by their respective legs such that

$$
\begin{aligned}
\mathbf{p}_i &= \mathbf{r} + \xi_i \mathbf{e}_i + \eta_i \mathbf{n}_i \\
&= \left[x^1 + \xi_i \left(\cos \varphi^i \right) - \eta_i \left(\sin \varphi^i \right) \right] \mathbf{i} + \left[x^2 + \xi_i \left(\sin \varphi^i \right) + \eta_i \left(\cos \varphi^i \right) \right] \mathbf{j} \qquad (1) \\
&= \left[x^1 \left(\cos \varphi^i \right) + x^2 \left(\sin \varphi^i \right) + \xi_i \right] \mathbf{e}_i + \left[-x^1 \left(\sin \varphi^i \right) + x^2 \left(\cos \varphi^i \right) + \eta_i \right] \mathbf{n}_i,
\end{aligned}
$$

where ξ_i and η_i are the local coordinates w.r.t. the pivot point. The area of the two rectangles can be simply given as:

$$A = \sum_{i=1}^{2} (\xi_{i2} - \xi_{i1})(\eta_{i2} - \eta_{i1}).\tag{2}$$

We define the energy of the described system as the sum of the energies of the individual legs $E = \sum_{i=1}^{2} E_i$:

$$E\left(x^1, x^2, \varphi^1, \varphi^2\right) \doteq \sum_{i=1}^{2} \frac{1}{A} \int_{\eta_i=\eta_{i1}}^{\eta_{i2}} \int_{\xi_i=\xi_{i1}}^{\xi_{i2}} f(\xi_i, \eta_i) I(\mathbf{p}_i)\, d\xi_i d\eta_i,\tag{3}$$

where f is an appropriately chosen function representing any filter that rotates with the legs. Note that upper indices indicate variables on which the energy depends and are not powers. The easiest way to understand the energy function is to consider f to be identical to 1. In this case the energy is low when the mean image intensity under the rectangles defined by the Pipette Hunter is low. The components of the energy gradient w.r.t. the coordinates of the pivot point are:

$$\frac{\partial E}{\partial x^1} = \frac{\partial E}{\partial \mathbf{r}} \cdot \mathbf{i}$$
$$\frac{\partial E}{\partial x^2} = \frac{\partial E}{\partial \mathbf{r}} \cdot \mathbf{j},\tag{4}$$

where $\frac{\partial E}{\partial \mathbf{r}} \cdot \mathbf{b}$ is the scalar (dot) product of the gradient vector $\frac{\partial E}{\partial \mathbf{r}} \equiv E\nabla$ with one of the standard basis vectors $\mathbf{b} \in \{\mathbf{i}, \mathbf{j}\}$. The gradient vector itself is a sum of two vectors (*i.e.* the coordinates of the pivot point dependent on the energies of both legs), each of them can be decomposed in the directions of its own leg, such that the integration boundaries become constants:

$$\frac{\partial E}{\partial \mathbf{r}} = \sum_{i=1}^{2} \left[\left(\frac{\partial E_i}{\partial \mathbf{r}} \cdot \mathbf{e}_i \right) \mathbf{e}_i + \left(\frac{\partial E_i}{\partial \mathbf{r}} \cdot \mathbf{n}_i \right) \mathbf{n}_i \right].\tag{5}$$

The energy gradient w.r.t. the rotations φ^1 and φ^2 are:

$$\frac{\partial E}{\partial \varphi^i} = \sum_{j=1}^{2} \frac{1}{A} \int_{\eta_j=\eta_{j1}}^{\eta_{j2}} \int_{\xi_j=\xi_{j1}}^{\xi_{j2}} f(\xi_j, \eta_j) I\nabla(\mathbf{p}_j) \cdot \frac{\partial \mathbf{p}_j}{\partial \varphi^i}\, d\xi_j d\eta_j.\tag{6}$$

From Eq. (1), the derivatives of the position vector $\mathbf{p}_j$ are:

$$\frac{\partial \mathbf{p}_j}{\partial \varphi^i} = \delta_{ij} \left\{ \xi_j \left[-(\sin\varphi^i)\,\mathbf{i} + (\cos\varphi^i)\,\mathbf{j} \right] + \eta_j \left[-(\cos\varphi^i)\,\mathbf{i} - (\sin\varphi^i)\,\mathbf{j} \right] \right\}$$
$$= \delta_{ij} (\xi_j \mathbf{n}_i - \eta_j \mathbf{e}_i).\tag{7}$$

where δ_{ij} is the Kronecker delta function, indicating that (unlike the pivot coordinates) the rotations of the legs contribute to the system energy independently.

Using Eqs. (4–7) and (a) the following identities $\mathbf{i} \cdot \mathbf{e}_i = \cos\varphi^i$, $\mathbf{i} \cdot \mathbf{n}_i = -\sin\varphi^i$, $\mathbf{j} \cdot \mathbf{e}_i = \sin\varphi^i$, $\mathbf{j} \cdot \mathbf{n}_i = \cos\varphi^i$, (b) the definitions of the directional derivatives $I_{\xi_i} = I\nabla \cdot \mathbf{e}_i$, $I_{\eta_i} = I\nabla \cdot \mathbf{n}_i$, the complete system is written as:

$$\frac{\partial E}{\partial x^1} = \frac{1}{A}\sum_{i=1}^{2}\int_{\eta_{i1}}^{\eta_{i2}}\int_{\xi_{i1}}^{\xi_{i2}} f(\xi_i,\eta_i)\left[\cos\varphi^i I_{\xi_i}(\mathbf{p}_i) - \sin\varphi^i I_{\eta_i}(\mathbf{p}_i)\right]d\xi_i d\eta_i$$

$$\frac{\partial E}{\partial x^2} = \frac{1}{A}\sum_{i=1}^{2}\int_{\eta_{i1}}^{\eta_{i2}}\int_{\xi_{i1}}^{\xi_{i2}} f(\xi_i,\eta_i)\left[\sin\varphi^i I_{\xi_i}(\mathbf{p}_i) + \cos\varphi^i I_{\eta_i}(\mathbf{p}_i)\right]d\xi_i d\eta_i \qquad (8)$$

$$\frac{\partial E}{\partial\varphi^i} = \frac{1}{A}\int_{\eta_{i1}}^{\eta_{i2}}\int_{\xi_{i1}}^{\xi_{i2}} f(\xi_i,\eta_i)\left[\xi_i I_{\eta_i}(\mathbf{p}_i) - \eta_i I_{\xi_i}(\mathbf{p}_i)\right]d\xi_i d\eta_i, \quad i = 1,2.$$

2.3 Simplification

Consider the simple case when no filter function is used: $f(\xi_i,\eta_i) \equiv 1$. Then the calculations will be limited to the boundaries of the rectangles. By using filters, the calculations can be expanded to the internal regions of the rectangles. Our simplified model uses no filter function. Let the primed ξ', η' variables note the variables measured from the origin of the standard basis in the directions of the respective local systems $\mathbf{e}_i$, $\mathbf{n}_i$, i.e. $(\xi_i',\eta_i') = (\mathbf{e}_i \cdot \mathbf{r} + \xi_i, \mathbf{n}_i \cdot \mathbf{r} + \eta_i)$. Note that the primed variables ξ', η' differ from their unprimed counterparts only by a displacement, hence $d\xi = d\xi'$, $d\eta = d\eta'$. The gradient components of the energy w.r.t. the pivot point (i.e. the first two lines of the extreme value Eqs. (8)) become single integrals:

$$\frac{\partial E}{\partial x^1} = \frac{1}{A}\sum_{i=1}^{2}\left(\cos\varphi^i \int_{\eta_i=\eta_{i1}}^{\eta_{i2}} I(\xi_{i2}',\eta_i') - I(\xi_{i1}',\eta_i')\, d\eta_i \right.$$

$$\left. - \sin\varphi^i \int_{\xi_i=\xi_{i1}}^{\xi_{i2}} I(\xi_i',\eta_{i2}') - I(\xi_i',\eta_{i1}')\, d\xi_i \right)$$

$$\frac{\partial E}{\partial x^2} = \frac{1}{A}\sum_{i=1}^{2}\left(\sin\varphi^i \int_{\eta_i=\eta_{i1}}^{\eta_{i2}} I(\xi_{i2}',\eta_i') - I(\xi_{i1}',\eta_i')\, d\eta_i \right. \qquad (9)$$

$$\left. + \cos\varphi^i \int_{\xi_i=\xi_{i1}}^{\xi_{i2}} I(\xi_i',\eta_{i2}') - I(\xi_i',\eta_{i1}')\, d\xi_i \right).$$

Note, that the integrands are the differences of the image intensity values on the regions' opposite boundaries (that is, the opposite edges of the rectangles).

Similarly, the gradient components of the energy w.r.t. the angles φ^1, φ^2 (the third line of the extreme value Eqs. (8)) become:

$$
\frac{\partial E}{\partial \varphi^1} = \frac{1}{A} \left(\int_{\xi_1 = \xi_{11}}^{\xi_{12}} \xi_1 \left[I\left(\xi_1', \eta_{12}'\right) - I\left(\xi_1', \eta_{11}'\right) \right] d\xi_1 \right.
$$

$$
\left. - \int_{\eta_1 = \eta_{11}}^{\eta_{12}} \eta_1 \left[I\left(\xi_{12}', \eta_1'\right) - I\left(\xi_{11}', \eta_1'\right) \right] d\eta_1 \right)
$$

$$
\frac{\partial E}{\partial \varphi^2} = \frac{1}{A} \left(\int_{\xi_2 = \xi_{21}}^{\xi_{22}} \xi_2 \left[I\left(\xi_2', \eta_{22}'\right) - I\left(\xi_2', \eta_{21}'\right) \right] d\xi_2 \right. \tag{10}
$$

$$
\left. - \int_{\eta_2 = \eta_{21}}^{\eta_{22}} \eta_2 \left[I\left(\xi_{22}', \eta_2'\right) - I\left(\xi_{21}', \eta_2'\right) \right] d\eta_2 \right) .
$$

Note that unlike in the case of the pivot equations, the integrands (of the single integrals) are the weighted differences of the image intensity values on the regions' opposite boundaries (that is, the opposite edges of the rectangles).

2.4 Solving the Equations

One way to minimize the energy $E\left(q^i\right)$ of a system that depends on general variables q^i, $i = 1, 2, ...n$ is to find the stationary solution for the gradient descent evolution equation: $\frac{\partial q^i}{\partial \tau} = -\frac{\partial E}{\partial q^i}$, where τ is the 'artificial' time, and at the stationary point $\frac{\partial q^i}{\partial \tau} = 0$ (hence $\frac{\partial E}{\partial q^i} = 0$).

In our case, the dimensions of the pivot point Eqs. (9) and the rotation angle Eqs. (10) are different. The first two is expressed in length units, while the second two is in radians. Thus we perform a normalization. The complete system, using the local coordinate system, consists of four coupled differential equations:

$$
\frac{\partial x^1}{\partial \tau} = -\frac{\partial E}{\partial x^1}
$$

$$
\frac{\partial x^2}{\partial \tau} = -\frac{\partial E}{\partial x^1}
$$

$$
\frac{\partial \varphi^1}{\partial \tau} = -\frac{2}{(\xi_{11} + \xi_{12})} \frac{\partial E}{\partial \varphi^1} \tag{11}
$$

$$
\frac{\partial \varphi^2}{\partial \tau} = -\frac{2}{(\xi_{21} + \xi_{22})} \frac{\partial E}{\partial \varphi^2} .
$$

The quantities on the right hand side are defined in (9) and (10).

2.5   Properties and Notes

The Pipette Hunter is neutral (does nothing) in a homogeneous environment. Weak 'external forces' (constants to Eq. (11)) may be added to avoid freezing in these regions.

In the simplified case, the integrals are calculated only on the edges of the rectangle shaped region, hence only the image intensity distribution at the boundary is taken into account. The intensity function can freely vary inside. This is not always acceptable. To impose some regularity requirements inside the rectangles, an appropriate filter function can be applied, hence the general equations in (8) need to be used. Minimizing the most significant Fourier coefficients of the intensity function (*i.e.* using a combination of the Fourier basis functions as a filter) allows minor variance inside the rectangles.

3   Results

We have implemented (11) in MATLAB. If some part of the legs are outside the image boundaries, we use the median value of the image for calculations, which is usually very close to the background intensity. For ξ_{i1} we use 0, and for the length of the legs $(\xi_{i2} - \xi_{i1})$ we use half of the longer side of the image. We find this value to work well when the pipette covers about half of the image (it can be the shorter side as well) or even goes over it.

The η values should be chosen such that if a rectangle is fit on a pipette edge, its sides that are aligned to $\mathbf{e}_i$ lie on the opposite sides of the dark region. For our tests we have empirically set 15 pixels for both η values which satisfied the above requirement, and thus the sides are symmetric to the corresponding leg. Note, that the distance between the sides $(\eta_{i1} + \eta_{i2})$ becomes 30 pixels. If this distance is too short, the algorithm is not able to fit the model on dark regions. Similarly, if the distance is too high, the result can be inaccurate.

We start multiple instances of the Pipette Hunter mechanism in swarm to cover the whole image, which is not possible by using only 2 legs. A simple swarm setup is to place a few mechanisms with different rotations on grid points over the image. This number can be minimized to keep the runtime low. As we require the pipette to be around or over the center of the image and set the length of the legs to be half of the longer side of the image, putting mechanisms to the sides and the center of the image will be enough to find the pipette region with at least one instance. Note, that this is a specific case and the search can fail if the requirements are not satisfied. In the general case where no assumption is made on the pipette's position, the runtime can be much higher as it depends on the number of instances in the swarm. Furthermore, we use a 2-phase run of the algorithm. In the first phase, we only update the angles and apply a force that pulls the two legs towards each other. The phase ends if the angles' changes are small or the legs get closer to each other than 0.1 rad. This allows the initialization of an instance with high angle difference (even $\pi/2$ rad or more). In the second phase we turn off the pulling force, update the pivot point as well and keep the restriction that does not let the legs get too close to each other.

180   K. Koos et al.

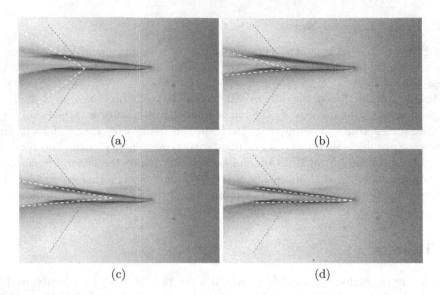

(a)

(b)

(c)

(d)

Fig. 4. Images captured during the runtime of the algorithm. Red dashed lines are the starting position of the legs, which is kept in all the images for comparison. White dashed lines are the current state of the legs. (**a**) During the first phase, both the introduced force and the image intensities pull the legs towards the dark regions. (**b**) End of the first phase. (**c**) In the second phase the pivot point is also updated. The legs are moved and rotated to cover darker image regions. (**d**) The result of the detection.

A few iterations of the algorithm on an example image using one mechanism is shown in Fig. 4.

3.1 Comparisons

To compare the detection of the Pipette Hunter to a reliable solution, we have manually determined the pipette tip positions in 31 stack images. Figure 5 contains images where the pipette orientation and the starting points of the algorithm differs. The average absolute difference between the algorithm's result and the hand-picked focus points is 3.53 ± 2.47 µm, which is 32.97 ± 23.10 pixels and the image size was 1388 by 1040 pixels. This error is acceptable in automatic patch-clamping. A cell's diameter is 10–20 µm. If the pipette tip is aimed at the cell's centroid, given the above error it will still reliably hit the cell. Furthermore, our results are better than the value reported in [8] for final tip-target distance in *in vitro* experiments, which was 12.06 ± 4.30 µm.

We have developed a simple baseline algorithm which can be compared to our method. The baseline model also searches for dark image regions. The model works if the pipette orientation is 0 rad. First, the algorithm searches for minimum point-pairs in the y direction in every slice, then fits two lines on them. The intersection of the two lines will be considered as the pipette tip. The algorithm has a linear runtime, but poor in quality. The baseline model often over-detects

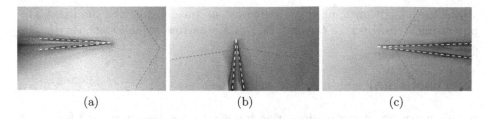

(a)   (b)   (c)

Fig. 5. Example configurations and results.

the pipette tip, which is illustrated in Fig. 6 and compared to the proposed model. On 15 appropriate image stacks (where the pipette orientation is 0 rad.) the mean absolute difference to the hand-picked solutions is 17.92 ± 9.49 μm (167.52 ± 88.68 pixels).

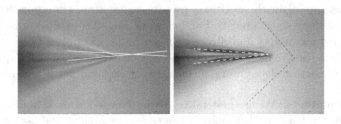

Fig. 6. Comparison to the baseline algorithm (left) on the same image stack.

3.2   Application on Fluorescent Images

Patch-clamping is sometimes performed in two-photon (fluorescent) imaging mode. The proposed pipette detection method can be applied on fluorescent images after applying an edge detection filter (e.g. Sobel detector in our case) and inverting the image. The edges of the pipette will be dark regions. Figure 7

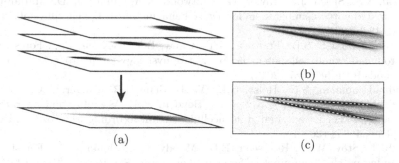

(a)   (c)

Fig. 7. Example fluorescent images and the detection result. **(a)** Images of a fluorescent pipette in different z-levels and the projected fluorescent image. **(b)** The projected image filtered with the Sobel operator. **(c)** Result of the pipette detection algorithm.

shows the algorithm applied on a fluorescent image after the discussed pre-processing steps. Because the dark regions are narrow, we used smaller η values and longer legs.

4   Discussion

In this paper an energy minimization framework has been proposed for patch-clamp pipette detection. The method works on minimum intensity projection of DIC images. The main idea is to fit rectangles on dark image regions that are the sides of the pipette. The algorithm can be applied in automated patch-clamping, where the pipette has to be changed often and slight changes in the pipette's length and shape can be detected in a robust way. The result includes the pipettes orientation and tilt angle as well. The steps of the algorithm are presented on example images. The results are compared to hand-picked pipette tip positions and to a baseline algorithm. After a preprocessing step, the approach can also be applied to fluorescent images. Further work includes extension to 3D, which will work on stack images and directly return an (x,y,z) point, both on DIC and fluorescent images. We believe our method can be extended to other applications, e.g. road intersection detection, neuron network segmentation and more.

Acknowledgements. We are grateful to Attila Ozsvár for his help in the imaging. We also thank to Gáspár Oláh, Márton Rózsa, Gábor Molnár and Gábor Tamás for their support in the automation of the microscopes. We acknowledge the financial support from the National Brain Research Programme (MTA-SE-NAP B-BIOMAG), the TEKES FiDiPro Programme, the European Union and the European Regional Development Funds (GINOP-2.3.2-15-2016-00001, GINOP-2.3.2-15-2016-00006).

References

1. Amuzescu, B., Scheel, O., Knott, T.: Novel automated patch-clamp assays on stem cell-derived cardiomyocytes: will they standardize in vitro pharmacology and arrhythmia research? J. Phys. Chem. Biophys. **4**, 153 (2014)
2. Desai, N.S., Siegel, J.J., Taylor, W., Chitwood, R.A., Johnston, D.: Matlab-based automated patch-clamp system for awake behaving mice. J. Neurophysiol. **114**(2), 1331–1345 (2015)
3. Kodandaramaiah, S.B., Franzesi, G.T., Chow, B.Y., Boyden, E.S., Forest, C.R.: Automated whole-cell patch-clamp electrophysiology of neurons in vivo. Nat. Methods **9**(6), 585–587 (2012)
4. Kodandaramaiah, S.B., Holst, G.L., Wickersham, I.R., Singer, A.C., Franzesi, G.T., McKinnon, M.L., Forest, C.R., Boyden, E.S.: Assembly and operation of the autopatcher for automated intracellular neural recording in vivo. Nat. Protoc. **11**(4), 634–654 (2016)
5. Kolb, I., Stoy, W.A., Rousseau, E.B., Moody, O.A., Jenkins, A., Forest, C.R.: Cleaning patch-clamp pipettes for immediate reuse. Sci. Rep. **6**, 35001 (2016)
6. Lacoste, C., Descombes, X., Zerubia, J.: Point processes for unsupervised line network extraction in remote sensing. IEEE Trans. Pattern Anal. Mach. Intell. **27**(10), 1568–1579 (2005)

7. Lafarge, F., Descombes, X., Zerubia, J., Pierrot-Deseilligny, M.: Automatic building extraction from dems using an object approach and application to the 3D-city
modeling. J. Photogrammetry Remote Sens. **63**(3), 365–381 (2008)
8. Long, B., Li, L., Knoblich, U., Zeng, H., Peng, H.: 3D image-guided automatic
pipette positioning for single cell experiments in vivo. Sci. Rep. **5**, 18426 (2015)
9. Molnar, Cs., Jermyn, I.H., Kato, Z., Rahkama, V., Östling, P., Mikkonen, P.,
Pietiäinen, V., Horvath, P.: Accurate morphology preserving segmentation of overlapping cells based on active contours. Sci. Rep. **6**, 32412 (2016)
10. Perin, R., Markram, H.: A computer-assisted multi-electrode patch-clamp system.
J. Visualized Exp. **80**, e50630 (2013)
11. Perrin, G., Descombes, X., Zerubia, J.: Tree crown extraction using marked point
processes. In: Proceedings of the European Signal Processing Conference, University of Technology, Vienna, Austria (2004)
12. Rochery, M., Jermyn, I.H., Zerubia, J.: Higher order active contours. Int. J. Comput. Vis. **69**(1), 27–42 (2006)
13. Thevenaz, P., Delgado-Gonzalo, R., Unser, M.: The ovuscule. IEEE Trans. Pattern
Anal. Mach. Intell. **33**(2), 382–393 (2011)
14. Thevenaz, P., Unser, M.: Snakuscules. IEEE Trans. Image Process. **17**(4), 585–593
(2008)
15. Wu, Q., Kolb, I., Callahan, B.M., Su, Z., Stoy, W., Kodandaramaiah, S.B., Neve,
R., Zeng, H., Boyden, E.S., Forest, C.R., Chubykin, A.A.: Integration of autopatching with automated pipette and cell detection in vitro. J. Neurophysiol. **116**(4),
1564–1578 (2016)

Non-reference Image Quality Assessment for Fingervein Presentation Attack Detection

Amrit Pal Singh Bhogal, Dominik Söllinger, Pauline Trung,
Jutta Hämmerle-Uhl, and Andreas Uhl$^{(\boxtimes)}$

Visual Computing and Security Lab (VISEL),
Department of Computer Sciences, University of Salzburg,
Salzburg, Austria
uhl@cosy.sbg.ac.at

Abstract. Non-reference image quality measures are used to distinguish real biometric data from data as used in presentation/sensor spoofing attacks. An experimental study shows that based on a set of 6 such measures, classification of real vs. fake fingervein data is feasible with an accuracy of 99% on one of our datasets. However, we have found that the best quality measure (combination) and classification setting highly depends on the target dataset. Thus, we are unable to provide any other recommendation than to optimise the choice of quality measure and classification setting for each specific application setting. Results also imply, that generalisation to unseen attack types might be difficult due to dataset dependence of the results.

1 Introduction

Biometric authentication techniques have emerged to replace or at least complement the traditional authentication methods (e.g. passwords). Consequently, various attacks have been increasingly observed threatening the reliability of this authentication approach. In particular, artifacts mimicking real biometrics traits or captured and displayed image or video footage of real biometric traits have been used to deceive biometric sensors and systems in so-called "presentation"- or "sensor-spoofing"- attacks. In general, counter-measures to such presentation attacks (or anti-spoofing [1]) in biometrics can be categorised into (1) liveness-based, (2) motion-based, and (3) texture-based methods. Liveness-based methods use signs of vitality to ensure that the image is captured from a living human being. In contrast, motion-based methods utilise unnatural movements on scenes as indication of spoofing, e.g. caused by hand motion when presenting a photo or a display to the sensor. Texture-based methods aim to explore textural artifacts in the images captured by the sensor (e.g. caused by recapturing artifacts). While liveness-based techniques are of course specific for the modality under investigation, texture-based methods often employ general purpose texture descriptors in a machine learning setting to discriminate real biometric data from spoofed variants. For example, [2] compares the attack detection performance of certain local descriptors on collections of spoofed iris, fingerprint, and face data. In order

© Springer International Publishing AG 2017
P. Sharma and F.M. Bianchi (Eds.): SCIA 2017, Part I, LNCS 10269, pp. 184–196, 2017.
DOI: 10.1007/978-3-319-59126-1_16

to circumvent the question which texture descriptors to choose, also generative deep learning techniques employing convolutional neural networks have been successfully used to identify spoofed data [3].

An entirely different approach is to consider the quality of the imagery in biometric anti-spoofing which can be interpreted as a specific form of texture-based technique. While this can be done in an approach entirely agnostic of the underlying modality by employing general purpose image quality measures (IQM) [4], a possible alternative is to consider specific properties of the target modality in the quality considerations (see e.g. [5] for quality assessment for face recognition spoofing detection). In this paper we revisit general purpose non-reference IQM (also termed "blind") for their suited-ness in presentation attack detection. In particular, while applying a similar methodology as in [4], we (i) apply non-reference IQM for the first time in biometric *fingervein spoofing detection*, (ii) aim at a different and larger set of non-reference IQM (6 instead of 2) compared to [4], and (iii) do not fuse the results with full-reference IQM but focus on blind IQM as a stand-alone technique (eventually also employing a single metric contrasting to [4] where most results given correspond to fusing a considerable amount of IQM also resulting in significant computational effort).

Section 2 reviews the state-of-the-art in fingervein (FV) spoofing detection while the blind IQM as used in this paper are explained in Sect. 3. Experimental results, including a description of the dataset used in this study, are presented in Sect. 4. Section 5 provides the conclusions of this paper.

2 Fingervein Spoofing Detection

One biometric trait enjoying more and more popularity are veins. One advantage of veins over other biometric traits is the fact that they are embedded *inside* the human body, as opposed to traits like fingerprints or faces. Moreover, vein images can be acquired in an unintrusive manner which is not the case for other biometric traits, such as fingerprint acquisition. However, despite being resistant to tampering, vein-based authentication is vulnerable to presentation attacks [6].

Contrasting to all subsequent techniques (which are texture-based), a first FV presentation attack detection technique based on liveness detection has been proposed [7], requiring FV video data to apply motion magnification techniques to classify into real and fake data.

In 2015, the first competition on counter-measures to fingervein spoofing attacks took place [8] (providing a dataset of real and fake FV images). The competition baseline algorithm looks at the frequency domain of FV images, exploiting the bandwidth of vertical energy signal on real fingervein images, which is different for fakes ones. Three teams participated in this competition. The first team (GUC) uses binarised statistical images features (BSIF). They represent each pixel as a binary code. This code is obtained by computing the pixel's response to a filter that are learnt using statistical properties of natural images [8]. The second team (B-Lab) uses monogenic scale space based global descriptors employing the Riesz transform. This is motivated by the fact that local object appearance & shape within an image can be represented as a distribution of local energy and local orientation information. The best approach (team GRIP-PRIAMUS) utilises local descriptors, i.e., local binary patterns (LBP), and local phase quantisation (LPQ) and Weber local descriptors (WLD). They distinguish between full and cropped images. LBPs and LPQ/WLD are used to classify full and cropped images, respectively.

However, counter-measures to finger vein spoofing attacks were/are already developed prior or independent to this competition. In 2013, the authors of [9] introduced a fake finger vein image detection based upon Fourier, and Haar and Daubechies wavelet transforms. For each of these features, the score of spoofing detection was computed. To decide whether a given finger vein image is fake or real, an SVM was used to combine the three features.

The authors of [10] propose windowed dynamic mode decomposition (W-DMD) to be used to identify spoofed finger vein images. DMD is a mathematical method to extract the relevant modes from empirical data generated by non-linear complex fluid flows. While DMD is classically used to analyse a set of image sequences, the W-DMD method extracts local variations as low rank representation inside a single still image. It is able to identify spoofed images by capturing light reflections, illuminations and planar effects.

Texture-based presentation attack detection techniques have been proven to be applicable to the imagery in the FV-Spoofing-Attack database [8] independent of the above-referenced competition, in particular baseline LBP [11]. In a recent paper [12], inspired by the success of basic LBP techniques [8,13] in finger vein spoofing detection and the availability of a wide variety of LBP extensions and generalisations in literature, we have empirically evaluated different features obtained by using these more recent LBP-related feature extraction techniques for finger vein spoofing detection. Also the steerable pyramid is used to extract features subsequently used for FV spoofing detection [11]. Steerable pyramids are a set of filters in which a filter of arbitrary orientation is synthesised as a linear combination of a set of basis functions. This enables the steerable pyramids scheme to compute the filter response at different orientations. This scheme shows consistent high performance for the finger vein spoofing detection problem and outperforms many other texture-classification-based techniques. It is compared to techniques from [8], including two LBP variants, and to quality-based approaches computing block-wise entropy, sharpness, and standard deviation.

Finally, a detection framework based on singular value decomposition (SVD) is proposed in a rather confused paper [13]. The authors utilise the fact that one is able to extract geometrical finger edge information from infrared finger images. Fingevein images are classified based on image quality assessment (IQA) without giving any clear indication about the actual IQA used and any experimental results.

3   Non-reference Image Quality Metrics

Current state-of-the-art non-reference Image Quality Assessment (NR IQM) algorithms are based on models that can learn to predict human judgments from

databases of human-rated distorted images. These kinds of IQM models are necessarily limited, since they can only assess quality degradations arising from the
distortion types that they have been trained on. However, it is also possible to
contemplate sub-categories of general-purpose NR IQM models having tighter
conditions. A model is said to be opinion-aware (OA) if it has been trained on a
database(s) of human rated distorted images and associated subjective opinion
scores.

Algorithms like DIIVINE, BIQI, BLIINDS-2 and BRISQUE are OA IQM
measures. However, IQM like NIQE, and BIQAA are opinion-unaware (OU) and
they make only use of measurable deviations from statistical regularities observed
in natural images without being trained on human-rated distorted images and
indeed without any exposure to distorted images.

Systematic comparisons of the NR IQM as used in this paper have been
published [14,15]. Both, in non-trained [14] as well as in specifically trained
manner [15] the correspondence to human vision turns out to be highly dependent on the dataset considered and the type of distortion present in the data.
Thus, there has been no "winner" identified among the techniques considered
with respect to correspondence to subjective human judgement and objective
distortion strength.

3.1 NIQE - Natural Image Quality Evaluator

A NR OU-DU IQM (no reference, opinion unaware & distortion unaware) is
based on constructing a collection of quality aware features and fitting them to
a multivariate Gaussian (MVG) model. The quality aware features are derived
from a simple, but highly regular natural scene statistic (NSS) model. NIQE [16]
only uses the NSS features from a corpus of natural images while BRISQUE (see
below) is trained on features obtained from both natural and distorted images
and also on human judgments of the quality of these images.

The classical spatial NSS model begins with preprocessing: local mean
removal and divisive normalisation. Once the new image pixels calculated by
the preprocessing have been computed, the image is partitioned into P×P image
patches. Specific NSS features are then computed from the coefficients of each
patch. Then the sharpness of each patch is determined and only patches with
higher sharpness are selected. A simple model of the NSS features computed from
natural image patches can be obtained by fitting them with an MVG density.

NIQE is applied by computing the 36 identical NSS features from patches
of the size P×P from the image to be quality analysed, fitting them with the
MVG model, then comparing its MVG fit to the natural MVG model. The NIQE
Index delivers performance comparable to top performing NR IQA models that
require training on large databases of human opinions of distorted images.

3.2 BLIINDS-2 - Blind Image Integrity Notator

BLIINDS-2 [17] uses natural scene statistics models of discrete cosine transform
(DCT) coefficients. The algorithm can be divided into four stages. At the first

stage the image is subjected to local 2-D DCT coefficient computation. At this point the image is partitioned into equally sized n×n blocks, then computing a local 2-D DCT on each of the blocks. The DCT coefficient extraction is performed locally in accordance with the HVS (Human Visual System) property of local spatial visual processing (i.e., in accordance with the fact that the HVS processes the visual space locally), thus, this DCT decomposition is accomplished across several spatial scales.

The second stage applies a generalised Gaussian density model to each block of DCT coefficients, as well as for specific partitions within each DCT block. In order to capture directional information from the local image patches, the DCT block is partitioned directionally into three oriented subregions. A generalised Gaussian fit is obtained for each of the oriented DCT coefficient subregions. Another configuration for the DCT block partition reflects three radial frequency subbands in the DCT block. The upper, middle and lower partitions correspond to the low-frequency, mid-frequency, and high-frequency DCT subbands, respectively. A generalised Gaussian fit is obtained for each of the radial DCT coefficient subregions as well.

The third step computes functions of the derived generalised Gaussian model parameters. These are the features used to predict image quality scores. The fourth and final stage is a simple Bayesian model that predicts a quality score for the image. Here the training is required. The prediction model is the only element of BLIINDS-2 that carries over from BLIINDS-1. The Bayesian approach maximises the probability that the image has a certain quality score given the model-based features extracted from the image. The posterior probability that the image has a certain quality score from the extracted features is modelled as a multidimensional generalised Gaussian density.

3.3 BIQAA - Blind Image Quality Assessment Through Anisotropy

BIQAA [18] is based on measuring the variance of the expected entropy of a given image upon a set of predefined directions. Entropy can be calculated on a local basis by using a spatial/spatial-frequency distribution as an approximation for a probability density function. The generalised Renyi entropy and the normalised pseudo-Wigner distribution (PWD) have been selected for this purpose. As a consequence, a pixel-by-pixel entropy value can be calculated, and therefore entropy histograms can be generated as well. The variance of the expected entropy is measured as a function of the directionality, and it has been taken as an anisotropy indicator. For this purpose, directional selectivity can be attained by using an oriented 1-D PWD implementation. So, the method is based on measuring the averaged anisotropy of the image by means of a pixel-wise directional entropy. BIQAA aims to show that an anisotropy measure can be used to assess both, the fidelity and quality of images.

3.4 BRISQUE

BRISQUE [19] does not require any transformation to another coordinate frame like DCT used by BLIINDS-2. BRISQUE has very low computational complexity, making it well suited for real time applications. The two main stages of BRISQUE are natural scene statistics in the spatial domain and quality evaluation. At the first stage an image is locally normalised (via local mean subtraction and divisive normalisation). Subsequently, 2 parameters are estimated $(\alpha, \sigma2)$ from a GGD fit of the normalised pixel data. These form the first set of features that will be used to capture image distortion. To show that pristine and distorted images are well separated in GGD parameter space, a set of pristine images from the Berkeley image segmentation database was taken. Similar kinds of distortions as present in the LIVE image quality database were introduced in each image at varying degrees of severity to form the distorted image set: JPEG 2000, JPEG, white noise, Gaussian blur, and fast fading channel errors. A model for the statistical relationships between neighboring pixels is also built. While normalised coefficients are definitely more homogeneous for pristine images, the signs of adjacent coefficients also exhibit a regular structure, which gets disturbed in the presence of distortion. To model this structure the empirical distributions of pairwise products of neighboring normalised coefficients along four orientations are used.

At the second stage a mapping is learned from feature space to quality scores using a regression module, yielding a measure of image quality. For that purpose a support vector machine (SVM) regressor (SVR) is used. SVMs are popular as classifiers since they perform well in high-dimensional spaces, avoid over-fitting and have good generalisation capabilities. In contrast to algorithms like NIQE and BLIINDS-2, BRISQUE requires training (a pre-trained version is employed in the experiments).

3.5 DIIVINE - Distortion Identification-Based Image Verity and Integrity Evaluation

DIIVINE [20] is based on a 2-stage framework involving distortion identification followed by distortion-specific quality assessment. Most present-day NR IQM algorithms assume that the distorting medium is known - for example, compression, loss induced due to noisy channel etc. Based on this assumption, distortions specific to the medium are modelled and quality is assessed. By far the most popular distorting medium is compression which implies that blockiness and blurriness should be evaluated. DIIVINE targets three common distortion categories, i.e. JPEG compression, JPEG2000 compression, and blur.

In order to extract statistics from distorted images the steerable pyramid decomposition is utilised. The steerable pyramid is an over-complete wavelet transform that allows for increased orientation selectivity. Since NR IQM algorithms are generally trained and tested on various splits of a single dataset (as described above), it is natural to wonder if the trained set of parameters are database specific. However, the training process of DIIVINE is simply a calibration, and once such training is performed, DIIVINE is capable of assessing the quality of any distorted image, since the performance of the algorithm was evaluated on an alternate database. A support vector machine (SVM) regressor (SVR) is used for the classification into the distortion categories.

DIIVINE was actually not developed under the constraint of real-time analysis of images, given that the performance of DIIVINE is as good as leading full-reference quality assessment (FR QA) algorithms.

3.6   BIQI - Blind Image Quality Index

BIQI [21] is also based on a 2-stage framework like DIIVINE. The two steps are image distortion classification based on a measure of how the natural scene statistic (NSS) are modified, followed by quality assessment, using an algorithm specific to the decided distortion. Once trained, an algorithm of the proposed framework does not require further knowledge of the distortion affecting the images to be assessed. The framework is modular in that it can be extended to any number of distortions.

BIQI starts with wavelet transforming an image over three scales and three orientations using the Daubechies 9/7 wavelet basis. The subband coefficients so obtained are parametrised using a generalised Gaussian distribution (GGD). An 18-D vector is formed and it is the representative feature vector for each image.

Given a training and test set of distorted images, a classifier is based on the feature vector to classify the images into five different distortion categories, based on the distortion type JPEG, JPEG2000, WN (white noise), Blur, and FF (fast fading). DIIVINE in contrast only classifies the distortion images into 3 categories. The classifier used is a support vector machine (SVM), which is also utilised in DIIVINE and BRISQUE. BIQI works well for images corrupted by WN and blur and to some extent for JPEG2000 and FF. However, the performance for JPEG compression is less impressive.

4   Experiments

4.1   Experimental Settings

The **The Spoofing-Attack Finger Vein Database** as used in the "1st Competition on Counter Measures to Finger Vein Spoofing Attacks" as described above and provided by IDIAP Research Institute consists of 440 index finger vein images of both real authentications and fake ones (i.e. attack attempts) to 110 different identities. The samples are split into two different categories (as shown in Fig. 1): *Full* (printed) images and *cropped* images where the resolution of the *full* images is 665 × 250 and that of the cropped images is 565 × 150 pixel, respectively.

For each image in the database quality scores were calculated with the IQM described in Sect. 3. We used the MATLAB implementations from the developers

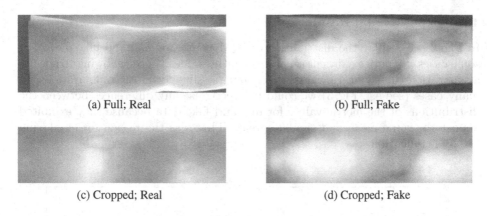

(a) Full; Real

(b) Full; Fake

(c) Cropped; Real

(d) Cropped; Fake

Fig. 1. Finger Vein DB samples

of BIQI, BLIINDS-2, NIQE, DIIVINE, BRISQUE[1] and BIQAA[2]. In all cases, we used the default settings. We normalised the result data with the result that 0 represents a good quality and 100 the bad one which is already the default result in all cases except BIQAA. Originally the data of BIQAA is between 0 and 1. However, the values are so small that we had to define our own limits for the normalisation. A thorough analysis shows that the values output by the software are all between 0.00005 and 0.05 therefore we used these figures as our limits. Moreover we had to change the "orientation" of the BIQAA quality scores to be conforming to our definition. Summarising, the following formula (1) was built:

$$x' = 100 - \frac{x - 0.05}{0.00005 - 0.05} \cdot 100 \tag{1}$$

In the first experimental stage we consider the distribution of the quality scores only. Our aim was to eventually find a threshold between the values of the real data and the fake ones for the various IQM.

Afterwards, in the second stage, we used the quality scores for a leave-one-out cross validation to get an exact assertion about the classification possibility with NR IQM. To classify our data we used k-nearest neighbours (kNN) classification. Our used k were 1, 3, 5, 7 and 9 for this experiment according to first pre-results. First, we only used one quality score for the classification. In the next step, we combined several quality scores of the different measures into one vector and used this for the kNN-classification. This method allowed us to test all possible combinations of IQM in a simple way. The distance for the kNN-classification was in the first case the difference between the two values and in the second case the distance between the two vectors. At the end, we got the classification accuracy for discriminating real from fake images for all IQM combinations.

---

[1] All available from http://live.ece.utexas.edu/research/quality/.

[2] Available at https://www.mathworks.com/matlabcentral/fileexchange/30800-blind-image-quality-assessment-through-anisotropy.

4.2   Experimental Results

In Fig. 2, we display the distribution of IQM values for real and fake data. For some cases, we notice a decent separation of the values almost allowing to specify a separation threshold. However, this is not possible for most configurations. In many cases (see e.g. Fig. 3) we could not recognise any differences between the distributions of the metric values for real and fake data because they exhibited almost the same mean and similar spread. That was the reason for employing kNN-Classification.

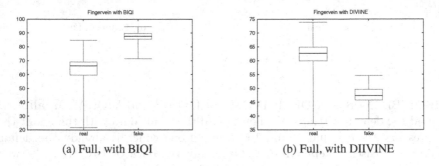

(a) Full, with BIQI   (b) Full, with DIIVINE

Fig. 2. Quality score distribution (positive examples)

Note that in Fig. 3 we see DIIVINE exhibiting highly overlapping value distributions for real and fake versions of cropped fingervein data, while for the full images a decent separation could be observed. This already indicates that the performance of the metrics in discriminating real from fake data obviously highly depends on the actual dataset under investigation.

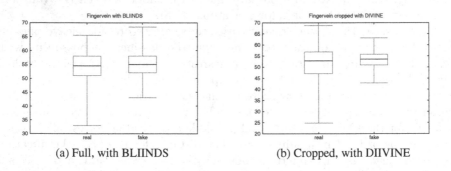

(a) Full, with BLIINDS   (b) Cropped, with DIIVINE

Fig. 3. Quality score distribution (negative examples)

In the case of kNN-classification with only one IQM, we already obtain good results for the full images. In Table 1 we can see that we got over 99% classification accuracy for BIQI. In this case we already could see the differences of the distributions of the real and the fake values of the quality scores (see Fig. 2a). For this reason, a high accuracy with kNN-classification was already expected. In good correspondence to the value distributions, we also get decent distinction accuracy with DIIVINE (>95%) while BLIINDS is only slightly superior to guessing.

Table 1. Best results for kNN-classification for full (left) and cropped (right) images (depending on k)

Algorithm	k	Accuracy		Algorithm	k	Accuracy
BIQI	9	99.17%		BIQI	9	72.71%
BLIINDS	5	56.04%		BLIINDS	7	52.50%
NIQE	7	72.29%		NIQE	7	59.17%
DIIVINE	9	95.42%		DIIVINE	9	62.71%
BRISQUE	9	68.96%		BRISQUE	7	60.63%
BIQAA	9	74.38%		BIQAA	7	80.21%

The result is somewhat less convincing for the cropped images as also displayed in Table 1 (right table). Surprisingly, BIQAA, an IQM not well perceived in literature clearly delivers the best result with >80% distinction accuracy while BLIINDS is again the worst IQM for our purpose. Our result conforms well to the competition results [8] where also different texture-based descriptors were used to discriminate real from fake images considering the full images and the cropped ones, respectively.

It seems that the different distortions present in the spoofed data are quite specific in terms of the nature and characteristic of the distortions, which is the only explanation of different IQM performing best on different datasets. In fact, our results confirm the general results on IQM quality prediction performance [14,15] in that it is highly dataset and distortion dependent which IQM provides the best results.

A further increase in classification accuracy was obtained by the combination of several IQM (a feature vector of IQM values is used instead of a scalar IQM value). Table 2 shows the best combinations for the considered databases from an exhaustive search. For the full images, we have finally found many metric combination configurations with a distinction accuracy of 99.79% while only that one with the lowest k and the lowest number of combined metrics is shown in the table.

For the cropped images, we could improve our results by 5% compared to the single measure results and the best result is over 85% and many results are over 80%.

From the latter table we are not able to confirm the trend of getting best results when combining a larger number of IQM [4]. In order to look into this effect more thoroughly (and to clarify the role of the k-parameter in

Table 2. Best metric combinations for fingervein images

Combination	k	Accuracy
Full		
BIQI, BRISQUE	3	99.38%
BIQI, DIIVINE, BRISQUE	1	99.79%
Cropped		
BIQI, BRISQUE	3	76.88%
BIQI, BLIINDS, DIIVINE	9	82.50%
BIQI, BLIINDS, NIQE, BRISQUE, BIQAA	9	84.42%
BIQI, BLIINDS, NIQE, DIIVINE, BRISQUE, BIQAA	7	84.58%
BIQI, BLIINDS, NIQE, BRISQUE	9	85.63%

kNN-classification) we have systematically plotted the results of the exhaustive classification scenarios.

We average all classification results by keeping the number of combined metrics fixed (Fig. 4b) and and by keeping the parameter k fixed (Fig. 4a). Combining more metrics and choosing k large leads to better results on average, where a large number of metrics has a higher influence than a large value for k. Top results in Table 2 for full images do not conform to these average observations, neither in terms of metrics combined, nor in terms of the k value, while for the cropped images the average behaviour is reflected in the best results.

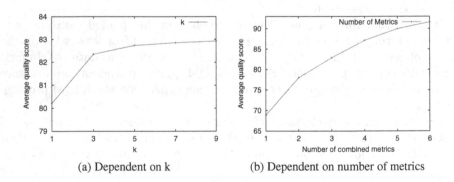

(a) Dependent on k   (b) Dependent on number of metrics

Fig. 4. Average quality score

5   Conclusion

We have found a high dependency on the actual dataset under investigation when trying to answer the question about the optimal choice of an image quality measure, even though the cropped images are just a part of the full images.

BIQI and BIQAA are found to be the top performing IQM for full and cropped images, respectively. Therefore, we are not able to identify a clear "winner" among the IQM based on the results analysed while BLIINDS seems to be the "looser". Still, BLIINDS is used in the best-performing IQM combination for cropped images, while BIQAA (the best-performing IQM for these images) is not.

Since the optimal choice of IQM is dependent on the dataset, it is probably also the nature of attack type that plays a certain role (e.g. if the attack is based on replayed data or if actual artifacts are being used). Thus, the generalisation of the results to unseen attack types might be not straightforward. A similar observation has been made recently in the context of open set spoofing detection where most detector types exhibited extremely degraded attack detection performance on unseen attack data (i.e. data, they have not been trained on) [22].

Acknowledgments. This work has been partially supported by the Austrian Science Fund, project no. 27776.

References

1. Marcel, S., Nixon, M., Li, S. (eds.): Handbook of Biometric Anti-Spoofing. Springer, New York (2014)
2. Gragnaniello, D., Poggi, G., Sansone, C., Verdoliva, L.: An investigation of local descriptors for biometric spoofing detection. IEEE Trans. Inf. Forensics Secur. **10**(4), 849–861 (2015)
3. Menotti, D., Chiachia, G., Pinto, A., Schwartz, W., Pedrini, H., Falcao, A.X., Rocha, A.: Deep representations for iris, face, and fingerprint spoofing detection. IEEE Trans. Inf. Forensics Secur. **10**(4), 864–879 (2015)
4. Galbally, J., Marcel, S., Fierrez, J.: Image quality assessment for fake biometric detection: application to iris, fingerprint, and face recognition. IEEE Trans. Image Process. **23**(2), 710–724 (2014)
5. Wen, D., Han, H., Jain, A.: Face spoof detection with image distortion analysis. IEEE Trans. Inf. Forensics Secur. **10**(4), 746–761 (2015)
6. Tome, P., Vanoni, M., Marcel, S.: On the vulnerability of finger vein recognition to spoofing. In: 2014 International Conference of the Biometrics Special Interest Group (BIOSIG), pp. 1–10, September 2014
7. Raghavendra, R., Avinash, M., Marcel, S., Busch, C.: Finger vein liveness detection using motion magnification. In: 2015 IEEE 7th International Conference on Biometrics Theory, Applications and Systems (BTAS), pp. 1–7, September 2015
8. Tome, P., Raghavendra, R., Busch, C., Tirunagari, S., Poh, N., Shekar, B.H., Gragnaniello, D., Sansone, C., Verdoliva, L., Marcel, S.: The 1st competition on counter measures to finger vein spoofing attacks. In: 2015 International Conference on Biometrics (ICB), pp. 513–518, May 2015
9. Nguyen, D.T., Park, Y.H., Shin, K.Y., Kwon, S.Y., Lee, H.C., Park, K.R.: Fake finger-vein image detection based on fourier and wavelet transforms. Digit. Signal Proc. **23**(5), 1401–1413 (2013)
10. Tirunagari, S., Poh, N., Bober, M., Windridge, D.: Windowed DMD as a microtexture descriptor for finger vein counter-spoofing in biometrics. In: 2015 IEEE International Workshop on Information Forensics and Security (WIFS), pp. 1–6, November 2015

11. Raghavendra, R., Busch, C.: Presentation attack detection algorithms for finger vein biometrics: a comprehensive study. In: 2015 11th International Conference on Signal-Image Technology Internet-Based Systems (SITIS), pp. 628–632, November 2015
12. Kocher, D., Schwarz, S., Uhl, A.: Empirical evaluation of IBP-extension features for finger vein spoofing detection. In: Proceedings of the International Conference of the Biometrics Special Interest Group (BIOSIG'16), Darmstadt, Germany, p. 8 (2016)
13. Mythily, B., Sathyaseelan, K.: Measuring the quality of image for fake biometric detection: application to finger vein. In: National Conference on Research Advances in Communication, Computation, Electrical Science and Structures (NCRAC-CESS), pp. 6–11 (2015)
14. Nouri, A., Charrier, C., Saadane, A., Fernandez-Maloigne, C.: Statistical comparison of no-reference images quality assessment algorithms. In: Proceedings of the Colour and Visual Computing Symposium (CVCS 2013) (2013)
15. Charrier, C., Saadane, A., Fernandez-Maloigne, C.: Comparison of no-reference image quality assessment machine learning-based algorithms on compressed images. In: Image Quality and System Performance XII. Prooceedings of SPIE, vol. 9396 (2015)
16. Mittal, A., Soundararajan, R., Bovik, A.C.: Making image quality assessment robust. In: Proceesings of the 46th Asilomar Conference on Signals, Systems and Computers (ASILOMAR) (2012)
17. Saad, M., Bovik, A.C., Charrier, C.: Blind image quality assessment: a natural scene statistics approach in the DCT domain. IEEE Trans. Image Process. 21(8), 3339–3352 (2012)
18. Gabarda, S., Cristobal, G.: Blind image quality assessment through anisotropy. J. Opt. Soc. Am. A 24, B42–B51 (2007)
19. Mittal, A., Moorthy, A.K., Bovik, A.C.: No-reference image quality assessment in the spatial domain. IEEE Trans. Image Process. 21(12), 4695–4708 (2012)
20. Moorthy, A.K., Bovik, A.C.: Blind image quality assessment: from natural scene statistics to perceptual quality. IEEE Trans. Image Process. 20(12), 3350–3364 (2011)
21. Moorthy, A.K., Bovik, A.C.: A two-step framework for constructing blind image quality indices. IEEE Signal Process. Lett. 17(5), 513–516 (2010)
22. Rattani, A., Scheirer, W., Ross, A.: Open set fingerprint spoof detection across novel fabrication materials. IEEE Trans. Inf. Forensics Secur. 10(11), 2447–2460 (2015)

Framework for Machine Vision Based Traffic Sign Inventory

Petri Hienonen[1,2], Lasse Lensu[1], Markus Melander[2,3],
and Heikki Kälviäinen[1(✉)]

[1] Machine Vision and Pattern Recognition Laboratory (MVPR),
School of Engineering Science, Lappeenranta University of Technology (LUT),
Skinnarilankatu 34, 53850 Lappeenranta, Finland
{lasse.lensu,heikki.kalviainen}@lut.fi
[2] Vionice Ltd., Brahenkatu 4, 53100 Lappeenranta, Finland
{petri.hienonen,markus.melander}@vionice.fi
[3] Finnish Transport Agency (FTA), Laserkatu 6, 53850 Lappeenranta, Finland

Abstract. Automatic traffic sign inventory and simultaneous condition analysis can be used to improve road maintenance processes, decrease maintenance costs, and produce up-to-date information for future intelligent driving systems. The goal of this research is to combine automatic traffic sign detection and classification with traffic sign inventory and condition analysis. This paper proposes a complete machine vision framework for the purpose and presents the results of its performance evaluation with three datasets: Traffic Signs Dataset, and two datasets collected for this research. The experimental results show that the system is able to detect, locate, and classify almost all the traffic signs, and is a suitable platform for traffic sign condition analysis.

Keywords: Traffic sign inventory · Detection · Classification · Localization · Distributed asset management · Condition analysis · Machine vision · Image processing

1 Introduction

Road maintenance operations and contracting require up-to-date traffic sign inventories, and the importance of accurate inventory information will even increase with the widespread deployment of intelligent transportation systems. Such inventories contain information on the sign type, location, direction, and condition. Compiling and keeping the inventories up-to-date requires a considerable amount of work. For example, in Finland the traffic signs are supposed to be inventoried every five years since missing or damaged signs cause danger to road users. The Finnish Transport Agency (FTA)[1] is responsible for approximately 78,000 km of highways. In total the Finnish road network is approximately 454,000 km long. FTA aims to automate the inventory and condition

---

[1] http://www.liikennevirasto.fi/web/en.

© Springer International Publishing AG 2017
P. Sharma and F.M. Bianchi (Eds.): SCIA 2017, Part I, LNCS 10269, pp. 197–208, 2017.
DOI: 10.1007/978-3-319-59126-1_17

analysis of the traffic signs to be performed continuously during the normal road maintenance. The system would improve services to citizens and decrease road maintenance costs by real-time machine vision based monitoring of the signs. Currently, traffic sign inventory is maintained manually which is slow, laborious, and error prone.

A machine vision based solution that updates the inventory during normal road maintenance offers benefits that cannot be achieved with current processes: close to real-time asset management, increased road security, improved competitive bidding processes, objective evaluation of contracts, and efficient information management for intelligent driver assistance systems. This article presents the machine vision methods and implementation of an automatic system for traffic signs inventory and simultaneous traffic sign condition analysis. The first study by the authors has been published in [1].

A practical solution for data collection is to mount a digital camera into in-service road maintenance vehicles to provide an efficient method for monitoring the entire road network and its traffic signs on a daily basis and with relatively low costs. Automatic traffic sign detection and classification have been previously studied from the viewpoint of self-driving vehicles and driver assistance systems, but systems combining the inventory and condition analysis have not been presented. In addition, condition analysis of traffic signs has not been comprehensively studied before, with the exception of automatic reflectance assessment. The work differentiates between traffic signs and sign posts, the latter of which are not considered in this study. Moreover, two novel datasets were collected in winter conditions and annotated for the evaluation and for traffic sign condition analysis.

2   Machine Vision Based System

Road maintenance vehicles traverse the roads often, especially in the wintertime. Therefore, the approach harnessing the vehicles for data collection can provide comprehensive analysis of the road network, and at the same time, acquire multiple images of individual signs for further analysis. There are two constraints affecting the proposed system: the system should be easy to install on existing road maintenance vehicles, and it should be operable on low-cost mobile hardware.

Automatic Traffic Sign Inventory (TSI) and condition analysis consists of three high level tasks: (i) Traffic Sign Recognition (TSR): Detection (TSD) and Classification (TSC), (ii) sign location estimation, and (iii) condition analysis. TSD is the problem of finding the traffic sign from an image which is a binary classification between traffic signs and the background. TSC is a multi-class classification problem where the previously detected traffic sign patches are classified as sign types to determine the class label. Common-use cases for TSR are autonomous driving, assisted driving, and mobile traffic sign mapping.

2.1   Related Work

Research mainly focuses on TSR [2] or TSD only [3,4], usually on a certain subclass of signs, for example, speed limit signs [5]. The survey [5] conducted into TSR shows that comparing the methods is difficult. Different studies typically use different comparison metrics and data, consider the subsequent tasks of TSD [6], TSC and tracking, or only part of the processing tasks. A system for inventory purposes in Spain was proposed in [7], containing 51 sign types. In [8] 176 different sign types were considered exploiting street-level panoramic images where promising results for large-scale automated surveying were reported. These studies differ from our work since they do not consider GPS positioning, sign condition analysis, and winter weather conditions.

Recently, three large and publicly available traffic sign datasets for the detection and classification have been released: Belgian KUL traffic sign detection and KUL traffic sign classification benchmark 2011 datasets [9], Traffic Signs Dataset [10,11], and German GTSDB and GTSRB datasets [2,4], but no datasets related to the traffic sign condition analysis exist. The German datasets were introduced for two competitions that benchmark different TSR methods which were used for selecting a TSR algorithm for the system presented in this article. The camera and installation setup need to be selected: either a single camera, a dual camera [9,12], or specialized equipment such as infrared cameras [13].

2.2   Environment

Traffic signs are standardized, and white or yellow is the second color of prohibitory signs. In Finland, Sweden, Poland and Iceland, the color is yellow for better visibility in snowy landscapes. In Finland, traffic signs come in three sizes (width of a side): small (400 mm), medium (640 mm) and large (900 mm).

TSR algorithms have to cope with natural, complex, and dynamic scenes (see Fig. 1). The appearance of traffic signs is affected by variations in lighting conditions due to, e.g., shadows, clouds and direct sunlight. Colors in captured images depend on the ambient light spectrum (daylight, headlights, infrastructure lighting) and viewing geometry (angle, distance). The color and reflectivity of traffic signs fade with time, and signs can be damaged, misaligned, or obstructed. Other

Fig. 1. Operating environment. (Color figure online)

objects with colors and shapes similar to traffic signs, such as advertisements, certain parts of vehicles, and buildings, make the recognition task more difficult.

2.3 Proposed System

The proposed system is presented in Fig. 2. A single camera setup was chosen for this implementation. The core components of the system are marked as gray. At the general level, object detection, object classification, and condition analysis all contain feature extraction, feature post-processing, and classification. The core components of the system and their purposes are as follows:

1. **Camera with known location**: A camera captures either still images or video, and the corresponding GPS location data is stored. For example, the camera can be integrated into a mobile phone with a built-in GPS receiver.
2. **Image pre-processing**: The captured images are processed to better suit to the next steps, including, e.g., motion de-blurring and color correction.
3. **Object detection**: The task of the detection module is to locate the objects (traffic signs) in the 2D image. The module outputs the location of possible signs in the image and the reliability of the detection, given by the classifier.
4. **Object classification**: The located signs (objects) are classified based on a predefined set of sign types.
5. **Localization**: The detections are combined with known camera parameters to estimate the sign location in with respect to the camera location.
6. **Trajectory prediction**: Information on the localized signs is further refined by predicting the space-time trajectories, used as a priori information for the next detection round. The relationship between the trajectories and the detections is asymmetric, and new detections can occur and the old ones vanish from the view in matching between the time steps.
6. **Global location assessment**: The sign locations are mapped onto the world coordinate system using interpolated/extrapolated GPS coordinates and 3D-localized signs. The GPS locations are estimated with ellipsoid projections on a surface.
7. **Condition evaluation**: The condition of the signs is analyzed. The sign image is segmented from the background, image features related to the sign condition are extracted, and the condition category is determined.

3 Traffic Sign Recognition and Localization

The algorithms chosen for the system are shown in Fig. 3. Object detection has to be done fast whereas there is more time for object classification, considering variations of the environment affecting the appearance of the traffic signs. During TSD, a binary classifier is used to discriminate traffic signs from the background. TSC is a multi-class classification problem where the previously detected traffic sign patches are classified as sign types to determine the class label.

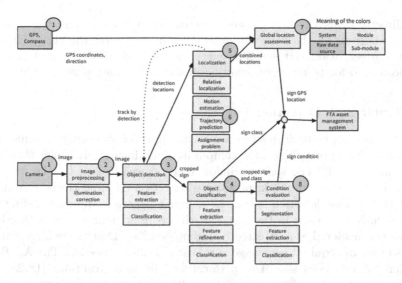

Fig. 2. The proposed system and the relations between different components.

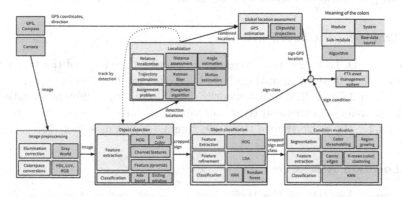

Fig. 3. Algorithms of the proposed system.

The TSR approaches utilize three prominent features: color, inner shape, and contour shape. Due to diverse natural lighting conditions, the use of color information is difficult and many heuristics have been proposed to make the color useful feature [14,15]. Related to shape, there are two general approaches: (i) parametrized methods such as Hough Transform (HT) for different shapes and (ii) model-based methods utilizing Haar-like features [16].

Dense HOG features have been used to capture the shape features of traffic signs [2,4,17]. The research conducted in [18] shows a comparison of HOG feature parameters, different scales, and their performance as basic shape features for TSR. It has been demonstrated that the color is an important feature in detection, but less important in classification. Dealing with high dimensionality reduction techniques or sparse modeling can be used as PCA and LDA [19].

LDA directly deals with the classes, whereas PCA tries to find the principal components from all the data without taking into account the class structure. Sparse representation and encoding methods, such as SPM [20] and LLC [21] have been used for traffic signs resulting in state-of-the-art performance.

3.1   Traffic Sign Detection

The detection can be performed in two ways: the computationally complex sliding window [18] approach and the computationally inexpensive color thresholding method [22–24]. The sliding window method was chosen for the proposed system since of good results in pedestrian detection [25] and traffic sign detection [4,18,26]. The detection is done by defining a score at different sliding window positions and scales in the image. If the score is greater than a threshold, the area is considered as a detected bounding box [26]. Due to the large number of classifications required per image, a robust classifier is needed. The AdaBoost classifier [27] has been shown to perform well in such situations [18,28], and is fast to train, and performs especially well with large feature sets. The ACF approach [25] is used for feature pyramids in the proposed system.

3.2   Traffic Sign Classification

Similar to TSD, HOG features are used to describe the shape of the traffic sign. Numerous classification algorithms have been used for TSC, including KNN [18], Random Forests [2,17], Neural Networks (NN) [2,29], and different variations of SVM [18]. Image features and their representation have been shown to be more important for the performance than selecting a specific classifier [18].

3.3   Location Assessment

The accurate location of the traffic sign is determined in two steps: (i) The location of the sign is first estimated relative to the car and (ii) the relative location is transformed into the global coordinate system. In the localization, the location of the sign relative to the camera has to be determined. The side length of a traffic sign is known which can be used for the determination of the location accuracy. After the distance to the sign is known, the estimate has to be refined using the geometry of the scene. For global location assessment, Karney's implementation for geodesic calculations [30] was used. The accuracy of GPS can be improved to be within 1–2 m in urban settings using sensor fusion, and below 0.10 m by using the Real Time Kinematics correction signal [31].

3.4   Traffic Sign Condition Analysis

The machine vision approach can be divided into three parts: (i) definition of the exact location of the sign surface (segmentation), (ii) selection and extraction of features that correlate with traffic sign condition, and (iii) determination of the condition of the traffic sign with classification. Next, the current human-performed condition analysis criteria are presented, followed by a machine vision solution for the same criteria. The machine vision approach for condition analysis was introduced in [1] and the comprehensive results are to be reported in the future.

4   Experimental Results

4.1   Datasets and Evaluation

The implemented system was tested with three datasets: Traffic Signs Dataset published by the Computer Vision Laboratory at Linköping University, Sweden [10,11] (Dataset 1), Finnish Winter Dataset (Dataset 2) and Lappeenranta Road Signs Dataset (Dataset 3). The two latter ones were collected in this research.

Dataset 1 was used in the TSD and TSC experiments since the Swedish and Finnish traffic signs are very similar. Dataset 1 consists of 20,000 images from video sequences of which around 20% have been annotated. The dataset consists of continuous video sequences, recorded during the summer and a single tour, in daytime in varying illumination conditions and different driving scenarios (rural, urban, and highway). The dataset contains 16 sign classes as shown in Table 1. Annotations smaller than 25×25 pixels were ignored due to too low resolution.

Dataset 2 contains approximately 20 h of video material for the system testing in difficult environmental conditions and to demonstrate the functionality of the

Table 1. Annotated signs available in the Traffic Signs Dataset.

Name	Sign category	Count
Priority road	Priority	496
Give way	Priority	145
Stop and give way	Priority	4
No standing or parking	Prohibitory	79
No parking	Prohibitory	102
Allowed direction left or right	Mandatory	27
Pass this left of right	Mandatory	347
Speed limit 30	Prohibitory	21
Speed limit 50	Prohibitory	163
Speed limit 60	Prohibitory	21
Speed limit 70	Prohibitory	223
Speed limit 80	Prohibitory	112
Speed limit 90	Prohibitory	30
Speed limit 100	Prohibitory	100
Speed limit 110	Prohibitory	29
Speed limit 120	Prohibitory	23
Total		1922

system shown in Fig. 5. The dataset was collected since suitable videos captured in snowy winter conditions with GPS location information were not available. The videos were recorded in an urban environment using a Garmin Virb Elite camera (with GPS) attached to a road maintenance vehicle inside the driver's cabin. The camera contains a CMOS sensor with a spatial resolution of 1980 × 1080 pixels and records 30 fps. The higher resolution, the better for the condition analysis. The camera's horizontal field of view is 151°. The exposure time for the video is controlled automatically based on average lighting of the view.

Dataset 3 consists of 325 still images for traffic sign condition analysis. The dataset contains 397 condition and class-annotated traffic signs. The dataset was introduced in [1].

4.2 Traffic Sign Detection

The approach was evaluated using Dataset 1. Common evaluation criteria include the number of True Positives (TP), False Negatives (FN), and False Positives (FP). The accuracy of detections can be measured based on the relative overlap between the Ground Truth (GT) bounding boxes (BBs) and the detected BBs. This measure is known as Intersection of Union (IoU) [26]. The performance of the detection accuracy is measured using the two main indicators False Positives Per Image (FPPI) and the miss rate. FPPI describes how many false positives are found on the average per image. The miss rate describes what percentage of the real signs are missed (FN) during the detection: 1 - True Positive Rate (TPR) where $TPR = TP/(TP + FN)$. In the tests, the detection is considered to be successful when IoU is more than 0.5 which is considered to be good enough [26].

The results are shown in Fig. 4 and Table 2. The detectors were trained for each sign category separately and for all the three sign categories. Figure 4

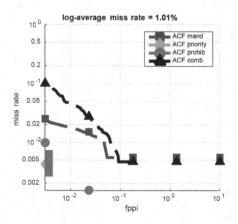

Fig. 4. Logarithmic FPPI/miss rate curve of the sign detection on Dataset 1 using Adaboost, and the HSV and HOG detectors for three subcategory detectors and the combined detector with IoU 0.7.

Table 2. TSD using Dataset 1 and three separately trained detectors.

Type	Mand.	Prior.	Prohib.	Combined
Training	193	353	520	1066
Test	181	292	383	856
FN	1	0	0	4
Error	0.006	0.014	0.024	

presents the miss rates on a logarithmic scale, showing very good detection performance.

4.3 Classification

The evaluation was also done using Dataset 1. The classification performance was computed by comparing the classification results to GT. The performance was evaluated using the HOG and grayscale features. Grayscale features are just normalized gray-level values of the image patch. Dimension reduction techniques tested were LDA and PCA, and the KNN, Naive Bayes and Random Forests classifiers. For KNN, $k = 5$ was selected by experimentation. PCA was used so that the set of features explains 95% of the variance in the features. The depth of the forest was limited to 64. The classification results, computed by using the 10-fold cross validation, are presented in Table 3.

Table 3. TSC using Dataset 1.

Feature	Dim. red.	Classifier	Mean error
Gray-scale	PCA	KNN	0.1957
Gray-scale	LDA	KNN	0.1251
HOG	None	KNN	0.0295
HOG	PCA	KNN	0.0295
HOG	LDA	KNN	0.0146
Gray-scale	None	Random Forest	0.1387
HOG	None	Random Forest	0.0424
HOG	LDA	Random Forest	0.0493
HOG	PCA	Random Forest	0.1215
HOG	LDA	Naive Bayes	0.0187
HOG	PCA	Naive Bayes	0.0586

4.4   Distance and Location Evaluation

During the collection of data, no GT data for the assessment of localization accuracy was collected. The distance estimation to the sign is one of the error-prone parts of traffic sign localization. The distance estimation accuracy was evaluated by comparing six independent separate images taken on six different traffic signs specifically for the purpose of computing distance estimation accuracy with GT measured using a laser distance meter. Images were taken with the same Garmin Virb camera as described earlier. The camera manufacturer does not give the exact focal length or width of view parameters for the camera, and they were approximated by using calibration. The camera performs lens correction for images automatically. The BBs for calculating the heights were placed manually, and the averages of the height and the width were computed for the projection. All the tested signs were 640 mm wide. The absolute and relative average errors were −0.31 m and −1,9% only.

5   Discussion

The framework for traffic sign inventory with simultaneous sign condition analysis were proposed and tested using three different datasets. Two datasets were collected during the research project, the one for condition analysis and the other for the detection, localization, and classification in difficult winter conditions. TSD uses the rigid, HOG+color feature detector which reached detection of 96.00% of the signs and runs around 15 fps on a 640 × 480 frame on a mobile i5-4200u processor. The best results for TSC were obtained using a HOG+LDA+KNN combination, classifying 98.55% of the signs correctly. The

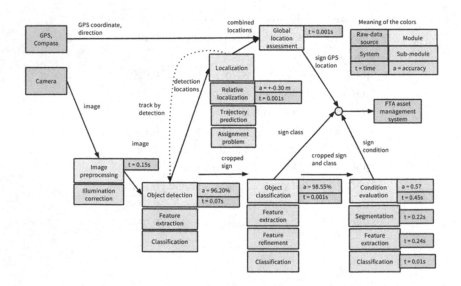

Fig. 5. Implemented system performance on a i5-4200 cpu.

mean error of the condition analysis phase per sign was 0.583 [1]. GPS localization is accurate, mostly because the signs are detected correctly. Figure 5 shows the performance of the proposed system.

In the future, system performance has to be verified in practice, including challenging weather conditions. To evaluate the performance in extreme conditions (fog, snowfall, low light, and rain) and to estimate the localization error more accurately, a new dataset is needed. This can be developed for the proposed system by extending Dataset 2. The TSI process can be further improved by adding more information specific to the environment and traffic signs, such as the movement of the vehicle and scene understanding. The TSD and TSC results can be combined with information from multiple frames. An improved classification model could produce appropriate a posteriori probabilities for TSC. Moreover, experiments can be extended by utilizing, for example, Convolutional Neural Networks (CNN) [32].

6 Conclusion

This research was a first step in automating and combining traffic sign condition analysis with TSI and, thus, reducing road maintenance costs and increasing quality of the traffic sign inventory data. The research shows that machine vision solutions are accurate enough for implementing a TSI system for automatic asset management. The sign detector and classifier performed very well and close to real-time. The best results were obtained using HOG and color features in the detection, and the HOG-LDA-KNN combination in the classification. In the future, the system should be tested more in challenging winter conditions.

Acknowledgements. The authors would like to thank the Finnish Transport Agency for funding of the TrafficVision research project.

References

1. Hienonen, P.: Automatic traffic sign inventory and condition analysis. Master's thesis, Lappeenranta University of Technology, Finland (2014)
2. Stallkamp, J., Schlipsing, M., Salmen, J., Igel, C.: Man vs. computer: benchmarking machine learning algorithms for traffic sign recognition. NN **32**, 323–332 (2012)
3. Baro, X., Escalera, S., Vitria, J., Pujol, O., Radeva, P.: Traffic sign recognition using evolutionary adaboost detection and forest-ECOC classification. IEEE ITS **10**(1), 113–126 (2009)
4. Houben, S., Stallkamp, J., Salmen, J., Schlipsing, M., Igel, C.: Detection of traffic signs in real-world images: the German traffic sign detection benchmark. In: Proceedings of IJCNN (2013)
5. Mogelmose, A., Trivedi, M., Moeslund, T.: Vision-based traffic sign detection and analysis for intelligent driver assistance systems: perspectives and survey. IEEE ITS **13**, 1484–1497 (2012)
6. Bahlmann, C., Zhu, Y., Ramesh, V., Pellkofer, M., Koehler, T.: A system for traffic sign detection, tracking, and recognition using color, shape, and motion information. In: Proceedings of IEEE IV (2005)

7. Maldonado-Bascon, S., Lafuente-Arroyo, S., P. Siegmann, H.G.M., Acevedo-Rodriguez, F.: Traffic sign recognition system for inventory purposes. In: Proceedings of IEEE IVS (2008)
8. Hazelhoff, L., Creusen, I.M.: Exploiting street-level panoramic images for large-scale automated surveying of traffic signs. MVA **25**, 1893–1911 (2014)
9. Timofte, R., Zimmermann, K., Van Gool, L.: Multi-view traffic sign detection, recognition, and 3D localisation. MVA **25**, 633–647 (2011)
10. Larsson, F., Felsberg, M.: Using Fourier descriptors and spatial models for traffic sign recognition. In: Heyden, A., Kahl, F. (eds.) SCIA 2011. LNCS, vol. 6688, pp. 238–249. Springer, Heidelberg (2011). doi:10.1007/978-3-642-21227-7_23
11. Larsson, F., Felsberg, M., Forssen, P.E.: Correlating Fourier descriptors of local patches for road sign recognition. IET Comput. Vis. **5**, 244–254 (2011)
12. Gu, Y., Yendo, T., Tehrani, M., Fujii, T., Tanimoto, M.: Traffic sign detection in dual-focal active camera system. In: Proceedings of IEEE IV (2011)
13. Gonzalez, A., Garrido, M., Llorca, D., Gavilan, M., Fernandez, J., Alcantarilla, P., Parra, I., Herranz, F., Bergasa, L., Sotelo, M., Revenga de Toro, P.: Automatic traffic signs and panels inspection system using computer vision. IEEE ITS **12**, 485–499 (2011)
14. Houben, S.: A single target voting scheme for traffic sign detection. In: Proceedings of IEEE IV (2011)
15. Fleyeh, H.: Color detection and segmentation for road and traffic signs. In: Proceedings of IEEE CIS (2004)
16. Viola, P., Jones, M.J.: Robust real-time face detection. IJCV **57**, 137–154 (2004)
17. Zaklouta, F., Stanciulescu, B., Hamdoun, O.: Traffic sign classification using k-d trees and random forests. In: Proceedings of IJCNN (2011)
18. Mathias, M., Timofte, R., Benenson, R., Gool, L.J.V.: Traffic sign recognition - how far are we from the solution? In: Proceedings of IJCNN (2013)
19. Martinez, A.M., Kak, A.C.: PCA versus LDA. IEEE PAMI **23**, 228–233 (2001)
20. Zhu, Y., Wang, X., Yao, C., Bai, X.: Traffic sign classification using two-layer image representation. In: Proceedings of ICIP (2013)
21. Lu, K., Ding, Z., Ge, S.: Sparse-representation-based graph embedding for traffic sign recognition. IEEE ITS **13**, 1515–1524 (2012)
22. de la Escalera, A., Moreno, L., Salichs, M., Armingol, J.: Road traffic sign detection and classification. IEEE IE **44**, 848–859 (1997)
23. Fleyeh, H.: Shadow and highlight invariant colour segmentation algorithm for traffic signs. In: Proceedings of CCIS (2006)
24. Broggi, A., Cerri, P., Medici, P., Porta, P., Ghisio, G.: Real time road signs recognition. In: Proceedings of IEEE IV (2007)
25. Dollar, P., Appel, R., Belongie, S., Perona, P.: Fast feature pyramids for object detection. IEEE PAMI **36**, 1532–1545 (2014)
26. Dollar, P., Wojek, C., Schiele, B., Perona, P.: Pedestrian detection: an evaluation of the state of the art. IEEE PAMI **34**, 743–761 (2012)
27. Breiman, L.: Random forests. Mach. Learn. **45**, 5–32 (2001)
28. Dollar, P., Tu, Z., Perona, P., Belongie, S.: Integral channel features. In: Proceedings of BMVC (2009)
29. Sermanet, P., LeCun, Y.: Traffic sign recognition with multi-scale convolutional networks. In: Proceedings of IJCNN (2011)
30. Karney, C.F.: Algorithms for geodesics. J. Geod. **87**, 43–55 (2013)
31. Martí, E.D., Martín, D., García, J., de la Escalera, A., Molina, J.M., Armingol, J.M.: Context-aided sensor fusion for enhanced urban navigation. Sensors **12**, 16802–16837 (2012)
32. LeCun, Y., Bengio, Y., Hinton, G.: Deep learning. Nature **521**, 436–444 (2015)

Copy-Move Forgery Detection Using the Segment Gradient Orientation Histogram

Ali Retha Hasoon Khayeat[1,2]([✉]), Paul L. Rosin[1], and Xianfang Sun[1]

[1] School of Computer Science and Informatices,
Cardiff University, Cardiff, UK
{KhayeatAR,RosinPL,SunX2}@Cardiff.ac.uk
[2] Computer Science Department, College of Science,
Kerbala University, Karbala, Iraq
aliretha@gmail.com

Abstract. The ready availability of image-editing software makes ensuring the authenticity of images an important issue. The most common type of image tampering is cloning, or Copy-Move Forgery (CMF), in which part(s) of the image are copied and pasted back into the same image. One possible transformation is where an object is copied, rotated and pasted; this type of forgery is called Copy-Rotate-Move Forgery (CRMF). Applying post-processing can be used to produce more realistic doctored images and thus can increase the difficulty of forgery detection. This paper presents a novel segmentation-based Copy-Move forgery detection method. A new method has been developed to segment the Copy-Move objects in a consistent way that is more efficient than Simple Linear Iterative Clustering (SLIC) segmentation for CMF/CRMF. We propose a new method to describe irregular shaped blocks (segments). The Segment Gradient Orientation Histogram (SGOH), is used to describe the gradient distribution of each segment. The quality of initial matches is improved by applying hysteresis to grow the primary detection regions. We show that the proposed method can effectively detect forgery involving translation and rotation. Moreover, the proposed method can detect forgery in images with blurring, brightness change, colour reduction, JPEG compression, variations in contrast and added noise.

Keywords: Copy-Move Forgery · Segment Gradient Orientation Histogram (SGOH) · Simple Linear Iterative Clustering (SLIC) · Otsu thresholding

1 Introduction

Copy-move is the most common type of image forgery (copy and paste), where regions of the image are cloned to cover objects in the scene. If this is done with care, visual detection of cloning will be difficult. Moreover, because the cloned regions can be in any location or can have any shape, searching all possible image portions in different sizes and locations is computationally infeasible.

© Springer International Publishing AG 2017
P. Sharma and F.M. Bianchi (Eds.): SCIA 2017, Part I, LNCS 10269, pp. 209–220, 2017.
DOI: 10.1007/978-3-319-59126-1_18

In Copy-Move Forgery (CMF), part(s) of the image are copied and pasted into the same image but in different places, possibly after a rotation. Moreover, because the copied-pasted region is from the same image, its characteristics (e.g. colour and noise) are compatible with that image. This type of forgery is more challenging to detect than other types, such as splicing and retouching. This is because the usual methods of detecting incompatibilities, using statistical measurements to compare different parts of the image, will be useless for CMF detection [6].

The most common approach to detect CMF consists of many steps, the most important step is the feature extraction. There are two different methods to extract features, either by tiling the image into blocks and then extracting the features from each block, or from interest points which are distributed over the image in different ways (e.g. SIFT, MSER, FAST, etc.). The block-based methods usually need a long time to extract features from the image. The keypoint-based methods are much faster than block-based methods, but can only detect part(s) of the duplicated region(s) because keypoints are distributed sparsely over the image.

We have considered a segmentation approach as a potential solution to overcome the problems of the block-based and keypoint-based methods. However, the main problem of the standard approach is: *there is no reliable method to segment identical objects consistently*. So, even with state-of-the art segmentation methods, there is no guarantee of segmenting Copy-Move objects in the same manner.

This paper presents a novel method to detect Copy-Move Forgery which performs a consistent image segmentation and extracts features from each segment. The image has been quantized into seven labels using the Rolling Guidance filter [18] followed by Otsu's thresholding [11]. Each quantized area (segment) has been described using a 3D colour histogram, the Segment Weighted Gradient Orientation Histogram (SWGOH), and its size. The 2nd Approximate Nearest Neighbour was used to detect the forgery at the segment level followed by a hysteresis technique to extend the primary detection. The proposed method is robust to rotation and the effects of post-processing method, and is relatively fast.

2   Related Works

Many papers have been proposed to detect CMF, and are mostly either block based or use keypoint based techniques. Recently two papers have used segmentation to detect CMF [8,13].

The authors of [13] tested four different image segmentation methods and used superpixels – the Simple Linear Iterative Clustering (SLIC) algorithm [1] – to over-segment the images. Then they extracted SIFT features from each segment, built a K-d tree for these features and used KNN to find the matches between patches. They computed the number of the matched feature points and identified suspicious pairs of patches which have many similar keypoints.

RANSAC was applied to estimate the transformation matrix between each pair of patches. Each pixel was represented by a dense SIFT descriptor. Then, the patch level matches were refined by applying matching to all the pixels in the matched patches and applying RANSAC to remove outliers.

The authors found that the segmentation method did not significantly influence the Copy-Move Forgery detection. Their approach is similar to finding matches between different images using SIFT features, considering each segment as a different image. Segmentation was used to divide the original image into various patches (small images). Their approach depends on extracting SIFT features from segments instead of from the whole image.

Li et al. [8] also used the SLIC method to segment the images. They used different scales of segmentation depending on the image contents itself. They set a large initial superpixel size for smooth images, and a small initial size for detailed images. The Discrete Wavelet Transformation (DWT) was used to analyse the frequency distribution of the image. According to their approach, the image is smooth when the majority of the energy of the host image is low-frequency; otherwise, the image is considered to be detailed. Then they extracted SIFT features from each segment and computed the Euclidean distance between features. If the number of matched points is more than a threshold, a correlation coefficient map generated to find the matched patches. They used SLIC to segment each matched patch to a smaller size and measured the local colour feature for each sub-patch. Neighbouring sub-regions (patches) are merged when the colour features are similar and the morphological close operation applied to generate detected forgery regions.

The above two methods [8,13] rely on using the keypoint-based approach (SIFT) in addition to image segmentation in order to detect CMF. In homogeneous regions there will be few SIFT features detected, causing matches to be missed. Moreover, the computational complexity of the two proposed methods is high.

3 Background

As the keypoint based techniques are likely to fail to detect interest points in relatively homogeneous regions, we instead use a dense descriptor to represent each segment. However, this will require repeatable segmentation, i.e. instances of the copied objects should be segmented in the same way, since otherwise, their descriptors will not match. As demonstrated in Sect. 5.2, SLIC is not repeatable: variations in an object's surroundings prevents the consistent segmentation of duplicated objects. We will present a new segmentation method that can segment the image consistently.

3.1 Threshold Selection Using Otsu's Method

The most common and simplest method to segment an image is using thresholding. Otsu [11] suggested a method to find the best threshold to binarize the grayscale image.

The binary image which is generated from applying a single Otsu threshold on the grayscale image contains a lot of small segments, and these would be unstable for matching in CMF detection, see the binary image in Fig. 1. Also, using the multi-level thresholding version of Otsu does not produce a better result, see the coloured images in Fig. 1.

Fig. 1. (from left to right) The grayscale input image, the binary image generated by Otsu's method, the segmented image using 7 Otsu thresholds, zoomed CMF areas.

3.2   Rolling Guidance Filter

In order to produce more meaningful segments, we will filter the image to remove noise and unnecessary details. Images contain significant structures and edges over a range of scale [18]. Many filtering techniques have been proposed to smooth the image while maintaining those structures. Edge-aware filters have been used to remove the low-contrast edges (gradual changes) and preserve the high-contrast edges, e.g. bilateral filter [15], guided filter [4] and weighted median filters [18]. In this work we use the Rolling Guidance Filter because it has been shown to be effective at removing small-scale structures while preserving large-scale structures by the use of scale-aware local operations. It can cope with irregular shapes and furthermore has low computation cost.

Fig. 2. (from left to right) The rolling guidance grayscale image, the rolling guidance binary image generated by Otsu's method, the rolling guidance segmented image using 7 Otsu thresholds, zoomed CMF areas in the rolling guidance image after threshing.

The rolling guidance method includes two main steps:

1. Remove small structures: In this step, the Gaussian filter is used. However, as well as removing the edges of structures smaller than the smoothing scale, it also blurs large-scale structures instead of preserving them.
2. Edge recovery: A joint bilateral filtering of the given input image and the image from the previous iteration is used to recover the edges. This can be understood as a filter that smooths the input image guided by the structure of the previous iteration image.

The binary image which is generated from applying the single Otsu threshold to the Rolling Guidance filtered image (smoothed image) produces a reasonably segmented image. As shown in the binary image of Fig. 2, the detrimental or unwanted content has been removed and the pixels have been clustered appropriately.

However, this does not adequately segment the Copy-Move objects in the image. Under-segmentation has caused the objects to become merged with the background. Therefore, we use the multi-level thresholding version of Otsu applied after the Rolling Guidance filtering, see the coloured images of Fig. 2.

4   Methodology

4.1   Segment Gradient Orientation Histogram (SGOH)

SIFT/DSIFT is restricted to regular blocks. Here, we have developed the Segment Gradient Orientation Histogram (SGOH) to describe the gradient for each segment (irregular block). The SIFT descriptor [9] has a 128 element feature vector for each keypoint. It considers a 16×16 neighbourhood around each keypoint and divides it into 4×4 sub-regions. For each sub-region, an eight-bin histogram of gradient magnitude weighted orientations is computed. DSIFT [2] follows the same approach however it considers all the pixels in the image as keypoints.

The steps to build the Segment Gradient Orientation Histogram (SGOH) are as follows:

1. The moment method (intensity centroid measure) [14] is used to find the canonical orientation for each segment.
2. Rotate each segment according to its canonical orientation to make the descriptor rotation invariant.
3. Construct a gradient magnitude weighted orientation histogram containing $18 = 360°/20°$ bins.
4. Normalize the generated feature vector (SGOH) between zero and one.

4.2   Proposed Algorithm

The Rolling Guidance filter [18] is used to smooth the image and preserve the strong edges, then the Otsu method [11] is used to find 7 different thresholds on the filtered image. We have tried a different number of thresholds (5, 7, 9, 11 and 13), and experimentally we found that using 7 thresholds segments the Copy-Move objects in the most consistent way. The 7 thresholds have been used to quantize the Rolling Guidance filtered image into 8 different labels. Next, connected component labelling is applied in each different intensity threshold range and properties (e.g. area, pixel list, etc.) are computed for each object. All segments of size less than 70 pixels ($T_1 = 70$) are removed; this threshold has been chosen experimentally, and is fixed for all experiments. A 3D colour (3DRGB) histogram has been used to describe the colour distribution of each segment, and a Segment Gradient Orientation Histogram (SGOH) has been built to represent the gradient of each segment, see Sect. 4.1. The SGOH is concatenated with 3DRGB and the segment area to form a feature vector. A K-d tree is built from the segment feature vectors, and for each segment its 2ANN is found. Matched segments are kept if the Euclidean distance between their concatenated feature vectors is less than a threshold ($T_2 = 0.002$) and the ratio between their sizes is less than a threshold ($T_3 = 1.5$). Make the size of the two matched segments is equal, save its coordinates in two separated lists and call RANSAC to remove the outliers. Finally, Hysteresis technique is applied to grow the detected Copy-Move regions.

4.3 Hysteresis Technique

To produce the best possible result, we use a hysteresis technique in the CMF/CRMF detection. Hysteresis thresholding is based on using two thresholds, one low and one high, and it considers the spatial information to improve the result. This technique is commonly employed in edge detection [3]. Recently, hysteresis has been used in forgery detection [5]; the "strong" matches detected using the high threshold, and the low threshold were rejected very "weak" matches. The main drawbacks of Jaberi et al. [5] are that they used a window to search for the new matched features which may contain parts outside of the CMF areas. Also, they recompute all the MIFT feature for all pixels in the detection window with each iteration.

To use the hysteresis thresholding in CMF/CRMF detection, we developed the following approach. Find the 2nd Approximate Nearest Neighbour (2ANN) for each feature vector (SGOH) within the strict low threshold (T_2) [5,19] and this will decrease the false matches. The low threshold is used to detect "strong" similar features, which represent the pixels from the original and the duplicated regions (segments). Apply RANSAC to remove the outliers and find the coordinate's transformation of the matched features. For each coordinate in the transformation list, recolour the block that takes this coordinate as a center. In the next step, dilate each region using a disk with a one-pixel radius size. For each of the newly added pixels, compute the improved DSIFT [6]. Build a K-d tree and find the 2ANN for each new feature vector. If the Euclidean distance between the matched features is less than the high threshold ($T_4 = 2T_2$), we store the coordinates of these features. Apply RANSAC to remove any new outliers and keep the new coordinates within the previously found transformation. Add the new pixels

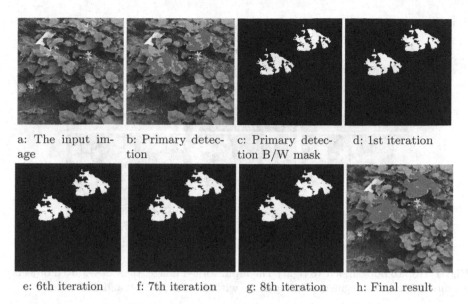

a: The input image | b: Primary detection | c: Primary detection B/W mask | d: 1st iteration

e: 6th iteration | f: 7th iteration | g: 8th iteration | h: Final result

Fig. 3. An example of growing the detection regions with hysteresis thresholding.

to the matching list and update the transformation matrix. Grow the detection regions by adding a new block located at the center of the new matched pixel. Repeat this process until no more pixels can be added to the primary detection. This region growing technique depends on the primary detection of the strong features matching and the spatial information, to iteratively add one block to the edges of the primary detection. As illustrated on Fig. 3 the detection regions have been grown, which increased the F-measure from 0.72 to 0.88.

5 Experiments

5.1 Dataset and Evaluation Method

We tested our method using the image database for Copy-Move Forgery Detection (CoMoFoD) [16]. CoMoFoD consists of 260 forged images categorized into two categories (small 512×512, and large 3000×2000). The small category consists of 200 original images with different types of forgery. We considered only the small images in our work. In the small category, images are divided into 5 different groups according to the applied manipulations, as follows: translation, rotation, scaling, distortion and a combination of all previous manipulations. Moreover, various types of post-processing methods (e.g. blurring, brightness change, color reduction, JPEG compression, contrast adjustments and added noise), are applied to all forged images in each group. The total number of images in the small group is 10400 with different types of manipulations. We used the F-measure [20] at the pixel level to evaluate the accuracy of our results.

Fig. 4. (top to bottom, left to right) The input forged image, the Copy-Moved objects, zoomed CMF areas segmented using SLIC with $K = 30, 55, 100$ and 300.

5.2   Experiment to Detect CMF Using SLIC

In our initial work we tried to use superpixel segmentation as the basis for detecting CMF; SLIC was applied, and a set of features was densely extracted from each segment.

The size of duplicated objects can vary from one image to another, as they can form a small or large part of an image. SLIC divides the image into irregular blocks which exhibit state-of-the-art boundary adherence [1]. The required number of approximately similar-sized superpixels (K) is the parameter to control the SLIC algorithm.

Problems with this approach are evident in Fig. 4. When the SLIC method is applied to this image it does not segment the Copy-Move objects (i.e. the two ladies) consistently, due to differences between their backgrounds, which reveals the unreliability of this approach.

So, instead of using SLIC to segment the image, we have used our proposed method to segment the Copy-Move objects, see Sect. 4.2, that is more consistent than SLIC and produces better CMF detection results, see Figs. 4 and 5.

5.3   Experiment to Detect CMF with Translation and Post-processing

We used our suggested method to test 40 different images with plain CMF (without post-processing), and obtained an F-measure $= 0.79$, see Table 1 and Fig. 5. Moreover, the proposed method is less complicated than the other suggested methods [8, 13] which use segmentation; it takes about 45 s to process one image.

We tested our proposed method on 280 images with different types of post-processing (image blurring, brightness change, colour reduction, JPEG compression, contrast adjustments and added noise), see Table 1.

a: input image   b: Rolling Guidance image   c: quantized image

d: primary detection mask   e: RANSAC result   f: final result after hysteresis

Fig. 5. An example of plain CMF detection (True Positive (TP), True Negative (TN), False Positive (FP), False Negative (FN)).

5.4 Experiment to Detect CRMF and Post-processing

Using rotation invariant features is the primary requirement of the Copy-Rotate-Move Forgery (CRMF) detection. The Segment Gradient Orientation Histogram (SGOH) is rotation invariant as each segment is rotated to its canonical orientation before computing the weighted histogram.

The experimental work illustrated that the suggested algorithm can detect rotated duplicated objects with acceptable performance. The algorithm detected forgery on 35 images out of 40 with F-measure = 0.71, see Table 1.

280 images with rotated duplicated objects and different types of post-processing (image blurring, brightness change, colour reduction, JPEG compression, contrast adjustments and added noise) have been tested. The suggested algorithm successfully detect forgery on 243 images. We have shown experimentally that our proposed method is not affected by the post-processing methods and it can detect forgery even on compressed or noisy images.

We found that the under-segmentation is the main reason that the proposed algorithm cannot detect forgery on some images (Fig. 6).

We compared our proposed method with Zernike moments (ZM) [6,21] to demonstrate the robustness of our method. Table 1 shows that Zernike moments

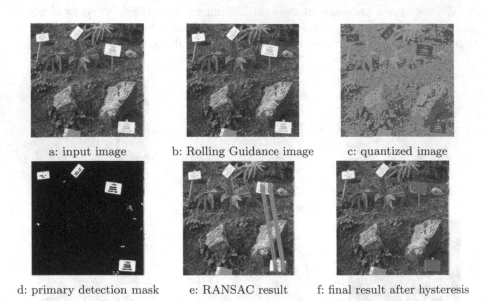

a: input image   b: Rolling Guidance image   c: quantized image

d: primary detection mask   e: RANSAC result   f: final result after hysteresis

Fig. 6. An example of Copy-Rotate-Move Forgery detection (True Positive (TP), True Negative (TN), False Positive (FP), False Negative (FN)).

Table 1. CMF/CRMF detection with post-processing.

Post-processing	Translation		Rotation	
	F-measure (SGOH)	F-measure (ZM) [6]	F-measure (SGOH)	F-measure (ZM) [6]
Without post-processing	0.79	0.88	0.71	0.68
Brightness change range (0.01, 0.8)	0.78	0.45	0.70	0.40
Contrast adjustment range (0.01, 0.8)	0.78	0.48	0.70	0.38
Colour reduction (32 intensity levels)	0.78	0.49	0.70	0.40
Image blurring (5 × 5 average filter)	0.79	0.56	0.70	0.49
Adding noise ($\mu = 0$, $\sigma^2 = 0.005$)	0.69	0.63	0.64	0.61
JPG compression (quality factor = 40)	0.78	0.60	0.69	0.62

are not robust to post-processing, which confirms the results of Fig. 9 in [21]. In contrast, our proposed method produces consistent results with post-processing.

6   Conclusions

In this paper, we considered Copy-Move forgery incorporating translation and rotation. A new segmentation method was suggested to segment the Copy-Move objects in a more consistent way than SLIC. We obtained good results on translation and reasonable results with rotation.

The Segment Gradient Orientation Histogram (SGOH), which was inspired by SIFT [9], was used to describe the gradient for each segment (irregular block).

The hysteresis technique was used to grow the detection region(s) and improve the primary detection result. Also, our method can detect CMF in images with blurring, brightness change, color reduction, JPEG compression, variations in contrast and added noise.

References

1. Achanta, R., Shaji, A., Smith, K., Lucchi, A., Fua, P., Süsstrunk, S.: SLIC superpixels compared to state-of-the-art superpixel methods. IEEE Trans. Pattern Anal. Mach. Intell. **34**(11), 2274–2282 (2012)
2. Bosch, A., Zisserman, A., Munoz, X.: Image classification using random forests and ferns. In: 11th International Conference on Computer Vision, pp. 1–8. IEEE (2007)
3. Canny, J.: A computational approach to edge detection. IEEE Trans. Pattern Anal. Mach. Intell. (PAMI) **8**(6), 679–698 (1986)
4. He, K., Sun, J., Tang, X.: Guided image filtering. IEEE Trans. Pattern Anal. Mach. Intell. **35**(6), 1397–1409 (2013)
5. Jaberi, M., Bebis, G., Hussain, M., Muhammad, G.: Accurate and robust localization of duplicated region in copy-move image forgery. Mach. Vis. Appl. **25**(2), 451–475 (2014)
6. Khayeat, A.R.H., Sun, X., Rosin, P.L.: Improved DSIFT Descriptor Based Copy-Rotate-Move Forgery Detection. In: Image and Video Technology. LNCS, vol. 8334, pp. 642–655. Springer (2016)
7. Klette, R.: Concise Computer Vision. Springer, London (2014)
8. Li, J., Li, X., Yang, B., Sun, X.: Segmentation-based image copy-move forgery detection scheme. IEEE Trans. Inf. Forensics Secur. **10**(3), 507–518 (2015)
9. Lowe, D.G.: Distinctive Image Features from Scale-Invariant Keypoints. Int. J. Comput. Vis. **60**(2), 91–110 (2004)
10. Muja, M., Lowe, D.G.: Fast approximate nearest neighbors with automatic algorithm configuration. In: International Conference on Computer Vision Theory and Applications, pp. 331–340. INSTICC Press (2009)
11. Otsu, N.: A threshold selection method from gray-level histograms. IEEE Trans. Syst. Man Cybern. **9**(1), 62–66 (1979)
12. Pietikäinen, M., Hadid, A., Zhao, G., Ahonen, T.: Computer Vision Using Local Binary Patterns. Springer, London (2011)

13. Pun, C.-M., Yuan, X.-C., Bi, X.-L.: Image forgery detection using adaptive over-segmentation and feature point matching. IEEE Trans. Inf. Forensics Secur. **10**(8), 1705–1716 (2015)
14. Rosin, P.L.: Measuring corner properties. Comput. Vis. Image Underst. **73**(2), 291–307 (1999)
15. Tomasi, C., Manduchi, R.: Bilateral filtering for gray and color images. In: Sixth International Conference on Computer Vision, pp. 839–846 (1998)
16. Tralic, D., Zupancic, I., Grgic, S., Grgic, M.: CoMoFoD - new database for copy-move forgery detection. In: Proceedings of 55th International Symposium ELMAR, pp. 25–27. IEEE (2013)
17. Yang, J., Jiang, Y.-G., Hauptmann, A.G., Ngo, C.-W.: Evaluating bag-of-visual-words representations in scene classification. In: Proceedings of the International Workshop on Multimedia Information Retrieval, vol. 63, p. 197. ACM Press (2007)
18. Zhang, Q., Shen, X., Xu, L., Jia, J.: Rolling guidance filter. In: Fleet, D., Pajdla, T., Schiele, B., Tuytelaars, T. (eds.) ECCV 2014. LNCS, vol. 8691, pp. 815–830. Springer, Cham (2014). doi:10.1007/978-3-319-10578-9_53
19. Zhang, T., Boult, T.E., Johnson, R.: Two thresholds are better than one. In: Conference on Computer Vision and Pattern Recognition, pp. 1–8. IEEE (2007)
20. DM Powers: Evaluation: from precision, recall and f-measure to ROC, informedness, markedness & correlation. J. Mach. Learn. Technol. **2**(1), 37–63 (2011)
21. Ryu, S.-J., Lee, M.-J., Lee, H.-K.: Detection of copy-rotate-move forgery using Zernike moments. In: Böhme, R., Fong, P.W.L., Safavi-Naini, R. (eds.) IH 2010. LNCS, vol. 6387, pp. 51–65. Springer, Heidelberg (2010). doi:10.1007/978-3-642-16435-4_5

BriefMatch: Dense Binary Feature Matching for Real-Time Optical Flow Estimation

Gabriel Eilertsen[1]([✉]), Per-Erik Forssén[2], and Jonas Unger[1]

[1] Department of Science and Technology, Linköping University, Linköping, Sweden
{gabriel.eilertsen,jonas.unger}@liu.se
[2] Department of Electrical Engineering, Linköping University, Linköping, Sweden
per-erik.forssen@liu.se

Abstract. Research in optical flow estimation has to a large extent focused on achieving the best possible quality with no regards to running time. Nevertheless, in a number of important applications the speed is crucial. To address this problem we present BriefMatch, a real-time optical flow method that is suitable for live applications. The method combines binary features with the search strategy from PatchMatch in order to efficiently find a dense correspondence field between images. We show that the BRIEF descriptor provides better candidates (less outlier-prone) in shorter time, when compared to direct pixel comparisons and the Census transform. This allows us to achieve high quality results from a simple filtering of the initially matched candidates. Currently, Brief-Match has the fastest running time on the Middlebury benchmark, while placing highest of all the methods that run in shorter than 0.5 s.

Keywords: Optical flow · Feature matching · Real-time computation

1 Introduction

Optical flow estimation is a fundamental problem within computer vision. It is useful in a wide range of applications, from temporal filtering to structure-from-motion. Due to its applicability, a huge body of work has been devoted to the topic. However, the great majority of methods do not focus on real-time, and it still remains a difficult challenge to determine robust and high quality flow fields in live applications. Real-time optical flow is essential e.g. for detecting moving objects on moving platforms, obstacle avoidance and gesture recognition. It can also be used in live video streaming, in order to perform frame interpolation, video stabilization, rolling-shutter correction etc.

In this work, we use local pixel matching, with binary robust independent elementary features (BRIEF) [11] as similarity criterion. We match these using principles from the PatchMatch algorithm [5]. Running on the GPU, this combination can efficiently estimate a candidate optical flow field. The field contains few outliers as compared to using direct patch comparisons or a Census transform [35]. This makes it possible to use only low-level filtering to obtain a high quality optical flow field, and thus avoid expensive global optimization.

© Springer International Publishing AG 2017
P. Sharma and F.M. Bianchi (Eds.): SCIA 2017, Part I, LNCS 10269, pp. 221–233, 2017.
DOI: 10.1007/978-3-319-59126-1_19

The main contributions of this paper can be summarized as follows:

- We propose to combine BRIEF and PatchMatch, for efficient dense feature matching in order to find a candidate flow field with few outliers.
- We show that BRIEF features improve both matching quality and efficiency, compared to the Census transform and direct pixels comparisons.
- By comparing to existing real-time and offline methods we show that our robust optical flow exhibits a very good trade-off between quality and running time. It is thus well-suited for real-time applications.

2 Background

While classical methods for optical flow estimation generally optimize globally in a coarse to fine setting, modern approaches tend to increasingly use local pixel matching. This means that fine detailed motion at large displacements can be better recovered. The matching can be performed using a sparse set of features [10,28,33,34], and then propagated to neighboring pixels to give a per-pixel flow estimation. Alternatively, the matching can be done densely, comparing per-pixel local patch distances or dense features [2,4,12,21,23]. However, a dense correspondence field (CF) contains many outliers, and some post-processing is inevitably needed in order to find a final optical flow. For example, the method proposed by Bailer et al. [2] performs a hierarchical correspondence field search, followed by outlier filtering. We also make use of a per-pixel correspondence search. However, in order to relieve the need for post-processing we perform the matching using feature descriptors that makes for relatively few outliers.

The first use of binary patch descriptors were the local binary pattern descriptors (LBP) [26], also known as the Census transform [35] which is the name we will use. The Census transform encodes the local properties around a pixel by comparing it to the pixels in a local neighborhood. Census matching is common in stereo where it often is used together with a Semi Global Matching (SGM) in order to find disparities between images [13,17,18]. The Census transform has also been used in recent state-of-the-art for optical flow estimation [2], and in order to promote real-time performance [25,31]. It has a number of favorable features that makes it well-suited for determining flow vector candidates [16,32]. A different formulation of binary features is used in the BRIEF descriptor [11], where random pixel-pairs are compared in a local neighborhood. This descriptor has been extensively used as an efficient alternative to e.g. SIFT or SURF. There is also at least one example of it being used for stereo matching [36]. However, to our knowledge BRIEF has not been used in the context of optical flow estimation, and not in combination with the PatchMatch algorithm. We will show that using the BRIEF descriptor for real-time per-pixel matching has significant advantages over the Census transform, both in terms of robustness and speed.

The PatchMatch algorithm [5] is an efficient and well-known method to estimate the correspondence field between two images. The method alters random searches and a propagation scheme, in order to efficiently find approximate nearest neighbors. There are examples of the method being used both for optical flow evaluation [4,21,23,34] and stereo matching [7,13]. By combining BRIEF and PatchMatch we are able to achieve a robust optical flow estimation in real-time.

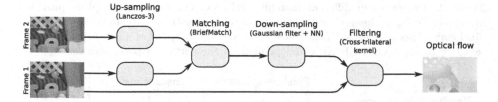

Fig. 1. BriefMatch optical flow pipeline.

3   Method

The proposed pipeline for BriefMatch optical flow estimation is outlined in Fig. 1. It has two main components: (1) per-pixel matching between images using BRIEF descriptors and PatchMatch, to obtain an approximate correspondence field (CF), and (2) outlier filtering of the CF using a cross-trilateral filter kernel. This section describes the pipeline, as well as implementation specific details.

3.1   Correspondence Field Search

A BRIEF descriptor is constructed by performing N pair-wise comparisons in a $S \times S$ pixels local neighborhood [11]. A binary feature vector is then built as a bit string, setting the bits based the comparisons. Denoting local coordinates as q and q', the descriptor at pixel p in the image I can be formulated as

$$F(p) = \sum_{i=1}^{N} 2^{i-1} \left(0.5 \, \text{sign}(I(p + q_i) - I(p + q_i')) + 0.5\right). \tag{1}$$

In BRIEF, the sampling of point pairs for the descriptor is done using a normal distribution, $q, q' \sim \mathcal{N}(0, \sigma)$, where the standard deviation is set to $\sigma = S/5$. The Census transform instead compares all pixels of a patch to the central pixel [35]. This can also be described using (1), if we define q, q' as

$$\begin{aligned} q &= (0, 0), \\ q' &= (i, j), \quad i, j = \{-S/2, -S/2 + 1, \ldots, S/2 - 1, S/2\}, \quad q \neq q'. \end{aligned} \tag{2}$$

Matching on descriptor fields $F(p)$ is performed by comparing the descriptors with the Hamming distance. This can be done efficiently using a `bitcnt` operation on the bitwise exclusive or (XOR) of two descriptors. In order to find a correspondence field between two images, we use the search strategy from PatchMatch [5]. This was originally designed to search for approximate nearest neighbors, comparing in terms of mean absolute pixel differences over patches. PatchMatch uses the fact that the offset to a nearest neighbor of a pixel is also often a good candidate for its adjacent pixels. By iterating random searches and a propagation scheme, the algorithm is able to find a good correspondence field in about 4–5 iterations. The algorithm is also directly applicable for finding nearest neighbors in terms of binary feature distances.

Table 1. Properties of different matching criteria. The binary descriptors provide invariance to light changes and a greater outlier robustness as compared to direct pixel comparisons. Census depends heavily on the central pixel of a patch, and is thus sensitive to noise. Only BRIEF is independent of the particular patch size used.

	Pixel comparisons	Census transform	BRIEF
Handles light changes	✗	✓	✓
Outlier robust	✗	✓	✓
Robust to pixel noise	✓	✗	✓
Scales well with patch size	✗	✗	✓

3.2 Properties of Comparison Criteria

In the previous subsection three different comparison criteria were mentioned for performing a correspondence field search – direct pixel comparisons, a Census transform and BRIEF descriptors. Figure 2 shows examples of the resulting correspondence fields from matching in terms of these criteria. The main differences are due to four distinct properties, which are summarized in Table 1. These are:

Light Changes: The most obvious advantage of using binary features, such as Census and BRIEF, is the invariance to global changes in intensity. Direct pixel comparisons on the other hand, depend on absolute pixel values.

Outlier Robustness: Encoding the local information around a pixel with binary features also results in a more robust matching that is less sensitive to outliers. For example, an inherent problem with patch comparisons is matching on edges where the background is moving relative to the foreground (see Fig. 2(a)). With binary features, outlier pixels have less influence compared to direct pixel comparisons. In order to further improve matching on edges, we also tested to include edge-aware binary matching, that only uses comparisons on either foreground or background [36]. However, this did not improve on the result to a large extent, so given the limited time budget for real-time performance we did not include this in the algorithm.

Robustness to Pixel Noise: A weakness of the Census transform is that it relies heavily on the value of the center pixel of the patch, and it is thus sensitive to that pixel's variance. This is not the case for BRIEF and direct patch comparisons.

Patch Size: While the complexity of direct patch comparisons and the Census transform are directly related to the size of the local neighborhood, BRIEF is only dependent on the number of binary feature comparisons (N in Eq. 1).

3.3 Flow Refinement

In order to improve on the correspondence field search described in Sect. 3.1, we make two modifications. First, we up-sample the input frames in the spatial

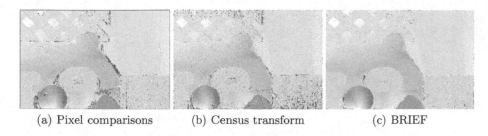

(a) Pixel comparisons   (b) Census transform   (c) BRIEF

Fig. 2. Examples of correspondence fields estimated using different comparison criteria, where the images have been up-sampled by a factor 3 before matching (see Sect. 3.3). Using BRIEF results in the least amount of outliers.

dimension, as illustrated in Fig. 1, to achieve sub-pixel accuracy and to provide a higher number of flow vector candidates. For the up-sampling the interpolation strategy is crucial, where a Lanczos-3 kernel improves significantly on bicubic or bilinear interpolation. After matching has been performed on the up-sampled frames, the resulting correspondence field is down-sampled to the original size. This is done by first filtering the flow vectors with a Gaussian filter, followed by a nearest neighbor down-sampling.

While the up-sampling strategy is able to refine the flow candidates, there are still many outliers in the flow field. Many existing patch-based methods for optical flow estimation use higher level optimization in order to refine the candidates, e.g. determining motions in segmented regions with RANSAC [12]. However, since we have a tight time budget to allow for real-time performance, we rely purely on local filtering of the correspondence field. While a median filter is able to remove many outliers, it also results in over-smoothed edges. Instead, we propose to use a cross-trilateral filter that in addition to spatial distance incorporates distances both in terms of flow vectors and image intensity,

$$\bar{u}(\boldsymbol{p}) = \text{medianFilter}(u(\boldsymbol{p})),$$

$$d_{EPE}(\boldsymbol{p}, \boldsymbol{q}) = \sqrt{(u(\boldsymbol{q}) - \bar{u}(\boldsymbol{p}))^2 + (v(\boldsymbol{q}) - \bar{v}(\boldsymbol{p}))^2},$$

$$d_I(\boldsymbol{p}, \boldsymbol{q}) = I(\boldsymbol{q}) - I(\boldsymbol{p}), \tag{3}$$

$$\bar{\bar{u}}(\boldsymbol{p}) = \frac{1}{W} \sum_{\boldsymbol{q} \in \Omega} u(\boldsymbol{q}) G_{\sigma_1}\left(d_{EPE}(\boldsymbol{p}, \boldsymbol{q})\right) G_{\sigma_2}\left(d_I(\boldsymbol{p}, \boldsymbol{q})\right) G_{\sigma_3}(\boldsymbol{q} - \boldsymbol{p}).$$

Here, $\boldsymbol{q}$ runs in a neighborhood Ω of the pixel $\boldsymbol{p}$. The distance d_{EPE} incorporates the differences in the flow field. It is computed by comparing flow vectors to the median filtered version of the flow, $\bar{u}(\boldsymbol{p})$, to increase outlier resistance. The distance d_I is a weighting term computed from the original image I. $G_{\sigma_{\{1,2,3\}}}$ are Gaussian kernels, which are normalized through the weight W. The filtering is performed in the same manner for both u and v. The final optical flow estimation $\bar{\bar{\boldsymbol{r}}} = (\bar{\bar{u}}, \bar{\bar{v}})$ has some advantages over a separate median filter, where it better preserves corners and boundaries of the flow as can bee seen in Fig. 3.

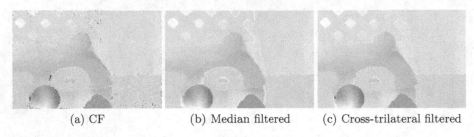

(a) CF   (b) Median filtered   (c) Cross-trilateral filtered

Fig. 3. The result of filtering the correspondence field (CF) using (b) a median filter and (c) the cross-trilateral filter in Eq. 3. The up-sampling ratio is 3, and BRIEF-64 features have been used, resulting in a total running time of about 65 ms.

3.4 Temporal Propagation

Since the aim is to provide a real-time optical flow algorithm for processing frames in a video sequence, we can explore between-frame correlations. One simple modification is to initialize the nearest neighbor search using the information from the previous matching,

$$r_t(p)_0 = r_{t-1}(p + r_{t-1}(p)). \tag{4}$$

Here, t is the current frame and 0 indicates the initial correspondences. Now, the number of iterations of the correspondence search can be cut in half without sacrificing performance. This makes for a significant reduction in running time.

3.5 Implementation

The described method is well-suited for parallel implementation. The only exception is the serial propagation of nearest neighbors in the PatchMatch algorithm. In order to approximate this on the GPU we use a jump flooding scheme [29]. All the steps in Fig. 1 have been implemented using CUDA, and the performance we report throughout this paper has been evaluated running on an Nvidia Geeforce GTX 980. For a typical setup, the running times of the different stages in Fig. 1 are given in Table 2.

For the filtering step (Eq. 3) we use a 13×13 pixels median filter. This involves sorting an array of 169 values for each pixel, which is expensive. Instead we choose to approximate the filter with a separable median computation, which significantly reduces running time without sacrificing quality to a large extent.

Table 2. Running time for the different steps in Fig. 1, given a typical parameter calibration (this can be changed to trade-off quality and speed). Times are estimated using a 640×480 resolution sequence, and running on a Geeforce GTX 980.

Up-sampling	Matching	Down-sampling	Filtering	Total time	Framerate
6.1 ms	33.1 ms	1.9 ms	26.5 ms	68.1 ms	14.7 fps

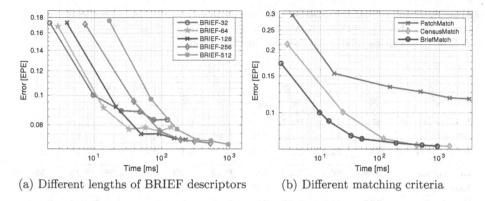

(a) Different lengths of BRIEF descriptors   (b) Different matching criteria

Fig. 4. Correspondence field search time vs. final error for the *RubberWhale* sequence. Sampling points have been estimated at different up-sampling factors before matching.

4   Results

In order to validate the performance of BriefMatch we perform a set of comparisons on the Middlebury training and test data [3]. In order to measure quality we use the average endpoint error (EPE), where the EPE is defined as the distance between estimated and ground truth flow vectors, $d_{EPE}(\boldsymbol{p}) = ||\boldsymbol{r}(\boldsymbol{p}) - \boldsymbol{r}_{gt}(\boldsymbol{p})||$.

Impact of Feature Length: With BriefMatch we have the option to trade off quality for running time by specifying the length N of the binary features in Eq. 1. Figure 4(a) shows the error on the *RubberWhale* sequence for a selection of descriptor sizes. The times specified are only for the matching, while the error is after performing the filtering in Eq. 3. For each descriptor length the up-sampling factor has been set in the range $[1, 6]$. Since the method has a random search component, the outcome may be slightly different between runs, and thus the results are averaged over 30 separate runs. From the results it is clear that the optimal descriptor length depends on the up-sampling factor.

Comparison to Census: In order to show that the BRIEF descriptor performs better than direct pixel comparisons or a Census transform, Fig. 4(b) shows the same comparison as in Fig. 4(a). However, it is now made between different comparison criteria (see Sect. 3). All criteria have been matched with the same settings of the correspondence search, and the results have been filtered with the same calibration of the filter in Eq. 3. For the direct pixel comparisons and the Census transform, the patch size has been tuned to achieve the best possible performance. For the BRIEF comparisons the descriptor length has been chosen for best performance, so that the plot is the lower envelope of the plots in Fig. 4(a). It is clear that the performance when using binary descriptors is improved to a large extent, compared to direct pixel comparisons. The difference between using Census and BRIEF is smaller, although significant for shorter running times. For example, it takes about 3 times as long time for Census to reach the quality of BRIEF, when BRIEF runs in the range of 20–60 ms.

Table 3. Results from the Middlebury benchmark. The list includes the top 3 performers and the six methods that run in under 0.5 s. The colors indicate how well BriefMatch performs relative to the compared methods.

Method	Time (s)	Avg rank	Army	Mequon	Schefflera	Wooden	Grove	Urban	Yosemite	Teddy
NNF-Local [12]	673	**3.2**	0.07	0.15	0.18	0.10	0.41	0.23	0.10	0.34
OFLAF [19]	1530	9.5	0.08	0.16	0.19	0.14	0.51	0.31	0.11	0.42
MDP-Flow2 [34]	**342**	10.2	0.08	0.15	0.20	0.15	0.63	0.26	0.11	0.38
BriefMatch	**0.068**	66.2	0.09	0.21	0.25	0.20	0.93	1.69	0.25	1.25
Rannacher [27]	0.12	74.8	0.11	0.25	0.57	0.24	**0.91**	1.49	**0.15**	0.69
Bartels [6]	0.15	79.3	0.12	0.22	0.35	0.28	0.97	1.20	0.20	0.91
FlowNet [14]	0.5	81.5	0.11	0.30	0.62	0.27	1.04	**0.46**	0.17	0.75
FlowNet2 [20]	0.091	82.2	0.22	0.67	0.61	0.28	0.97	0.59	0.19	**0.60**
PGAM+LK [1]	0.37	118.6	0.37	1.08	0.94	1.40	1.37	2.10	0.36	1.89

Middlebury Benchmark: The Middlebury online benchmark[1] currently comprises 125 methods. In terms of the average EPE, BriefMatch places in the middle of these, while being the fastest of all. Table 3 lists 3 of the top-performing methods from the benchmark, as well as the ones that run in <0.5 s. BriefMatch performs very well for 3 of the sequences (green), with an error that is not very far from the top-performing methods while being about 4 orders of magnitude faster. However, for 2 sequences (yellow) the result is approximately equivalent to the best of the fast methods, and for 3 of the sequences (red) the quality is worse than many of the fast methods. In Sect. 5 we try to analyze why this is the case.

Comparison to Real-Time Methods: A comparison with existing real-time methods for optical flow is listed in Table 4. The table includes three methods from OpenCV's CUDA library. The first is a Lucas-Kanade solver [24] in a pyramidal implementation [8]. The second is Farnebäck's method [15], that is based on polynomial expansion to approximate the neighborhood of each pixel. The third is the method proposed by Brox et al. [9], using a variational model and a warping technique. We also include the recent real-time method presented by Kroeger et al. [22], which uses a dense inverse search (DIS) for finding patch correspondences. All methods have been executed with constant parameters over the sequences, selected in an effort to give the best quality in approximately 50 ms. However, the pyramidal LK and the polynomial expansion methods do not provide viable options to trade quality for time. For these, increasing the number of iterations or pyramid scaling increases time, but quality does not scale well, and it is a better trade-off to run in shorter time. The times reported are estimated on a machine equipped with an Intel Xeon X5680 (3.33 GHz) CPU and a Geeforce GTX 980 GPU. From the results we can see that BriefMatch reduces

---

[1] http://vision.middlebury.edu/flow/eval/results/results-e1.php.

Table 4. Comparison to real-time methods using the Middlebury training data. The colors indicate how well BriefMatch performs relative to the compared methods.

Method	Time (ms)	Avg EPE	R.Whale	Hydrangea	Grove2	Grove3	Urban2	Urban3
BriefMatch	49	**0.514**	**0.079**	**0.152**	**0.164**	**0.687**	0.606	1.396
Pyr. LK [8]	25	1.312	0.306	0.563	0.394	1.367	2.301	2.940
Pol. exp. [15]	**22**	0.823	0.250	0.323	0.354	1.175	0.913	1.921
Warp OF [9]	54	0.565	0.175	0.232	0.326	0.911	**0.580**	**1.167**
DIS [22]	57	0.800	0.230	0.261	0.379	0.976	1.344	1.607

the error to 45–75% for four of the sequences (green), as compared to the second best method. The only sequences where another method yields slightly better quality are the *Urban2* and *Urban3* datasets (yellow). We discuss the reason for this in the next section.

5   Limitations

Looking at the results in Tables 3 and 4, BriefMatch is not consistent in how well it performs relative to other methods. For some sequences it performs very well (marked with green), and for others there are many outliers (marked with red). Elaborating on the cause for this, we can discern the following reasons:

1. **Repetitive patterns:** Image structures that occurs repetitive can cause many outliers in the correspondence field. This is for example the case in the *Urban* sequences in Tables 3 and 4. These sequences are also computer generated, which potentially may increase the problem. In order to successfully deduce the motion in areas of repetitive patterns, a global optimization is inevitably needed, which would make real-time performance difficult.
2. **Occlusion:** Many outliers can be created in areas that are occluded from one frame to the next and vice versa, e.g. close to image boundaries in a sequence with camera motion. This is the case for the *Urban* and *Teddy* sequences. To alleviate this problem a global optimization would also be needed.
3. **Z-motion:** In comparing patches between images – either directly or in terms of binary features – there is no invariance to image scale. This is a problem if objects or the camera are moving perpendicular to the image plane, as in the *Yosemite* sequence. In order overcome this problem, matching may need to be performed at multiple scales.

Figure 5 exemplifies the two first problems, using the *Urban3* sequence. In the mid regions of the images the building facades are highly repetitive, causing a large number of outliers. The problem with occlusion can e.g. be seen close to the top and bottom image borders, caused by a vertical camera motion.

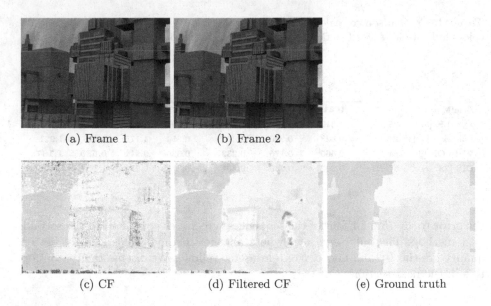

| (a) Frame 1 | (b) Frame 2 |

| (c) CF | (d) Filtered CF | (e) Ground truth |

Fig. 5. Repetitive patterns and occluded areas result in a large fraction of outliers in the correspondence field (CF). These are difficult to completely remove with a local outlier filtering technique.

6   Discussion

We have shown that for real-time optical flow estimation BRIEF has a significant advantage over the Census transform, and that binary features in general are much better suited for the problem than direct comparisons of pixels. Furthermore, our optical flow algorithm offers a substantial increase in quality compared to existing real-time methods. Also, comparing to offline state-of-the-art methods it can in some circumstances perform on a par with these in terms of quality, while being about 4 orders of magnitude faster.

Although BriefMatch shows promising results, problems occur for repetitive patterns and occlusions. These problems would be the main focus for improving the method, investigating how they can be alleviated without using an expensive global optimization formulation. Another possibility is to use BriefMatch in offline applications. Since a number of the best performing methods use direct patch comparisons, from our investigation we expect that using BRIEF for these methods has the potential of increasing quality and/or reducing running time.

Other straightforward improvements include bidirectional matching, color matching, multiple neighbors in the correspondence search, improved temporal considerations, etc. The up/down-sampling strategy may also be subject to improvement, exploring other interpolation kernels and sampling schemes. We also expect that the current implementation can be improved on, for example using Halide[2] and by adapting the implementation to better use shared memory

---

[2] http://halide-lang.org/.

and thread cooperation. Finally, for the BRIEF descriptor we have used random patch comparisons, were pixel pairs are drawn from a normal distribution, as in the original BRIEF formulation. Similar performance could probably be obtained with smaller descriptors, by using mining of feature pairs, as was done for e.g. the ORB descriptor [30].

Acknowledgments. This project was funded by the Swedish Foundation for Strategic Research (SSF) through grants IIS11-0081 Virtual Photo Sets and RIT 15-0097 SymbiCloud, and by the Swedish Research Council through grants 2015-05180, 2014-5928 (LCMM) and 2014-6227 (EMC2).

References

1. Alba, A., Arce-Santana, E., Rivera, M.: Optical flow estimation with prior models obtained from phase correlation. In: Bebis, G., et al. (eds.) ISVC 2010. LNCS, vol. 6453, pp. 417–426. Springer, Heidelberg (2010). doi:10.1007/978-3-642-17289-2_40
2. Bailer, C., Taetz, B., Stricker, D.: Flow fields: dense correspondence fields for highly accurate large displacement optical flow estimation. In: Proceedings of the ICCV 2015, December 2015
3. Baker, S., Scharstein, D., Lewis, J.P., Roth, S., Black, M.J., Szeliski, R.: A database and evaluation methodology for optical flow. IJCV **92**(1), 1–31 (2011)
4. Bao, L., Yang, Q., Jin, H.: Fast edge-preserving patchmatch for large displacement optical flow. In: Proceedings of the CVPR 2014, June 2014
5. Barnes, C., Shechtman, E., Finkelstein, A., Goldman, D.B.: PatchMatch: a randomized correspondence algorithm for structural image editing. ACM Trans. Graph. **28**(3), 24:1–24:11 (2009)
6. Bartels, C., de Haan, G.: Smoothness constraints in recursive search motion estimation for picture rate conversion. IEEE TCSVT **20**(10), 1310–1319 (2010)
7. Bleyer, M., Rhemann, C., Rother, C.: PatchMatch stereo - stereo matching with slanted support windows. In: Proceedings of the BMVC, pp. 14.1–14.11 (2011)
8. Bouguet, J.-Y.: Pyramidal implementation of the affine Lucas Kanade feature tracker description of the algorithm. Intel Corp. **5**(1–10), 4 (2001)
9. Brox, T., Bruhn, A., Papenberg, N., Weickert, J.: High accuracy optical flow estimation based on a theory for warping. In: Pajdla, T., Matas, J. (eds.) ECCV 2004. LNCS, vol. 3024, pp. 25–36. Springer, Heidelberg (2004). doi:10.1007/978-3-540-24673-2_3
10. Brox, T., Malik, J.: Large displacement optical flow: descriptor matching in variational motion estimation. IEEE Trans. PAMI **33**(3), 500–513 (2011)
11. Calonder, M., Lepetit, V., Strecha, C., Fua, P.: BRIEF: binary robust independent elementary features. In: Daniilidis, K., Maragos, P., Paragios, N. (eds.) ECCV 2010. LNCS, vol. 6314, pp. 778–792. Springer, Heidelberg (2010). doi:10.1007/978-3-642-15561-1_56
12. Chen, Z., Jin, H., Lin, Z., Cohen, S., Wu, Y.: Large displacement optical flow from nearest neighbor fields. In: Proceedings of the CVPR 2013, June 2013
13. Cho, J.H., Humenberger, M.: Fast PatchMatch stereo matching using cross-scale cost fusion for automotive applications. In: Proceedings of the IEEE IV 2015, pp. 802–807, June 2015
14. Dosovitskiy, A., Fischer, P., Ilg, E., Hausser, P., Hazirbas, C., Golkov, V., van der Smagt, P., Cremers, D., Brox, T.: FlowNet: learning optical flow with convolutional networks. In: Proceedings of the ICCV 2015 (2015)

232   G. Eilertsen et al.

15. Farnebäck, G.: Two-frame motion estimation based on polynomial expansion. In: Bigun, J., Gustavsson, T. (eds.) SCIA 2003. LNCS, vol. 2749, pp. 363–370. Springer, Heidelberg (2003). doi:10.1007/3-540-45103-X_50
16. Hafner, D., Demetz, O., Weickert, J.: Why is the census transform good for robust optic flow computation? In: Kuijper, A., Bredies, K., Pock, T., Bischof, H. (eds.) SSVM 2013. LNCS, vol. 7893, pp. 210–221. Springer, Heidelberg (2013). doi:10.1007/978-3-642-38267-3_18
17. Hirschmuller, H., Scharstein, D.: Evaluation of stereo matching costs on images with radiometric differences. IEEE Trans. PAMI 31(9), 1582–1599 (2009)
18. Humenberger, M., Engelke, T., Kubinger, W.: A census-based stereo vision algorithm using modified semi-global matching and plane fitting to improve matching quality. In: Proceedings of the CVPR 2010 Workshops, pp. 77–84, June 2010
19. Hyun Kim, T., Seok Lee, H., Mu Lee, K.: Optical flow via locally adaptive fusion of complementary data costs. In: Proceedings of the ICCV 2013, December 2013
20. Ilg, E., Mayer, N., Saikia, T., Keuper, M., Dosovitskiy, A., Brox, T.: FlowNet 2.0: evolution of optical flow estimation with deep networks. Technical report, arXiv:1612.01925, December 2016
21. Jith, O.U.N., Ramakanth, S.A., Babu, R.V.: Optical flow estimation using approximate nearest neighbor field fusion. In: Proceedings of the ICASSP 2014, pp. 673–677, May 2014
22. Kroeger, T., Timofte, R., Dai, D., Gool, L.: Fast optical flow using dense inverse search. In: Leibe, B., Matas, J., Sebe, N., Welling, M. (eds.) ECCV 2016. LNCS, vol. 9908, pp. 471–488. Springer, Cham (2016). doi:10.1007/978-3-319-46493-0_29
23. Lu, J., Yang, H., Min, D., Do, M.N.: Patch match filter: efficient edge-aware filtering meets randomized search for fast correspondence field estimation. In: Proceedings of the CVPR 2013, June 2013
24. Lucas, B.D., Kanade, T.: An iterative image registration technique with an application to stereo vision. In: Proceedings of the IJCAI 1981, vol. 81, pp. 674–679 (1981)
25. Müller, T., Rabe, C., Rannacher, J., Franke, U., Mester, R.: Illumination-robust dense optical flow using census signatures. In: Mester, R., Felsberg, M. (eds.) DAGM 2011. LNCS, vol. 6835, pp. 236–245. Springer, Heidelberg (2011). doi:10.1007/978-3-642-23123-0_24
26. Ojala, T., Pietikainen, M., Harwood, D.: Performance evaluation of texture measures with classification based on Kullback discrimination of distributions. In: Proceedings of the ICPR 1994, vol. 1, pp. 582–585, October 1994
27. Rannacher, J.: Realtime 3D motion estimation on graphics hardware. Undergraduate thesis, Heidelberg University (2009)
28. Revaud, J., Weinzaepfel, P., Harchaoui, Z., Schmid, C.: EpicFlow: edge-preserving interpolation of correspondences for optical flow. In: Proceedings of the CVPR 2015, June 2015
29. Rong, G., Tan, T.-S.: Jump flooding in GPU with applications to Voronoi diagram and distance transform. In: Proceedings of the I3D 2006, pp. 109–116 (2006)
30. Rublee, E., Rabaud, V., Konolige, K., Bradski, G.: ORB: an efficient alternative to SIFT or SURF. In: Proceedings of the ICCV 2011, pp. 2564–2571, November 2011
31. Stein, F.: Efficient computation of optical flow using the census transform. In: Rasmussen, C.E., Bülthoff, H.H., Schölkopf, B., Giese, M.A. (eds.) DAGM 2004. LNCS, vol. 3175, pp. 79–86. Springer, Heidelberg (2004). doi:10.1007/978-3-540-28649-3_10

32. Vogel, C., Roth, S., Schindler, K.: An evaluation of data costs for optical flow. In: Weickert, J., Hein, M., Schiele, B. (eds.) GCPR 2013. LNCS, vol. 8142, pp. 343–353. Springer, Heidelberg (2013). doi:10.1007/978-3-642-40602-7_37
33. Weinzaepfel, P., Revaud, J., Harchaoui, Z., Schmid, C.: DeepFlow: large displacement optical flow with deep matching. In: Proceedings of the ICCV 2013, December 2013
34. Xu, L., Jia, J., Matsushita, Y.: Motion detail preserving optical flow estimation. IEEE Trans. PAMI **34**(9), 1744–1757 (2012)
35. Zabih, R., Woodfill, J.: Non-parametric local transforms for computing visual correspondence. In: Eklundh, J.-O. (ed.) ECCV 1994. LNCS, vol. 801, pp. 151–158. Springer, Heidelberg (1994). doi:10.1007/BFb0028345
36. Zhang, K., Li, J., Li, Y., Hu, W., Sun, L., Yang, S.: Binary stereo matching. In: Proceedings of the ICPR 2012, pp. 356–359, November 2012

Robust Data Whitening as an Iteratively Re-weighted Least Squares Problem

Arun Mukundan[(✉)], Giorgos Tolias, and Ondřej Chum

Visual Recognition Group, Czech Technical University in Prague,
Prague, Czech Republic
{arun.mukundan,giorgos.tolias,chum}@cmp.felk.cvut.cz

Abstract. The entries of high-dimensional measurements, such as image or feature descriptors, are often correlated, which leads to a bias in similarity estimation. To remove the correlation, a linear transformation, called whitening, is commonly used. In this work, we analyze robust estimation of the whitening transformation in the presence of outliers. Inspired by the Iteratively Re-weighted Least Squares approach, we iterate between centering and applying a transformation matrix, a process which is shown to converge to a solution that minimizes the sum of ℓ_2 norms. The approach is developed for unsupervised scenarios, but further extend to supervised cases. We demonstrate the robustness of our method to outliers on synthetic 2D data and also show improvements compared to conventional whitening on real data for image retrieval with CNN-based representation. Finally, our robust estimation is not limited to data whitening, but can be used for robust patch rectification, *e.g.* with MSER features.

1 Introduction

In many computer vision tasks, visual elements are represented by vectors in high-dimensional spaces. This is the case for image retrieval [3,14], object recognition [17,23], object detection [9], action recognition [20], semantic segmentation [16] and many more. Visual entities can be whole images or videos, or regions of images corresponding to potential object parts. The high-dimensional vectors are used to train a classifier [19] or to directly perform a similarity search in high-dimensional spaces [14].

Vector representations are often post-processed by mapping to a different representation space, which can be higher or lower dimensional. Such mappings or embeddings can be either non-linear [2,5] or linear [4,6]. In the non-linear case, methods that directly evaluate [2] or efficiently approximate [5] non-linear kernels are known to be perform better. Typical applications range from image classification [5] and retrieval [4] to semantic segmentation [8]. Examples of the linear kind are used for dimensionality reduction in which dimensions carrying the most meaningful information are kept. Dimensionality reduction with Principal

© Springer International Publishing AG 2017
P. Sharma and F.M. Bianchi (Eds.): SCIA 2017, Part I, LNCS 10269, pp. 234–247, 2017.
DOI: 10.1007/978-3-319-59126-1_20

Component Analysis (PCA) is very popular in numerous tasks [4,6,15]. In the same vein as PCA is data whitening, which is the focus of this work[1].

A whitening transformation is a linear transformation that performs correlation removal or suppression by mapping the data to a different space such that the covariance matrix of the data in the transformed space is identity. It is commonly learned in an unsupervised way from a small sample of training vectors. It is shown to be quite effective in retrieval tasks with global image representations, for example, when an image is represented by a vector constructed through the aggregation of local descriptors [13] or by a vector of Convolutional Neural Network (CNN) activations [11,22]. In particular, PCA whitening significantly boosts the performance of CNN compact image vectors, *i.e.* 256 to 512 dimensions, due to handling of inherent co-occurrence phenomena [4]. Principal components found are ordered by decreasing variance, allowing for dimensionality reduction at the same time [12]. Dimensionality reduction may also be performed in a discriminative, supervised fashion. This is the case in the work by Cai *et al.* [6], where the covariance matrices are constructed by using information of pairs of similar and non-similar elements. In this fashion, the injected supervision performs better separation between matching and non-matching vectors and has better chances to avoid outliers in the estimation. It has been shown [10] that an unsupervised approach based on least squares minimization is likely to be affected by outliers: even a single outlier of high magnitute can significantly deviate the solution.

In this work, we propose an unsupervised way to learn the whitening transformation such that the estimation is robust to outliers. Inspired by the Iteratively Re-weighted Least Squares of Aftab and Hartley [1], we employ robust M-estimators. We perform minimization of robust cost functions such as ℓ_1 or Cauchy. Our approach iteratively alternates between two minimizations, one to perform the centering of the data and one to perform the whitening. In each step a weighted least squares problem is solved and is shown to minimize the sum of the ℓ_2 norms of the training vectors. We demonstrate the effectiveness of this approach on synthetic 2D data and on real data of CNN-based representation for image search. The method is additionally extended to handle supervised cases, as in the work of Cai *et al.* [6], where we show further improvements. Finally, our methodology is not limited to data whitening. We provide a discussion on applying it for robust patch rectification of MSER features [18].

The rest of the paper is organized as follows: In Sect. 2 we briefly review conventional data whitening and give our motivation, while in Sect. 3 we describe the proposed iterative whitening approach. Finally, in Sects. 4 and 5 we compare our method to the conventional approach on synthetic and real data, respectively.

[1] The authors were supported by the MSMT LL1303 ERC-CZ grant, Arun Mukundan was supported by the SGS17/185/OHK3/3T/13 grant.

2   Data Whitening

In this section, we first briefly review the background of data whitening and then give a geometric interpretation, which forms our motivation for the proposed approach.

2.1   Background on Whitening

A whitening transformation is a linear transformation that transforms a vector of random variables with a known covariance matrix into a set of new variables whose covariance is the identity matrix. The transformation is called "whitening" because it changes the input vector into a white noise vector.

We consider the case where this transformation is applied on a set of zero centered vectors $\mathcal{X} = \{\mathbf{x}_1, \ldots, \mathbf{x}_i, \ldots, \mathbf{x}_N\}$, with $\mathbf{x}_i \in \mathbb{R}^d$, where $\Sigma = \sum_i \mathbf{x}_i \mathbf{x}_i^\top$. The whitening transformation P is given by

$$P^\top P = \Sigma^{-1}. \tag{1}$$

In Fig. 1 we show a toy example of 2D points and their whitened counterpart.

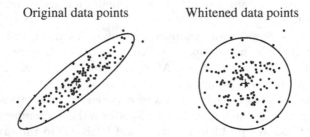

Original data points   Whitened data points

Fig. 1. Left: Points in 2D and their covariance shown with an ellipse. Right: The corresponding whitened 2D point set.

Assumption. In the following text, we assume that the points of $\mathcal{X}$ do *not* lie in a linear subspace of dimensionality $d' < d$. If this is the case, a solution is to first identify the d'-dimensional subspace and perform the proposed algorithms on this subspace. The direct consequence of the assumption is that the sample covariance matrix Σ is full rank, in particular $\det(\Sigma) > 0$.

It is clear from (1) that the whitening transformation is given up to an arbitrary rotation $R \in \mathbb{R}^{d \times d}$, with $R^\top R = I$. The transformation matrix P of the whitening is thus given by

$$P = R\Sigma^{-1/2}. \tag{2}$$

2.2   Geometric Interpretation

We provide a geometric interpretation of data whitening, which also serves as our motivation for the proposed method in this work.

Observation. Assuming zero-mean points, the whitening transform P in (2) minimizes the sum of squared ℓ_2 norms among all linear transforms T with $\det(T) = \det(\Sigma)^{-1/2}$.
Proof.

$$
\begin{aligned}
C_{\ell_2}(P) &= \sum_i \|P\mathbf{x}_i\|^2 \\
&= \sum_i tr\left(\mathbf{x}_i^\top P^\top P\mathbf{x}_i\right) \\
&= \sum_i tr\left((\mathbf{x}_i\mathbf{x}_i^\top) P^\top P\right) \\
&= tr\left(\left(\sum_i \mathbf{x}_i\mathbf{x}_i^\top\right) P^\top P\right) \\
&= tr\left(\Sigma P^\top P\right) \\
&= \sum_{j=1}^{d} \lambda_j,
\end{aligned}
\tag{3}
$$

where λ_i are the eigenvalues of $\Sigma P^\top P$ and $\|\cdot\|$ is denoting ℓ_2 norm. Upon imposing the condition $\det(T) = \det(\Sigma)^{-1/2}$, we get that $\det(\Sigma P^\top P) = \prod_{j=1}^{d} \lambda_j$ is constant with respect to P. It follows from the arithmetic and geometric mean inequality, that the sum in (3) is minimized when $\lambda_i = \lambda_j, \forall i = j$. Equality of all eigenvalues allows us to show that

$$
\begin{aligned}
\Sigma P^\top P &= I \\
P^\top P &= \Sigma^{-1} \\
P &= R\Sigma^{-1/2}
\end{aligned}
\tag{4}
$$

which is exactly the solution in (2) that also minimizes (3). The need for the existence of Σ^{-1} justifies the stated full rank assumption.

We have just shown that learning a whitening transformation reduces to a least squares problem.

3   Robust Whitening

In this section we initially review the necessary background on the the iteratively re-weighted least squares (IRLS) method recently proposed by Aftab and Hartley [1], which is the starting point for our method. Then, we present the robust whitening and centering procedures, which are posed as weighted least squares problems and performed iteratively. Finally, the extension to the supervised case is described.

3.1   Background on IRLS

In the context of distance minimization the IRLS method minimizes the cost function

$$C_h(\theta) = \sum_{i=1}^{N} h \circ f(\theta, \mathbf{x}_i), \tag{5}$$

where f is a distance function that is defined on some domain, h is a function that makes the cost less sensitive to outliers, and $\mathbf{x}_i \in \mathcal{X}$. Some examples of robust h functions are ℓ_1, Huber, pseudo-Huber, *etc.* as described in [1]. For instance, assume the case of the geometric median of the points in $\mathcal{X}$. Setting $f(\boldsymbol{\mu}, \mathbf{x}_i) = ||\boldsymbol{\mu} - \mathbf{x}_i||$ and $h(z) = z$, we get the cost (5) as the sum of ℓ_2 norms. The minimum of this cost is attained when $\boldsymbol{\mu}$ is equal to the geometric median.

It is shown [1] that a solution for $\operatorname{argmin}_\theta C_h(\theta)$ may be found by solving a sequence of weighted least squares problems. Given some initial estimate θ^0, the parameters θ are iteratively estimated

$$\theta^{t+1} = \operatorname*{argmin}_{\theta} \sum_{i=1}^{N} w(\theta^t, \mathbf{x}_i) f(\theta, \mathbf{x}_i)^2, \tag{6}$$

where for brevity $w(\theta^t, \mathbf{x}_i)$ is denoted w_i^t in the following. Provided $h(\sqrt{z})$ is differentiable at all points and concave, for certain values of w_i^t and conditions on f this solution minimizes $C_h(\theta)$. In some cases, it may even be possible to find a simple and anlytic solution.

Given that the iterative procedure indeed converges to a minimum cost of (5), we get the following condition on the weights:

$$\begin{aligned} \nabla_\theta (h \circ f(\theta, \mathbf{x}_i)) &= 0, \\ \nabla_\theta (w_i^t f(\theta, \mathbf{x}_i)^2) &= 0. \end{aligned} \tag{7}$$

This results in the following weights

$$w_i^t = \frac{h'(f(\theta^t, \mathbf{x}_i))}{2f(\theta^t, \mathbf{x}_i)}. \tag{8}$$

Geometric median. The geometric median $\boldsymbol{\mu}$ of a set of points $\{\mathbf{x}_i\}$ is the point that minimizes the sum of ℓ_2 distances to the points. As shown in one of the cases in the work by Aftab and Hartley [1], the problem of finding the geometric median can be cast in an IRLS setting for certain value of weights. Setting $f(\boldsymbol{\mu}, \mathbf{x}_i) = ||\boldsymbol{\mu} - \mathbf{x}_i||$ and $h(z) = z$, the IRLS algorithm minimizes the sum of distances at each iteration, thus converging to the geometric median.

3.2   Method

From the observation in Sect. 2.2, we know that there is a closed-form solution to the problem of finding a linear transformation P so that $\sum_i ||P\mathbf{x}_i||^2$ is minimized subject to a fixed determinant $\det(P)$. The idea of the robust whitening is to use this least squares minimizer in a framework similar to the iterative re-weighted least squares to minimize a robust cost.

Robust transformation estimation. In contrast to the conventional whitening and the minimization of (3), we now propose the estimation of a whitening transform (transformation matrix P) in a way that is robust to outliers. We assume zero mean points and seek the whitening transformation that minimizes the robust cost function of (5). We set $f(P, \mathbf{x}_i) = ||P\mathbf{x}_i||$ and use the ℓ_1 cost function $h(z) = z$. Other robust cost functions can be used, too[2].

We seek to minimize the sum of ℓ_2 norms in the whitened space

$$C_{\ell_1}(P) = \sum_{i=1}^{N} f(P, \mathbf{x}_i) = \sum_{i=1}^{N} ||P\mathbf{x}_i||. \tag{9}$$

The corresponding iteratively re-weighted least squares solution is given by

$$P^{t+1} = \underset{P}{\text{argmin}} \sum_{i=1}^{N} w_i^t ||P\mathbf{y}_i^t||^2, \tag{10}$$

where $\mathbf{y}_i^t = P^t \mathbf{y}_i^{t-1}$ and $\mathbf{y}_i^0 = \mathbf{x}_i$. This means that each time transformation P^t is estimated and applied to whiten the data points. In the following iteration, the estimation is performed on data points in the whitened space. The effective transformation at iteration t with respect to the initial points $\mathbf{x}_i$ is given by

$$\hat{P}^t = \prod_{i=1}^{t} P^i. \tag{11}$$

Along the lines of proof (3) we find a closed form solution that minimizes (9) as

$$\sum_i w_i^t ||P\mathbf{y}_i^t||^2$$

$$= tr\left(\left(\sum_i w_i^t \mathbf{y}_i^t \mathbf{y}_i^{t^\top}\right) P^\top P\right) \tag{12}$$

$$= tr\left(\tilde{\Sigma} P^\top P\right)$$

where $\tilde{\Sigma} = \sum_i w_i^t \mathbf{y}_i^t \mathbf{y}_i^{t^\top}$ is a *weighted covariance*. Therefore, P is given, up to a rotation, as

$$P = R\tilde{\Sigma}^{-1/2}. \tag{13}$$

---

[2] We also use Cauchy cost in our experiments. It is defined as $h(z) = b^2 log(1 + z^2/b^2)$.

Joint centering and transformation matrix estimation. In this section we describe the proposed approach for data whitening. We propose to jointly estimate a robust mean $\boldsymbol{\mu}$ and a robust transformation matrix P by alternating between the two previously described procedures: estimating the geometric median and estimating the robust transformation. In other words, in each iteration, we first find $\boldsymbol{\mu}$ keeping P fixed and then find P keeping $\boldsymbol{\mu}$ fixed. In this way the assumption for centered points when finding P is satisfied. Given that each iteration of the method outlined above reduces the cost, and that the cost must be non-negative, we are assured convergence to a local minimum.

We propose to minimize cost

$$C_{\ell_1}(P, \boldsymbol{\mu}) = \sum_{i=1}^{N} ||P(\mathbf{x}_i - \boldsymbol{\mu})||. \tag{14}$$

In order to reformulate this as an IRLS problem, we use $h(z) = z$, and $f(P, \boldsymbol{\mu}, \mathbf{x}_i) = ||P(\mathbf{x}_i - \boldsymbol{\mu})||$. Now, at iteration t the minimization is performed on points $\mathbf{y}_i^t = \hat{P}^t(\mathbf{x}_i - \hat{\boldsymbol{\mu}}^t)$ and the conditions for convergence with respect to $\boldsymbol{\mu}$ (skipping t and notation for effective parameters for brevity) are

$$\nabla_{\boldsymbol{\mu}}(h \circ f) = \nabla_{\boldsymbol{\mu}} ||P(\mathbf{x}_i - \boldsymbol{\mu})||$$
$$= \nabla_{\boldsymbol{\mu}} \sqrt{(\mathbf{y}_i - \boldsymbol{\mu})^\top P^\top P(\mathbf{y}_i - \boldsymbol{\mu})}$$
$$= \frac{1}{2||P(\mathbf{y}_i - \boldsymbol{\mu})||} \cdot \nabla_{\boldsymbol{\mu}} M \tag{15}$$

$$\nabla_{\boldsymbol{\mu}}(w_i \cdot f^2) = w_i \cdot \nabla_{\boldsymbol{\mu}} M$$

where we have $M = (\mathbf{y}_i - \boldsymbol{\mu})^\top P^\top P(\mathbf{y}_i - \boldsymbol{\mu})$. This gives the expression for the weight

$$w_i^t = \frac{1}{2||\hat{P}^t(\mathbf{x}_i - \hat{\boldsymbol{\mu}}^t)||}. \tag{16}$$

A similar derivation gives us the weights for the iteration step of P. Therefore in each iteration, we find the solutions to the following weighted least squares problems,

$$\boldsymbol{\mu}^{t+1} = \underset{\boldsymbol{\mu}}{\operatorname{argmin}} \sum_{i=1}^{N} w_i(P^t, \boldsymbol{\mu}^t) ||P^t(\mathbf{y}_i - \boldsymbol{\mu})||^2, \tag{17}$$

$$P^{t+1} = \underset{P}{\operatorname{argmin}} \sum_{i=1}^{N} w_i(P^t, \boldsymbol{\mu}^{t+1}) ||P(\mathbf{y}_i^t - \boldsymbol{\mu}^{t+1})||^2. \tag{18}$$

The effective centering and transformation matrix at iteration t are given by

$$\hat{\boldsymbol{\mu}}^t = \sum_{i=1}^{t} \left(\prod_{j=1}^{i-1} P_j^{-1} \right) \boldsymbol{\mu}^i \quad , \quad \hat{P}^t = \prod_{i=1}^{t} P^i. \tag{19}$$

Algorithm 1. Robust Whitening

1: **procedure** ROBUST WHITENING($\mathcal{X}$)
2:    $\mathbf{z}_0 \leftarrow \mathcal{X}$
3:    $\boldsymbol{\mu}_0 \leftarrow$ Initialize centre to mean of $\mathbf{z}_0$
4:    $P_0 \leftarrow$ Initialize transform to identity matrix
5:    **for** t$\leq$niter **do**
6:        $\boldsymbol{\mu}^t \leftarrow \frac{1}{N} \sum_{i=1}^{N} w_i(P^{t-1}, \boldsymbol{\mu}^{t-1})\mathbf{z}_i^{t-1}$
7:        $\tilde{\Sigma}^t \leftarrow \sum_{i=1}^{N} w_i(P^{t-1}, \boldsymbol{\mu}^t)(\mathbf{z}_i^{t-1} - \boldsymbol{\mu}^t)(\mathbf{z}_i^{t-1} - \boldsymbol{\mu}^t)^{\top}$
8:        $P^t \leftarrow \frac{\text{chol}(\tilde{\Sigma}^t)}{\det\left(\text{chol}(\tilde{\Sigma}^t)\right)^{1/d}}$
9:        $\mathbf{z}^t \leftarrow P^t \left(\mathbf{z}^{t-1} - \boldsymbol{\mu}^t\right)$
10:        $\hat{\boldsymbol{\mu}}^t \leftarrow \sum_{i=1}^{t} \left(\prod_{j=1}^{i-1} P^{j^{-1}}\right) \boldsymbol{\mu}^i$
11:        $\hat{P}^t \leftarrow \prod_{i=0}^{t} P^i$
12:    **end for**
13:    **return** $\hat{\boldsymbol{\mu}}^t, \hat{P}^t$
14: **end procedure**

The whole procedure is summarized in Algorithm 1, where chol is used to denote the Cholesky decomposition.

3.3   Extension with Supervision

We firstly review the work of Cai *et al.* [6] who perform supervised descriptor whitening and then present our extension for robust supervised whitening.

Background on linear discriminant projections [6]. The linear discriminant projections (LDP) are learned via supervision of pairs of similar and dissimilar descriptors. A pair (i, j) is similar if $(i, j) \in \mathcal{S}$ while dissimilar if $(i, j) \in \mathcal{D}$. The projections are learned in two parts. Firstly, the whitening part is obtained as the square-root of the intra-class covariance matrix $C_{\mathcal{S}}^{-1/2}$, where

$$C_{\mathcal{S}} = \sum_{(i,j \in \mathcal{S})} (x_i - x_j)(x_i - x_j)^{\top}. \tag{20}$$

Then, the rotation part is given by the PCA of the inter-class covariance matrix which is computed in the space of the whitened descriptors. It is computed as $\text{eig}\left(C_{\mathcal{S}}^{-1/2} C_{\mathcal{D}} C_{\mathcal{S}}^{-1/2}\right)$, where

$$C_{\mathcal{D}} = \sum_{(i,j \in \mathcal{D})} (x_i - x_j)(x_i - x_j)^{\top}. \tag{21}$$

Algorithm 2. Supervised Robust Whitening

1: **procedure** SUPERVISED ROBUST WHITENING($\mathcal{X}, \mathcal{S}$)
2:      $\mathcal{X}_\mathcal{S} = \{d : d = x_i - x_j, x_i \in \mathcal{X}, x_j \in \mathcal{X}, (i,j) \in \mathcal{S}\}$
3:      $\mathcal{X}_\mathcal{S} = \{\mathcal{X}_\mathcal{S} \cup -\mathcal{X}_\mathcal{S}\}$
4:      $\boldsymbol{\mu}_1, P_1 \leftarrow$ Robust Whitening($\mathcal{X}_\mathcal{S}$)
5:      $\boldsymbol{\mu} \leftarrow$ Geometric Median($\mathcal{X}$)
6:      $\bar{\mathcal{X}} \leftarrow \mathcal{X} - \boldsymbol{\mu}$
7:      $\boldsymbol{\mu}_2, P_2 \leftarrow$ Robust Whitening($P_1 \bar{\mathcal{X}}$)
8:      $R_2 \leftarrow$ eig($(P_2^\top P_2)^{-1}$)
9:      $\hat{\boldsymbol{\mu}} \leftarrow \boldsymbol{\mu} + \boldsymbol{\mu}_2$
10:     $\hat{P} \leftarrow P_1 R_2$
11:     **return** $\hat{P}, \hat{\boldsymbol{\mu}}$
12: **end procedure**

The final whitening is performed by $P_{\mathcal{SD}}^\top(x - m)$, where m is the mean descriptor and $P_{\mathcal{SD}} = C_\mathcal{S}^{-1/2} \cdot \text{eig}\left(C_\mathcal{S}^{-1/2} C_\mathcal{D} C_\mathcal{S}^{-1/2}\right)$. It is noted [6] that, if the number of descriptors is large compared to the number of classes (two in this case), then $C_\mathcal{D} \approx C_{\mathcal{S} \cup \mathcal{D}}$ since $|\mathcal{S}| \ll |\mathcal{D}|$. This is the approach we follow.

Robust linear discriminant projections. The proposed method uses the provided supervision in a robust manner by employing the method introduced in Sect. 3.2. The whitening is estimated in a robust manner by Algorithm 1 on the intra-class covariance. In this manner, small weights are assigned to pairs of descriptors that are found to be outliers. Then, the mean and covariance are estimated in a robust manner in the whitened space. The whole procedure is summarized in Algorithm 2. Mean μ_1 is zero due to the including the pairs in a symmetric manner.

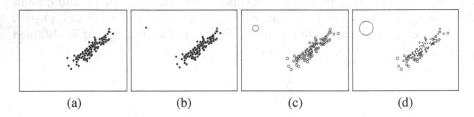

| (a) | (b) | (c) | (d) |

Fig. 2. (a) Set of 2D points drawn from a Gaussian distribution with zero mean. (b) Same set as (a) with an additional point (outlier) placed at a distance equal to 2 times the maximum distance from the center of the initial set. (c) Visualization of the weights assigned in the set of (b) with the robust whitening which uses the ℓ_1 cost function. Note that the size of the circles is inversely proportional to the weight. (d) Same as (c), but using the Cauchy cost.

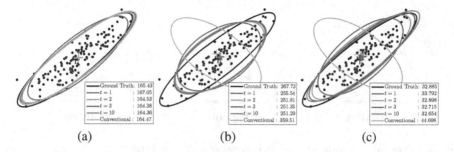

(a)   (b)   (c)

Fig. 3. Visualization of the covariance (ellipse) and center (cross) of the estimated whitening transformation at iteration t and the conventional estimate. The example is performed using the set of 2D points of Fig. 2. The ground truth distribution that created the data points is shown in black. The conventional estimate is shown in cyan. We show the effective estimate of the tth iteration. The two approaches are compared without an outlier in (a) or with an outlier using ℓ_1 in (b) or Cauchy cost function in (c). The outlier is placed at a distance equal to 10 times the maximum inlier distance. The outlier is not plotted to keep the scale of the figure reasonable. The ℓ_1 (or Cauchy) cost is shown in the legend. (Color figure online)

4   Examples on Synthetic Data

We compare the proposed and the conventional whitening approaches on synthetic 2D data in order to demonstrate the robustness of our method to outliers. We sample a set of 2D points from a normal distribution, which is shown in Fig. 2(a) and then add an outlier and show the result in Fig. 2(b). In the absence of outliers, both methods provide a similar estimation as shown in Fig. 3. It is also shown how the iterative approach reduces the cost at each iteration. With the presence of an outlier, the estimation of the conventional approach is largely affected, while the robust method gives a much better estimation, as shown in Fig. 3. Using the Cauchy cost function the estimated covariance is very close to

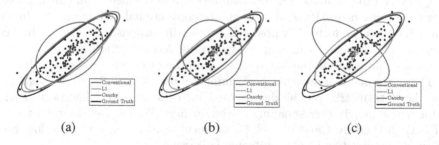

(a)   (b)   (c)

Fig. 4. Visualization of the covariance (ellipse) and center (cross) of the estimated whitening transformation using the conventional approach and ours. The example is performed using the set of 2D points of Fig. 2. The two approaches are compared for the case of an outlier placed at distance equal to 3 (a), 5 (b) and 10 (c) times the maximum inlier distance. The outlier is not shown to keep the resolution high.

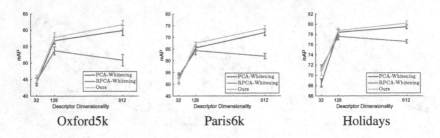

Oxford5k   Paris6k   Holidays

Fig. 5. Retrieval performance comparison using mAP on 3 common benchmarks. Comparison of the conventional PCA whitening, RPCA whitening and our approach for descriptors of varying dimensionality. The training set contains a small subset of 512 vectors randomly selected. The experiment is performed 10 times and mean performance is reported while standard deviation is shown on the curves. Descriptors extracted using VGG.

that of the ground truth. The weights assigned to each point with the robust approach are visualized in Fig. 2 and show how the outlier is discarded in the final estimation. Finally, in Fig. 4, we compare the conventional way with our approach for outlier of increasing distance.

5   Experiments

In this section, the robust whitening is applied to real-application data. In particular, we test on SPOC [4] descriptors, which are CNN-based image descriptors constructed via sum pooling of network activations in the internal convolutional layers. We evaluate on 3 popular retrieval benchmarks, namely Oxford5k, Paris6k and Holidays (the upright version), and use around 25 k training images to learn the whitening. We use VGG network [21] to extract the descriptors and, in contrast to the work of Babenko and Lempitsky [4], we do not ℓ_2-normalize the input vectors. The final ranking is obtained using Euclidean distance between the query and the database vectors. Evaluation is performed by measuring mean Average Precision (mAP). As in the case of conventional whitening, the dimension reduction is performed by preserving those dimensions that have the highest variance. This is done by finding an eigenvalue decomposition of the estimated covariance and ordering the eigenvectors according to decreasing eigenvalue.

There are many approaches performing robust PCA [7,24,25] by assuming that the data matrix can be decomposed into the sum of a low rank matrix and a sparse matrix corresponding to the outliers. We employ the robust PCA (RPCA) method by Candès et al. [7] to perform a comparison. The low rank matrix is recovered and PCA whitening is learned on this.

We present results in Table 1, where the robust approach offers a consistent improvement over the conventional PCA whitening [4]. Especially in the case where the whitening is learned on few training vectors, the improvement is larger as outliers will heavily influence the conventional whitening, as shown in Fig. 5.

Table 1. Retrieval performance comparison using mAP on 3 common benchmarks. Comparison of retrieval using the initial sum-pooled CNN activations, post-processing using the baselines and our methods for unsupervised and supervised whitening. Results for descriptors of varying dimensionality. The full training set is used. Descriptors extracted using VGG. S: indicates the use of supervision.

Dataset		Oxford5k			Paris6k			Holidays		
Method	S	32D	128D	512D	32D	128D	512D	32D	128D	512D
Raw		–	–	51.4	–	–	61.6	–	–	78.8
PCA whitening		44.7	56.6	66.7	53.4	67.0	77.1	69.6	78.4	80.6
RPCA whitening		44.0	52.4	55.6	**55.9**	61.1	65.1	70.5	75.8	77.4
Ours		**45.8**	**58.5**	**67.7**	50.0	**68.3**	78.4	70.7	78.8	81.8
LDP	×	39.4	59.9	68.8	56.1	70.2	76.6	67.5	77.7	80.8
Ours	×	**49.9**	**62.3**	**70.3**	**57.6**	**72.0**	**78.0**	69.0	78.6	**82.1**

Our approach is also better than RPCA whitening for large dimensionalities. It seems that RPCA underestimates the rank of the matrix and does not offer any further improvements for large dimensions.

6   Discussion

The applicability of the proposed method goes beyond robust whitening. Consider, for example, the task of affine-invariant descriptors of local features, such as MSERs [18]. A common approach is to transform the detected feature into a canonical frame prior to computing a robust descriptor based on the gradient map of the normalized patch (SIFT [17]). To remove the effect of an affine transformation, a centre of gravity and centered second-order moment (covariance matrix) are used. It can be shown that both the centre of gravity and the covariance matrix are affine-covariants, *i.e.* if the input point set is transformed by an affine transformation A, they transform with the same transformation A.

The proposed method searches μ and P by minimization over all possible affine transformations with a fixed determinant. In turn, μ is fully affine covariant and P is affine covariant up to an unknown scale (and rotation, $P^\top P$ cancels the rotation). To the best of our knowledge, this type of robust-to-outliers covariants have not been used.

7   Conclusions

We cast the problem of data whitening as minimization of robust cost functions. In this fashion we iteratively estimate a whitening transformation that is robust to the presence of outliers. With the use of synthetic data, we show that our estimation is almost unaffected even with extreme cases of outliers, while it also offers improvements when whitening CNN descriptors for image retrieval.

References

1. Aftab, K., Hartley, R.: Convergence of iteratively re-weighted least squares to robust M-estimators. In: IEEE Winter Conference on Applications of Computer Vision (2015)
2. Arandjelović, R., Zisserman, A.: Three things everyone should know to improve object retrieval. In: CVPR (2012)
3. Arandjelovic, R., Zisserman, A.: All about VLAD. In: CVPR (2013)
4. Babenko, A., Lempitsky, V.: Aggregating local deep features for image retrieval. In: ICCV (2015)
5. Bo, L., Sminchisescu, C.: Efficient match kernel between sets of features for visual recognition. In: NIPS (2009)
6. Cai, H., Mikolajczyk, K., Matas, J.: Learning linear discriminant projections for dimensionality reduction of image descriptors. IEEE Trans. PAMI **33**(2), 338–352 (2011)
7. Candès, E.J., Li, X., Ma, Y., Wright, J.: Robust principal component analysis? J. ACM **58**(3), 11 (2011)
8. Carreira, J., Caseiro, R., Batista, J., Sminchisescu, C.: Semantic segmentation with second-order pooling. In: Fitzgibbon, A., Lazebnik, S., Perona, P., Sato, Y., Schmid, C. (eds.) ECCV 2012. LNCS, vol. 7578, pp. 430–443. Springer, Heidelberg (2012). doi:10.1007/978-3-642-33786-4_32
9. Dalal, N., Triggs, B.: Histograms of oriented gradients for human detection. In: CVPR (2005)
10. De la Torre, F., Black, M.J.: Robust principal component analysis for computer vision. In: ICCV (2001)
11. Gordo, A., Almazan, J., Revaud, J., Larlus, D.: Deep image retrieval: learning global representations for image search. In: arXiv (2016)
12. Huber, P.J.: Projection pursuit. In: The annals of Statistics (1985)
13. Jégou, H., Chum, O.: Negative evidences and co-occurences in image retrieval: the benefit of PCA and Whitening. In: Fitzgibbon, A., Lazebnik, S., Perona, P., Sato, Y., Schmid, C. (eds.) ECCV 2012. LNCS, pp. 774–787. Springer, Heidelberg (2012). doi:10.1007/978-3-642-33709-3_55
14. Jegou, H., Perronnin, F., Douze, M., Sánchez, J., Perez, P., Schmid, C.: Aggregating local image descriptors into compact codes. IEEE Trans. PAMI **34**(9), 1704–1716 (2012)
15. Ke, Y., Sukthankar, R.: PCA-SIFT: a more distinctive representation for local image descriptors. In: CVPR (2004)
16. Lim, J.J., Zitnick, C.L., Dollár, P.: Sketch tokens: a learned mid-level representation for contour and object detection. In: CVPR (2013)
17. Lowe, D.G.: Object recognition from local scale-invariant features. In: ICCV (1999)
18. Matas, J., Chum, O., Urban, M., Pajdla, T.: Robust wide-baseline stereo from maximally stable extremal regions. Image Vis. Comput. **22**(10), 761–767 (2004)
19. Perronnin, F., Dance, C.: Fisher kernels on visual vocabularies for image categorization. In: CVPR (2007)
20. Poppe, R.: A survey on vision-based human action recognition. In: Image and Vision Computing (2010)
21. Simonyan, K., Zisserman, A.: Very deep convolutional networks for large-scale image recognition. In: arXiv (2014)

22. Tolias, G., Sicre, R., Jégou, H.: Particular object retrieval with integral max-pooling of CNN activations. In: arXiv (2015)
23. Turk, M.A., Pentland, A.P.: Face recognition using eigenfaces. In: CVPR (1991)
24. Wright, J., Ganesh, A., Rao, S., Peng, Y., Ma, Y.: Robust principal component analysis: exact recovery of corrupted low-rank matrices via convex optimization. In: NIPS (2009)
25. Xu, H., Caramanis, C., Sanghavi, S.: Robust PCA via outlier pursuit. In: NIPS (2010)

DEBC Detection with Deep Learning

Ian E. Nordeng$^{(\boxtimes)}$, Ahmad Hasan, Doug Olsen,
and Jeremiah Neubert

University of North Dakota, Grand Forks, ND, USA
{ian.nordeng, ahmadjarjis.hasan}@und.edu,
olsen@aero.und.edu, jeremiah.neubert@engr.und.edu

Abstract. This work presents a novel system utilizing state of the art deep convolutional neural networks to detect dead end body component's (DEBC's) to reduce costs for inspections and maintenance of high tension power lines. A series of data augmenting techniques were implemented to develop 2,437 training images which utilized 146 images from a sensor trade study, and a test flight using UAS for inspections. Training was completed using the Python implementation of Faster R-CNN's object detection network with the VGG16 model. After testing the network on 111 aerial inspection photos captured with an UAS, the resulting convolutional neural network (CNN) was capable of an accuracy of 83.7% and precision of 91.8%. The addition of 270 training images and inclusion of insulators increased detection accuracy and precision to 97.8% and 99.1% respectively.

Keywords: Convolutional neural networks · CNN · Inspections · Machine learning

1 Introduction

Infrastructure maintenance in the United States is a multi-trillion-dollar industry, of which the electrical grid makes up a significant portion [1]. Maintenance of the electrical grid poses substantial costs. To help reduce these costs, we developed a deep learning neural network capable of detecting the dead-end body component (DEBC) from high tension power lines to allow for further analysis of the part due to wear.

The DEBC is a full tension device that is used to attach the conductor to the power line structure while maintaining electrical current. This component is comprised of an outer aluminum sleeve with a four-bolt pad welded to one end, a steel forging with a steel eye, and an aluminum insert. The outer aluminum sleeve grips the aluminum strands of the power line, while the inner aluminum inserts grip the inner aluminum matrix core wires separately. The eye of the steel forging is connected to the insulator string on the dead-end tower or substation. Jumper connectors attach to the outer sleeve pad and are used to connect pairs of powerline conductors. To aid inspection and maintenance of the DEBC, the component was detected with a bounding box annotation allowing for segmentation and analysis of possible wear or failures of the DEBC.

Deep convolutional neural networks (CNN) were chosen for the task of detecting the DEBC as they have made a resurgence in visual recognition tasks in recent years,

© Springer International Publishing AG 2017
P. Sharma and F.M. Bianchi (Eds.): SCIA 2017, Part I, LNCS 10269, pp. 248–259, 2017.
DOI: 10.1007/978-3-319-59126-1_21

overtaking other methods in image classification challenges [2]. This is further exemplified by the Pascal VOC challenge, a yearly challenge from 2005–2012, with the goal of recognizing objects from several visual object classes through a supervised learning process [4]. The challenge has commonly been used as a comparison between different object detection networks. Convolutional neural networks have consistently outperformed other methods and increased precision of detection in the Pascal Visual Object Classes (VOC) challenge [3].

Faster R-CNN was chosen for the purposes of aiding inspections as it was the state of the art in object detection available, as evidenced by the Pascal VOC 2007 challenge. The pretrained deep VGG16 model was implemented due to its high precision and public availability, as shown in [5]. All training in this work was completed using a Titan X GPU.

1.1 Background

To determine the most effective object detection CNN available, four different object detection CNN's were considered including R-CNN, SPPnet, YOLO, and Faster R-CNN. Due to the post processing focus of the proposed algorithm, accuracy and precision were the main considerations for each network. Table 1 lists a direct comparison of each network on the Pascal VOC 2007 dataset as represented by the mean average precision (mAP) attained on the challenge by each network.

Table 1. mAP of each detection network considered.

CNN considered	mAP in Pascal VOC 2007
R-CNN	58.5%
SPPnet	60.9%
YOLO	63.4%
Faster R-CNN	73.2%

R-CNN was the first successful object detection algorithm utilizing a CNN, and increased mAP of the previous state of the art method by over 30% [6]. This was done by utilizing region proposals. Region proposals are specific regions in the image determined as an object, provided by a separate algorithm, and performing a convolutional network forward pass for each proposed region. Although originally successful, there are many drawbacks to the region-based convolutional neural network. First, training is a multi-staged pipeline requiring finetuning the CNN on object proposals generated separately, then fitting a support vector machine (SVM) to the CNN features, and performing bounding box regression. Second, training is expensive in both hard drive space and time spent during training. The SVM and bounding box regression training requires storing features extracted from each object proposal to disc which may require storing hundreds of gigabytes of data. Third, detection is slow as features are extracted from each object proposal in each test image. Methods considered below improve both speed and precision over this implementation.

Spatial Pyramid Pooling (SPP-net) was proposed to speed up R-CNN by sharing computation [7]. SPP-net first computes a convolutional feature map for the entire input image, then classifies each object proposal using a feature vector from the shared feature map. Features are extracted through maxpooling the portion of the feature map inside the proposal into a fixed output size. Training is still multi staged, as it must extract feature vectors, fine tune the network with log loss, train a SVM, and finally fit bounding box regressors. SPPnet cannot update the convolutional layers preceding the spatial pyramid pooling limiting accuracy of deep networks.

You only look once (YOLO), is a real-time object detection CNN [8]. This method applies a single neural network to the full image at test time to provide global context, divides the image into equally spaced regions, and predicts bounding boxes and probabilities for each region. This method provides a fast object detection that runs in real time, up to 45 FPS, for the more computationally expensive and accurate model. As stated in [8], this network provides a mAP of approximately 10% less than Faster R-CNN.

Faster R-CNN is the latest improvement over R-CNN and introduced Region Proposal Networks (RPN) that share convolutional layers with the object detection network [5]. The region proposals are created by adding two additional convolutional layers. The first layer encodes each convolutional feature map position into a feature vector. The second layer outputs an objectness score and regressed bounds for k region proposals, relative to various scales and aspect ratios, at each convolutional map position. These added layers create a fully-convolutional network that can be trained end-to-end for the task of generating detection proposals. Faster R-CNN also developed a training scheme that alternates between fine-tuning for the region proposal task, and fine-tuning for object detection with the proposals fixed. Faster R-CNN was chosen for our purposes as it achieved the highest mAP of all methods considered at 73.2%.

2 Procedure

To properly train Faster R-CNN for detection of DEBC's, several tasks were completed with the following processes. The Python version of Faster R-CNN was obtained via [5] utilizing the VGG16 model as explained in Sect. 2.1. Section 2.2 expands on data collection of DEBC's. Section 2.3 provides methods for data augmentation to enhance the training dataset. Section 2.4 details the creation of the ground truth annotations for the training data. Section 2.5 provides the methods for collection of test data. Section 2.6 explains the process for training the different networks considered.

2.1 Faster R-CNN

The Python reimplementation of Faster R-CNN was obtained from [5]. As previously mentioned, Faster R-CNN incorporates a small region proposal network that shares a common set of convolutional layers with a standard detection network and was built using the Caffe framework [9]. This region proposal network takes an image of any size as input, and outputs rectangular object proposals with objectness scores, or in other words, a measurement of belonging to a set of object classes *vs.* the background. Training the region proposal network layers is accomplished through assigning a binary object or not object to anchors that are predicted from k region proposals. These anchors are assigned as positive while training when they either have the highest Intersection-over-Union (IoU) overlap with a ground truth box or IoU overlap greater than 0.7 with any ground truth box. Non-positive anchors did not contribute to training. The loss function for an image,

$$L(\{p_i\}, \{t_i\}) = \tfrac{1}{N_{cls}} \sum_i L_{cls}(p_i, p_i^*) + \lambda \tfrac{1}{N_{reg}} \sum_i p_i^* L_{reg}(t_i, t_i^*),$$ (1)

is minimized to generate a trained network. In the loss function i is the anchor index in a mini batch, p_i the predicted probability of the i'th anchor being an object. p_i^* represents the ground-truth label and is set to 1 for a positive anchor and 0 if negative. t_i provides the vector representation of the four parameterized coordinates of the predicted bounding box with t_i^* the ground-truth box associated with the positive anchor. The classification loss L_{cls} represents the log loss over the two object *vs.* non object classes. The regression loss $L_{reg}(t_i, t_i^*) = R(t_i, t_i^*)$ where R represents the robust loss function. The L_{cls} and L_{reg} functions are normalized with N_{cls}, N_{reg}, and the balancing weight λ. The classification of the region can then be performed once the regions for the object detected are determined. In this work, the VGG16 model was used to perform this classification.

The VGG16 model utilizes a total of 16 convolutional and max pooling layers with max pooling occurring after every three convolutional layers [10]. After each convolutional layer a ReLU layer was applied to increase nonlinearity. The response-normalized activity $b_{x,y}^i$ is given by

$$b_{x,y}^i = \frac{a_{x,y}^i}{\left(k + \alpha \sum_{j=\max\left(0, i-\frac{n}{2}\right)}^{\min\left(N-1, i+\frac{n}{2}\right)} (a_{x,y}^j)^2\right)}$$ (2)

in which $(a_{x,y}^j)$ denotes the activity of a neuron computed by applying kernel I at (x, y), the sum runs over n adjacent kernel maps, and N represents the total number of kernels in the layer. The last three layers of the VGG16 CNN are the fully connected layers to perform the classification.

A Titan X GPU was used to train Faster R-CNN with the complex VGG16 network, which required ~ 11 Gb of GPU memory. Training was performed using the alternating optimization method per [5]. This method performed a 4-step training to learn shared features between the region proposal network with a separate detection network. First, the RPN was trained as above. Second, the detection network was trained separately without sharing layers. Third, the detection network was used to initialize the region proposal network training but fixed the shared layers, fine-tuning only the region proposal network layers. The final stage shared the convolutional layers, keeping them fixed, and fine-tuned the fully connected layers, which formed a unified network.

2.2   DEBC Data Collection

The original data was provided from two local businesses. The first provided data from a sensor trade study [11], and the second a UAS test flight. The trade study collected data to determine the optimal camera sensor, viewing angle, and distance for a human to be able to identify potential maintenance concerns.

In the trade study, the cameras used to collect the data include the Sony NEX 7, and a Sony α6000 sensor converted to perform as a multispectral camera. The converted Sony α6000 sensor provided 700–800 nm wavelength light along with the standard visible light. Each image was of size 6000 × 4000 pixels. Two different DEBC's were attached to a forklift and lifted to the test height ranging from 4 m to 12 m from the camera height, and tested under various weather conditions. A total of 111 images from the sensor trade study were provided.

From the UAS test flight, images were collected on live high voltage power lines. The UAS flew approximately 15–20 m from the DEBC's imaged. A sensor comparable to the Sony NEX 7 from Field of View was used for data collection. For our purposes, only images with the DEBC within view were considered. From the UAS test flight, 30 training images were collected.

With a total of 146 images (Fig. 1), the training data set required a large amount of data augmentation to be performed to allow for a sufficiently large training set. As a comparison, the Pascal VOC training dataset provides close to 5,000 images per class to be considered. An additional 270 images from another inspection flight were later included and unaltered to improve the network accuracy. In addition to the 270 images, another class of images considering insulators was added to help the network differentiate between insulators and DEBC's.

Fig. 1. Example images from training data. Top images from sensor trade study, bottom images from test flight.

2.3   DEBC Data Augmentation

To generate enough data to properly train the network to accurately detect the DEBC, a series of simple image processing techniques were performed (Fig. 2). All image processing methods were completed using OpenCV functions [12].

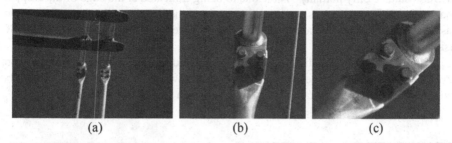

| (a) | (b) | (c) |

Fig. 2. Sample augmented images. (a) Original image. (b) Cropped image of left DEBC. (c) Image rotated 45°.

The first method involved manually cropping the DEBC from each original image. Due to the Faster R-CNN algorithm automatically resizing all input images to 1000 pixels on the larger side, and 600 pixels on the smaller side, the cropped images allowed for increasing the robustness of the network to differences of scale. The cropped images were also used for multiple data augmentations below.

To account for various viewing angles the DEBC may be oriented, each cropped image from before was manipulated to create a series of rotations. Rotations were performed using an SO(3) rotation matrix in degrees. The rotation matrix as

$$Rotation\,Matrix = \begin{bmatrix} \cos(\theta) & -\sin(\theta) & 0 \\ \sin(\theta) & \cos(\theta) & 0 \\ 0 & 0 & 1 \end{bmatrix} \quad (3)$$

was applied to the image. Rotations were applied using inverse warping where each intended pixel location in the rotated image is computed, then the corresponding location in the original image is sampled. The cropped images were rotated from their original position in 15° intervals to a total of 60°. The images were then cropped to remove the resulting black corners of the image from the rotations.

To account for possible out of focus or grainy images, two different morphological operations were performed on the images. These morphological operations include a slight dilation and erosion. This process is done by convolving a kernel (β) over the image I. β has a defined anchor point at the center of the kernel. For dilation, kernel β is convolved over the image and the maximal pixel value overlapped by β is computed and replaced by the image pixel in the anchor points position with that maximal value. This causes bright regions within an image to expand. Erosion is done similarly but instead uses the minimal pixel value for the anchor point causing bright regions. For our purposes, β was chosen to be of size [3 × 3] with only a single pass for slight erosion and dilation operations. Each of the above images were then flipped horizontally.

The last method cropped 1000 × 1000 pixel sized patches in a raster scan pattern with 50% overlap from the original 6000 × 4000 images. This technique was performed to create translations of the DEBC, as well as allow for edge cases where the DEBC would be only partially visible due to being truncated at the edge of the cropped image. The 50% overlap was used to ensure parts of the image would always be visible. Lastly, the total images had to be manually sorted to remove images without the DEBC visible.

With all image processing techniques completed, a total of 2,437 images were developed. With many images to train the network now available, the data needed to be annotated for the CNN to train on the object in each image to be detected.

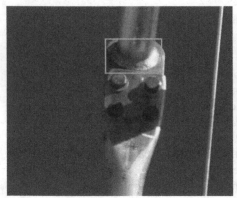 

Fig. 3. Example image with ground truth bounding box annotation.   **Fig. 4.** Example Image from validation set.

2.4   Annotations

Training a Faster R-CNN model required all cases of the intended object categories to be annotated. For our purposes, one class was considered outside the background class, a catch all for all non-defined objects, as the DEBC. Each annotation was created to encompass a portion of the DEBC while limiting all other features. A second class which considered the insulators was annotated and added to the training set later. With Faster R-CNN developed to train on the Pascal VOC dataset, the annotations performed matched the format using the LabelImg software [13]. The Pascal VOC format stores each bounding box annotations location, class, and image location on file in xml format. All annotations were manually selected and a sample of an annotation is provided in Fig. 3.

2.5   Test Data

Images to test the network performance were needed to evaluate the trained networks. Figure 4 provides sample images that were utilized and provided by the company which collected the test flight training data, and provided a newer set of true inspection images for the DEBC's. This test data included 111 images which include 115 total DEBCs taken from a much closer range and viewing angle in which the DEBC was easier to view. These test images were kept separate from the training data.

2.6   Training

The networks evaluated were fine-tuned from an ImageNet pre-trained VGG16 model over differing numbers of iterations and numbers of images following the methods in [14]. The learning rate was set as 0.0001 for 60k mini-batches and 0.00001 for the next 20k mini-batches, momentum as 0.9, and weight decay as 0.0005 as provided by the VGG16 model. The numbers of iterations varied from 40,000 to 150,000, often alternating the higher number of iterations in the first and third stage, and the lower number of iterations in the second and fourth. These alternating numbers of iterations were done as the default iterations were set to alternate from 80,000 to 40,000. The most visually accurate networks based on the number of iterations run were chosen for a full evaluation of the network. Time training the network was heavily dependent on the number of iterations, but took approximately three to seven days of training.

3   Results

The 111 test images were used to evaluate several different trained networks while varying both the number of iterations the network was trained with, and the number of training images. ROC and Precision-Recall curves were developed for each trained network to determine network accuracy (Fig. 5). Data for the curves was generated by setting the threshold for detection to the value of 0.1, and storing all detections and confidence intervals.

With all data available, true positives, true negatives, false positives, and false negatives were recorded while varying the threshold from 0.1 to 1 in 0.05 intervals. Figure 6 provides examples of how the network could accurately locate the DEBC in a variety of positions and poses, including one image where most the DEBC was outside the image. True positives were classified as matches if the detection appeared visually correct allowing for the user to easily view and evaluate the DEBC condition within the bounding box. True negatives were only considered from the list of false positives. As the threshold increased and the false positives were no longer detected, they became true negatives. False positives were defined as any detections that were not of a DEBC. False negatives were tallied for any DEBC not detected in the dataset. Both ROC and Precision Recall curves were developed by calculating the recall/true positive rate (TPR), the false positive rate (FPR), and precision as $TPR = \frac{TP}{TP+FN}$, $FPR = \frac{FP}{TN+FP}$, and $Precision = \frac{TP}{TP+FP}$ respectively. The ROC curve for all three networks considered

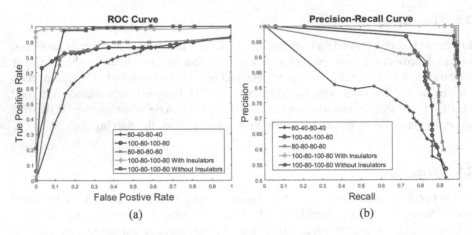

Fig. 5. (a) ROC curve of five different networks, numbers list the iterations of each stage of training for that network (in thousands). The curve with insulators includes additional images and a separate class considering insulators, the curve without insulators includes the additional images, but not the separate insulator class. (b) Precision Recall curve for the same five networks in (a)

was plotted using the TPR, and FPR, whereas the Precision Recall curve was plotted using the precision and recall at each threshold value. After finding a high rate of false positives due to detecting insulators as DEBC's, an additional 270 images were included in the network with insulators as a separate class. The network was retrained with the more successful number of iterations, 100,000, 80,000, 100,000 and 80,000.

As per [15], the network trained with 100,000, 80,000, 100,000, and 80,000 iterations dominates the ROC space of the first three trained networks at higher confidence interval thresholds as evidenced in Fig. 5. Taking the closest point to a TPR of 1 and FPR of 0 in the ROC curve while closer to a precision of 1 and recall of 1 in the precision recall curve, a threshold of 0.9 was found using the network trained with 100,000, 80,000, 100,000 and 80,000. From this data, the accuracy, $acc. = \frac{TP+TN}{Total}$, was also found. Accuracy was determined as 83.7% while maintaining a precision of 91.8%. A second point was also considered with a threshold of 0.85, but resulted in a slightly lower precision of 90.01% and the same accuracy. The confusion matrix at the 0.9 threshold is listed in Table 2.

Inclusion of the additional 270 images, along with the new insulator class, substantially improved the detections of the DEBC. Adding the 270 images increased the TPR, while the addition of the insulator class decreased the FPR on the ROC curve. The Precision Recall curve demonstrates inclusion of the additional images and including the additional insulator class dominates in the ROC space. Due to a preference for higher recall, the F_2 measure was calculated. The highest score was found as $F_2 = 0.6168$ at a threshold of 0.85. The accuracy and precision of this network at a threshold of 0.85 is 97.8% and 99.1% respectively with the confusion matrix for this threshold listed in Table 3.

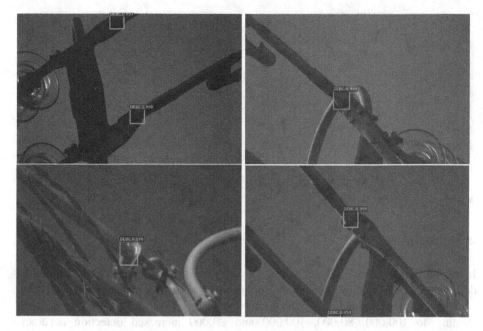

Fig. 6. Example results of successful DEBC detections.

Table 2. Confusion matrix for network 100, 80, 100, 80 (in thousands) with threshold 0.9.

	Predicted negative	Predicted positive	Total
Actual negative	TN: 85	FP: 8	93
Actual positive	FN: 26	TP: 89	115
Total	111	97	208

Table 3. Confusion matrix for network 100, 80, 100, 80 (in thousands) with additional images and insulator class, threshold 0.85.

	Predicted negative	Predicted positive	Total
Actual negative	TN: 22	FP: 1	23
Actual positive	FN: 2	TP: 113	115
Total	24	114	138

Failures fell mostly into two categories, false positives with high confidence, and false negatives (Fig. 7). False positives occurred primarily due to insulators in the background, and are believed to occur due to a lack of insulators in the training set. Inclusion of the additional images and the insulator class diminished these false positives. False negatives were attributed to both variation in lighting conditions and viewing angle, likely due to a lack of variability in the original training data set. Inclusion of additional training images helped in these detections.

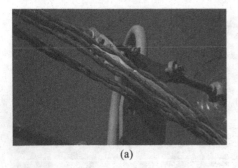 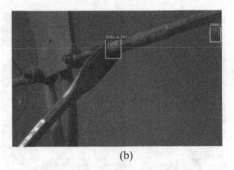

(a)   (b)

Fig. 7. Failures in detection. (a) Occluded DEBC not detected. (b) False positive DEBC detection

4 Conclusion

In this work, deep convolutional neural networks were applied and evaluated based on detection of DEBC's from dead end high tension power lines using Faster R-CNN. It was demonstrated that increasing the number of iterations for each of the four stages of training to 100,000, 80,000, 100,000 and 80,000 increased detection accuracy. Including 270 additional images, subsequently increasing the variability of the training set, increased the TPR. After finding a common object that caused a high number of false positives, inclusion of that object as a separate class decreased the false positive rate.

Future improvements to the network are suggested. Per [16], removing difficult training images can increase the precision. The training data from the sensor trade study provided several images where the DEBC was difficult to see due to situations such as the sun in the direct background. Most false positives were found to be other round shaped objects in the image. Additional classes and training images may help to reduce these false positives further. Increasing the annotation size for the DEBC to include more features, for example the four bolts on the DEBC, may also help reduce or eliminate this problem. Additionally, adding more classes can increase the utility of the network.

References

1. McNichol, E.: It's time for states to invest in infrastructure (2016). http://www.cbpp.org/research/state-budget-and-tax/its-time-for-states-to-invest-in-infrastructure
2. Krizhevsky, A., Sutskever, I., Hinton, G.: ImageNet classification with deep convolutional neural networks. In: NIPS 2012
3. Girshick, R.: Fast R-CNN object detection with Caffe. From Caffe CVPR presentation (2015). http://tutorial.caffe.berkeleyvision.org/caffe-cvpr15-detection.pdf
4. Everingham, M., Van Gool, L., Williams, C.K.I., Winn, J., Zisserman, A.: The PASCAL Visual Object Classes (VOC) challenge. IJCV **88**(2), 303–338 (2010)

5. Ren, S., He, K., Girshick, R., Sun, J.: Faster R-CNN: towards real-time object detection with region proposal networks. In: NIPS 2015. https://github.com/rbgirshick/py-faster-rcnn

6. Girshick, R., Donahue, J., Darrell, T., Malik, J.: Rich feature hierarchies for accurate object detection and semantic segmentation. In: CVPR 2014

7. He, K., Zhang, X., Ren, S., Sun, J.: Spatial pyramid pooling in deep convolutional networks for visual recognition. In: Fleet, D., Pajdla, T., Schiele, B., Tuytelaars, T. (eds.) ECCV 2014. LNCS, vol. 8691, pp. 346–361. Springer, Cham (2014). doi:10.1007/978-3-319-10578-9_23

8. Redmon, J., Divvala, S., Girshick, R., Farhadi, A.: You only look once: unified, real-time object detection. arXiv 2016

9. Jia, Y., et al.: Caffe: Convolutional Architecture for Fast Feature Embedding. In: ACM 2014. http://caffe.berkeleyvision.org/installation.html

10. Simonyan, K., Zisserman, A.: Very deep convolutional networks for large-scale image recongition. In: ICLR 2015

11. Lemler, K., Heichel, J., Dvorak, D.: Powerline Sensor Trade Study Sensor Test Plan (2016)

12. OpenCV Dev Team: OpenCV Documentation (2016). Accessed http://opencv.org/documentation.html

13. Tzutalin. LabelImg. Git code (2013). https://github.com/tzutalin/labelImg

14. Beaucorps, P.: Train Py-Faster-RCNN on another dataset (2015). https://github.com/deboc/py-faster-rcnn/tree/master/help

15. Davis, J., Goadrich, M.: The relationship between precision-recall and ROC curves. In: ICML 2006

16. Girshick, R.: Fast R-CNN. In: ICCV 2015

Object Proposal Generation Applying the Distance Dependent Chinese Restaurant Process

Mikko Lauri$^{(\boxtimes)}$ and Simone Frintrop

Department of Informatics, University of Hamburg, Hamburg, Germany
{lauri,frintrop}@informatik.uni-hamburg.de

Abstract. In application domains such as robotics, it is useful to represent the uncertainty related to the robot's belief about the state of its environment. Algorithms that only yield a single "best guess" as a result are not sufficient. In this paper, we propose object proposal generation based on non-parametric Bayesian inference that allows quantification of the likelihood of the proposals. We apply Markov chain Monte Carlo to draw samples of image segmentations via the distance dependent Chinese restaurant process. Our method achieves state-of-the-art performance on an indoor object discovery data set, while additionally providing a likelihood term for each proposal. We show that the likelihood term can effectively be used to rank proposals according to their quality.

1 Introduction

Image data in robotics is subject to uncertainty, e.g., due to robot motion, or variations in lighting. To account for the uncertainty, it is not sufficient to apply deterministic algorithms that produce a single answer to a computer vision task. Rather, we are interested in the full Bayesian posterior probability distribution related to the task; e.g., given the input image data, how likely is it that a particular image segment corresponds to a real object? The posterior distribution enables quantitatively answering queries on relevant tasks which helps in decision making. For example, the robot more likely succeeds in a grasping action targeting an object proposal with a high probability of corresponding to an actual object [7,13].

In this paper, we propose a method for object discovery based on the distance dependent Chinese restaurant process (ddCRP). In contrast to other approaches, we do not combine superpixels deterministically to generate object proposals, but instead place a ddCRP prior on clusters of superpixels, and then draw samples from the posterior given image data to generate proposals. This firstly increases the diversity of object proposals, and secondly enables calculation of a likelihood term for each proposal. We show that the likelihood term may be used to rank proposals according to their quality. Additionally, the likelihood term might be exploited by a mobile robot to plan its actions.

An overview of our approach is shown in Fig. 1. We begin with a superpixel oversegmentation of the input image, and then place a ddCRP prior on

© Springer International Publishing AG 2017
P. Sharma and F.M. Bianchi (Eds.): SCIA 2017, Part I, LNCS 10269, pp. 260–272, 2017.
DOI: 10.1007/978-3-319-59126-1_22

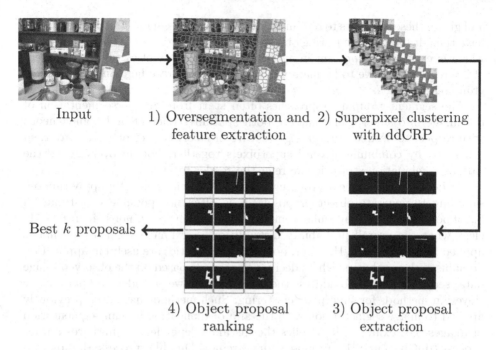

Fig. 1. Overview of the object discovery approach. Superpixels in an initial overseg-mentation (1) are grouped applying the distance dependent Chinese restaurant process (ddCRP) (2). Multiple segmentation samples are drawn from the ddCRP posterior distribution. Object proposals are extracted from the set of segmentation samples (3), and ranked according to how likely they correspond to an object (4).

clusters of superpixels. The ddCRP hyperparameters are selected to encourage object proposal generation: clusters of superpixels with high internal similarity and external dissimilarity are preferred. We apply Markov chain Monte Carlo (MCMC) to draw samples of the posterior distribution on clusterings of super-pixels. We extract all unique clusters which form our set of object proposals. We rank the object proposals according to the Gestalt principles of human object perception [14]. We propose to include the likelihood term, i.e., how often each proposal appears in the set of samples, as part of the ranking, and show that this effectively improves the quality of the proposals.

The paper is organized as follows. Section 2 reviews related work and states our contribution w.r.t. the state-of-the-art. In Sects. 3, 4 and 5, we present in detail the steps involved in the overall process shown in Fig. 1. Section 6 describes an experimental evaluation of our approach. Section 7 concludes the paper.

2   Related Work

Object discovery methods include window-scoring methods (e.g. [2]) that slide a window over the image which is evaluated for its objectness, and segment-grouping methods (e.g. [10]), that start with an oversegmentation of the image and group these segments to obtain object proposals. Segment-grouping methods have the advantage of delivering object contours instead of only bounding boxes, which is especially important in applications such as robotics where the object of interest might have to be manipulated. We concentrate here on the segment-grouping approach.

The segment-grouping approaches often start from an oversegmentation of the image into superpixels that are both spatially coherent and homogeneous with respect to desired criteria, e.g., texture or color. Object proposals are then generated by combining several superpixels together. For an overview of the various combination strategies we refer the reader to [9].

Although some segment-grouping approaches such as e.g. [10] apply random sampling to generate object proposals, it is often not possible to estimate a likelihood value for a particular combination of superpixels, nor is it intuitively clear what the overall probability distribution over image segments is that is applied in the sampling. However, both these properties are useful in application domains such as robotics, where decisions are made based on the observed image data, see, e.g., [7,13]. To address these limitations, we consider non-parametric Bayesian methods for superpixel clustering. Such methods have been previously applied to image segmentation with the aim of replicating human segmentation of images. For example, [6] applies the distance dependent Chinese restaurant process (ddCRP) and [12] proposes a hierarchical Dirichlet process Markov random field for the segmentation task. In [5], multiple segmentation hypotheses are produced applying the spatially dependent Pitman-Yor process. Recent work applies a Poisson process with segment shape priors for segmentation [4].

In our work, similarly to [6], we apply Markov chain Monte Carlo (MCMC) sampling from a ddCRP posterior to generate clusters of superpixels. However, in contrast to earlier work our main aim is object discovery. We tune our method especially towards this aim by setting the model hyperparameters to produce clusters of superpixels that have a strong link to human object perception as described by the Gestalt principles of human object perception [14].

3   The Distance Dependent Chinese Restaurant Process

We first oversegment the input image into superpixels (step 1 in Fig. 1). For each superpixel, we compute a feature vector x_i that we define later. We generate object proposals by grouping superpixels together applying the distance dependent Chinese restaurant process (ddCRP) [3], a distribution over partitions.

The ddCRP is illustrated by an analogy where data points correspond to customers in a restaurant. Every customer links to another customer with whom they will sit at the same table. A partitioning is induced by this set of customer links: any two customers i and j are seated at the same table if i can be reached from j traversing the links between customers (regardless of link direction). Applied to object proposal generation, the image is the restaurant, the customers are superpixels, and the assignment of customers to tables corresponds to a

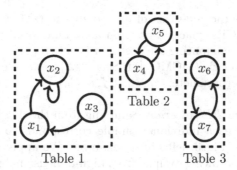

Fig. 2. The distance dependent Chinese restaurant process. Customers corresponding to superpixels in the input image are denoted by the nodes x_i. The links between customers induce a table assignment which corresponds to a segmentation of the image.

segmentation of the image, with each table forming one object proposal – see Fig. 2 for an illustration.

In the ddCRP, the probability that a customer links to another is proportional to the distance between customers. Let c_i denote the index of the customer linked to by customer i, d_{ij} the distance between customers i and j, and D the set of all such distances. The customer links are drawn conditioned on the distances,

$$p(c_i = j \mid D, f, \alpha) \propto \begin{cases} \alpha & \text{if } j = i \\ f(d_{ij}) & \text{if } j \neq i \end{cases}, \tag{1}$$

where α is a parameter defining the likelihood of self-links, and $f : [0, \infty) \to \mathbb{R}^+$ is a decay function that relates the distances between customers to the likelihood of them connecting to each other. We require f to be non-increasing and $f(\infty) = 0$.

We next define the posterior over customer links. Let $\boldsymbol{x} = x_{1:N}$ denote the collection of all data points. Denote by $\boldsymbol{c} = c_{1:N}$ the vector of customer links, and by $\boldsymbol{z}(\boldsymbol{c})$ the corresponding vector of assignments of customers to tables. Denote by $K \equiv K(\boldsymbol{c})$ the number of tables corresponding to link assignment $\boldsymbol{c}$. Furthermore, write $\boldsymbol{z}^k(\boldsymbol{c})$ for the set of all customers i that are assigned to table $k \in \{1, \ldots K\}$. For each table k, we assume that the data x_i, $i \in \boldsymbol{z}^k(\boldsymbol{c})$, is generated from $p(\cdot \mid \theta_k)$. The parameter θ_k is assumed to be drawn from a base measure G_0, which may be considered a prior on θ. Thus, the posterior is

$$p(\boldsymbol{c} \mid \boldsymbol{x}, D, f, \alpha, G_0) \propto \left(\prod_{i=1}^{N} p(c_i \mid D, f, \alpha) \right) p(\boldsymbol{x} \mid \boldsymbol{z}(\boldsymbol{c}), G_0). \tag{2}$$

The first term on the right hand side above is the ddCRP prior, and the second likelihood term is conditionally independent between the tables k:

$$p(\boldsymbol{x} \mid \boldsymbol{z}(\boldsymbol{c}), G_0) = \prod_{k=1}^{K} p(\boldsymbol{x}_{\boldsymbol{z}^k(\boldsymbol{c})} \mid G_0), \tag{3}$$

where $x_{z^k(c)}$ denotes the collection of data points in table k under link configuration c. As the ddCRP places a prior on a combinatorial number of possible image segmentations, computing the posterior is not tractable. Instead, we apply Markov chain Monte Carlo (MCMC) [11, Sect. 24.2] to sample from the posterior given the model hyperparameters $\eta = \{D, f, \alpha, G_0\}$.

Sampling from the ddCRP posterior: Sampling from the ddCRP corresponds to step 2 of Fig. 1, and each individual sample corresponds to a segmentation of the input image - see Fig. 3, left, for an example. We apply Gibbs sampling, a MCMC algorithm for drawing samples from high-dimensional probability density functions, introduced for the ddCRP in [3]. The idea is to sample each variable sequentially, conditioned on the values of all other variables in the distribution. Denote by c_{-i} the vector of link assignments excluding c_i. We sequentially sample a new link assignment c_i^* for each customer i conditioned on c_{-i} via

$$p(c_i^* \mid c_{-i}, x, \eta) \propto p(c_i^* \mid D, f, \alpha)p(x \mid z(c_{-i} \cup c_i^*), G_0). \tag{4}$$

The first right hand side term is the ddCRP prior of Eq. (1), and the second term is the marginal likelihood of the data under the partition $z(c_{-i} \cup c_i^*)$. The current link c_i is first removed from the customer graph which may either cause no change in the table configuration, or split a table (c.f. Fig. 2). Then, reasoning about the effect that a potential new link c_i^* would have on the table configuration, it can be shown that [3]

$$p(c_i^* \mid c_{-i}, x, \eta) \propto \begin{cases} \alpha & \text{if } c_i^* = i \\ f(d_{ij}) & \text{if } c_i^* = j \text{ does not join two tables} \\ f(d_{ij})L(x, z, G_0) & \text{if } c_i^* = j \text{ joins tables } k \text{ and } l, \end{cases} \tag{5}$$

where

$$L(x, z, G_0) = \frac{p(x_{z^k(c_{-i}) \cup z^l(c_{-i})} \mid G_0)}{p(x_{z^k(c_{-i})} \mid G_0)p(x_{z^l(c_{-i})} \mid G_0)}. \tag{6}$$

The terms in the nominator and denominator can be computed via

$$p(x_{z^k(c)} \mid G_0) = \int \left(\prod_{i \in z^k(c)} p(x_i \mid \theta) \right) p(\theta \mid G_0)d\theta. \tag{7}$$

Recall that we interpret the base measure G_0 as a prior over the parameters: $G_0 \equiv p(\theta)$. If $p(\theta)$ and $p(x \mid \theta)$ form a conjugate pair, the integral is usually straightforward to compute.

4   Object Proposal Generation and Likelihood Estimation

We extract a set of object proposals (step 3 in Fig. 1) from samples drawn from the ddCRP posterior. Furthermore, we associate with each proposal an estimate of its likelihood of occurrence. As proposals are clusters of superpixels, we use here notation s_i to refer to superpixels instead of their feature vectors x_i.

To sample a customer assignment c from the ddCRP posterior, we draw a sample from Eq. (5) for each $i = 1, \ldots, N$. Denote by c_j the jth sample, and by $K_j \equiv K(c_j)$ the number of tables in the corresponding table assignment. We can view c_j as a segmentation of the input image, $\bigcup_{k=1}^{K_j} S_{j,k}$, where $S_{j,k} = \{s_i \mid i \in z^k(c_j)\}$ is the set of superpixels assigned to table k by c_j. E.g., in Fig. 2, we would have $S_{j,1} = \{s_1, s_2, s_3\}$, $S_{j,2} = \{s_4, s_5\}$, and $S_{j,3} = \{s_6, s_7\}$.

We sample M customer assignments c_j, $j = 1, \ldots, M$, and write $S_j = \{S_{j,1}, S_{j,2}, \ldots, S_{j,K_j}\}$ as the set of segments in the jth customer assignment. E.g., for the case of Fig. 2, we have $S_j = \{S_{j,1}, S_{j,2}, S_{j,3}\} = \{\{s_1, s_2, s_3\}, \{s_4, s_5\}, \{s_6, s_7\}\}$. The set O of object proposals is obtained by keeping all unique segments observed among the sampled customer assignments: $O = \bigcup_{j=1}^{M} S_j$.

Each proposal $o \in O$ appears in at least one and in at most M of the assignments S_j, $j = 1, \ldots, M$. We estimate the likelihood of each proposal by

$$P(o) = \left[\sum_{j=1}^{M} \mathbb{1}\,(o \in S_j) \right] \Big/ \left[\sum_{j=1}^{M} |S_j| \right], \tag{8}$$

where $\mathbb{1}(A)$ is an indicator function for event A, and $|\cdot|$ denotes set cardinality. Figure 3 illustrates the likelihood values for the proposals.

Fig. 3. Left: an example of a segmentation result from the ddCRP. Each segment is a proposal o. Right: The corresponding proposal likelihood estimates $P(o)$.

5   Gestalt Principles for Object Discovery

We select the hyperparameters $\eta = \{D, f, \alpha, G_0\}$ to promote two important principles: objects tend to have *internal consistency* while also exhibiting *contrast* against their background. This ensures that the proposal set O contains segments that are likely to correspond to objects. As O contains segments from all parts of the image, there are certainly also segments that belong to the background and contain no objects. To mitigate this drawback, we rank the proposals in O and output them in a best-first order. For ranking, we calculate a set of scores from the proposals based on properties such as convexity and symmetry, that have also been shown to have a strong connection to object perception [14]. Next, we describe the superpixel feature extraction, the selection of the ddCRP hyperparameters, and the ranking of object proposals (step 4 of Fig. 1).

Feature extraction: We compute three feature maps from the input image as in: the grayscale intensity I, and the red-green and blue-yellow color contrast maps RG and BY, respectively. The feature vector x_i for superpixel i is

$$x_i = \begin{bmatrix} x_{i,I} & x_{i,RG} & x_{i,BY} & x_{i,avg} \end{bmatrix}^T, \tag{9}$$

where $x_{i,I}$, $x_{i,RG}$, and $x_{i,BY}$ are the 16-bin normalized histograms of the intensity, red-green, and blue-yellow contrast maps, respectively, and $x_{i,avg}$ is the average RGB color value in the superpixel.

Hyperparameter selection: We incorporate contrast and consistency via the distance function d and the base measure G_0, respectively. The distance function d and the decay function f determine how likely it is to link two data points. We impose a condition that only superpixels that share a border may be directly linked together. Also, superpixels with similar contrast features should be more likely to be linked. We define our distance function as

$$d(i,j) = \begin{cases} \infty & \text{if } s_i \text{ and } s_j \text{ are not adjacent} \\ \sum_{n \in \{I,RG,BY\}} w_n \cdot v(x_{i,n}, x_{j,n}) & \text{otherwise} \end{cases}, \tag{10}$$

where $v(x,y) = \frac{1}{2}\|x - y\|_1$ is the total variation distance, and w_n is a weight for feature $n \in \{I, RG, BY\}$, s.t. $\sum_n w_n = 1$. The distance function d has values in the range $[0,1]$, or the value ∞. The weights w_n may be tuned to emphasize certain types of contrasts, but in our experiments we set all to $1/3$. We set an exponential decay function $f(d) = \exp(-d/a)$, where $a > 0$ is a design hyperparameter, to make it more likely to link to similar superpixels.

We encourage internal consistency in the segments by setting the base measure G_0. For the likelihood terms in Eq. (7), we only consider the average RGB color feature $x_{i,avg}$ of the superpixels[1], which is a 3-dimensional vector. We set a multivariate Gaussian cluster likelihood model $p(x_{i,avg} \mid \theta) = N(x_{i,avg}; \mu, \Sigma)$. The model parameters are $\theta = \{\mu, \Sigma\}$, where μ and Σ are the mean vector and covariance matrix, respectively. We apply the Normal-inverse-Wishart distribution as a conjugate prior [11, Sect. 4.6.3], i.e. $p(\theta \mid G_0) = NIW$ $(\theta \mid m_0, \kappa_0, v_0, S_0) = N(\mu \mid m_0, \frac{1}{\kappa_0}\Sigma) \cdot IW(\Sigma \mid S_0, v_0)$. Here, m_0, κ_0, indicate our prior mean for μ and how strongly we believe in this prior, respectively,

---

[1] The other elements of the feature vector are considered via the distance function d.

and S_0 is proportional to the prior mean for Σ and v_0 indicates the strength of this prior. With this choice, adjacent superpixels with similar average RGB colors have a high likelihood of belonging to the same table in the ddCRP.

Object proposal ranking: Similarly as in [15], for each object proposal $o \in O$, we compute the following Gestalt measures that have been shown to have a relation to human object perception [14]:

- symmetry, calculated by measuring the overlaps l_1 and l_2 between the object proposal o and its mirror images along both of its principal axes, i.e., eigenvectors of its scatter matrix. We use the symmetry measures $\frac{\lambda_1 l_1 + \lambda_2 l_2}{\lambda_1 + \lambda_2}$ and $\max\{l_1, l_2\}$, where λ_i are the eigenvalues of the scatter matrix,
- solidity, the ratio of the area of the convex hull of o to the area of o itself,
- convexity, the ratio of the proposal's boundary length and the boundary length of its convex hull,
- compactness, the ratio of the area of o to the squared distance around the boundary of o, i.e., its perimeter,
- eccentricity, the ratio of the distance between the foci of the ellipse encompassing o and its major axis length, and
- centroid distance, the average distance from the centroid of the proposal to its boundary.

As in [15], we apply the first sequence of the KOD dataset [8] to train a support vector machine (SVM) regression model [11, Sect. 14.5] from the Gestalt measures of a proposal o to the intersection-over-union (IoU) of o with the ground truth objects.

Applying the SVM, we can predict a score $s(o)$ for any object detection proposal in O. The proposals with the highest score are deemed most likely to correspond to an actual object. We propose a weighted variant of this score taking into account the likelihood (Eq. (8)):

$$s_w(o) = P(o)s(o). \tag{11}$$

The rationale for this definition is that we would like to give higher priority to object proposals that (1) have a high score $s(o)$ and (2) appear often in the segmentations, indicating robustness with respect to internal consistency and external contrasts as defined via our model hyperparameters. For example in Fig. 3, the scores of proposals with high $P(o)$, i.e., proposals that appear in many samples from the ddCRP, are given higher priority.

As an optional step, we add non-maxima suppression (NMS) for duplicate removal: iterating over all object proposals o in descending order of score, all lower ranked proposals with an IoU value greater than 0.5 with o are pruned.

6   Evaluation

We evaluate our object proposal generation method on the Kitchen Object Discovery (KOD) dataset [8]. We select this dataset as it contains sequences from challenging cluttered scenes with many objects (approximately 600 frames and 80 objects per sequence). This makes it more suitable for our envisioned application area of robotics than other datasets consisting mostly of single images. Ground truth labels indicate the true objects for every 30th frame.

We tuned our method and trained the proposal scoring SVM on the first sequence of the data set, and apply it to the remaining four sequences, labeled Kitchen A, B, C, and D, for testing. For superpixel generation, we apply the SLIC algorithm [1] with a target of 1000 superpixels with a compactness of 45. Features for superpixels are computed as described in Sect. 5. We set a self-link likelihood as $\log \alpha = 0$. For the exponential decay function $f(d) = \exp(-d/a)$, we set $a = 0.05$. For the base measure, we set $m_0 = \begin{bmatrix} 1 & 1 & 1 \end{bmatrix}^T$ with a low confidence $\kappa_0 = 0.1$, and $S_0 = 10 \cdot I_{3 \times 3}$ with $v_0 = 5$.

For each image, we draw $M = 50$ samples of segmentations applying the ddCRP. Samples from a burn-in period of 50 samples were first discarded to ensure the underlying Markov chain enters its stationary distribution. We rank the proposals applying the score $s(o)$ or the likelihood-weighted score $s_w(o)$, and return up to 200 proposals with the highest score. Before ranking we removed proposals larger than 10% or smaller than 0.1% of the image size.

We compare our method to the saliency-guided object candidates (SGO) of [15], the objectness measure (OM) of [2], and the randomized Prim's algorithm (RP) of [10]. SGO is a recent method that performs well on the KOD dataset. The other two methods are representatives of the window-scoring (OM) and segment-grouping (RP) streams of object discovery methods. We measure precision and recall in terms of the number of valid object proposals that have IoU ≥ 0.5 with the ground truth. As OM outputs proposals as bounding boxes, we evaluate all methods with bounding boxes for a fair comparison. We define the bounding box of a proposal as the smallest rectangle enclosing the whole proposal.

Table 1. Area under curve (AUC) values for precision and recall averaged over all frames on the test data sequences labeled A through D, and averaged over all test sequences. "Weighted" refers to using the score $s_w(o)$, "plain" to using the score $s(o)$. Non-maxima suppression (NMS) was applied in all cases. The greatest values for each sequence are shown in a bold font.

	Kitchen A		Kitchen B		Kitchen C		Kitchen D		Average	
	Prec.	Rec.	Prec.	Rec.	Prec.	Rec.	Prec.	Rec.	Prec.	Rec.
Ours (weighted)	**19.4**	**93.3**	25.2	**86.0**	12.1	**86.7**	26.7	47.4	**20.8**	**78.3**
Ours (plain)	16.8	83.1	22.2	79.1	11.8	85.3	**27.9**	49.2	19.7	74.2
SGO [15]	9.8	60.9	**25.3**	85.5	9.6	81.6	**27.9**	**51.9**	18.2	70.0
OM [2]	11.5	45.7	14.7	44.4	**18.1**	83.8	8.6	17.2	13.2	47.8
RP [10]	11.1	61.2	12.3	46.0	12.0	70.0	11.8	25.2	11.8	50.6

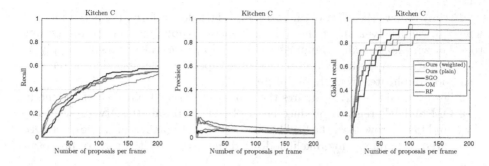

Fig. 4. From left to right: recall, precision, and global recall (fraction of all objects in the sequence detected) averaged over all frames in the Kitchen C sequence. The results are shown as a function of the number of best-ranked proposals considered.

The results are summarized in Table 1. As shown by the average column, the proposed method with likelihood weighting performs best both in terms of precision and recall. With the plain scoring we still slightly outperform SGO, OM, and RP. On individual sequences, we reach the performance of SGO on sequences B and D, while outperforming it on A and C. OM has better precision and similar recall as our method and SGO on sequence C, but does not perform as well on other sequences. On sequences A, B, and C, applying our likelihood-weighted proposal scoring improves performance compared to the plain scoring method. Thus, the likelihood is useful for ranking proposals, providing complementary information not available with the plain score.

For sequence C, the recall, precision, and global recall (fraction of all objects in the sequence detected over all frames) as a function of the number of best-ranked proposals considered are shown in Fig. 4. We achieve higher precision and global recall than SGO for a low number of proposals (<50) per frame. We achieve greater global recall than all the other methods, detecting a greater fraction of all objects over the whole sequence.

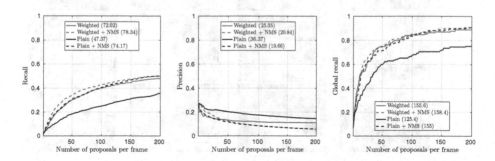

Fig. 5. Evaluation of the ranking methods. Plain refers to the score $s(o)$, weighted is the likelihood weighted score $s_w(o)$, while NMS indicates applying non-maxima suppression (duplicate removal). The numbers in parenthesis show the AUC values for each curve.

Figure 5 shows the effect of ranking method on the performance of our method when averaging over all of the four sequences. Applying likelihood-weighting together with non-maxima suppression (NMS) improves the results over applying the plain score. Applying NMS decreases the reported precision, since it removes also good duplicates from the set of proposals.

Figure 6 qualitatively compares the 5 best proposals from each of the methods. OM and RP tend to produce large object proposals (last two rows). The third and fourth row show the likelihood weighted and plain scoring, respectively. Compared to plain scoring, likelihood weighting increases the rank of proposals

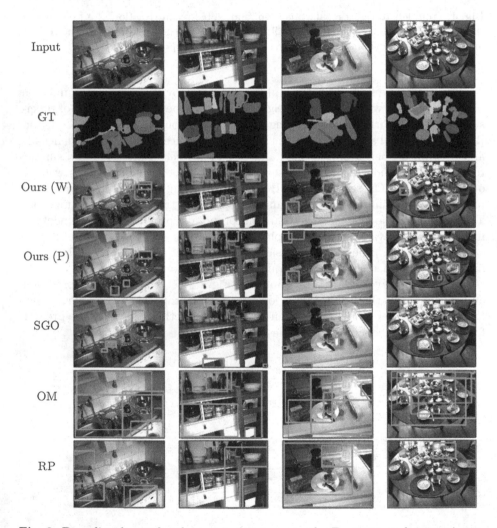

Fig. 6. Bounding boxes for the top 5 object proposals. From top to bottom: input image, ground truth labels, ours (likelihood weighted), ours (plain score), SGO [15], OM [2], and RP [10]. From left to right: one frame from sequence A, B, C, or D.

that appear often in the ddCRP samples. For example, in the last column, fourth row, the plain score gives a high rank for the patch of floor in the lower left corner and the patch of table covering in the lower middle part of the image. These proposal rarely appear in the ddCRP samples. With likelihood weighting (last column, third row), the often appearing proposals on the coffee cup in the middle left part and near the glass in the top left part of the image are preferred as they have a higher likelihood, as also seen from Fig. 3.

7   Conclusion

We introduced object proposal generation via sampling from a distance dependent Chinese restaurant process posterior on image segmentations. We further estimated a likelihood value for each of the proposals. Our results show that the proposed method achieves state-of-the-art performance, and that the likelihood estimate helps improve performance. Further uses for the likelihood estimates may be found, e.g., in robotics applications. Other future work includes extending the method to RGB-D data, and an analysis of the parameter dependency.

References

1. Achanta, R., Shaji, A., Smith, K., Lucchi, A., Fua, P., Süsstrunk, S.: SLIC superpixels compared to state-of-the-art superpixel methods. IEEE Trans. PAMI **34**(11), 2274–2282 (2012)
2. Alexe, B., Deselaers, T., Ferrari, V.: Measuring the objectness of image windows. IEEE Trans. PAMI **34**(11), 2189–2202 (2012)
3. Blei, D.M., Frazier, P.I.: Distance dependent Chinese restaurant processes. J. Mach. Learn. Res. **12**, 2461–2488 (2011)
4. Ghanta, S., Dy, J.G., Niu, D., Jordan, M.I.: Latent Marked Poisson Process with Applications to Object Segmentation. Bayesian Analysis (2016)
5. Ghosh, S., Sudderth, E.B.: Nonparametric learning for layered segmentation of natural images. In: Proceedings of CVPR, pp. 2272–2279 (2012)
6. Ghosh, S., Ungureanu, A.B., Sudderth, E.B., Blei, D.M.: Spatial distance dependent Chinese restaurant processes for image segmentation. In: Advances in Neural Information Processing Systems, pp. 1476–1484 (2011)
7. van Hoof, H., Kroemer, O., Peters, J.: Probabilistic segmentation and targeted exploration of objects in cluttered environments. IEEE Trans. Robot. **30**(5), 1198–1209 (2014)
8. Horbert, E., Martín García, G., Frintrop, S., Leibe, B.: Sequence-level object candidates based on saliency for generic object recognition on mobile systems. In: Proceedings of the International Conference on Robotics and Automation (ICRA), pp. 127–134, May 2015
9. Hosang, J., Benenson, R., Dollár, P., Schiele, B.: What makes for effective detection proposals? IEEE Trans. PAMI **38**(4), 814–830 (2016)
10. Manén, S., Guillaumin, M., Van Gool, L.: Prime object proposals with randomized Prim's algorithm. In: Proceedings of ICCV (2013)
11. Murphy, K.P.: Machine Learning: A Probabilistic Perspective. The MIT Press, Cambridge (2012)

12. Nakamura, T., Harada, T., Suzuki, T., Matsumoto, T.: HDP-MRF: a hierarchical nonparametric model for image segmentation. In: Proceedings of the 21st International Conference on Pattern Recognition (ICPR), pp. 2254–2257, November 2012
13. Pajarinen, J., Kyrki, V.: Decision making under uncertain segmentations. In: Proceedings of International Conference on Robotics and Automation (ICRA), pp. 1303–1309, May 2015
14. Wagemans, J., Elder, J.H., Kubovy, M., Palmer, S.E., Peterson, M.A., Singh, M., von der Heydt, R.: A century of Gestalt psychology in visual perception: I. Perceptual grouping and figure-ground organization. Psychol. Bull. **138**(6), 1172–1217 (2012)
15. Werner, T., Martín-García, G., Frintrop, S.: Saliency-guided object candidates based on Gestalt principles. In: Nalpantidis, L., Krüger, V., Eklundh, J.-O., Gasteratos, A. (eds.) ICVS 2015. LNCS, vol. 9163, pp. 34–44. Springer, Cham (2015). doi:10.1007/978-3-319-20904-3_4

Object Tracking via Pixel-Wise and Block-Wise Sparse Representation

Pouria Navaei, Mohammad Eslami$^{(\boxtimes)}$, and Farah Torkamani-Azar

Cognitive Telecommunications Research Group,
Department of Electrical Engineering,
Shahid Beheshti University, Evin, 1983963113 Tehran, Iran
{p_navaei,m_eslami,f-torkamani}@sbu.ac.ir

Abstract. Object tracking is an important task within the field of computer vision. In this paper, a new and robust method for target tracking in video sequences is proposed based on sparsity representation. Also, in order to increase the accuracy of the tracking, the proposed method uses both group and individual sparse representations. The appearance changes of the target are considered by an on-line subspace training and the appearance model is updated in a procedure which is modified by considering both global and local analysis which brings more accurate appearance model. The proposed appearance representation model is exploited along with the particle filter framework to estimate the target's state and our particle filter uses a modified observation model too. This method is evaluated on several tracking benchmark videos with some different tracking challenges. The results show the robustness of the proposed method in dealing with challenges such as occlusions, changes in illuminations and poses with respect to other related methods.

Keywords: Appearance representation model · Target tracking · Sparse representation · Particle filter

1 Introduction

Target tracking in videos or visual tracking plays a key role in many fields of computer vision applications such as intelligent surveillance, intelligent transportation, activity recognition and etc. Although many algorithms have been proposed, but still some challenges have remained in which researchers have interest to solve them. Most of the visual tracking algorithms consist of three components

- *Motion model:* is used to predict the state of the target in the frame.
- *Appearance model:* represents the appearance of the target according to its visual characteristics.
- *Search method:* considers the appearance and motion models to select the most likely target's state.

© Springer International Publishing AG 2017
P. Sharma and F.M. Bianchi (Eds.): SCIA 2017, Part I, LNCS 10269, pp. 273–284, 2017.
DOI: 10.1007/978-3-319-59126-1_23

The main challenge in designing a robust tracking algorithm is changes in the appearance of the target caused by blurring, non-uniform illuminating, size changing and partial occlusions. Therefore, appearance model is the one of the key components in the robust tracking that has received more attention in the recent years [1].

In the most of previous tracking methods, appearance model was based on the templates [1–3] or subspaces [4]. However, these methods are not suitable when occlusions or drastic changes occur in the target appearance. Recently, a some of appearance model techniques is presented based on the sparse representation, which have more desirable performance in dealing with appearance corruptions and especially occlusions [5–9]. Sparsity representation has many attractive applications such as compressive sensing, dimension reduction, source separation, super resolution [10] in computer vision and also in other subjects of signal processing such as classification [11], cognitive radios [12] and etc. First of sparse representation based tracking method was proposed in [5] by Mei and Ling, which has some unsolved problems such as high computational cost, low number of templates in the dictionary and occlusion effects in the updated dictionary. Therefore further efforts were done to solve these problems e.g. articles [6,7] have been able to address the problems of the paper [5]. However, as the results of experiments show both methods have low accuracy in some scenarios yet. In this paper inspired from [6,7], an effective tracking method is proposed based on both block and pixel based sparsity representations which its results represent the more accuracy and tracking stability with them.

The rest of this paper is organized as follow. The sparse tracking method is described in Sect. 2 and the proposed method is suggested in Sect. 3. Section 4 presents the experimental results and finally the paper is concluded in Sect. 5.

2   Tracking Based on Sparse Representation

In this section, the basics of tracking based on sparsity representation is introduced. In addition, two relevant recent well cited methods called as Sparse Prototypes Tracker (SPT) [6] and Structured Sparse Representation Tracker (SSRT) [7] are discussed too.

2.1   Original Sparse Representation Model

Mei and Ling proposed sparsity representation based tracking method [5]. In this type, target appearance is modeled by a sparse linear combination of target and trivial templates as shown in Fig. 1. In fact, they propose an algorithm (l_1 tracker) by casting the tracking problem as finding the most likely patch with sparse representation and handling partial occlusion with trivial templates. Trivial templates is an identity matrix and is exploited to model occlusion and noise in the real-world observation data. More precisely $y \in R^d$ could be the observation vector as:

$$y \cong Ta + e = \begin{bmatrix} T & I \end{bmatrix} \begin{bmatrix} a \\ e \end{bmatrix} = Dc \tag{1}$$

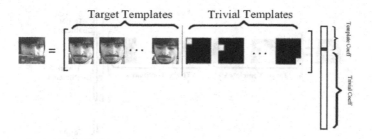

Fig. 1. Original sparse representation model for target tracking [5].

where $T = [t_1, t_2, ..., t_m] \in R^{d \times m}$ $(d \gg m)$ is the set of training templates and $I \in R^{d \times d}$ is the trivial templates, which $\begin{bmatrix} T & I \end{bmatrix}$ can be assumed as a dictionary of representation. Vector $a \in R^m$ is the coefficients vector and $e \in R^d$ is the error vector in which indicates the partial occlusion. The occlusion only covers a portion of the target appearance and therefore it is possible to assume that the error vector e and consequently vector $c \in R^{d+m}$ are sparse [5]. To find the sparse vector c, the following minimization problem should be solved,

$$\min_{c} \frac{1}{2} \|Dc - y\|_2^2 + \lambda \|c\|_1 \tag{2}$$

where $\|\|_2$ and $\|\|_1$ denote norms l_2 and l_1 respectively.

2.2 Sparse Prototypes Tracker (SPT)

In article [6], Wang et al. proposed an extension named as *Sparse Prototypes Tracker (SPT)* for target representing. They exploit the strength of both subspace learning and sparse representation for modeling object appearance. For object tracking, they model target appearance with PCA basis vectors U, and account for occlusion with trivial templates I by

$$y \cong Uz + e = \begin{bmatrix} U & I \end{bmatrix} \begin{bmatrix} z \\ e \end{bmatrix} \tag{3}$$

where z indicates the coefficients of basis vectors. In their formulation, the prototypes consist of just a small number of PCA basis vectors, therefore the z will be dense and the appearance problem can be modified as follow. Figure 2 shows the difference in representations of [5] and SPT [6] which target templates are replaced by PCA basis. Prototypes consist of PCA basis vectors and trivial templates.

$$\min_{z,e} \frac{1}{2} \|y - Uz - e\|_2^2 + \lambda \|e\|_1 \tag{4}$$

It is obvious that the number of used basis vectors in matrix U could be effective on accuracy.

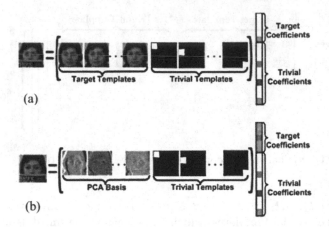

Fig. 2. Sparse representation models for target tracking. (a) Original [5] (b) Sparse Prototypes Tracker (SPT) [6].

2.3   Structured Sparse Representation Tracker (SSRT)

In SPT [6], authors only use information from individual pixels and do not exploits any predetermined assumptions about the structure of the sparse coefficients. But the performance of using the group sparsity or structured sparsity is higher than using just original sparsity [9]. In other words, having previous knowledge of the signal's structure and exploiting it can yield the better results. *Structured Sparse Representation Tracker (SSRT)* is proposed in [7] by Bai and Li with assuming continuous occlusion and previous knowledge of the dictionary structure. As shown in Fig. 3, authors first partition the observed sample and also each of the training templates into R local parts which makes contiguous occlusion (highlighted with red) can be stacked (grouped) as a block sparse vector that has clustered nonzero entries. Then the partitioned regions are stacked into $1 - D$ vectors y. Also, Corresponding structuring should be considered for PCA or subspace templates. More details can be found in [7].

3   Proposed Tracking Algorithm

In this section, our proposed method based on SPT and SSRT methods is explained. The proposed appearance model is defined first and then the particle filter tracking framework is adjusted for coping the model. Finally, procedure for updating the appearance model is discussed.

3.1   Proposed Appearance Model

Based on tracking with structured sparse representation model, since the occlusion geometry is unknown, therefore regardless of the occlusion geometry, the sample is partitioned into predefined blocks. In cases where occlusion does not

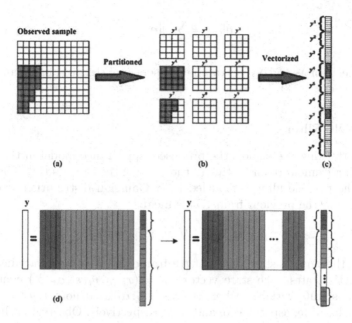

Fig. 3. A simple illustration of structured sparse representation. (a) Observed holistic sample or template, (b) Partition the sample into the local areas, (c) Convert local areas into vectors and putting them in an observed vector, (d) Block structured basis. (Color figure online)

completely fill a block (e.g. 7th block in Fig. 3), the block may be determined as clean (without occlusion) or in contrast full of occlusion, and this simple decision criterion leads to a weak accuracy in tracking procedure.

In order to solve this problem, we propose to represent the appearance model of the target by using original sparse representation of pixels and group sparse representation simultaneously. In this model, $l_{2,1}$ and l_1 norms are used to represent group and individual pixel sparsity, respectively. The proposed sparse tracking model is:

$$\min_{z,e} \frac{1}{2} \|\bar{y} - Uz - e\|_2^2 + \lambda_1 \|e\|_1 + \lambda_2 \|e\|_{2,1} \qquad (5)$$

where U is the PCA subspace extracted from target templates. Also, e is the error vector that includes $e = \left[e^{1^T}, e^{2^T}, \cdots, e^{J^T} \right]$ where J is the total number of blocks and e^j is the error vector for the jth block. In this fashion, the lower size of data can model more states of the object. The vector $\bar{y}$ is the centered observation vector, i.e. $\bar{y} = y - \mu$ which μ is the average vector of the training space. The subspace coefficients z and sparse error vector e should be found while vector e is considered regarded to both pixel based and group based sparseness properties. Pixel based sparseness of the error vector is considered by $\|e\|_1$. The sparseness in groups is computed by

$$\|e\|_{2,1} = \sum_{j=1}^{J} \|e^j\|_2. \tag{6}$$

The coefficients of λ_1 and λ_2 control the sparseness of the pixel based and groups based sparseness.

3.2   Particle Filter

For robust tracking, we exploit the proposed appearance model in the particle filter tracking framework and estimate the state of the target's [13]. The motion model in the particle filter is modeled by a Gaussian distribution around the target's state in the previous frame. This means

$$p(\boldsymbol{x}_t | \boldsymbol{x}_{t-1}) = \mathcal{N}(\boldsymbol{x}_t : \boldsymbol{x}_{t-1}, \boldsymbol{\Psi}) \tag{7}$$

where $\boldsymbol{x}_t$ is the target's state vector at tth frame and $\boldsymbol{\Psi}$ is the covariance matrix of the target's states. The state vector $\boldsymbol{x}_t = (x_t, y_t, \theta_t, s_t, \alpha_t, \phi_t)$ contains six parameters as state variables where $x_t, y_t, \theta_t, s_t, \alpha_t, \phi_t$ denote x, y translations, rotation angle, scale, aspect ratio, and skew respectively. Observation likelihood function is calculated as:

$$p(\boldsymbol{y}_t | x_t) = \exp(- \|\boldsymbol{y}_t - \hat{\boldsymbol{y}}_t\|_2^2) \tag{8}$$

where $\hat{\boldsymbol{y}}_t$ is prediction of the observed sample in the tth frame based on state $\boldsymbol{x}_t$. The formula $\hat{\boldsymbol{y}}_t = \boldsymbol{T}\boldsymbol{a}$ is used in literature of tracking for particle filtering. However, we propose a modified observation model which is inspired by [6] as follow.

$$p(\boldsymbol{y}_t | x_t) = \exp\left(- \left[\|\boldsymbol{y}_t - \boldsymbol{U}\boldsymbol{z} - \boldsymbol{e}\|_2^2 + \lambda_2 \times NOEB \right]\right)$$
$$= \exp(- [term1 + term2]) \tag{9}$$

As mentioned before, similar criterion is proposed in the SPT method of [6], but their reconstruction error ($term1$) was just calculated over the pixels without occlusions. In Eq. (9), we also consider the number of occlusion blocks in $term2$ as $NOEB$, which is the sum of the *Error Number* of each block in an observed sample. Figure 3 shows the concept of the observed sample which contains some blocks. Suppose that, the number of blocks in each observed sample is J, then the $NOEB$ is $NOEB = \sum_{j=1}^{J} \gamma_j$ where γ_j is the *Error Number* of each block and is computed as follow.

$$\gamma_n = \frac{number\ of\ occluded\ pixels\ in\ the\ nth\ block}{number\ of\ pixels\ in\ the\ nth\ block} \tag{10}$$

In addition, two thresholds tr_L and tr_H are used to define three types of Error Number as follow.

- If $\gamma_j \leq tr_L$, the block is considered as error-free and Error Number will be set to $\gamma_j = 0$.
- If $\gamma_j \geq tr_H$, the block is considered as completely error block and Error Number will be set to $\gamma_j = 1$.
- If $tr_L \geq \gamma_j \leq tr_H$, some of the pixels in the block have errors and Error Number will be set to γ_j.

3.3   Updating Appearance Model

Because of changes in the appearance of the targets during tracking sequences, it is not logical to use a fixed subspace for target appearance representation. Therefore updating the appearance model dynamically could improve the tracking performance. It is important to update by using the *correct templates* which has no errors such as occlusion and background. So the first step of the updating procedure is to select the *correct templates*. We propose to use a local analysis along with global analysis. Suppose that observed sample y is selected by the particle filter and the corresponding error vector is computed as e by (5).

In global analysis if the number of occlusion blocks in the selected error vector e, is greater than a certain threshold, then the sample y will be rejected and not be used for updating. Otherwise this sample will be used to update the subspace after local analysis as follows. For each block j:

- If $\gamma_j \leq tr_L$, the entire block pixels remains unchanged.
- If $\gamma_j \geq tr_H$, the entire block pixels are replaced with the mean vector of the subspace μ.
- If $tr_L \geq \gamma_j \leq tr_H$, only occlusion pixels are replaced by corresponding values in the mean vector μ and other pixels left unchanged.

After determining the *correct templates* based on above mentioned procedure and collects them (e.g. 5 corrected templates), they will be used to update the subspace U and the mean vector μ by exploiting the incremental learning algorithm presented in [4].

4   Experimental Results

The proposed tracking algorithm was simulated in the Matlab platform while CVX is used to solve (5) [14,15]. In order to evaluate the performance of the proposed algorithm, four different sequences are selected in which have different tracking challenges as shown in Table 1.

The results of the proposed algorithm are compared with two other sparse tracking algorithms, SPT-2013 [6] and SSRT-2012 [7]. While the SPT simulation codes have been written by its author and are available to use, the simulation code for SSRT algorithm are provided by ourselves.

Table 1. Dataset characteristics: length and challenges

Dataset	Frames	Challenges
David	1–470	Pose and illumination variation, occlusion
Faceocc2	1–819	In-plane rotation and occlusion
Car6	1–705	Heavy occlusion
Jumping	1–313	Fast motion and motion Blur

Each observed sample is resized to size 32×32 for SPT and our method and 15×12 for SSRT. Each observed sample is partitioned to 64 and 6 blocks for our method and SSRT, respectively. The number of 600 particles are selected for particle filter. In all experiments, $\lambda_1 = 0.02$, $\lambda_2 = 0.27$, $tr_L = 0.25$ and $tr_H = 0.75$. The number of basis for PCA subspace is 10.

In order to evaluate and compare the proposed method with other algorithms quantitatively, the overlapping diagrams are drawn. *Overlap Rate (OLR)* [16,17], the overlap area between the detected target and the area specified by the ground truth is defined as

$$OLR = \frac{area\left(ROI_{TR} \cap ROI_{GT}\right)}{area\left(ROI_{TR} \cup ROI_{GT}\right)} \tag{11}$$

where the ROI_{TR} is the target's ROI which is the result of the tracking algorithm and ROI_{GT} is the corresponding correct area in the Ground truth. In addition, the *Center Location Error (CLE)*, the Euclidean distance between the centers of the found target and Ground truth is computed.

Figure 4 shows the results of the "David" sequence which contains both changing in illumination and target state. The Overlap Rate diagrams are illustrated in Fig. 4(b) and show that the performance of the proposed algorithm is better than other algorithms.

Figure 5 shows the results of the "faceocc2" sequence. In this sequence partial occlusions occur along with rotation of the target. The performance of the proposed algorithm is same as other algorithms until the 700th frame. But after a large occlusion (700th frame), the proposed algorithm brings much better performance than the two other algorithms.

Figure 6 is about sequence of the "car6" where the vehicle moves and a large occlusion occurs at 280th frame. While SPT and SSRT fail to track, the proposed algorithm tracks the target in all frames of this sequence as well!

Figure 7 shows the results of the "jumping" sequence. The target have fast motion and blurred in the most of the frames. The Overlap Rate diagrams are illustrated in Fig. 7(b) and show that the performance of different trackers. It can be inferred from Fig. 7 that, while the proposed algorithm performs almost similar to SPT, but is very better than SSRT algorithm.

Furthermore, the quantity measurements are reported here for each sequence and algorithm. The average of CLE and OLR of the algorithms for all frames are

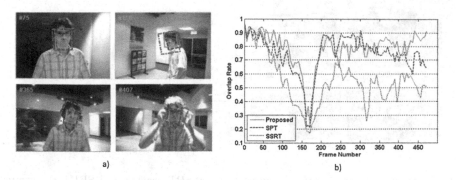

Fig. 4. Tracking results of the proposed tracker, SPT tracker and SRRT tracker on "David" sequence. (a) Quality evaluation, (b) Overlapping rate diagram.

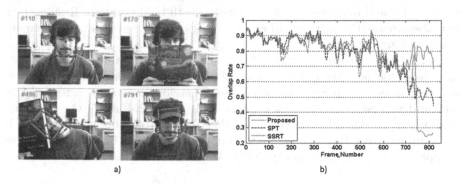

Fig. 5. Tracking results of the proposed tracker, SPT tracker and SRRT tracker on "Faceocc2" sequence. (a) Quality evaluation, (b) Overlapping rate diagram.

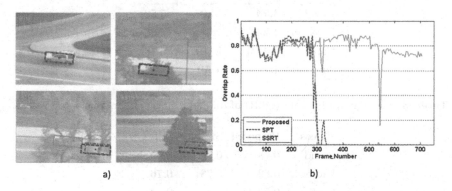

Fig. 6. Tracking results of the proposed tracker, SPT tracker and SRRT tracker on "Car6" sequence. (a) Quality evaluation, (b) Overlapping rate diagram.

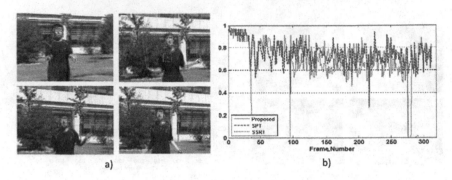

a)   b)

Fig. 7. Tracking results of the proposed tracker, SPT tracker and SRRT tracker on "jumping" sequence. (a) Quality evaluation, (b) Overlapping rate diagram.

reported in the Tables 2 and 3 for each sequence. As shown in the table, the lowest average CLE and maximum OLR are denoted with bold notation. Results show that, the proposed algorithm and criteria brings better performance of sparsity based tracking.

Finally, in this paper, new criteria are proposed in (5) and (9) to represent and track objects more precisely and robustly. But, for practical usage, it is necessary to deal with real-time sequences. Therefore, we try to extended this work and propose a new and fast solution algorithm in future. On the other hand, using just one adaptive subspace (PCA based) to represent the objects is not comprehensive and it is better to train more different or nonlinear spaces to improve the representation capability. Also, extracting the background infor-

Table 2. Average center location error (CLE) of different methods for considered videos.

Videos	Proposed	SPT [6]	SSRT [7]
David	**3.24**	4.25	5.52
Faceocc2	**4.24**	6.17	8.85
Car6	**3.46**	78.01	86.10
Jumping	4.0	**3.69**	22.04

Table 3. Average overlap rate (OLR) of different methods for considered videos.

Videos	Proposed	SPT [6]	SSRT [7]
David	**0.78**	0.74	0.55
Faceocc2	**0.82**	0.78	0.76
Car6	**0.79**	0.34	0.33
Jumping	**0.78**	0.75	0.46

mation of the image and considering it can be useful to model and find the new representation spaces.

5   Conclusion

This paper proposed a robust and fast tracking algorithm using sparse representation along with particle filtering. In order to represent the target's appearance, simultaneous pixel based and block based sparse representations are considered. Based on blocking and grouping concepts which are used to develop the appearance model, a new observation model in particle filter is suggested too. Finally, a simple additional criterion is proposed to select and modify the correct templates which will be used for PCA subspace updating. Experiments show the robustness of the proposed tracking algorithm according to major challenges such as occlusion, illumination changes, resizing, rotating and also brings better performance in comparison with recent SPT and SSRT algorithms.

References

1. Li, X., Hu, W., Shen, C., Zhang, Z., Dick, A., Hengel, A.V.D.: A survey of appearance models in visual object tracking. ACM Trans. Intell. Syst. Technol. (TIST) **4**, 58 (2013)
2. Comaniciu, D., Ramesh, V., Meer, P.: Kernel-based object tracking. IEEE Trans. Pattern Anal. Mach. Intell. **25**, 564–577 (2003)
3. Chateau, T., Laprest, J.T.: Realtime Kernel based tracking. ELCVIA Electron. Lett. Comput. Vis. Image Anal. **8**(1), 27–43 (2009)
4. Ross, D.A., Lim, J., Lin, R.-S., Yang, M.-H.: Incremental learning for Robust visual tracking. Int. J. Comput. Vis. **77**, 125–141 (2008)
5. Mei, X., Ling, H.: Robust visual tracking using l1 minimization. In: IEEE 12th International Conference on Computer Vision, pp. 1436–1443 (2009)
6. Wang, D., Lu, H., Yang, M.-H.: Online object tracking with sparse prototypes. IEEE Trans. Image Process. **22**, 314–325 (2013)
7. Bai, T., Li, Y.: Robust visual tracking with structured sparse representation appearance model. Pattern Recogn. **45**, 2390–2404 (2012)
8. Zhuang, B., et al.: Visual tracking via discriminative sparse similarity map. IEEE Trans. Image Process. **23**(4), 1872–1881 (2014)
9. Huang, J., Zhang, T.: The benefit of group sparsity. Ann. Stat. **38**, 1978–2004 (2010)
10. Baraniuk, R.G., et al.: Applications of sparse representation and compressive sensing [scanning the issue]. In: Proceedings of the IEEE, vol. 98, no. 6, pp. 906–909 (2010)
11. Zhang, L., Yang, M., Feng, X.: Sparse representation or collaborative representation: which helps face recognition? In: IEEE International Conference on Computer Vision (ICCV) (2011)
12. Eslami, M., Torkamani-Azar, F., Mehrshahi, E.: A centralized PSD map construction by distributed compressive sensing. IEEE Commun. Lett. **19**(3), 355–358 (2015)

13. Arulampalam, M.S., Maskell, S., Gordon, N., Clapp, T.: A tutorial on parti-
    cle filters for online nonlinear/non-Gaussian Bayesian tracking. IEEE Trans. Sig.
    Process. **50**, 174–188 (2002)
14. Grant, M., Boyd, S.: CVX: Matlab software for disciplined convex programming,
    version 2.0 beta, September 2013. http://cvxr.com/cvx,
15. Grant, M.C., Boyd, S.P.: Graph implementations for nonsmooth convex programs.
    In: Blondel, V.D., Boyd, S.P., Kimura, H. (eds.) Recent Advances in Learning and
    Control. LNCIS, vol. 371, pp. 95–110. Springer, London (2008)
16. Everingham, M., Van Gool, L., Williams, C.K., Winn, J., Zisserman, A.: The pascal
    visual object classes (VOC) challenge. Int. J. Comput. Vis. **88**, 303–338 (2010)
17. Čehovin, L., Leonardis, A., Kristan, M.: Visual object tracking performance mea-
    sures revisited. IEEE Trans. Image Process. **25**(3), 1261–1274 (2016)

Supervised Approaches for Function Prediction of Proteins Contact Networks from Topological Structure Information

Alessio Martino[1]([✉]), Enrico Maiorino[1], Alessandro Giuliani[2],
Mauro Giampieri[1], and Antonello Rizzi[1]

[1] Department of Information Engineering, Electronics and Telecommunications,
University of Rome La Sapienza, Via Eudossiana 18, 00184 Rome, Italy
{alessio.martino,enrico.maiorino,mauro.giampieri,
antonello.rizzi}@uniroma1.it
[2] Department of Environment and Health, Istituto Superiore di Sanitá,
Viale Regina Elena 299, 00161 Rome, Italy
alessandro.giuliani@iss.it

Abstract. The role performed by a protein is directly connected to its physico-chemical structure. How the latter affects the behaviour of these molecules is still an open research topic. In this paper we consider a subset of the Escherichia Coli proteome where each protein is represented through the spectral characteristics of its residue contact network and its physiological function is encoded by a suitable class label. By casting this problem as a machine learning task, we aim at assessing whether a relation exists between such spectral properties and the protein's function. To this end we adopted a set of supervised learning techniques, possibly optimised by means of genetic algorithms. First results are promising and they show that such high-level spectral representation contains enough information in order to discriminate among functional classes. Our experiments pave the way for further research and analysis.

Keywords: Pattern recognition · Supervised learning · Support Vector Machines · Protein contact networks · Normalised Laplacian matrix

1 Introduction

A protein is a biological macromolecule that is at the basis of every biological process, e.g. enzyme catalysis, DNA replication, response to stimuli, molecules transport, cell structures, and the like. A protein is composed by one or more long chains of amino-acids residues linked in a chain by peptide bonds. There are 20 different kinds of amino-acid residues and the particular sequence of amino-acids that composes a protein is called *primary structure*.

When in solution, protein molecules assume their specific 3D structure by a process called *protein folding*. The particular 3D shape of a protein is at the basis of its physiological role, moreover this configuration undergoes (slight but

© Springer International Publishing AG 2017
P. Sharma and F.M. Bianchi (Eds.): SCIA 2017, Part I, LNCS 10269, pp. 285–296, 2017.
DOI: 10.1007/978-3-319-59126-1_24

crucial) changes to adapt to its micro-environment. Indeed, the deformation affects the interaction potentials between the protein's atoms and the external environment, allowing it to carry out a specific function. In this regard, a protein can be conceived as a nano-machine equipped with sensors and actuators, and engineered—i.e. evolved—to be, from a chemical point of view, as stable as possible.

There is a deep relation between the function and the structure of a protein, and investigating the latter is a fundamental step in understanding the former. A thorough comprehension of how a protein works is in turn of great significance for a variety of practical settings, like drug design and the diagnosis of diseases. In this work we approach this problem from a topological point of view by a minimalist representation of the protein structure, called *protein contact network* (PCN) [1].

The main objective of this work is to investigate how the structure of a protein is related to its function by exploiting supervised machine learning techniques, building upon the spectral properties of the relative PCN. It is worth noting until now that there has been no consistent effort in relating functional and structural properties of proteins in a systematic way. This work takes into account a set of proteins of the Escherichia Coli proteome [2] represented as PCNs. Within this set we consider two classes, i.e. the subset of enzymes and its complement, non-enzymes, where each element of the first class is associated with an Enzyme Commission number [3], that describes the chemical reactions it catalyses, as the ground-truth class label.

This problem is then reformulated as a classification task. Specifically, the target of the classification task is to predict the particular class of each protein starting from a spectral representation of the corresponding protein contact network.

The remainder of this paper is structured as follows: in Sect. 2 we will discuss some essential concepts and definitions regarding graphs and their properties, along with PCNs and their graph-based representation; in Sect. 3 we will present the set of algorithms we used for our analysis, along with the pre-processing stage in order to map PCNs in suitable real-valued features vectors; in Sect. 4 we will show the obtained results and, finally, in Sect. 5 we will draw some conclusions, along with interesting extensions and future works.

2   Definitions

2.1   Fundamentals of Graph Theory and Graph Spectra

Graphs are objects capable of describing data and structures both under a topological and semantic point of view, often used to represent conveniently a set of objects and their relations in many data science fields and applications.

Formally, a graph $G = (V, E)$ is composed by a set of nodes (or vertices) V and a set of edges (or links) E, where $|V| = N$ and $|E| = M$ with N not necessarily equal to M; an edge $e = (v_i, v_j) \in E$ is a link between nodes v_i and v_j.

A graph can be described by means of the adjacency matrix $\mathbf{A}$, a binary matrix defined as:

$$\mathbf{A}_{i,j} = \begin{cases} 1 & \text{if } (v_i, v_j) \in E \\ 0 & \text{otherwise} \end{cases} \tag{1}$$

and if the graph is undirected (that is, if $e = (v_i, v_j) \in E$, then $e = (v_j, v_i) \in E$), such matrix is symmetric by definition. The *degree* D of node i is defined as the number of nodes connected to it:

$$D(i) = \sum_{j=1}^{N} \mathbf{A}_{i,j} \tag{2}$$

Starting from (2), the degree matrix $\mathbf{D}$ has the form

$$\mathbf{D}_{i,j} = \begin{cases} D(i) & \text{if } i = j \\ 0 & \text{otherwise} \end{cases} \tag{3}$$

which is diagonal by definition. Given $\mathbf{D}$ and $\mathbf{A}$, the Laplacian matrix $\mathbf{L}$ of a graph is defined as

$$\mathbf{L} = \mathbf{D} - \mathbf{A} \tag{4}$$

and its normalised version has the form

$$\mathcal{L} = \mathbf{D}^{-\frac{1}{2}} \mathbf{L} \mathbf{D}^{-\frac{1}{2}} \tag{5}$$

The *normalised graph Laplacian* matrix has an interesting property [4] in case of unweighted and undirected graphs, which will turn useful for our analysis:

Property. The set of eigenvalues of $\mathcal{L}$ (i.e. its spectrum) $S = \{\lambda_i\}_{i=1}^{N}$ lies in range $[0; 2]$, independently of the number of eigenvalues of $\mathcal{L}$.

2.2 Protein Contact Networks and Kernel Density Estimator

A protein can effectively be described as a 3D object defined by the location (i.e. 3D coordinates) of the amino-acids which compose the protein itself [1]. Amino-acids, being the monomers of the protein (polymer) are also called *residues*. Inter-residue interactions determine the unique spatial arrangement of the protein and therefore a graph is a convenient representation for such a configuration, where residues are the nodes of the graph and edges indicate spatial proximity between different residues.

Specifically, if the distance between two nodes is below a given threshold (typically 8Å), the two nodes can be considered adjacent. However, some authors (e.g. [5,6]) consider two nodes as adjacent if their distance in the 3D space is between 4 and 8 Å. The lower threshold is set in order to ignore first-neighbour contacts on the protein's linear chain, since they are expected in every protein and provide no additional information on its spatial organisation. In this work, we adopt this convention.

Since proteins can be described as graphs, we can evaluate PCNs adjacency matrices and spectra as in Sect. 2.1. Notice that in this unlabelled graph representation, the different chemical properties of amino-acids have been deliberately neglected.

The property stated in Sect. 2.1 provides the following, precious, insight: since the aforementioned $[0; 2]$ range in which eigenvalues lie is independent from N, one can think of processing graphs (e.g. evaluating dissimilarity) having different number of nodes. However, the number of eigenvalues is still function of N and in order to overcome this problem, we estimate [6] the graph spectral density $p(x)$ by means of a Kernel Density Estimator (KDE) [7]. Amongst the several kernel functions available, the Gaussian kernel is one of the mostly used:

$$p(x) = \frac{1}{N} \sum_{i=1}^{N} \frac{1}{\sqrt{2\pi\sigma^2}} e^{\frac{-(x - \lambda_i)^2}{2\sigma^2}} \tag{6}$$

where σ is the kernel bandwidth. We define the distance between two graphs (G_1 and G_2) as the squared difference between their corresponding spectral densities ($p_1(x)$ and $p_2(x)$, respectively) all over the $[0; 2]$ range:

$$d(G_1, G_2) = \int_0^2 (p_1(x) - p_2(x))^2 dx \tag{7}$$

2.3 Enzyme Commission Number

The Enzyme Commission number (hereinafter EC) is a numerical coding scheme utilised for classifying the physiological role of enzymes. In particular, the EC number of an enzyme encodes the chemical reaction it catalyses. An EC number is a sequence of four digits, separated by dots, in which the first digit (1–6) indicates one of the six major groups[1] and the latter three digits represent a progressively finer functional classification of the enzyme. As we will deal with supervised machine learning algorithms (Sect. 3.2), it is easy to map each protein in our dataset with its group which will serve as the label. However, not all proteins are enzymes and therefore for some of them the EC number might not exist: in this case, such proteins will have label 7, which means *not-enzyme*.

3   Proposed Approach

3.1   Preprocessing

We start considering our proteins as a set of plain text files, each of which describes a given graph. All graphs are undirected and we do not consider weights on edges between amino-acids. In order to feed this dataset to our algorithms (which take as input N_F-dimensional real-valued vectors, Sect. 3.2, where N_F is the number of features) a mandatory pre-processing stage is performed on the basis of Sect. 2. Indeed, for each graph:

---

[1] EC 1: Oxidoreductases; EC 2: Transferases; EC 3: Hydrolases; EC 4: Lyases; EC 5: Isomerases; EC 6: Ligases.

1. the adjacency matrix is evaluated according to (1)
2. the degree matrix is evaluated according to (3)
3. Laplacian and normalised Laplacian matrices are evaluated according to (4) and (5), respectively
4. normalised Laplacian matrix eigenvalues are evaluated

The set of eigenvalues represents our pattern which, to this stage, is a vector in $\mathbb{R}^N$ where the number of nodes N might be different from protein to protein.

3.2 Supervised Algorithms

Support Vector Machines. Amongst the chosen algorithms, the first competitor will be a One-Against-All non-linear Support Vector Machines (SVMs) [10] ensemble with Gaussian Radial Basis Function (GRBF) kernel.

$$GRBF(\mathbf{a}, \mathbf{b}) = e^{-\gamma \cdot d^2(\mathbf{a},\mathbf{b})} \qquad (8)$$

where $d^2(\cdot)$ is the squared Euclidean distance.

The two main parameters, namely the regularisation term C and the kernel shape γ, will be tuned according to a grid search ($log_2 C = [-20; 20] \times log_2 \gamma = [-20; +20]$) with cross-validation.

K-Nearest Neighbours. Conversely to SVMs, K-Nearest Neighbours (K-NN) is an *instance-based* algorithm [9] and therefore it does not require any training phase. The only parameter to be tuned is K, the number of neighbours to be considered in the classification stage. As our dataset does not have prohibitive dimensions (in terms of number of patterns), we will gather the optimal K using a bruteforce approach; that is, trying every K from 1 up to the number of patterns in the Training Set and select the best K as the value that leads to the minimum error rate on the Validation Set.

CURE Support Vector Machines. The CURE[2] Support Vector Machines is an optimised and extended version of the plain Support Vector Machines described above. Specifically:

(a) albeit the Gaussian kernel (6) is widely used, it might not be the most suitable choice for the problem at hand and how to select the right bandwidth σ deserves some attention

(b) the very same SVMs parameter (C and γ) can be tuned in a smarter way, if compared to a grid-search approach

(c) some features (i.e. KDE samples, Sect. 4.1) might be more important than others, thus the dissimilarity measure can be tuned according to a weights vector

---

[2] *Choose yoUR own Estimator.*

The CURE Support Vector Machines overcome these problems thanks to an optimisation/tuning procedure orchestrated by a genetic algorithm (GA) [8] in which the genetic code which identifies the generic individual from a given population has the form:

$$[C \; \gamma \; \mathbf{w} \; KT \; \sigma] \tag{9}$$

where $\mathbf{w}$ is the weights vector which tunes the dissimilarity measure in the GRBF kernel function (8). The latter can therefore be restated as

$$wGRBF(\mathbf{a}, \mathbf{b}) = e^{-\gamma \cdot d_w^2(\mathbf{a}, \mathbf{b}, \mathbf{w})} \tag{10}$$

where in turn

$$d_w^2(\mathbf{a}, \mathbf{b}, \mathbf{w}) = \sum_{i=1}^{N_F} \mathbf{w}_i (\mathbf{a}_i - \mathbf{b}_i)^2 \tag{11}$$

Moreover, in (9), KT is an integer in range $[1; 4]$ which indicates the kernel type (1 = Gaussian, 2 = Epanechnikov, 3 = rectangular box, 4 = triangular) and σ is the bandwidth used by kernel KT.

Each SVM will be trained and optimised independently in order to separate a given class (marked as positive) from all other classes (marked as negatives). To do so, each individual from the genetic population will train such SVM on the Training Set by using the set of parameters written in its genetic code: C will regularise the penalty value in the SVMs convex optimisation problem, γ and $\mathbf{w}$ will tune the dissimilarity/kernel function (10)–(11), KT and σ will select the KDE and its bandwidth in order to extract the set of samples which represent a given PCN.

However, separate tuning of such SVMs, due to heavy labels unbalancing (Table 1), might lead the GA towards apparently good solutions, if the error rate is selected as (part of) the linear convex combination[3], i.e. the fitness function. In order to overcome this problem, the fitness function (to be minimised) has been re-stated as well as the linear convex combination between the complement of the F-score[4] and the percentage of patterns elected as Support Vectors[5].

4   Experimental Results

4.1   Dataset Description

In order to validate our algorithms, we used the 454-patterns Escherichia Coli dataset introduced in [11, 12], named DS-G-454. Such dataset has been introduced

---

[3] E.g. let us suppose we have 100 patterns in our Validation Set, equally distributed amongst 10 different classes; thus, 10 patterns will have positive labels and 90 patterns will have negative labels. If our SVM predicts all patterns as negatives, we will have a 10% error rate - a rather good value - which might lead the genetic algorithm to believe this is a good solution whereas, obviously, it is not.

[4] Defined as the harmonic mean between precision and recall.

[5] In order to avoid overfitting.

in [2], where the Authors collected the whole Escherichia Coli proteome. However, of the 3173 proteins collected, only 454 have their 3D structure available from [13], starting from which we were able to build their respective graphs. Moreover, we processed such graphs according to Sect. 3.1 with the following caveats:

1. we generated a first dataset (hereinafter SCOTT454) by evaluating the Gaussian KDE (6) with bandwidth (i.e. parameter σ) according to the Scott's rule[6] [14]
2. we generated a second dataset (hereinafter HSCOTT454) by setting σ as half the Scott's rule

Finally, $N_F = 100$ samples linearly spaced in $[0; 2]$ have been extracted from the density function evaluated with Eq. (6). Such final 100 samples unambiguously identify our pattern which, to this stage, is a vector in $\mathbb{R}^{100}$ and in turn the dissimilarity measure between patterns, formerly (7), collapses into the Euclidean distance.

However, both HSCOTT454 and SCOTT454 will not substitute in any case the original dataset in which each record is the set of eigenvalues for a given protein since the CURE SVMs will be free to evaluate different KDEs; indeed, such SVMs will basically repeat the above steps of evaluating the KDE with a given bandwidth and extracting 100 samples, where the KDE does not necessarily has to be Gaussian and the bandwidth does not necessarily has to be (a function of) the Scott's rule.

We split the 454-patterns dataset into three non-overlapping sets, namely Training Set, Validation Set and Test Set. Roughly, the Training Set contains 50% of the total number of patterns (229 patterns), whereas the Validation and Test Sets contain 25% of the remaining patterns (111 and 114 patterns, respectively) and such split has been done in a stratified fashion; that is, preserving proportions amongst labels. For the sake of completeness, Table 1 summarises labels distribution in the aforementioned three splits:

4.2   Test Results

Coherently with the CURE SVMs approach and in order to ensure a fair comparison, each of the algorithms described in Sect. 3.2 has been restated in a One-Against-All fashion; that is, there will be as many classifiers as there are labels and the ith classifier will be trained in order to separate the ith class (marked as positive) from all other classes (marked as negatives).

The set of parameters considered for comparison are:

1. $\text{accuracy} = \dfrac{TP + TN}{TP + TN + FP + FN}$

---

[6] The Scott's rule has been selected as a starting point from our analysis, as it is the optimal bandwidth value in case of normal distributions which, however, is a condition not properly respected by our PCNs.

Table 1. Labels distribution in Training, Validation and Test Sets. In brackets, the respective percentage value.

Class ID	Training set	Validation set	Test set
1	22 (10%)	10 (9%)	11 (9%)
2	49 (21%)	24 (21%)	24 (21%)
3	36 (16%)	18 (16%)	18 (16%)
4	18 (8%)	8 (7%)	9 (8%)
5	10 (4%)	4 (4%)	5 (4%)
6	8 (3%)	4 (4%)	4 (4%)
7	86 (38%)	43 (39%)	43 (38%)
Total	229 (100%)	111 (100%)	114 (100%)

2. sensitivity (or recall) $= \dfrac{TP}{TP + FN}$

3. specificity $= \dfrac{TN}{TN + FP}$

4. negative predictive value $= \dfrac{TN}{TN + FN}$

5. positive predictive value (or precision) $= \dfrac{TP}{TP + FP}$

where TP, TN, FP and FN are the true positives, true negatives, false positives and false negatives, respectively.

Tables 2 and 3 summarise the K-NN and SVM performances on HSCOTT454, respectively. In such Tables, the ith row corresponds to the ith classifier which, recall, has been trained to recognise the ith class as positive. Also values marked as "NaN" are the outcome of a 0-by-0 division.

Table 2. K-Nearest Neighbours results on HSCOTT454

Classifier	Accuracy	Sensitivity	Specificity	NPV	PPV
1	89%	0%	98%	90%	0%
2	81%	13%	99%	81%	75%
3	84%	0%	100%	84%	NaN
4	92%	0%	100%	92%	NaN
5	95%	0%	99%	96%	0%
6	96%	0%	100%	96%	NaN
7	73%	37%	94%	71%	80%

In a similar way, Tables 4 and 5 summarise the K-NN and SVM performances on SCOTT454, respectively.

Table 3. Support Vector Machines results on HSCOTT454

Classifier	Accuracy	Sensitivity	Specificity	NPV	PPV
1	90%	0%	100%	90%	NaN
2	81%	8%	100%	80%	100%
3	84%	0%	100%	84%	NaN
4	92%	0%	100%	92%	NaN
5	96%	0%	100%	96%	NaN
6	96%	0%	100%	96%	NaN
7	75%	42%	94%	73%	82%

Table 4. K-Nearest Neighbours results on SCOTT454

Classifier	Accuracy	Sensitivity	Specificity	NPV	PPV
1	90%	0%	100%	90%	NaN
2	76%	4%	96%	79%	20%
3	85%	6%	100%	85%	100%
4	92%	33%	97%	94%	50%
5	96%	0%	100%	96%	NaN
6	96%	0%	100%	96%	NaN
7	71%	28%	97%	69%	86%

Table 5. Support Vector Machines results on SCOTT454

Classifier	Accuracy	Sensitivity	Specificity	NPV	PPV
1	90%	9%	99%	91%	50%
2	79%	0%	100%	79%	NaN
3	84%	0%	100%	84%	NaN
4	92%	0%	100%	92%	NaN
5	96%	0%	100%	96%	NaN
6	96%	0%	100%	96%	NaN
7	73%	51%	86%	74%	69%

From Tables 2, 3, 4 and 5 it is clear that both algorithms, in both cases, tend to predict all patterns as negatives, as shown by NaNs in positive predictive value[7] (PPV) and very high negative predictive value (NPV) and specificity. Interestingly, the 7th classifier (for both algorithms in both cases) does not return such results and recalling that the 7th classifier is in charge of separating enzymes from not-enzymes, indicates an approximate spectrum/EC number mapping, encoded by the data-driven classifier function. This in turn suggests the existence

---

[7] A clear sign that no patterns have been predicted as positive, either true or false.

294 A. Martino et al.

of a relation between protein structure and function, that is preserved by the spectral representation employed in this work.

Let us further investigate by showing the CURE SVMs results in Table 6. Given the randomness in GAs, such results have been obtained by averaging five GA runs.

Table 6. The CURE Support Vector Machines results

Classifier	Accuracy	Sensitivity	Specificity	NPV	PPV	Kernel type
1	54%	82%	51%	96%	15%	4
2	46%	75%	39%	85%	25%	4
3	54%	82%	51%	96%	15%	2
4	70%	33%	73%	93%	10%	2
5	70%	60%	71%	97%	9%	1
6	94%	0%	97%	96%	0%	1
7	78%	53%	93%	77%	82%	1

As first observation, the SVMs are much more robust with respect to positive predictions, this as the result of choosing (a function of the) F-score as the fitness value in the GA; indeed, the F-score by considering both precision and recall intrinsically considers also false positives and false negatives, "stretching" the confusion matrix to be as much diagonal as possible. Second, the 7th SVM over-performs the other 7th classifiers, thanks to the optimisation procedure. Third, the (sub)-optimal kernel type as returned by the GA is the Gaussian kernel, which proves our first assumption in introducing (see (6)) and using such type of kernel for our first experiments.

5   Conclusions and Future Works

The classification task we face in this work is highly challenging and (at least to our knowledge) has never been faced in a systematic manner. It is worth noting that proteins are nano-machines whose basic structure has not a unique "optimisation target", such as performing a specific physiological function (like the catalysis of a given chemical reaction). Conversely, protein molecules must at the same time accommodate many chemico-physical constraints, the most demanding one being probably to be soluble in water [15].

One of the many constraints the particular 3D configuration of a functional protein molecule must obey is the efficient transmission of allosteric signals through the structure [16]. Allostery is the mechanism that allows the protein to sense its micro-environment and to transmit a relevant message, sensed by a different part of the molecule (allosteric site) through the entire structure, to reach the "active site" (in the case of an enzyme the part of the structure devoted to the catalytic work). This mechanism allows the molecule to modify active site configuration so to adapt the reaction kinetics according to the particular physiological needs. Network formalisation, while surely extremely minimalistic, is highly effective as for signal transmission efficiency description, being able to get rid of many aspects of allosteric mechanism [17].

In our opinion, the largely unexpected success of functional prediction from PCN, stems from the focus on signal transmission of the PCN formalisation. This is evident when considering the statistics of the dichotomic separation of non-enzymes (class 7) from all the other classes. This is a somewhat "semantically asymmetric" case, like the Alice in Wonderland not-birthdays, since there are many modes to be a not-enzyme (structural proteins, motor proteins, membrane pores, ...). This is why (see Table 6) we are not disturbed by the low sensitivity of the class 7 prediction task (sensitivity = 53%), but at the same time, the specificity is extremely high (93%). This means that the corresponding synthesised SVM (see Table 6) was pretty sure of 'what-is-not-a-not-enzyme' and, in more plain terms, the system is very effective in classifying a protein as an enzyme. While, at least in principle, all the proteins must sense their microenvironment and adapt to it [18], the allosteric properties are expected to be more prominent for enzymes than for non-enzymatic molecules. This is in line with the behaviour of the 7th CURE SVM prediction that recognises very well the enzymatic/non-enzymatic character of patterns.

Finer details (the recognition of specific ECs) of the proposed structure/function recognition are still difficult to interpret and need to enlarge the dataset, but the obtained results seem to go along a biophysically motivated avenue.

We will further study this machine learning-based way of predicting functional behaviour starting from proteins topological information and some further analyses can be carried out. Indeed, it is possible to check how the classifiers performances change as the aforementioned [4; 8]Å (Sect. 2.2) range changes. Moreover, several variants of the CURE SVMs can be applied, by considering linear classification or other different KDEs or different (dis)similarity measures in the Gaussian RBF kernel. Finally, a hierarchical classifier can be applied in order to improve between-enzymes classification.

References

1. Di Paola, L., De Ruvo, M., Paci, P., Santoni, D., Giuliani, A.: Protein contact networks: an emerging paradigm in chemistry. Chem. Rev. **113**, 1598–1613 (2013)
2. Niwa, T., Ying, B.W., Saito, K., Jin, W., Takada, S., Ueda, T., Taguchi, H.: Bimodal protein solubility distribution revealed by an aggregation analysis of the entire ensemble of Escherichia coli proteins. Proc. Natl. Acad. Sci. USA **106**, 4201–4206 (2009)
3. Webb, E.C.: Enzyme Nomenclature. Academic Press, San Diego (1992)
4. Jurman, G., Visintainer, R., Furlanello, C.: An introduction to spectral distances in networks. Front. Artif. Intell. Appl. **226**, 227–234 (2011)

5. Livi, L., Maiorino, E., Giuliani, A., Rizzi, A., Sadeghian, A.: A generative model for protein contact networks. J. Biomol. Struct. Dyn. **34**, 1441–54 (2016)
6. Maiorino, E., Rizzi, A., Sadeghian, A., Giuliani, A.: Spectral reconstruction of protein contact networks. Phys. A **471**, 804–817 (2017)
7. Parzen, E.: On estimation of a probability density function and mode. Ann. Math. Stat. **33**, 1065–1076 (1962)
8. Goldberg, D.: Genetic Algorithms in Search, Optimization and Machine Learning. Addison-Wesley Longman Publishing Co., Inc., Boston (1989)
9. Mitchell, T.: Machine Learning. McGraw-Hill, Boston (1997)
10. Bishop, C.M.: Pattern Recognition and Machine Learning. Springer, New York (2007)
11. Livi, L., Giuliani, A., Sadeghian, A.: Characterization of graphs for protein structure modeling and recognition of solubility. Curr. Bioinform. **11**, 106–114 (2016)
12. Livi, L., Giuliani, A., Rizzi, A.: Toward a multilevel representation of protein molecules: Comparative approaches to the aggregation/folding propensity problem. Inf. Sci. **326**, 134–145 (2016)
13. Berman, H.M., Westbrook, J., Feng, Z., Gilliland, G., Bhat, T.N., Weissig, H., Shindyalov, I.N., Bourne, P.E.: The protein data bank. Nucleic Acids Res. **28**(1), 235–242 (2000). http://www.rcsb.org/pdb/home/home.do
14. Scott, D.: On optimal and data-based histograms. Biometrika **66**, 605–610 (1979)
15. Giuliani, A., Benigni, R., Zbilut, J.P., Webber, C.L., Sirabella, P., Colosimo, A.: Nonlinear signal analysis methods in the elucidation of protein sequence-structure relationships. Chem. Rev. **102**(5), 1471–1492 (2002)
16. Changeux, J.P., Edelstein, S.J.: Allosteric mechanisms of signal transduction. Science **308**(5727), 1424–1428 (2005)
17. Di Paola, L., Giuliani, A.: Protein contact network topology: a natural language for allostery. Curr. Opin. Struct. Biol. **31**, 43–48 (2015)
18. Tsai, C.J., Del Sol, A., Nussinov, R.: Allostery: absence of a change in shape does not imply that allostery is not at play. J. Mol. Biol. **378**(1), 1–11 (2008)

Top-Down Deep Appearance Attention for Action Recognition

Rao Muhammad Anwer[1](✉), Fahad Shahbaz Khan[2], Joost van de Weijer[3],
and Jorma Laaksonen[1]

[1] Department of Computer Science, Aalto University School of Science,
Espoo, Finland
rao.anwer@aalto.fi
[2] Computer Vision Laboratory, Linköping University, Linköping, Sweden
[3] CS Department, Computer Vision Center,
Universitat Autonoma de Barcelona, Barcelona, Spain

Abstract. Recognizing human actions in videos is a challenging problem in computer vision. Recently, convolutional neural network based deep features have shown promising results for action recognition. In this paper, we investigate the problem of fusing deep appearance and motion cues for action recognition. We propose a video representation which combines deep appearance and motion based local convolutional features within the bag-of-deep-features framework. Firstly, dense deep appearance and motion based local convolutional features are extracted from spatial (RGB) and temporal (flow) networks, respectively. Both visual cues are processed in parallel by constructing separate visual vocabularies for appearance and motion. A category-specific appearance map is then learned to modulate the weights of the deep motion features. The proposed representation is discriminative and binds the deep local convolutional features to their spatial locations. Experiments are performed on two challenging datasets: JHMDB dataset with 21 action classes and ACT dataset with 43 categories. The results clearly demonstrate that our approach outperforms both standard approaches of early and late feature fusion. Further, our approach is only employing action labels and without exploiting body part information, but achieves competitive performance compared to the state-of-the-art deep features based approaches.

Keywords: Action recognition · CNNs · Feature fusion

1 Introduction

Action recognition is an active research problem in computer vision with applications in, e.g., real-time surveillance, security, video retrieval, human-computer interfaces and sports video analysis. Many approaches employ the popular bag-of-words framework for action recognition. Local descriptors based bag-of-words representation have shown promising results for action recognition [20,27]. Several local features, such as, 3D-SIFT [21] and motion boundary histograms [26]

© Springer International Publishing AG 2017
P. Sharma and F.M. Bianchi (Eds.): SCIA 2017, Part I, LNCS 10269, pp. 297–309, 2017.
DOI: 10.1007/978-3-319-59126-1_25

are utilized for video description. These features capture shape, appearance and motion information crucial for action recognition.

Recently, convolutional neural networks (CNNs) have shown promising results on a variety of vision applications including action recognition [22]. CNNs are typically trained on a large labeled data and consist of a series of convolution and pooling operations followed by one or more fully-connected (FC) layers. Initially, most deep learning based approaches rely on capturing appearance information by training the network on RGB patches. Recently, motion-based CNN features have been investigated for the problems of action classification and detection [7,22]. The motion-based CNNs operates on a dense optical flow signal to capture the motion patterns. The appearance and motion deep networks are combined in a late fusion manner. Features from the FC layers are then used for classification. The deep appearance and motion features are used both as holistic representations [22] and in combination with human pose estimation [3].

As discussed above, activations extracted from the output of the FC layers of the deep network are typically used as *features* for domain transfer in CNNs. Different to these approaches, recent studies [4,18] have shown that activations from the convolutional layers provide excellent performance for object and texture recognition. These layers can be exploited within a bag-of-features pipeline by employing them as dense local features. The convolutional layer based bag-of-features framework has been successfully used for the task of object and texture recognition [4]. The deeper convolutional layers are known to possess higher discriminative power [29] and mitigate the need to use a fixed input image size. In this work, we investigate the fusion of appearance and motion based local convolutional features for the problem of action recognition.

In the last decade, most approaches for image classification and action recognition relied on the popular bag-of-words (BOW) based representations. The BOW approach starts by feature detection followed by feature extraction stage. Several hand-crafted features such as SIFT [19] are used for image description. The feature extraction stage is followed by a vocabulary construction step where the local features are vector quantized into a fixed size visual codebook. Consequently, the final representation is obtained by encoding the local features to a visual vocabulary. Within the BOW framework, the fusion of multiple cues such as color and shape is a well studied problem [15,24]. The two standard strategies to fuse multiple cues within the BOW framework are early and late feature fusion. Early fusion fuses color and shape at the feature level as a result of which a joint multi-cue visual vocabulary is constructed. The second strategy, late fusion, fuses multiple cues at the feature encoding level by concatenating the explicit image representations of each visual cue. Early fusion possesses the property of feature binding since the spatial connection between color and shape is preserved at the feature level. Early fusion has been shown to provide improved results for natural scene categories [15]. Late fusion provides feature compactness since separate vocabularies are constructed for each visual cue and has been shown to provide superior performance for man-made categories [15].

As mentioned above, the two standard fusion approaches are only optimal for a specific type of object categories. The color attention based fusion approach [15] aims to combine the advantages of both early and late fusion. In color attention approach, color is used to modulate the shape features. The modulation can be applied both top-down and bottom-up and results in sampling more shape features from regions in an image that are likely to contain an object instance. Color attention possesses the feature binding property since color and shape are combined at the feature level. However, like late fusion, it also possesses the feature compactness property since separate vocabularies are constructed for color and shape. Color attention combines hand-crafted color and shape features for bag-of-words based object recognition. In this work, we re-visit the attention based fusion framework [15] to fuse motion and appearance local features obtained by the convolutional layers of the deep networks for bag-of-deep-features based action recognition.

Contributions: In this paper, we investigate the problem of fusing deep appearance and motion features for action recognition. We introduce an attention based bag-of-deep-features framework to combine deep appearance and motion based local convolutional features. Firstly, dense local appearance (RGB) and motion (flow) based local convolutional features are extracted from the spatial and temporal deep networks. Afterwards, a separate visual vocabulary is constructed for the deep motion and appearance features. Class-specific appearance information is then learned to modulate the weights of the deep motion features. Consequently, a category-specific histogram is constructed for each action class resulting in a discriminative video representation.

We validate our proposed approach by performing experiments on two challenging video datasets namely JHMDB with 21 categories and ACT with 43 action classes. On the JHMD dataset, the proposed approach provides a significant performance improvement of 4.6% and 4.1% compared to standard approaches of early and late fusion, respectively. Similarly on the ACT dataset, the proposed approach obtains a gain of 3.2% and 2.4% compared to standard approaches of early and late fusion, respectively. Furthermore, our approach, without exploiting body part information, achieves competitive performance compared to state-of-the-art approaches employing deep appearance and motion approaches.

2   Related Work

Recently, CNNs have shown significant improvement in performance over the state-of-the-art for various computer vision applications such as image classification and action recognition [17,22]. CNNs, also known as deep networks, comprise of a series of convolution and pooling layers followed by several fully connected (FC) layers and are trained using a large amount of labeled training data. Several recent works have proposed deep features based video representations for action recognition [3,7,22]. Simonyan and Zisserman [22] proposed a two-stream CNN architecture where separate deep networks are trained to capture spatial and temporal features. The spatial network operates on RGB images whereas the temporal stream takes optical flow signal as input. The work of [3] introduced pose based CNNs based on appearance and flow information for action classification.

As discussed above, state-of-the-art action recognition approaches [3,7,22] employ deep architectures where both appearance (RGB) and motion (optical flow) information is exploited. Generally, the appearance and motion based CNNs are trained separately and combined at the FC layers. Other than the FC layers, recent works [4,18] in image classification have demonstrated the effectiveness of convolutional layer activations instead of FC ones. The convolutional layers are discriminative while containing semantically meaningful information. The work of [4] proposed a bag-of-deep-features approach where convolutional layer activations are used as local descriptors. In this work, we employ the bag-of-deep-features framework and investigate the problem of fusing deep appearance and motion information for action recognition.

There exist two main approaches to fuse multiple cues within the bag-of-features framework. The first approach, called as early fusion, combines the visual cues before the vocabulary construction stage. This results in a single visual vocabulary with visual words representing multiple visual cues. Early fusion is shown to be especially suitable for natural object categories [15]. The second fusion strategy is termed as late fusion, where the two visual cues are processed separately and only combined at the final representation level. This implies that separate vocabularies are constructed for each visual cue and the final representation is the concatenation of all representations. Late fusion is shown to provide improved performance compared to early fusion for man-made object categories by [14,15]. Further, late fusion of color and shape have shown to provide improved performance compared to early fusion for texture recognition [11], object detection [10], and action recognition [9]. Different to early and late fusion, the work of [14,15] proposed an attention based fusion framework, to combine color and shape features, for object recognition. In color attention framework, color and shape are processed separately by constructing explicit vocabularies for both cues. Color is used to construct top-down attention maps, used to modulate the shape features. Color attention was shown to provide superior performance compared to both early and late fusion for object recognition.

Our Approach: Here, we re-visit the fusion framework of [14,15] to combine deep appearance and motion based local features, within the bag-of-deep-features framework, for action recognition. We train separate deep networks to capture appearance (RGB) and motion (optical flow) information. Activations from the last convolutional layer of the two networks are then used as local features for each video frame. We construct separate vocabularies for the deep motion and appearance features. Deep appearance is used to construct top-down attention maps, used to modulate the deep motion features. A fixed-length video level representation is then obtained by max aggregation over all video frames.

3   Deep Features for Action Recognition

We train two CNNs to capture spatial and temporal features. The spatial network is trained on the ImageNet ILSVRC-2012 dataset [5] and the temporal network is trained on the UCF101 dataset [23].

Appearance Features: The spatial network takes an RGB image as an input and captures the appearance information. We employ the VGG-F network [2] which is similar to AlexNet and is faster to train. The network consists of five convolutional and three fully-connected layers. The network takes an RGB image of 224 × 224 dimensions. The first convolutional layer comprises of stride of 4 pixels. The remaining four convolutional layers consists of a convolution stride of 1 pixel. The number of convolution filters is 64 in the first convolutional layer and 256 in the remaining four convolutional layers. During training, the learning rate is set to 0.001, the weight decay to 0.0005 and the momentum to 0.9.

Motion Features: The temporal network takes optical flow signal as input and captures the motion information. Similar to [3,7], we compute the optical flow from each consecutive pair of frames using the method of [1]. The values of the motion fields are transformed to the interval [0, 255]. The flow maps are saved as a 3-dimensional image by stacking the flow in $x-$ and $y-$ directions together with the flow magnitude. The network is trained on optical flow images using the UCF101 dataset [23] containing 13320 videos and 101 different classes. Similar to the spatial network, we employ the VGG-F architecture consisting of five convolutional and three FC layers. The work of [7] trains a temporal network using region proposals for action detection. Different to [7], the network is trained using the optical flow on the entire image for action classification. Figure 1 shows a few activations from the spatial (RGB) and temporal (flow) networks.

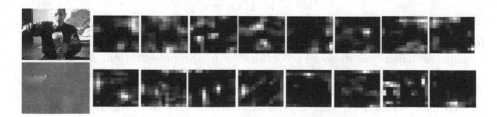

Fig. 1. Visualization of activations with the highest energy from the deepest (last) convolutional layer in the spatial (top row) and temporal (bottom row) networks. Appearance feature maps are computed from the RGB frame (top left) and motion feature maps from the corresponding flow image (bottom left).

4 Top-Down Appearance Attention for Action Recognition

Here, we investigate fusion strategies to combine deep appearance and motion based convolutional features for the problem of action recognition. We then propose an attention based framework where top-down appearance information is used to modulate deep motion features. Given a video, we extract dense local convolutional features f_{bj}, $j = 1, ..., M^b$, in each frame B^b, $b = 1, 2, ..., N$, where M^b is the total number of feature sites in frame b. We extract dense local features from the last convolutional layer of the deep spatial (RGB) and temporal (flow) networks, respectively. The extracted local convolutional appearance and motion features are then quantized into fixed-sized visual vocabularies. The visual vocabularies for appearance and motion cues are represented as, $W^k = \{w_1^k, ..., w_{V^k}^k\}$ with $k \in \{ap, mo, apmo\}$ for the two separate vocabularies for appearance and motion and for the joint visual vocabulary of appearance and motion, respectively. In the case of early fusion, the local features f_{bj} are quantized into a single vocabulary with joint appearance-motion words w_{bj}^{apmo}. In the case of late fusion, separate visual vocabularies are constructed for appearance and motion cues with visual-words (w_{bj}^{ap}, w_{bj}^{mo}). In both fusion cases, visual-words $w_{bj}^k \in W^k$ is the jth quantized convolutional feature of the bth frame of a video for a visual cue k.

In the standard bag-of-words framework, the final representation is a histogram constructed by counting the occurrence of each visual-word in a frame. In case of early fusion, a single histogram is constructed based on joint appearance-motion words:

$$h\left(w_n^{apmo}|B^b\right) \propto \sum_{j=1}^{M^b} \delta\left(w_{bj}^{apmo}, w_n^{apmo}\right) \tag{1}$$

with

$$\delta\left(x, y\right) = \begin{cases} 0 & \text{for } x \neq y \\ 1 & \text{for } x = y \end{cases} \tag{2}$$

In the case of late fusion, we construct separate histogram representations for appearance $h\left(w_n^{ap}|B^b\right)$ and motion $h\left(w_n^{mo}|B^b\right)$, respectively. The two histograms are then concatenated to obtain the final representation. As discussed earlier, both early and late feature fusion approaches are advantageous for a certain set of categories. Early fusion possesses the property of feature binding due to joint vocabulary whereas late fusion possesses the property of feature compactness due to separate visual vocabularies.

Next, we introduce an attention based bag-of-deep-features framework to combine deep appearance and motion based local convolutional features. In the attention framework, the visual cues are processed separately and combined at a later stage in the presence of top-down attention. We reformulate Eq. 1 to modulate the motion features with top-down appearance attention:

$$h\left(w_n^{mo}|B^b, class\right) \propto \sum_{j=1}^{M^b} a\left(\mathbf{x}_{bj}, class\right) \delta\left(w_{bj}^{mo}, w_n^{mo}\right), \tag{3}$$

where, $a(\mathbf{x}_{bj}, class)$ are the attention weights and describe attention of the jth local feature of the bth frame. The attention is top-down and induces spatial binding since it is dependent on both the location $\mathbf{x}_{bj}$ and the corresponding action $class$. The top-down attention component $a(\mathbf{x}_{bj}, class)$ is defined to be the probability of an action class given its deep appearance value, described as

$$a(\mathbf{x}_{bj}, class) = p\left(class | \mathbf{w}_{bj}^{ap}\right). \tag{4}$$

where $\mathbf{w}_{bj}^{ap}$ describes an appearance visual-word. We compute the appearance probabilities $p\left(class | \mathbf{w}_{bj}^{ap}\right)$ as,

$$p(class | \mathbf{w}^{ap}) \propto p(\mathbf{w}^{ap} | class)\, p(class) \tag{5}$$

where, $p(\mathbf{w}^{ap} | class)$ is the empirical distribution. The distribution is obtained by taking a summation over the indexes of the training frames belonging to the action category I^{class} as

$$p(\mathbf{w}_n^{ap} | class) \propto \sum_{I^{class}} \sum_{j=1}^{M^b} \delta\left(w_{bj}^{ap}, \mathbf{w}_n^{ap}\right), \tag{6}$$

To obtain the prior $p(class)$ over the action classes, we use the training data. The attention formulation in Eq. 3 reduces to standard bag-of-deep-features based motion histogram when the probabilities $p(class | \mathbf{w}^c)$ are uniform. In the attention framework, motion features are given more weights in regions with higher appearance attention compared to regions where attention is low. Note that due to the top-down attention, a different distribution is obtained for each action class using the same deep motion visual-words. The final representation is obtained by concatenating all action category-specific histograms. The proposed attention based representation combines the advantages of both late and early fusion. Similar to early fusion, the final representation possesses feature binding property since the appearance and motion features are binded spatially. Similar to late fusion, the attention based final representation possesses the property of feature compactness since separate vocabularies are constructed for appearance and motion features. Finally, an attention based video representation is obtained by max aggregating the frame-level attention histograms over all video frames.

5   Experiments

Here, we present the results of our experiments. We compare our approach with standard early and late fusion approaches. We also provide a comparison of our approach with state-of-the-art results reported in literature.

Datasets: We validate our approach on two challenging video datasets: JHMDB [8] and ACT [25] datasets. The JHMDB dataset consists of 21 human actions, such as *jump, golf, climb and swing-baseball*. The dataset consists of

Fig. 2. Example images from the JHMDB (top two rows) and ACT datasets (bottom row). The JHMDB dataset consists of 928 video clips of 21 different action categories, such as *jump, golf, shoot-gun, shoot-bow, kick-ball, brush hair and swing-baseball*. The ACT dataset consists of 11234 video clips of 43 different action categories, such as *swinging golf, swinging tennis, pouring-juice, jumping-high and cutting-apple*.

928 video clips where there are 36 to 55 clips per action category. There are 15 to 40 frames in each video clip. We use the train/test splits provided with the dataset. The ACT dataset contains 11234 high quality video clips. The dataset is divided into 7260 training videos and 3974 test videos. The dataset consists of 43 action classes such as *swinging golf, cutting-orange and cutting-apple*. We use the train/test splits provided with the dataset. On both datasets, the performance is evaluated in terms of mean accuracy over all action classes. Each test video clip is assigned the action category label of the classifier giving the highest response. Figure 2 shows example images from the JHMBD and ACT datasets.

Experimental Setup: As discussed in Sect. 3, we employ spatial and temporal deep networks to obtain appearance and motion features. We train the RGB and flow VGG-F networks using the Matconvnet library [2]. The RGB network is trained on ImageNet 2012 dataset whereas the flow network is trained on the UCF-101 dataset. The standard deep features are extracted from the FC7 layer of the spatial (RGB) and temporal (flow) networks, respectively. For the bag-of-deep-features representations, we extract the convolutional features from the output of the last convolutional layer of the two networks. For both RGB and flow, we construct vocabularies of 4096 words using the K-means algorithm. Since the final representation for late fusion is $4096 + 4096 = 8192$ dimensional, for a fair comparison, we construct a vocabulary of 8192 words for early fusion. For classification, we employ SVMs with linear kernels.

Attention Cue Evaluation: We first evaluate the role of appearance and motion features as the attention and modulated cues. As discussed earlier, the attention cue contains the prior knowledge about action categories and used to alter the weights of the histogram of modulated cue. We investigate the pairs of appearance-appearance, motion-motion, motion-appearance and appearance-motion attention model. Table 1 shows the results when different attention-modulated cues were used. The results do not change when the same visual cue is used both as the attention and modulated cue. The accuracy improves when using motion as attention cue compared to using appearance alone. The best results are obtained when deep appearance features are used as an attention cue to modulate the weights of motion features. On JHMDB, the top-down appearance provides a gain of 5.8% compared to the motion-motion attention.

Table 1. Attention cue evaluation (mean accuracy in %). We evaluate the role of appearance and motion as attention and modulated cues. The best results are obtained when appearance is used as an attention cue to modulate deep motion features.

Attention cue	Modulated cue	JHMDB	ACT
Appearance	Appearance	41.9	56.2
Motion	Motion	55.9	61.3
Motion	Appearance	44.0	60.2
Appearance	Motion	**61.7**	**71.9**

Table 2. Feature fusion evaluation (mean accuracy in %). We evaluate fusion strategies both with standard deep features (FC layer) and bag-of-deep-features framework. On the JHMDB dataset, our proposed fusion approach provides the best results with a gain of 1.8% compared to the late fusion of standard deep features (FC-A, FC-M). On the ACT dataset, our approach achieves the best performance with a gain of 2.4% compared to the late fusion of BOW deep features (A, M).

Method	JHMDB	ACT
FC-Appearance (FC-A)	36.8	56.7
FC-Motion (FC-M)	57.0	59.1
Late Fusion (FC-A, FC-M)	59.9	68.5
BOW Appearance (A)	41.7	56.1
BOW Motion (M)	55.8	61.0
BOW Early Fusion (A, M)	57.1	68.7
BOW Late Fusion (A, M)	57.6	69.5
BOW Attention (TD-A)	**61.7**	**71.9**

Feature Fusion Comparison: We compare our fusion approach with two standard fusion methods: early and late fusion. We further compare our approach with standard deep features extracted from the FC layers of the deep networks. Table 2 shows the results of different fusion approaches on the two action recognition datasets. On the JHMDB dataset, the standard FC based deep appearance features achieve the classification score of 36.8%. The standard FC based deep motion features obtain the classification score of 57.0%. The late fusion of FC based deep appearance and motion features improves the results by 2.9%

with recognition accuracy of 59.9%. The bag-of-deep-features based appearance and motion representations obtain the classification scores of 41.7% and 55.8%, respectively. Among the standard BOW based fusion approaches, late fusion provides slightly improved performance by obtaining the classification score of 57.6%, compared to early fusion. The best results are obtained with our attention based fusion framework which provides a significant gain of 4.1% compared to late fusion of appearance and motion (A, M).

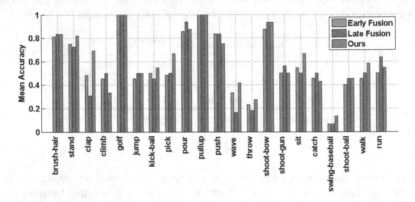

Fig. 3. Per-class comparison of our fusion approach with early and late fusion on the JHMDB dataset. Our approach improves the results on most of the action classes.

On the ACT dataset (Table 2), the standard FC based deep appearance features achieve the classification accuracy of 56.7%. The standard FC based motion features achieve a recognition rate of 59.1%. The late fusion of FC based motion and appearance features provide a score of 68.5%. The bag-of-deep-feature based motion and appearance representations obtain classification scores of 61.0% and 56.1%, respectively. The early feature fusion approach improves the classification results with a recognition accuracy of 68.7%. The late feature fusion obtains a score of 69.5%. Our fusion approach outperforms both early and late fusion by achieving a classification accuracy of 71.9%. Further, our approach outperforms late fusion of FC based standard deep appearance and motion features by 3.4%.

Table 3. State-of-the-art comparison on the JHMDB dataset with 21 action categories. The results are shown in terms of accuracy (%). Our approach provides superior results compared to existing methods.

	brush-hair	catch	clap	climb	golf	jump	kick-ball	pick	pour	pullup	push	run	shoot-ball	shoot-bow	shoot-gun	sit	stand	swing-baseball	throw	walk	wave	Acc
Action Tubes [7]	79.1	33.4	53.9	60.3	99.3	18.4	26.2	42.0	92.8	98.1	29.6	24.6	13.7	92.9	42.3	67.2	57.6	66.5	27.9	58.9	35.8	53.3
P-CNN Pose [3]	80.6	43.8	53.8	63.9	100	46.3	34.0	33.3	89.6	100	**97.2**	47.5	41.0	77.8	74.9	34.1	72.7	39.7	16.7	72.2	38.9	59.9
IDT-FV Pose [3]	77.8	47.9	74.4	69.4	91.7	35.7	50.1	41.7	91.7	100	86.1	**70.7**	**53.7**	80.0	66.4	**68.4**	72.7	59.4	16.7	77.8	25.0	64.7
Deep Actioness [28]	76.4	49.7	**80.3**	43.0	92.5	24.2	**57.7**	70.5	78.7	77.2	31.7	35.7	27.0	88.8	**76.9**	29.8	68.6	72.8	**31.5**	44.4	26.2	56.4
Ours	**83.3**	**50.0**	76.9	**83.3**	100	**66.7**	45.5	**75.0**	**93.8**	100	91.7	54.5	36.4	**100**	68.8	58.3	**81.8**	**93.3**	27.3	**91.7**	**41.7**	**70.1**

It is worthwhile to investigate the combination of our bag-of-deep-features based fusion approach and the standard FC based deep features, since they are potentially complementary. A further gain of 8.4% and 6.2% in accuracy is obtained (presented in Tables 3 and 4) on the JHMDB and ACT datasets respectively by combining our attention based fusion approach with the standard FC based deep features. This clearly suggests that our bag-of-deep-features based fusion representation is complementary to the standard deep features and combining them results in a significant improvement in performance.

Figure 3 shows the per-category performance comparison of our approach with late and early fusion methods on the JHMDB dataset. Our attention based method provides improved performance on nine classes and achieves similar results to the standard fusion approaches on six categories. A significant gain is achieved especially for *clap* (+21.0%), *pick* (+16.7%) and *sit* (+12.2%) categories, all in comparison to the two standard fusion methods.

Table 4. State-of-the-art comparison on the ACT dataset with 43 classes. The results are shown in terms of accuracy (%). Existing approaches are based on very deep (VGG16) architecture. Our approach provides competitive performance despite employing the shallow VGG-F network architecture, compared to state-of-the-art using very deep VGG16 network. It is worth to mention that our fusion approach is generic and can be used with any CNN architecture including very deep networks (VGG16).

Method	Two stream [22]	LSTM [6]	Siamese [25]	Ours
Acc	78.7	78.6	**80.6**	78.7

State-of-the-Art Comparison: Table 3 shows the state-of-the-art comparison on the JHMDB dataset. Our final representation is the combination of the proposed attention based fusion and the standard FC based deep features. The P-CNN based framework [3] combining body part information with appearance and motion based deep features, achieves a classification accuracy of 59.9%. The results are further improved to 64.7% when using improved dense trajectory features with pose information. Our approach, without exploiting any pose information, achieves a gain of 5.4% compared to IDT-FV Pose [3]. It is worth to mention that the IDT-FV Pose method [3] is complementary to our approach and their combination is expected to further improve the results.

Table 4 shows the comparison on the ACT dataset. The work of [22] proposes a two-stream CNN using the very deep (VGG16) architecture and obtains accuracy of 78.7%. The Siamese network based approach [25] that works by modeling the action as a transformation on a high-level feature space achieves a score of 80.6%. Our approach, employing the shallow VGG-F network provides competitive performance with a score of 78.7%. However, our fusion approach is generic and can be used with any CNN architecture including very deep networks (VGG16). The two-stream and Siamese network approaches are complementary to our method and can be combined to further improve the results.

6   Conclusions

We proposed an approach within the bag-of-deep-features framework to combine deep appearance and motion features. Appearance and motion based local features are extracted from the spatial and temporal networks, respectively. Separate vocabularies are constructed for the appearance and motion cues. Top-down deep appearance information is used to modulate the deep motion features. Experiments show that our approach provides significant improvements compared to the standard fusion approaches based same set of deep features. A promising future direction is to investigate the integration of semantic part based information [12, 13] within the proposed framework. Another research direction is to investigate integrating semantic information in a weakly supervised fashion [16] for real-world autonomous applications.

Acknowledgments. This work has been funded by the grant 251170 of the Academy of Finland, Projects TIN2013-41751- P, TIN2016-79717-R of the Spanish Ministry of Economy, Industry and Competitiveness, SSF through a grant for the project Symbi-Cloud, VR starting grant (2016-05543), through the Strategic Area for ICT research ELLIIT. The calculations were performed using computer resources within the Aalto University School of Science "Science-IT" project and NSC. We also acknowledge the support from Nvidia.

References

1. Brox, T., Bruhn, A., Papenberg, N., Weickert, J.: High accuracy optical flow estimation based on a theory for warping. In: Pajdla, T., Matas, J. (eds.) ECCV 2004. LNCS, vol. 3024, pp. 25–36. Springer, Heidelberg (2004). doi:10.1007/978-3-540-24673-2_3
2. Chatfield, K., Simonyan, K., Vedaldi, A., Zisserman, A.: Return of the devil in the details: delving deep into convolutional nets. In: BMVC (2014)
3. Cheron, G., Laptev, I., Schmid, C.: P-CNN: Pose-based CNN features for action recognition. In: ICCV (2015)
4. Cimpoi, M., Maji, S., Kokkinos, I., Vedaldi, A.: Deep filter banks for texture recognition, description, and segmentation. IJCV **118**(1), 65–94 (2016)
5. Deng, J., Dong, W., Socher, R., Li, L.-J., Li, K., Li, F.-F.: Imagenet: a large-scale hierarchical image database. In: Proceedings of the CVPR (2009)
6. Donahue, J., Hendricks, L., Guadarrama, S., Rohrbach, M., Venugopalan, S., Darrell, T., Saenko, K.: Long-term recurrent convolutional networks for visual recognition and description. In: CVPR (2015)
7. Gkioxari, G., Malik, J.: Finding action tubes. In: CVPR (2015)
8. Jhuang, H., Gall, J., Zuffi, S., Schmid, C., Black, M.: Towards understanding action recognition. In: ECCV (2013)
9. Khan, F.S., Anwer, R.M., van de Weijer, J., Bagdanov, A., Lopez, A., Felsberg, M.: Coloring action recognition in still images. IJCV **105**(3), 205–221 (2013)
10. Khan, F.S., Anwer, R.M., van de Weijer, J., Bagdanov, A.D., Vanrell, M., Lopez, A.M.: Color attributes for object detection. In: CVPR (2012)
11. Khan, F.S., Anwer, R.M., van de Weijer, J., Felsberg, M., Laaksonen, J.: Compact color-texture description for texture classification. PRL **51**, 16–22 (2015)

12. Khan, F.S., Anwer, R.M., Weijer, J., Felsberg, M., Laaksonen, J.: Deep semantic pyramids for human attributes and action recognition. In: Paulsen, R.R., Pedersen, K.S. (eds.) SCIA 2015. LNCS, vol. 9127, pp. 341–353. Springer, Cham (2015). doi:10.1007/978-3-319-19665-7_28
13. Khan, F.S., van de Weijer, J., Anwer, R.M., Felsberg, M., Gatta, C.: Semantic pyramids for gender and action recognition. TIP **23**(8), 3633–3645 (2014)
14. Khan, F.S., van de Weijer, J., Vanrell, M.: Top-down color attention for object recognition. In: ICCV (2009)
15. Khan, F.S., van de Weijer, J., Vanrell, M.: Modulating shape features by color attention for object recognition. IJCV **98**(1), 49–64 (2012)
16. Khan, F.S., Xu, J., van de Weijer, J., Bagdanov, A., Anwer, R.M., Lopez, A.: Recognizing actions through action-specific person detection. TIP **24**(11), 4422–4432 (2015)
17. Krizhevsky, A., Sutskever, I., Hinton, G.E.: Imagenet classification with deep convolutional neural networks. In: NIPS (2012)
18. Liu, L., Shen, C., van den Hengel, A.: The treasure beneath convolutional layers: cross-convolutional-layer pooling for image classification. In: CVPR (2015)
19. Lowe, D.: Distinctive image features from scale-invariant points. IJCV **60**(2), 91–110 (2004)
20. Oneata, D., Verbeek, J., Schmid, C.: Action and event recognition with Fisher vectors on a compact feature set. In: ICCV (2013)
21. Scovanner, P., Ali, S., Shah, M.: A 3-Dimensional sift descriptor and its application to action recognition. In: ACM MM (2007)
22. Simonyan, K., Zisserman, A.: Two-stream convolutional networks for action recognition in videos. In: NIPS (2014)
23. Soomro, K., Zamir, A.R., Shah, M.: UCF101: a dataset of 101 human actions classes from videos in the wild. arXiv preprint arXiv:1212.0402 (2012)
24. van de Sande, K.E.A., Gevers, T., Snoek, C.G.M.: Evaluating color descriptors for object and scene recognition. PAMI **32**(9), 1582–1596 (2010)
25. Wan, X., Farhad, A., Gupta, A.: Actions transformations. In: CVPR (2016)
26. Wang, H., Klaser, A., Schmid, C., Liu, C.-L.: Dense trajectories and motion boundary descriptors for action recognition. IJCV **103**(1), 60–79 (2013)
27. Wang, H., Ullah, M.M., Klaser, A., Laptev, I., Schmid, C.: Evaluation of local spatio-temporal features for action recognition. In: BMVC (2009)
28. Wang, L., Qiao, Y., Tang, X., Gool, L.V.: Actionness estimation using hybrid fully convolutional networks. In: CVPR (2016)
29. Zeiler, M.D., Fergus, R.: Visualizing and understanding convolutional networks. In: Fleet, D., Pajdla, T., Schiele, B., Tuytelaars, T. (eds.) ECCV 2014. LNCS, vol. 8689, pp. 818–833. Springer, Cham (2014). doi:10.1007/978-3-319-10590-1_53

Machine Learning

Soft Margin Bayes-Point-Machine Classification via Adaptive Direction Sampling

Karsten Vogt[(✉)] and Jörn Ostermann

Institut für Informationsverarbeitung,
Leibniz Universität Hannover, Hanover, Germany
{vogt,office}@tnt.uni-hannover.de

Abstract. Supervised machine learning is an important building block for many applications that involve data processing and decision making. Good classifiers are trained to produce accurate predictions on a training set while also generalizing well to unseen data. To this end, Bayes-Point-Machines (BPM) were proposed in the past as a generalization of margin maximizing classifiers, such as Support-Vector-Machines (SVM). For BPMs, the optimal classifier is defined as an expectation over an appropriately chosen posterior distribution, which can be estimated via Markov-Chain-Monte-Carlo (MCMC) sampling. In this paper, we propose three improvements on the original BPM classifier. Our new statistical model is regularized based on the sample size and allows for a true soft-margin formulation without the need to hand-tune any nuisance parameters. Secondly, this model can handle multi-class problems natively. Finally, our fast adaptive MCMC sampler uses Adaptive Direction Sampling (ADS) and can generate a sample from the proposed posterior with a runtime complexity quadratic in the size of the training set. Therefore, we call our new classifier the Multi-class-Soft-margin-Bayes-Point-Machine (MS-BPM). We have evaluated the generalization capabilities of our approach on several datasets and show that our soft-margin model significantly improves on the original BPM, especially for small training sets, and is competitive with SVM classifiers. We also show that class membership probabilities generated from our model improve on Platt-scaling, a popular method to derive calibrated probabilities from maximum-margin classifiers.

1 Introduction

Models of statistical learning and classification methods are vital components in many current applications, such as autonomous driving [13], natural language processing [2] and game AI [23]. A challenging aspect of machine learning concerns the balancing of classification accuracy and generalizability on unseen data, especially if only few training examples are available.

For classification problems, supervised learning has the aim to derive a decision function $y = h(x)$ from a labeled training set $Tr = (x_i, y_i)_{i=1}^N$, where $x \in \mathbb{R}^F$ are feature vectors from an F-dimensional feature space and $y \in C$ are labels chosen from a finite set of class labels. Different theoretical models and

© Springer International Publishing AG 2017
P. Sharma and F.M. Bianchi (Eds.): SCIA 2017, Part I, LNCS 10269, pp. 313–324, 2017.
DOI: 10.1007/978-3-319-59126-1_26

learning algorithms have been proposed in the past. Support-Vector-Machines (SVM), originally developed by Cortes and Vapnik [5], have retained widespread usage due to their excellent theoretical underpinnings and their competitive performance on many datasets. Other vector machine approaches were later proposed to alleviate some of the shortcomings of SVMs. These include, for example, extensions for multi-class problems [1, 26], probabilistic decision functions [25, 26] and highly sparse solutions [25].

Herbrich et al. [7] presented their own take of a vector-machine classifier based on the concept of a Bayesian point estimate of the optimal parametrized decision plane. This Bayesian-Point-Machine (BPM) ties maximum-margin classification into a larger framework of Bayesian decision making. As a nontrivial byproduct, learning a BPM constructs an approximation of the Bayesian posterior over all classification models. This posterior distribution can, for example, be used to inexpensively derive various statistics for use in more complex decision models or to compute calibrated class membership probabilities. In this paper, we propose three improvements to the BPM classifier. Firstly, the BPM is based on a regularized hard-margin model. Although the BPM has been proven to have good generalization capabilities for the hard-margin case, this was never conclusively shown for the soft-margin variant. Our experiments in Sect. 5 show that this may not be the case. The regularization also introduces an additional hyperparameter into the model, which must be carefully tuned. To solve these problems, we will substitute the statistical data model with a true soft-margin model that contains no nuisance parameters. Secondly, we extend this new formulation to handle multi-class problems natively. These changes necessitate the development of a new sampling approach. Therefore, we introduce a novel sampling algorithm that can create a sample from our posterior with a runtime complexity of $O(N^2|C| + N|C|^2)$. Our statistical classifier will subsequently be called the Multi-class-Soft-margin-Bayes-Point-Machine (MS-BPM).

Our paper is composed as follows. Section 2 provides a brief introduction to BPMs. Sections 3 and 4 then introduce our new soft-margin model and a fast multi-class sampling algorithm. We evaluate the generalization capabilities and class membership probabilities of the MS-BPM in Sect. 5 and conclude with Sect. 6.

2   Bayes-Point-Machines

The BPM utilizes a very simple statistical model. In the hard-margin case, all classifiers that manage to perfectly separate a training set Tr receive a uniform likelihood, while classifiers that generate at least one training error are discarded. The set of valid classifiers can be described by a convex polytope called the version space. Using a Bayes estimator with an assumed L_2-loss, the point estimate of an optimal decision plane is simply the center of mass of this version space. This point is also called the Bayes-point classifier. It was shown that the BPM generalizes the concept of maximum-margin classification and will often generalize at least as well as the SVM [7]. The soft-margin case, where some margin is sacrificed to mitigate the effects of outliers and overlapping class distributions, was handled for the kernelized version of the algorithm by regularizing the Gram matrix. In effect, this allows for some misclassified training examples near the decision boundary. This approach introduces a tunable dataset-dependent hyper-parameter whose value must be optimized, e.g. using cross-validation.

Since the Bayes-point can be formulated as an expectation over the posterior distribution of classification models, sampling methods based on the Markov-Chain-Monte-Carlo (MCMC) methodology can be an effective way of estimation [21]. In the original works, a billiard scheme was proposed to generate a sample from the uniformly distributed version space [7,22]. Later works improved the computational efficiency using the Expectation Propagation algorithm by approximating the posterior under the assumption of local Gaussianity [14].

In the next section, we present our new statistical model that directly models the soft-margin case without introducing an additional hyperparameter. Furthermore, our model can be straightforwardly extended to multi-class problems.

3   Statistical Model of Soft Margin Classification

Sampling from the distribution of decision boundaries requires the definition of a posterior distribution $p(\beta|\mathit{Tr})$. This section therefore introduces and eluci-dates the required components of our statistical multi-class soft-margin model. This includes a parametrization β of the decision boundaries, a data-dependent likelihood term $l(\mathit{Tr}|\beta)$ and a prior distribution $p(\beta)$ for the model parameters.

3.1   Parametrization

A non-probabilistic classifier can be parametrized using any partitioning function that subdivides the feature-space into $|C|$ partitions. To simplify the sampling, we will focus on linear partitionings. Non-linear decision boundaries can then be modeled via non-linear projections of the feature-space, e.g. using the kernel trick [8]. Following the example of generalized linear models [10], each class c has an associated linear predictor $f_c(x)$. Given a feature-vector x, we always choose the class which produces the maximum response:

$$h(x) = \arg\max_{c \in C} f_c(x) = \arg\max_{c \in C} \beta_c^T x + \beta_{c,0}. \tag{1}$$

The parameters β_c are the importance weights of the linear predictor for class c and $\beta_{c,0}$ its intercept. We will further call a specific instantiation parametrized by the vector β a configuration. Furthermore, the parameters of this model can be reduced by subtracting $\beta_1^T x + \beta_{1,0}$ from all predictor functions. In this formulation, the anchor class $c = 1$ will always produce a zero response, while the remaining functions model the relative predictions for each class compared to the anchor class.

The remaining model parameters are still redundant in regard to uniform scaling. Herbich et al. [7] solved this problem for the two-class case by reparame-terizing the model using a hyperspherical coordinate transform and normalizing the radius to 1. We argue that the original cartesian parametrization allows for a simpler MCMC sampling algorithm. We solve the redundancy in a more classical fashion by introducing appropriate priors on the model parameters.

3.2 Data Likelihood

The likelihood used by Herbrich et al. [7] is based on a simplified data model that is only valid for the hard-margin case. All configurations that achieve zero empirical training errors have a constant likelihood, while all configurations that produce at least one error are discarded. In case of outliers and overlapping class distributions, it may prove beneficial to admit at least some errors. In the original formulation, this is achieved by ignoring training errors that are geometrically close to the decision plane. For our soft-margin model, we would like to derive a likelihood that is more closely related to a well-defined data generating process. The likelihood of the entire training dataset Tr is usually defined by its log-loss:

$$l(Tr|\boldsymbol{\beta}) = \exp(\text{LogLoss}(Tr, \boldsymbol{\beta})) = \prod_{i=1}^{N} p(y_i|\boldsymbol{x}_i, \boldsymbol{\beta}). \tag{2}$$

The logistic regression [10], for example, substitutes the class label probabilities $p(y_i|\boldsymbol{x}_i, \boldsymbol{\beta})$ with the logistic function $(1 + \exp(\boldsymbol{x}_i^t \cdot \boldsymbol{\beta}))^{-1}$. The BPM, on the other hand, assumes a 0–1 loss. We define $p(y_i|\boldsymbol{x}_i, \boldsymbol{\beta}) = 1_{y_i=h(\boldsymbol{x}_i)}$. It can be easily seen that even a single misclassified example pulls the entire likelihood down to zero, which is highly problematic for the non-separable case. Intuitively, this can be interpreted as the BPM model placing infinite confidence on the decisions of the learned classifier. In order to handle overlapping class distributions, we propose to regularize the model by additionally estimating the classification confidences from the data. Our modified likelihood reads as

$$l(Tr|\boldsymbol{\beta}, \boldsymbol{\pi}) = \prod_{i=1}^{N} \pi_{y_i, h(\boldsymbol{x}_i)}, \tag{3}$$

where $\pi_{c,p} \in (0, 1)$ is the probability that an example x_i with a true class label $c = y_i$ is classified as class $p = h(\boldsymbol{x}_i)$. These parameters would require dataset-dependent tuning. We can improve the robustness of our model in regard to the parameters $\boldsymbol{\pi}$ by placing an appropriate prior distribution on them, thus creating a hierarchical model. In the Bayesian spirit, we then marginalize these parameters. Assuming Dirichlet priors with parameters $\boldsymbol{\alpha}$, this produces the likelihood

$$l_{\text{dm}}(Tr|\boldsymbol{\beta}, \boldsymbol{\alpha}) = \int l(Tr|\boldsymbol{\beta}, \boldsymbol{\pi}) \cdot p_{\text{Dir}}(\boldsymbol{\pi}|\boldsymbol{\alpha}) \, d\boldsymbol{\pi}$$

$$\propto \prod_{c=1}^{|C|} \frac{\prod_{p=1}^{|C|} \Gamma(M_{c,p} + \alpha_{c,p})}{\Gamma(\sum_{p=1}^{|C|} M_{c,p} + \alpha_{c,p})}, \tag{4}$$

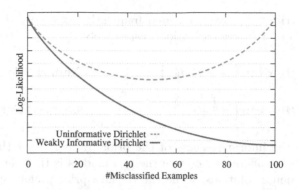

Fig. 1. Comparison of the log-likelihoods for a small training set ($N = 100$) plotted over the number of misclassified examples. Models under comparison are the Dirichlet-multinomial model with an uninformative Dirichlet prior and a weakly informative Dirichlet prior.

where $\Gamma(.)$ is the Gamma function and $M_{c,p}$ are the counts of how many training examples from class c were assigned to partition p. This model is also called a Dirichlet-multinomial or multivariate Pólya distribution [15]. In our model, the confidence we place on a classifier is largely based on the number of training examples it was derived from. In the separable case our regularized model will tend towards the BPM model for large N. Yet we still require a principled way of tuning the α parameters. General pointers of parametrizing Dirichlet distributions can be gleaned from the statistical literature. Generally, we get an uninformative flat prior by setting $\alpha_{c,p} = 1$. It turns out that this is not a sensible choice for classification models. As can be seen in Fig. 1, such a prior would place too much weight on models that exhibit high empirical errors. We need to guarantee that reductions in error always corresponds with increases in likelihood. This property trivially holds for the weakly informative prior with $\alpha_{c,p} = 1$, $c \neq p$ and $\alpha_{c,c} = 1 + N$. Furthermore, we will introduce the regularization parameter ν by setting $\alpha_{c,c} = 1 + \frac{N}{\nu}$. This way, setting $\nu \rightarrow \infty$ produces the uninformative prior while $\nu \rightarrow 0$ strongly penalizes misclassifications and corresponds in the limit with the original BPM model. Sensible choices for ν lie in the interval $(0, 1]$, but our model is largely robust to the specific choice of ν. For all experiments, we simply set it fixed to $\nu = 1$.

3.3 Feature Weight Prior

The parametrization introduced in Sect. 3.1 is redundant in regard to uniform scaling; that is $\beta \equiv t \cdot \beta$ for $t > 0$. This has the consequence that, given a uniform prior over the weights, the resulting posterior distribution will be improper. The typical solution involves replacing the uniform priors with proper ones. We would expect a good prior distribution to be zero-centered, symmetrical, weakly-informative and simple to compute. In past works, the normal distribution and the Laplace distribution have been used frequently, especially since they have strong ties to L_2 and L_1 regularization, respectively.

$$p(\boldsymbol{\beta}) \propto \exp\left(-\frac{1}{2\sigma^2}\|\boldsymbol{\beta}\|_2^2\right) \qquad \text{Normal Prior} \qquad (5)$$

$$p(\boldsymbol{\beta}) \propto \exp\left(-\frac{1}{\sigma}\|\boldsymbol{\beta}\|_1\right) \qquad \text{Laplace Prior} \qquad (6)$$

We can reduce the informativeness of the prior by increasing the scale parameter σ. The main difference between the two models is that the normal prior produces more dense solutions, while the Laplace prior prefers sparsity in the weight parameters. In more recent works, even more sparsity inducing prior distributions have been used [25]. The original BPM approach is restricted to dense models. Although, for computational reasons, our current implementation only uses dense normal priors, the MS-BPM method could be used in conjunction with any of these sparsity inducing priors.

4   Fast Multi-class MCMC Sampler

In this section, we will introduce an efficient sampling scheme for our proposed statistical model. Multivariate sampling is achieved by performing fast univariate sampling along randomized search directions. Quick convergence can then be reached by adapting the distribution of search directions to the local properties of the posterior distribution.

4.1   Univariate vs Multivariate Sampling

The optimization of such classification problems based on a 0–1 loss is known to be NP-hard [17]. Each training example splits the likelihood along $|C| - 1$ half-spaces. As such, the potential number of equivalence-classes of different solutions can be stated as $2^{N \cdot (|C|-1)}$. Direct sampling from the multivariate posterior distribution quickly becomes prohibitively expensive even for small training sets. As can be seen in Fig. 2, the posterior distribution tends to be highly discontinuous, which is a direct result of our choice of the 0–1 loss. The lack of useful gradient information also diminishes the effectiveness of a large class of MCMC algorithms, such as billiard schemes [16], Hamiltonian Monte Carlo [9] and covariance adaptive slice sampling [24]. One important observation is that arbitrary univariate sampling paths can only intersect at most $N \cdot (|C| - 1)$ discontinuities. This implies that a univariate sampling algorithm could be implemented with a much lower computational complexity than a multivariate one. We describe such a sampling method in Sect. 4.2. To facilitate fast convergence of the Markov chain, it is essential to select useful search directions with a high probability. Our univariate sampler can be directly embedded in a number of higher-level sampling methods, such as Gibbs-sampling [21], Hit-and-Run [4] and Adaptive Direction Sampling (ADS) [6].

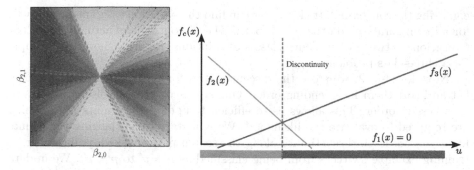

Fig. 2. A heatmap of the posterior distribution for a two-class toy problem with a 1D feature-space. Discontinuities are represented by gray stippled lines.

Fig. 3. Example of a three-class problem. The upper envelope of the linear predictor functions defines the partition membership intervals for one individual training example x along the search path. In this particular case, class 2 is selected left of the discontinuity and class 3 otherwise. Class 1 is never selected.

4.2   Efficient Univariate Sampling Along Arbitrary Search Paths

Our fast univariate sampler will start at a configuration β_t and be given a search direction d. The set of possible configurations

$$\beta_{t+1} = \beta_t + u \cdot d \tag{7}$$

forms our search path for the next configuration. It is important to realize that β_t and d are fixed. As such, u is the random variable that we are actually sampling from. The task can also be stated as a problem of sampling from the posterior distribution conditioned on the search path in Eq. (7).

Our approach to this problem can be broken down into the following four steps:

1. Find all discontinuities along the search path.
2. Construct a discrete distribution of all intervals spanned by two consecutive discontinuities.
3. Draw an interval from this distribution.
4. Finally, draw a new configuration from the selected interval.

The first step can be directly tackled by substituting (7) into (1) as follows:

$$h(x) = \arg\max_{c \in C} (\beta_{t,c} + u \cdot d_c)^T x_i + (\beta_{t,c,0} + u \cdot d_{c,0}).$$

Each example x_i in the training set will generate at most $|C| - 1$ discontinuities. These are situated at values for u, where the predictor functions of two classes become equal. By computing the upper envelope of all $|C|$ predictor functions, e.g. using the convex-hull trick [18], we can find the partition assignment intervals for all configurations on the search path. The discontinuities actually mark the transitions between equivalence classes of solutions. Figure 3 shows an example for a three-class problem.

Next for step 2, we store the discontinuities for all training examples in a list and sort them in ascending order. Our aim is to visit all discontinuities in a successive order. This allows us to efficiently update the counts $M_{c,p}$, which are required to evaluate the likelihood. We will start by initializing the counts for $u = -\infty$. Each discontinuity along the search path marks a point, where a training example switches from being classified as $p = p'$ to $p = p''$. We update the counts accordingly:

$$M_{y_i,p'} = M_{y_i,p'} - 1$$
$$M_{y_i,p''} = M_{y_i,p''} + 1.$$

To compute the interval probability for the discretized sampling problem, we have to integrate over the conditional posterior density:

$$p_j = \int_{u_j}^{u_{j+1}} l_{\mathrm{dm}}(M|\boldsymbol{\alpha}) \cdot p(\boldsymbol{\beta}_t + u \cdot \boldsymbol{d}) \, \mathrm{d}u$$

$$= l_{\mathrm{dm}}(M|\boldsymbol{\alpha}) \cdot \int_{u_j}^{u_{j+1}} p(\boldsymbol{\beta}_t + u \cdot \boldsymbol{d}) \, \mathrm{d}u. \qquad (8)$$

Notice that the likelihood remains constant over the entire interval, since it only depends on the counts M. Integrating over the prior distribution is also trivial for the case of an isotropic normal distribution.

After selecting an interval from the discretized interval distribution for step 3, all that remains is to draw a new configuration from the selected interval in step 4. Once again, we make use of the fact that the likelihood is constant. Therefore, the problem reduces to sampling from the prior distribution, conditioned on the selected interval. In our case, this means to draw a configuration $\boldsymbol{\beta}_{t+1}$ from an appropriately parametrized truncated normal distribution.

The runtime complexity of our sampling algorithm can be stated as $O(N^2|C| + N|C|^2)$ for the kernelized version. For typical datasets, where $N \gg |C|$, this is equivalent to the fast approximated BPM approach in [7] ($O(N^2|C|)$) and compares favorably with support-vector-machines ($O(N^3|C|)$), relevance-vector-machines ($O(N^3|C|)$) and import-vector-machines ($O(N^2q^2|C|)$) (we assumed a o-vs-r scheme for the non multi-class methods). Of course, sampling based methods will usually also incur a much higher constant factor compared to optimization based learning algorithms, so this advantage may only play out for very large datasets.

4.3   Choosing Good Search Directions

Reliably choosing good search directions is of great importance. Two non-adaptive methods are Gibbs sampling [21] and Hit-and-Run sampling [4]. Gibbs sampling proposes to only use search directions that are parallel to the axes of the parameter space. The sampler alternates between these directions using either a predefined schedule or random schedule. In the case that two or more of the parameters are highly correlated, the Markov chain may be required to temporarily assume a low-probability state in order to reach more promising parts of the posterior distribution. This property may cause slow convergence. Hit-and-Run sampling, on the other hand, chooses a uniformly sampled random search direction at each iteration. It is more robust and can often show surprisingly fast convergence [12]. An adaptive sampling scheme, which exploits knowledge about the local correlation structure between parameters, is expected to significantly improve convergence in most cases. One simple adaptive sampling method is the ADS scheme [6]. ADS works by sampling multiple Markov chains in parallel. At each step, one chain is randomly chosen to be iterated on. In contrast to Hit-and-Run sampling the search direction is however not chosen uniformly. Information from two other randomly selected chains is utilized in order to steer the search along the principal directions of the parameter-space. To avoid the sampler getting stuck in a particular subspace of the parameter-space, some precautions have to be made. Following the findings of Gilks et al. [6], it has proven effective to occasionally use a search direction generated by a non-adaptive method. The sampling behavior is typically very robust in regard to this selection probability. In our implementation, e.g., we arbitrarily fixed it to select ADS with 85% probability and Hit-and-Run sampling with 15% probability.

5 Evaluation

In this section, we evaluate our proposed MS-BPM method. We use the original BPM model with soft-margin regularization and the SVM as baseline methods. For all experiments, we simulated 200 independent MCMC chains of length 50, using 1000 iterations for the ADS sampler. We set $\nu = 1$ as described in Sect. 3.2. The hyperparameters for the SVM and BPM classifiers were optimized using a grid-search approach. All methods use the same RBF kernel using the kernel γ that was selected during the grid-search for the SVM runs. The kernel parametrizations for all methods were implemented exactly as in [3].

5.1 UCI Datasets

Our main evaluation is based on the commonly used supervised learning datasets from the UCI database [11]. These seven datasets cover a range of different classification problems of varying size, feature space and number of classes. We show the validity of our method for small and large training sets by training on 10% and 50% bootstrap samples for each dataset. The presented values in Table 1 show the out-of-bag accuracies for 100 independent runs and their standard deviations. As can be seen, our MS-BPM method displays similar performance characteristics as the baseline SVM classifier, yet it does not require hand-tuning of any regularization hyperparameters. The original BPM approach for soft-margin

Table 1. Out-of-bag accuracy estimates for the original regularized BPM, the SVM and our MS-BPM classifier on UCI datasets. Estimates were averaged over 100 bootstrap runs for simulated training sets of small (10%) and large (50%) size. For each experiment, the best result is printed in bold. Ties with the first place, as determined by a two-sample t-test with a 97.5% confidence interval, are also printed in bold.

Dataset	10% Bootstrap			50% Bootstrap		
	BPM	MS-BPM	SVM	BPM	MS-BPM	SVM
Diabetes	69.2 ± 4.5	**71.3 ± 3.8**	**70.5 ± 4.0**	71.5 ± 4.0	**73.4 ± 3.4**	**73.5 ± 3.5**
Ecoli	75.2 ± 14.4	**80.4 ± 3.2**	**80.8 ± 3.7**	76.1 ± 16.7	83.8 ± 2.6	**85.4 ± 2.3**
Image	78.2 ± 22.6	90.2 ± 1.6	**92.1 ± 1.5**	93.1 ± 5.8	94.9 ± 1.2	**96.0 ± 0.6**
Ionosphere	82.2 ± 10.3	**87.6 ± 4.1**	**86.2 ± 5.3**	92.9 ± 1.7	**93.8 ± 1.5**	**93.5 ± 1.7**
Sat-Images	86.5 ± 0.7	87.4 ± 0.6	**88.0 ± 0.5**	88.8 ± 0.3	90.4 ± 0.4	**90.8 ± 0.4**
Sonar	62.2 ± 8.7	**68.9 ± 4.6**	**69.7 ± 4.7**	78.7 ± 5.2	**80.5 ± 3.6**	**81.3 ± 3.4**
Votes	83.0 ± 12.0	**90.2 ± 2.3**	**90.6 ± 2.7**	92.0 ± 6.3	**93.0 ± 1.3**	**93.4 ± 1.4**

classification regularized the Gram-matrix to allow for some empirical errors on the training set. Our experiments show that this approach is not competitive with our improved statistical model on most datasets, and especially for small training sets. The large standard deviations also indicate some robustness problems that are not observable in our method. A Wilcoxon signed rank test shows with a 97.5% confidence level that our MS-BPM classifier significantly improves on the BPM classifier. The same test is inconclusive when used to compare the results of the MS-BPM and SVM classifiers.

5.2   Class Membership Probabilities

The class membership probabilities generated by our model often tend to better represent the true probabilities than classifiers that were calibrated subsequently after training, e.g. using Platt scaling [19]. This difference only gets amplified for small training sets, as any post-hoc calibration has to be based on a sub-sampling method, such as cross-validation. Figure 4 compares the membership probabilities for an SVM model and our classifier on the Ripley synthetic dataset [20]. This dataset features a two-class problem in a two-dimensional feature space. Both classes are mixtures of two Gaussians with distinct modes. This difference can be measured by comparing the log loss ($E[-\log(p(y_i|\boldsymbol{x}_i))]$) as estimated from a test sample. Our experiment gave the following results:

$$\text{LogLoss}_{SVM} = 0.15$$
$$\text{LogLoss}_{MS-BPM} = 0.08$$

Thus, our MS-BPM model improves over the SVM by approximately 53%. Most of the gains come from the improved estimation of membership probabilities in the higher-density regions of the dataset.

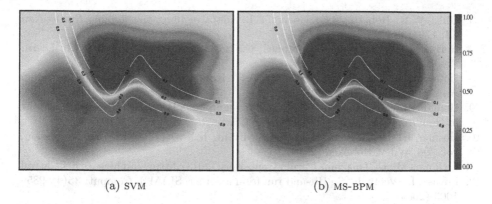

(a) SVM   (b) MS-BPM

Fig. 4. Posterior class membership probabilities for the Ripley synthetic dataset. Both classifiers use an RBF kernel with $\gamma = 7.5$ as selected via grid search. The heatmaps show the posterior plots of the learned classification models while the overlaid contour plots depict the true probabilities. The MS-BPM classifier (right) tends to generate more confident predictions than the SVM classifier (left) and produces smoother class boundaries.

6   Conclusion

In this paper, we presented our proposed improvements to the BPM classifier. The experiments demonstrated that our MS-BPM model exhibits similar performance to SVMs and significantly improves on the original BPM, especially for small training sets. Yet it requires less hand-tuning of hyperparameters while also supporting multi-class problems natively. We also showed that the class membership probabilities generated by our model are superior to post-hoc calibrated probabilities for maximum-margin models. The algorithmic complexity of our learning algorithm $(O(N^2|C| + N|C|^2))$ also compares favorably to other kernelized vector-machine classifiers.

Acknowledgments. This work was supported by the German Science Foundation (DFG) under grant OS 295/4-1.

References

1. Bordes, A., Bottou, L., Gallinari, P., Weston, J.: Solving multiclass support vector machines with LaRank. In: ICML, pp. 89–96. ACM (2007)
2. Cambria, E., White, B.: Jumping NLP curves: a review of natural language processing research. IEEE Comput. Intell. Mag. **9**(2), 48–57 (2014)
3. Chang, C.C., Lin, C.J.: LIBSVM: a library for support vector machines. ACM Trans. Intell. Syst. Technol. (TIST) **2**(3), 27 (2011)
4. Chen, M.H., Schmeiser, B.W.: General hit-and-run Monte Carlo sampling for evaluating multidimensional integrals. Oper. Res. Lett. **19**(4), 161–169 (1996)
5. Cortes, C., Vapnik, V.: Support-vector networks. Mach. Learn. **20**(3), 273–297 (1995)

6. Gilks, W.R., Roberts, G.O., George, E.I.: Adaptive direction sampling. Statistician **43**, 179–189 (1994)
7. Herbrich, R., Graepel, T., Campbell, C.: Bayes point machines. J. Mach. Learn. Res. **1**, 245–279 (2001)
8. Hofmann, T., Schölkopf, B., Smola, A.J.: Kernel methods in machine learning. Ann. Stat. **36**, 1171–1220 (2008)
9. Homan, M.D., Gelman, A.: The no-u-turn sampler: adaptively setting path lengths in Hamiltonian Monte Carlo. J. Mach. Learn. Res. **15**(1), 1593–1623 (2014)
10. Hosmer Jr., D.W., Lemeshow, S., Sturdivant, R.X.: Applied Logistic Regression, vol. 398. Wiley, Hoboken (2013)
11. Lichman, M.: UCI machine learning repository (2013). http://archive.ics.uci.edu/ml
12. Lovász, L., Vempala, S.: Hit-and-run from a corner. SIAM J. Comput. **35**(4), 985–1005 (2006)
13. Luettel, T., Himmelsbach, M., Wuensche, H.J.: Autonomous ground vehicles concepts and a path to the future. In: Proceedings of the IEEE 100 (Special Centennial Issue), pp. 1831–1839 (2012)
14. Minka, T.P.: Expectation propagation for approximate Bayesian inference. In: UAI, pp. 362–369. Morgan Kaufmann Publishers Inc., San Francisco (2001)
15. Mosimann, J.E.: On the compound multinomial distribution, the multivariate β-distribution, and correlations among proportions. Biometrika **49**(1/2), 65–82 (1962)
16. Neal, R.M.: Slice sampling. Ann. Stat. **31**, 705–741 (2003)
17. Nguyen, T., Sanner, S.: Algorithms for direct 0–1 loss optimization in binary classification. In: ICML, pp. 1085–1093 (2013)
18. PEGWiki: Convex hull trick (2016)
19. Platt, J., et al.: Probabilistic outputs for support vector machines and comparisons to regularized likelihood methods. Adv. Large Margin Classif. **10**(3), 61–74 (1999)
20. Ripley, B.D.: Pattern Recognition and Neural Networks. Cambridge University Press, Cambridge (2007)
21. Robert, C., Casella, G.: Monte Carlo Statistical Methods. Springer, New York (2013)
22. Ruján, P.: Playing billiards in version space. Neural Comput. **9**(1), 99–122 (1997)
23. Silver, D., Huang, A., Maddison, C.J., Guez, A., Sifre, L., Van Den Driessche, G., Schrittwieser, J., Antonoglou, I., Panneershelvam, V., Lanctot, M., et al.: Mastering the game of go with deep neural networks and tree search. Nature **529**(7587), 484–489 (2016)
24. Thompson, M., Neal, R.M.: Covariance-adaptive slice sampling. arXiv preprint arXiv:1003.3201 (2010)
25. Tipping, M.E.: Sparse Bayesian learning and the relevance vector machine. J. Mach. Learn. Res. **1**, 211–244 (2001)
26. Zhu, J., Hastie, T.: Kernel logistic regression and the import vector machine. In: NIPS, pp. 1081–1088 (2001)

ConvNet Regression for Fingerprint Orientations

Patrick Schuch[1,2](✉), Simon-Daniel Schulz[2], and Christoph Busch[1]

[1] NTNU, Gjøvik, Norway
{patrick.schuch2,christoph.busch}@ntnu.no
[2] Dermalog Identification Systems GmbH, Hamburg, Germany
simon.schulz@dermalog.com

Abstract. Estimation of orientation fields is a crucial task in fingerprint recognition. Many processing steps depend on their precise estimation and the direction of fingerprint minutiae is a valuable information. But especially for regions of low quality the task is not trivial and engineered approaches on local features may fail. Methods that combine local and global features learned from the data are state of the art and benchmarked with the framework FVC-ongoing. We propose to use Convolutional Neural Networks trained in a regression to estimate the orientation field (ConvNetOF). Regression is more accurate than classification in this case. Our approach achieves an RMSE of 8.53° on the Bad Quality Dataset of the FVC-ongoing benchmark. This is the best result reported so far.

Keywords: Fingerprint recognition · Orientation field · ConvNet · Regression

1 Motivation and Introduction

Fingerprint recognition is one of the most wide spread biometric modalities, when it comes to identification and verification of individuals. Recognition algorithms make use of the distinctive features in the fingerprints. Fingerprint minutiae are features, which are typically used for recognition. Minutiae are characteristic points of the papillary ridges, e.g. an ending and a bifurcation [13]. The spatial distribution and relations of positions and directions of minutiae are unique for every finger which allows to distinguish fingerprints.

The direction of a fingerprint minutia is one of its most valuable informations for recognition besides its type and position. It directly depends on the local orientation at its location. The orientation field (OF) of the papillary ridges (see Fig. 3a) is itself another important feature in fingerprint recognition [13].

Besides this, the OF is relevant information for image enhancement and many processing steps along the workflow of a biometric feature extraction [13]. Deviations between the estimation and the real OF have to be as small as possible for the whole fingerprint area [3]. Otherwise biometric features may not be extracted correctly or spurious features may be generated.

© Springer International Publishing AG 2017
P. Sharma and F.M. Bianchi (Eds.): SCIA 2017, Part I, LNCS 10269, pp. 325–336, 2017.
DOI: 10.1007/978-3-319-59126-1_27

Because of this, an accurate and reliable estimation of the OF is needed for fingerprint recognition. But an accurate estimation is challenging especially for low-quality fingerprint images.

Techniques and ideas for estimating the OF are vast. They can roughly be divided into local and global techniques [3]. Local techniques are based on the very vicinity of every point, e.g. by calculation of local gradients on grey-values in fingerprint images. Those techniques often are not reliable in areas of low quality [3]. In contrast, global techniques usually take benefit of models for the global OF (see Fig. 1a–c for typical patterns of OFs). The drawback in constructing OFs is that this tends to overly smooth local irregularities and regions of high curvature. In consequence hypothesized models are insufficiently representing the actual OF. Computational complexity in general is higher for global methods than for local ones [3]. Hybrid versions of both try to compensate the drawbacks. However, especially for images of low quality the estimation of the OF is still challenging. The Fingerprint Verification Contest (FVC-ongoing) is providing a benchmark area for fingerprint orientation extraction [9]. As results of this benchmark show, deviations between estimated and real OF are still significantly higher for low quality fingerprint images than for images of higher quality [8]. Closing this gap is one key factor for a more accurate and more reliable fingerprint recognition.

Recently, methods of machine learning, which combine local and global features and furthermore learn directly from the data seem to become a promising solution for OF estimation [20]. In general, techniques which learn from the data, have shown their superiority over engineered techniques in the last decade for various image processing tasks. Techniques from the domain of Deep Learning (DL) and especially Convolutional Neural Nets (ConvNets) are state of the art at numerous benchmarks, e.g. ILSVR [16]. Significant improvements have been achieved by DL in the domains of Speech Recognition, Signal Processing, Object Recognition, Natural Language Processing, and especially in Multi-Task and Transfer Learning [2].

The versatility of ConvNets and Deep Learning techniques enables them to estimate the OF of fingerprints. Our approach is to train a ConvNet as a regression. This allows to learn an estimation for the continuous valued OF directly from the data.

The rest of the paper is organized as follows: Related work in terms of OF estimation and benchmarking of proposed approaches is discussed in Sect. 2. Our suggested approach will be explained in Sect. 3. Section 4 summarizes the results and conclusions are made in Sect. 5. Section 6 adds remarks on the findings of this work and gives an outlook on future work.

2   Related Work

2.1   State of the Art: Benchmarking

Benchmarks are inevitable for a quantitative evaluation and comparison of approaches. The University of Bologna provides such a public benchmark framework for specific tasks in biometric recognition: *FVC-ongoing* [9]. It also contains a benchmark for *Fingerprint Orientation Extraction* (FOE). The benchmark is on-going and it allows to measure performance of algorithms for fingerprint orientation estimation. Implemented algorithms can be uploaded and tested. FVC-ongoing is the only benchmark offering independent measurements on common sequestered dataset and defined metrics for this task. Therefore we report our results based on the quantitative measurements provided by FVC-ongoing.

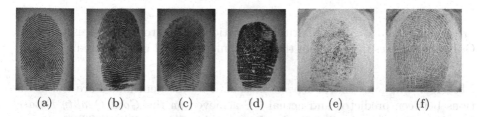

    (a)   (b)   (c)   (d)   (e)   (f)

Fig. 1. (a–c) show images of good quality. The orientations differ between fingers and form typical patterns. (d–f) show examples of images with lower quality representing typical challenges. Quality of a fingerprint image can be affected by the moisture of the finger and many other factors. (d) shows a sample with very moist skin, where the fingerprint in (e) is rather dry. In addition, (f) shows scars.

Data Set. The benchmark consists of two data sets. Dataset *FOE-TEST* is available for evaluation purposes by the contestants. Dataset *FOE-STD-1.0* is available only to the organizers of the benchmark. Both training and test set are divided into two categories: images of good and images of bad quality. According to their image's quality, the sets are called *Good Quality Dataset* and *Bad Quality Dataset* [7]. For the *Good Quality Dataset* 10 samples are provided, while for the *Bad Quality Dataset* 50 samples are provided (see Fig. 1 for examples). About 90,000 training data points are provided which represent the local orientation at a single pixel of an eight-fold sub-sampling grid (see Table 1).

The images are captured with fingerprint livescanners at a resolution of 500dpi. The *Bad Quality Dataset* shows typical challenges in processing fingerprint images. This set consists of images from fingers with different levels of skin moisture (compare the wet finger in Fig. 1d to the dry one in Fig. 1e) and the presence of scars in the fingerprint (see Fig. 1f). The data is close to what operational data of low quality may look like. The ground truth label data has been produced by manual labelling [8]. The orientation is sampled at an equidistant grid and angles are provided in 256 steps. Labelling is carried out with support of a tool introduced by Maltoni et al. [7].[1] Additionally to the finger-

---

[1] The workflow for labelling is roughly as follows (see [8] for details): A human expert selects a pixel location, which he wants to label. The tool estimates the local orientation by calculating the gradient. The expert may choose to accept the orientation estimate provided by the tool or do a manual correction. A Delaunay triangulation on all labelled points is performed. Each sampling point will be interpolated based on the supporting points of its surrounding Delaunay triangle.

Table 1. FVC provides datasets consisting of a Good and a Bad Quality dataset each.

Set	Name	Number of samples	Number of data points
FOE-TEST	Good Quality Dataset	10	18946
	Bad Quality Dataset	50	75812
FOE-STD	Good Quality Dataset	10	19260
	Bad Quality Dataset	50	89562

print images and the ground truth orientation, a foreground mask is provided. Only OF samples which are in the foreground area will be evaluated.

Metrics. The four central aspects measured by the benchmark are the deviations between predicted and actual OF achieved on the *Good Quality Dataset* (AvgErr$_{GQ}$) and on the *Bad Quality Dataset* (AvgErr$_{BQ}$) of *FOE-STD1.0*, memory consumption, and average processing time for each sample. The measure for the OF deviation is the average Root Mean Squared Error (RMSE) observed at all data points. RMSE averages over all sampling points in the fingerprint area in a single sample image. One may argue that deviation might be more important in highly curved regions than in regions of more or less constant orientation, e.g. regions around OF singularities are highly curved. Accurate estimation in those regions is necessary for localization of singular points. In contrast, the benchmark organizers argue that weighting all points equally is suited well for most of the other feature extraction tasks where orientation is needed [8]. The most important measure is AvgErr$_{BQ}$ since this metric quantifies the ability of algorithms to handle challenging images.

2.2 State of the Art: Algorithms

Many ideas for fingerprint OFs estimation have been proposed. A broad survey of OF estimations with qualitative assessments is given e.g. by Biradar et al. [3]. But only seven results have been published for FVC-ongoing so far. The two approaches *LocalDict* and *ROF* are performing best in terms of minimizing the deviation achieved on the *Bad Quality Dataset* of FVC-ongoing. Therefore, those methods are worth a closer inspection and will be described below.

Yang et al. provide the best performing algorithm yet called *LocalDict* [20]. They propose to learn dictionaries of OF patterns. The dictionary contains prototypes for local orientation patterns. In a second step, co-occurrence and spatial distribution of the prototypes is learned. Those aspects represent the global structure of fingerprint OFs. Thus, the proposed algorithm combines local and global information. The algorithm first learns a rough estimate of the OF. The locally best fitting prototype is assigned to each point. Finally, corrections of assigned prototypes are performed to optimize likelihood of spatial co-occurrence of the prototypes.

Cao et al. proposed an algorithm they call *ROF*. It extracts first an estimation of the OF by the gradient method applied to a root filtered image [6]. The OF is represented as the gradient vector field. In addition, the positions of singularities are estimated. The idea is to smooth the OF while keeping divergence and coherence of the orientation vector field. Intensity of smoothing is varied according to a specific local quality and the distance to a near-by singularity. Thus, areas of high quality and those close to singularities will be smoothed less.

Using Neural Networks and utilizing learning from the data for fingerprint recognition has been suggested previously. Baldi et al. already proposed to use a structure like modern Siamese ConvNets (without pooling layers) for fingerprint indexing already in 1993 [1]. Zhu et al. used a Multi-Layer Perceptron to estimate a 16 step quantization of the OF in 2006 [21]. Olsen et al. used self organizing networks to estimate fingerprint sample quality [14].

Using techniques from DL especially for OF estimation is a more recent development. Sahasrabudhe et al. proposed to use Restricted Boltzman Machines (RBM) to estimate fingerprint OFs [17]. An RBMs is probabilistic model which uses a bi-directional neural network. RBMs therefore are not straight feed-forward. An initial OF is estimated and the estimation is vectorized into x and y components. Each component is fed into a separate single-layer RBM. The trained weights of an RBM contain representations for the data used for training which form a basis. Trained RBMs try to approximate the input by this basis. The corresponding output can be interpreted as the best fit to the learned representations. Thus, the output is like a corrected version of the input, which fits best the learned data. The corrected OFs are used to enhance fingerprint images. Finally, performance is measured in terms of the number of spurious minutia extracted by a biometric feature extraction on the enhanced images and in term of the accuracy a biometric comparison algorithm achieved with such extracted features.

The most relevant work with respect to its methodology is an approach by Cao et al. which proposes to use a ConvNet trained as a classifier for orientation [5]. They propose to train a ConvNet for a classification task. Target labels for the classes are a selection of 128 characteristic OFs, which have to be selected beforehand. Cao et al. propose to use engineered noise to corrupt input images. This in turn shall simulate artefacts one in fingerprint images of bad quality.

3 Proposed Approach

3.1 Idea

We propose to train ConvNets as a regression to estimate the OF in fingerprint images. During the training for a regression, a ConvNet *model* $\mathcal{M}$ usually learns to minimize the quadratic error between its propagation $\mathcal{M}(inp)$ and the target value $T(inp)$ for a given input inp:

$$\min_{\mathcal{M}} ||T(inp) - \mathcal{M}(inp)||^2 \tag{1}$$

During testing, for an input $\hat{inp}$ the model $\mathcal{M}$ will create a prediction $\mathcal{M}(\hat{inp})$.

A ConvNet model $\mathcal{M}$ itself is assembled from multiple sets (layers) of trainable filter kernels. The output of each layer is fed into the next layer of filters. While such models learn simple local features in the first layers, the following layers learn more complex and more global features.

By doing so, our approach utilizes learned local and global characteristics at once. This turned out to be a successful strategy in the *LocalDict* approach. Our approach has three advantages over *LocalDict*. First, our approach does not need an initial estimation of the OF. Second, ConvNets use sparse representations for information. This enables a more flexible representation than the one-hot representation used by *LocalDict*. Third, our approach is an end-to-end solution, i.e. input is the raw grey-value image and the corresponding foreground mask and the output is an estimation of the OF. No separate processing steps need to be carried out. No special assumptions about the spatial distribution have to be separately modelled by learned data.

We train the model as a regression on a vectorization of the target orientations. Compared to Cao et al.'s classification approach, regression is a more natural approach for the estimation of continuous values. In addition, no selection of target patterns is necessary.

3.2   Model Architecture

The proposed ConvNet has been trained in the framework *Caffe* provided by Jia et al. [10]. Our approach combines three different types of layers provided in *Caffe*: Convolutional layers, Pooling layers and non-linear transfer layers. Gray values of the fingerprint images are normalized in the foreground area to have zero mean and unit standard deviation while the background is set to zero (Normalization). To enforce the same image dimensions for all training samples the images are embedded into a larger canvas.

Neurons in ConvLayer work like filter kernels (see Fig. 3b for trained filters). Pooling layers perform a reduce operation on the local neighbourhoods. They therefore work like sub-sampling. The pooling functions in this approach is the maximum over all local values. The layers are therefore called Max-Pooling. Non-linear transfer units simply apply a non-linear function to each input value, e.g. $ReLU(x) = \max(0, x)$.

Our ConvNet is designed for the special needs in OF estimation (see Fig. 2a). Accurate local estimations are needed as well as regional smoothness and global patterns. It therefore differs from typically very deep cascade of same 3×3 filters. The original fingerprint image is normalized in a Normalization layer. The normalized image is filled into a larger canvas of 576×464 pixels in an Embedding layer. In the following blocks of ConvLayers and ReLU layers are concatenated. Additionally, in the first three blocks MaxPooling layers are used to sub-sample the image dimensions to the provided target dimensions. Filter sizes are designed to cover half of the width of a typical fingerprint ridge. This is done to ensure good local estimations. In the next three blocks larger ConvLayers of dimension $13 \times 13 \times 49$ each are used for regional smoothness. All ConvLayers

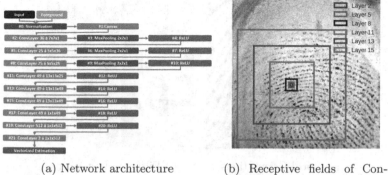

(a) Network architecture   (b) Receptive fields of ConvLayers

Fig. 2. Block diagram of the layout of the evaluated model $\mathcal{M}$ which calculates a vectorized estimation of the OF for a given input fingerprint sample and a given foreground mask. $\mathcal{M}$ consists of Normalization, canvas, Convolutional (ConvLayer), Pooling (MaxPooling), and Rectified Linear Unit (ReLU) layers. The receptive field, which is processed from the original image, increases with each layer. While the filters work locally in the first ConvLayers, the last ConvLayers work in a more global range. In this way, local and global features can be combined by a ConvNet.

have a striding of 1 and do padding to equalize height and width of input and output. The subsequent combination of MaxPooling and larger kernel leads to a larger receptive field for each output of layer 15, i.e. all pixels within the turquoise area contribute to the output value of layer 15 in Fig. 2b. This allows to combine the local features to more global ones. Usually, so-called fully-connected layers are used at the end of the cascade of layers. In a fully-connected layer each neuron is connected with every input as in classical Multi-Layer Perceptrons. We use ConvLayers with kernel height and width of 1 as a proxy for such fully-connected layers at the end of our layer cascade [12]. The final layer has two output channels which estimating the vectorized target orientation.

3.3 Training Algorithm

The ConvNet in our approach has been trained using a Stochastic Gradient Descent [4]. The cost function is formulated as a quadratic regression on the two-component representation of the orientation θ at input inp:

$$\min_{\mathcal{M}} \left\| \begin{pmatrix} \sin(2 \cdot \theta(inp)) \\ \cos(2 \cdot \theta(inp)) \end{pmatrix} - \mathcal{M}(inp) \right\|^2 \tag{2}$$

The images and ground truth orientation data from FOE-TEST are taken as training data. This data provides 94,758 labelled targets for training. Figure 3a visualizes a typical representative taken from the training data. The model $\mathcal{M}$ has 1,347,967 parameters. A cost function for large weights is added for a so-called *Weight Decay* to enforce generalisation [11]. Since large weights induce

332   P. Schuch et al.

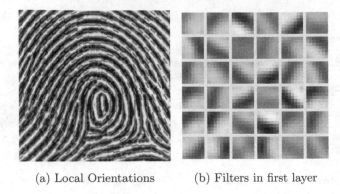

(a) Local Orientations   (b) Filters in first layer

Fig. 3. The training data provides fingerprint images and the hand-labelled ground truth orientation, indicated here as red lines. The filter kernels in the first ConvLayer work like edge filters and do a rough estimation of this orientation.

high costs, weight decay punishes over-specialisation of weights and therefore prevents over-training.

Parameters for training are the following: Weight decay factor is 10. Starting learning rate is 10^{-5} and adapted according to Inverse Decay policy with $\gamma = 10^{-4}$ and a power of 0.75.

Figure 3b visualizes the filter kernels of the first ConvLayer after training. The kernels of the first ConvLayer work similar to edge filters. The next ConvLayer recombines those features to more complex features. Figure 4 visualizes the output of all ConvLayers for a fingerprint sample after training. The outputs of some filters have only very little absolute values. This is an effect of Weight Decay reducing the energy of unnecessary filter kernels.

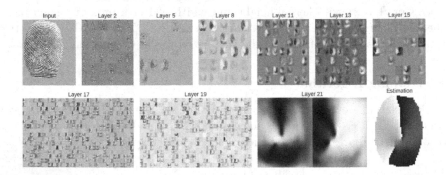

Fig. 4. The output of all ConvLayers for a single input fingerprint sample. Layer 21 represents the estimation for the vectorized OF of the input image. The vectorized OF can be used to calculate the final OF estimation. *Weight decay* can prevent a model $\mathcal{M}$ from over-fitting. As a result of weight decay the output of some kernels has low absolute values.

4   Results

As mentioned in Sect. 2.1, the four central aspects observed in the benchmark are the deviations achieved on the *Good Quality Dataset* and on the *Bad Quality Dataset* of *FOE-STD1.0*, processing time, and memory consumption. Figure 5 visualizes the four aspects for all reported results. The benchmark organizers do not provide a overall ranking based on the four aspects.

The reported deviations $AvgErr_{GQ}$ for all algorithms do not outperform the baseline algorithm significantly (see Table 2). Performing well for images of good quality therefore does not seem to be challenging even for simple algorithms. The error rates on this set range from $5.24°$ to $6.7°$ while the baseline algorithm achieves an error rate of $5.86°$. ConvNetOF achieves $5.80°$. The local information extracted by the baseline algorithm is just sufficient.[2] However, the deviation $AvgErr_{GQ}$ can be taken into account as a lower bound for the deviations $AvgErr_{BQ}$.

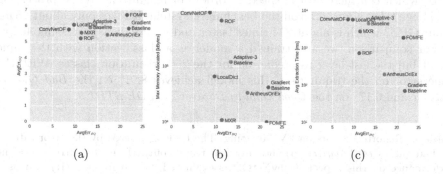

(a)   (b)   (c)

Fig. 5. Visualization of the reported results for the benchmark FVC-ongoing FOE. (a) shows a scatter plot for the deviations achieved on both datasets. Four algorithms outperform the baseline with respect to the deviation of the Bad Quality Dataset. (b) reveals that memory consumption varies significantly between these top four. A trade-off between Speed and Accuracy can be observed in (c).

To our mind, the most important aspect is the deviation achieved on the *Bad Quality Dataset* for *FOE-STD1.0*. Here deviations vary more than on the *Good Quality Dataset*: they range from $9.66°$ to $21.83°$. Our approach ConvNetOF achieves $8.53°$. This reduces the deviation to about 88% relative to the former

---

[2] Outperforming the baseline algorithm for $AvgErr_{GQ}$ might be challenging for a good reason: The *Gradient* algorithm is more or less used to generate the ground truth. One can assume that for good quality images the human editor might consider the initial estimation to be right even though it might show a systematic bias by the algorithm. In contrast, for bad images manual correction might be obvious to the human editor. However, FVC-ongoing still remains the best mean to compare OF estimators.

best result. Figure 5a visualizes the error rates for all reported results (compare to Table 2). On training data *FOE-TEST* ConvNetOF achieves 5.14°.

Timing and memory constraints strongly depend on the application. In Fig. 5b the memory consumption is plotted against the deviation on the *Bad Quality Dataset*. In general, memory consumption is seldom a limitation and may only be a critical issue for systems with very limited memory, e.g. for SmartCards. The consumption of memory varies by orders of magnitudes and ConvNetOF has the highest requirements for memory among all evaluated algorithms.

Figure 5c reveals a trade-off between deviation and average processing time: The longer the computation time, the more accurate the result. Like in memory consumption, the reported top results vary in their average processing time in an order of magnitude. However, all algorithms are way faster than the time limit of 60 s per sample allowed by the benchmark framework. Our approach takes the longest time to process an image. This is due to the evaluation performed on a CPU at FVC-ongoing. However, ConvNets are suited for operation on a GPU, which can increase the speed in orders of magnitude: While processing one image takes about 6.1 s on the benchmark system, it takes only about 25 ms of our GPU[3] which is about 244 times faster and would allow processing 40 fps.

For comparison, we also trained a model as a classification with the same layout but the last layer as prediction for the 256 orientation classes. With the same error on the training set, this model achieves 8.91° on the *Bad Quality Dataset* and 6.17° on the *Good Quality Dataset* for *FOE-STD1.0*.

Table 2. Reported results on FVC-ongoing. The table is ordered by the error rate on the bad quality set ($AvgErr_{BQ}$) and only four results outperform the current baseline performance on this aspect. ConvNetOF is evaluated as a regression (R) and as a classification (C). As a regression it performs best among all evaluated algorithms on this data set. Best algorithm per aspect is marked bold. For result of ConvNetOF see https://biolab.csr.unibo.it/FvcOnGoing/UI/Form/AlgResult.aspx?algId=5604

Algorithm	AvgErr$_{BQ}$ [°]	AvgErr$_{GQ}$ [°]	Avg. Time [ms]	Max Mem. [kBytes]	Ref
ConvNetOF (R)	**8.53**	5.80	6,096	939,212	
ConvNetOF (C)	**8.91**	6.17	6,257	943,888	
LocalDict	9.66	6.08	5,987	67,544	[20]
ROF	11.20	**5.24**	762	671,984	[6]
MXR	11.36	5.59	2,937	11,140	n/a
Adaptive-3	13.27	5.93	4,772	121,936	[18]
AntheusOriEx	17.06	5.46	205	34,176	n/a
FOMFE	21.44	6.70	1,996	**10,196**	[19]
Gradient	21.83	5.86	**74**	42,872	[15]

---

[3] An NVIDIA GTX 780 has been used for evaluation.

5   Conclusion

We have proposed to use ConvNets trained in a regression to estimate the OF of fingerprints. Our approach has been evaluated on the benchmark framework FVC-ongoing, which is the most relevant benchmark for estimation of OF. ConvNetOF achieves a deviation of 8.53° on the *Bad Quality Dataset*. Our approach therefore outperforms all other algorithms in this aspect. The deviation on bad quality images is lowered to about 88% relative to the second best result. This narrows the performance gap between the estimation of OF on images of good and those of bad quality. The performance of ConvNetOF on the *Good quality Dataset* is competitive to the other evaluated algorithms. The model trained as a regression outperforms the model trained as a classification. We found a generalization gap between training and testing.

In terms of memory consumption our approach has the highest requirements among all evaluated algorithms. Using a GPU it outperforms all other approaches in terms of speed.

6   Discussion and Outlook

The trained model is likely to be over-sized for this task. Inspection of the trained ConvNet reveals that some filter kernels may be obsolete. For application it would be reasonable to reduce the size of the ConvNet. This would not only make it faster and less memory consuming but it would also prevent over-training. However, runtime optimization is out of scope for this work.

Some remarks on the benchmark FOE of FVC-ongoing seem worth mentioning. The number of images (especially for the *Good Quality Dataset*) is small. In addition, the ground truth for the orientation may be biased since it has been edited by a human expert who manually corrects the output of an OF extraction algorithm. Both facts in combination are bad circumstances for learning from the data. Evaluations on larger datasets seem reasonable.

References

1. Baldi, P., Chauvin, Y.: Neural networks for fingerprint recognition. Neural Comput. **5**(3), 402–418 (1993)
2. Bengio, Y., Courville, A., Vincent, P.: Representation learning: a review and new perspectives. IEEE Trans. Pattern Anal. Mach. Intell. **35**(8), 1798–1828 (2013)
3. Biradar, V.G., Sarojadevi, H.: Article: fingerprint ridge orientation extraction: a review of state of the art techniques. Int. J. Comput. Appl. **91**(3), 8–13 (2014)
4. Bottou, L.: Large-scale machine learning with stochastic gradient descent. In: Lechevallier, Y., Saporta, G. (eds.) COMPSTAT 2010, pp. 177–186. Springer, Heidelberg (2010)
5. Cao, K., Jain, A.: Latent orientation field estimation via convolutional neural network. In: 2015 International Conference on Biometrics (ICB), pp. 349–356, May 2015

6. Cao, K., Liang, J., Tian, J.: A div-curl regularization model for fingerprint orientation extraction. In: 2012 IEEE Fifth International Conference on Biometrics: Theory, Applications and Systems (BTAS), pp. 231–236. IEEE (2012)
7. Cappelli, R., Maio, D., Maltoni, D.: Semi-automatic enhancement of very low quality fingerprints. In: Proceedings of 6th International Symposium on Image and Signal Processing and Analysis, 2009. ISPA 2009, pp. 678–683. IEEE (2009)
8. Cappelli, R., Maltoni, D., Turroni, F.: Benchmarking local orientation extraction in fingerprint recognition. In: 2010 20th International Conference on Pattern Recognition (ICPR), pp. 1144–1147. IEEE (2010)
9. Dorizzi, B., et al.: Fingerprint and on-line signature verification competitions at ICB 2009. In: Tistarelli, M., Nixon, M.S. (eds.) ICB 2009. LNCS, vol. 5558, pp. 725–732. Springer, Heidelberg (2009). doi:10.1007/978-3-642-01793-3_74
10. Jia, Y., Shelhamer, E., Donahue, J., Karayev, S., Long, J., Girshick, R., Guadarrama, S., Darrell, T.: Caffe: convolutional architecture for fast feature embedding. arXiv preprint arXiv:1408.5093 (2014)
11. Krogh, A., Hertz, J.A.: A simple weight decay can improve generalization. In: Advances in Neural Information Processing Systems, vol. 4, pp. 950–957 (1995)
12. Lin, M., Chen, Q., Yan, S.: Network in network. arXiv preprint arXiv:1312.4400 (2013)
13. Maltoni, D., Maio, D., Jain, A.K., Prabhakar, S.: Handbook of Fingerprint Recognition. Springer, London (2009)
14. Olsen, M.A., Tabassi, E., Makarov, A., Busch, C.: Self-organizing maps for fingerprint image quality assessment. In: 2013 IEEE Conference on Computer Vision and Pattern Recognition Workshops (CVPRW), pp. 138–145. IEEE (2013)
15. Ratha, N.K., Chen, S., Jain, A.K.: Adaptive flow orientation-based feature extraction in fingerprint images. Pattern Recognit. **28**(11), 1657–1672 (1995)
16. Russakovsky, O., Deng, J., Su, H., Krause, J., Satheesh, S., Ma, S., Huang, Z., Karpathy, A., Khosla, A., Bernstein, M., Berg, A.C., Fei-Fei, L.: ImageNet large scale visual recognition challenge. Int. J. Comput. Vis. (IJCV) **115**(3), 211–252 (2015)
17. Sahasrabudhe, M., Namboodiri, A.M.: Learning fingerprint orientation fields using continuous restricted Boltzmann machines. In: 2013 2nd IAPR Asian Conference on Pattern Recognition (ACPR), pp. 351–355. IEEE (2013)
18. Turroni, F., Maltoni, D., Cappelli, R., Maio, D.: Improving fingerprint orientation extraction. IEEE Trans. Inf. Forensics Secur. **6**(3), 1002–1013 (2011)
19. Wang, Y., Hu, J., Phillips, D.: A fingerprint orientation model based on 2D Fourier expansion (FOMFE) and its application to singular-point detection and fingerprint indexing. IEEE Trans. Pattern Anal. Mach. Intell. **29**(4), 573–585 (2007)
20. Yang, X., Feng, J., Zhou, J.: Localized dictionaries based orientation field estimation for latent fingerprints. IEEE Trans. Pattern Anal. Mach. Intell. **36**(5), 955–969 (2014)
21. Zhu, E., Yin, J.P., Zhang, G.M., Hu, C.F.: Fingerprint ridge orientation estimation based on neural network. In: Proceedings of the 5th WSEAS International Conference on Signal Processing, Robotics and Automation. ISPRA 2006, pp. 158–164 (2006)

Domain Transfer for Delving into Deep Networks Capacity to De-Abstract Art

Corneliu Florea[✉], Mihai Badea, Laura Florea, and Constantin Vertan

Image Processing and Analysis Laboratory,
Univeristy Politehnica of Bucharest, Bucharest, Romania
corneliu.florea@upb.ro

Abstract. Humans are capable of perceiving a natural scene at a glance and painters, through their representations, push this capacity to the limit by abstraction. Yet we still do it... What about machines? In this paper we address the problem of recognizing the theme (scene type) in digitized paintings. The approach is based on the Convolutional Neural Network framework and the chosen architecture is the recent Residual Network. In the first level of evaluation, we determine the recognition rate of a CNN given a database of 80,000 annotated digitized paintings. In the second level we evaluate the impact of extending the training database with photographs directly and via two domain adaptation functions and thus we are able to assess the abstraction level that CNN is capable to achieve.

1 Introduction

Pablo Picasso said "There is no abstract art. You must always start with something. Afterward you can remove all traces of reality." As art follows humanity through its entire history, if one integrates with respect to the level of abstraction he will note that the closer to the present moment we are, the more traces of reality have been removed.

In last period two trends favored the apparition of works similar to this one. First there were consistent efforts to digitize more and more paintings so that modern systems may learn from large databases. Two of such popular efforts are Your Paintings (now Art UK[1]) which contains more than 200.000 paintings and WikiArt[2] with around 100.000 paintings. The databases come with multiple annotations. For this work we are particulary interested in annotations about the painting's theme or scene type. From this point of view, a more complete database is the WikiArt collection, where the labelling category is named *genre*.

Secondly the development of the Deep Neural Networks allowed classification performance that was not imagined before. Here, we will inspire from the use of the more popular Convolutional Neural Networks (CNN). Given the achievable

---

C. Florea and M. Badea—Equally contributed.

[1] http://artuk.org/.
[2] http://www.wikiart.org/.

© Springer International Publishing AG 2017
P. Sharma and F.M. Bianchi (Eds.): SCIA 2017, Part I, LNCS 10269, pp. 337–349, 2017.
DOI: 10.1007/978-3-319-59126-1_28

performance, the focus may switch from improving the performance to its use in practical tasks.

Starting from the idea that the Deep Neural Networks share similarities with the human vision [7] and the fact that such networks are already proven to do great jobs in other perception inspired areas like object recognition or even in creating artistic images, we ask ourselves if they can pass the abstraction limit and correctly recognize the scene type of a painting. We will first compare the results of residual network (ResNet) on the standard WikiArt database with previous methods from state of the art. We will then test different domain transfer augmentations to see if they can help increase the achieved recognition rate and also if the network is capable to pass the abstraction limit and learn from different types of images that contain the same type of scenes. Furthermore, we introduce several alternatives for domain transfer to achieve a dual-task: improve the scene recognition performance and understand the abstraction capabilities of machine learning systems.

Regarding the deep networks, multiple improvements have been proposed. In many situations, if given database is smaller, better performance is reachable if the network parameters are previously found for a different task on large database such as ImageNet. Next, these values are updated to a given task. This is called fine-tuning and it is a case of transfer learning. As our investigation is related to a different domain transfer we will avoid to use both so to establish clearer conclusions. To compensate, we are relying on the recent architecture of the residual networks (Resnet [16]) that was shown to be able to overcome the problem of vanishing gradients, reaching better accuracy for the same number of parameters when compared to previous architectures.

The remainder of the paper is organized as follows: Sect. 2 presents previous relevant work, Sect. 3 summarizes the CNN choices made and Sect. 4 will discuss different aspects of painting understanding. Section 5 presents the used databases, while implementation details and results are in Sect. 6. The paper ends with discussions about the impact of the results.

2   Related Work

Object and Scene Recognition in Paintings. Computer based painting analysis has been in the focus of the computer vision community for a long period. A summary of various directions approached, algorithms and results for not-so-recent solutions are in the review of Bentowska and Coddington [5]. However the majority addressed style (art movement) or artist recognition. Object recognition has been in the focus of Crowley and Zisserman [9] while searching through YourPaintings dataset with learning on photographic data.

Scene recognition in paintings is also named genre recognition following the labels from the WikiArt collection. This topic was approached by Condorovici et al. [8] and by Agarwal et al. [1]; both works, using the classical feature + classifier approach, tested smaller databases with few (5) classes: 500 images - [8] and 1500 images - [1]. More extensive evaluation, using data from WikiArt, was performed by Saleh and Elgammal [22], which investigated an extensive list of visual features and metric learning to optimize the similarity measure between paintings and respectively by Tan et al. [25], which employed an AlexNet architecture [19] of CNN initialized (fine tuned) on ImageNet to recognize both style and genre of the painting.

The process of transferring knowledge from natural photography to art objects has also been previously addressed beyond the recent transfer from ImageNet to Wikiart [25]. 3D object reconstruction can be augmented if information from old paintings is available [3]. Classifiers (deep CNNs) trained on real data are able to locate objects such as cars, cows and cathedrals [10]. The problem of detecting/recognizing objects in any type of data regardless if it is real or artistic was named cross-depiction by Hall et al. [15]; however the problem is noted as being particular difficult and even in the light of dedicated benchmarks [6], as the results show a lot of place for improvement. Another comment is that all solutions that showed some degree of success did it for older artistic movements where scene depiction was without particular abstraction. To our best knowledge there isn't any significant success for more modern art.

Of particular interest for our work is the algorithm recently introduced by Gatys et al. [14], which used various layers of CNN trained for object recognition to separate the content from the style of an image and to enable style transfer between pairs; the most impressive results are in the transfer of artistic style to photographs rendering the later as being painted in rather abstract ways.

Scene Recognition in Photographs. Scene recognition in natural images is an intensively studied topic, but under the auspices of being significantly more difficult than object recognition or image classification [30]. We will refer the reader to a recent work [17] for the latest results on the topic. We merely note that the introduction of the SUN database [27] (followed by expansions) placed a significant landmark (and benchmark) on the issue and that it was shown that using domain transfer (e.g. from Places database), the performance may be improved [30].

Scene Recognition by Humans. While it is outside the purpose of this paper to discuss detailed aspects of the human neuro-mechanisms involved in scene recognition, following the integrating work of Sewards [23] we stress one aspect: compared to object recognition where localized structures are used, for scene recognition the process is significantly more tedious and complex. Object recognition, "is solved in the brain via a cascade of reflexive, largely feedforward computations that culminate in a powerful neuronal representation in the inferior temporal cortex" [11]. In contrast, scene recognition includes numerous and complex areas as the process starts with peripherical object recognition, continues with central object recognition, activating areas such entorhinal cortex, hippocampus and subiculum [23].

Concluding, there is consensus, from both neuro-scientist and from the computer vision community that scene recognition is a particularly difficult task.

This task becomes even harder when the subject images are heavily abstracted paintings from modern art.

3 Architecture and Training

In the remainder of the paper, we will use the Residual Network (ResNet) [16] architecture with 34 layers. All the hyper-parameters and the training procedure follows precisely the original ResNet [16]. Nominally, the optimization algorithm is 1-bit Stochastic Gradient Descent, the initialization is random (i.e. from scratch) and when the recognition accuracy on the validation set plateaus we decrease the learning rate by a factor of 10. The implementation is based on the CNTK library[3].

Database Augmentation. To improve the recognition performance various database augmentation scenarios have been tested. The ones that have produced positive effects are flipping and slight rotation. The flipped samples are all horizontal flips of the original images. Regarding the rotations, all images have been rotated either clockwise or counterclockwise with $3°$, $6°$, $9°$ or $12°$. We do not refer here to the domain transfer experiments.

4 Painting Understanding and Domain Transfer

A task that we undertake is to get an understanding of the machine learning systems (in our case deep CNN) grasp of art. For CNN, the favorite visualization tool has been proposed by Zeiler and Fergus [28] by introducing deconvolutional layers and visualizing activations maps onto features.

Attempts to visualize CNNs for scene recognition using this technique indicated that activation is related to objects, thus leading to the conclusion that multiple object detectors are incorporated in such a deep architecture [29]. In parallel, visualization of activation for genre [25] shown that, for instance, the landscape type of scene leads to activating almost the entire image, thus being less neat to draw any conclusion. Consequently, we tried a different approach to investigate the intrinsically mechanisms of deep CNN. Our approach exploits domain transfer.

Given the increased power of machine learning systems and the limited amount of data available to a specific task, a plethora of transfer learning techniques appeared [20]. The process of transfer learning is particular popular when associated with deep learning. First, let us recall that the lower layers of deep nets trained on large databases are extremely powerful features when coupled with powerful classifier (such as SVM) and maybe a feature selector, no matter the task [13]. Secondly, the process of *fine tuning* deep networks assumes taking a network that has been pre-trained on another database, and, using a small learning rate, only adapt it to the current task.

---

[3] Available at https://github.com/microsoft/cntk/.

In contrast, the concept of domain transfer or domain adaptation appeared as an alternative to increase the amount of information over which a learner may be trained directly (without fine tuning) in order to improve its prediction capabilities. Many previous solutions and alternatives have been introduced. We will refer to the work of Ben-David et al. [4] for theoretical insights on the process.

It has been shown that domain transfer is feasible and the resulting learner has improved performance if the two domains are adapted. Saenko et al. [21] showed that using a trained transformation, the domain transfer is beneficial. We investigate two alternatives. Firstly, we consider the Laplacian style transfer introduced by Aubry et al. [2]. They use a variant of bilateral filter to transfer the edginess from the reference, artistic image to the realistic photo. Secondly we consider the neural algorithm introduced by Gatys et al. [14]. Using a deep CNN they decompose an image into style and content. Intuitively the major difference between an artistic image and a photo is the style; doing style transfer, the second will be adapted to the first one's domain. Among several deep CNN architectures investigated, in the original work and confirmed by our experiments, only the VGG19 [24] leads to qualitative results.

5   Databases

For the various experiments undertaken, two databases have been employed. These are the WikiArt paintings dataset which was collected from Internet and used for the first time in this shape by Karayev et al. [18] and respectively the SUN database [27]. The former contains the bulk of the images used for training and testing, while the latter is used only as an auxiliary database for domain transfer experiments.

5.1   WikiArt Database

The WikiArt database contains approximately 80,000 digitized images of fine-art paintings. They are labelled within 27 different styles (cubism, rococo, realism, fauvism, etc.), 45 different genres (illustration, nude, abstract, portrait, landscape, marina, religious, literary, etc.) and belong to more than 1000 artists. To our knowledge this is the largest database currently available that contain genre annotations. Due to the fact that some classes are limited as number of examples, we chose to use only the ones that are well illustrated.

For our tests we considered a set that contains 79434 images of paintings. Some of the scene types were not well represented (i.e. less than 200 images) and we gathered them into a new class called "Others". This led to a division of the database into 26 classes. The names of the classes and the number of training and testing images in each class can be seen in Table 1.

We note that annotation is weak, as one may find arguable labels. For instance "literary" and "illustration" categories may in fact have "landscape" themes. Another observation is that there exists two "collector" classes: "others" and "genre". However, as this distribution matches practical situations, we used the database as it is, without altering the annotations.

5.2   Natural Scene Databases

As an additional source of real data we have relied on images from the SUN database [27]. In its original form, it contains 899 classes and more than 130,000 images. Yet only a few classes, which have a direct match with our database were selected. These classes and the number of images in each class added in the training process can be seen as the top, green segment in Fig. 1.

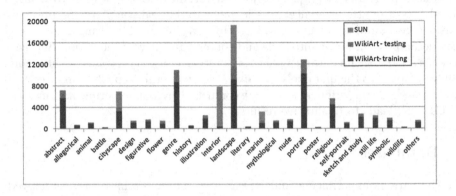

Fig. 1. Structure of the used databases

Table 1. Comparison with state of the art methods. The table is horizontally split to group solutions that have used databases with similar size. Acronyms: BOW - Bag of Words, ITML - or Iterative Metric Learning; pHoG - HoG pyramid as in [12]; pLBP - LBP pyramid implemented in VLFeat [26] DeCAF [13] assumes the first 7 levels of AlexNet trained on ImageNet.

Method	No. classes	No. images	Test ratio	Accuracy (%)
Agarwal et al. [1] - SIFT + BOW	5	1500	10%	82.53
Agarwal et al. [1] - ensemble				84.46
Saleh and Elgammal [22] - Classemes + Boost	10	63.691	33%	57.87
Saleh and Elgammal [22] - Classemes + ITML				60.28
Saleh and Elgammal [22] - Classemes + Fusion				60.28
Tan et al. [25] AlexNet - scratch			n/a	69.29
Tan et al. [25] CNN- finetune				**74.14**
ResNet 34 - scratch			20%	*73.74*
pHoG + SVM	26	79,434	20%	44.37
pLBP + SVM				39.58
DeCAF + SVM				59.05
AlexNet - scratch				53.02
ResNet 34 - scratch				**61.15**

6   Implementation and Results

It was our main interest to study the various ways in which the performance of the different systems tested can be increased. This included experiments on the classification methods themselves or various alterations brought to the database.

6.1   Comparison with State of the Art Methods

Agarwal et al. [1], Tan et al. [25] and Saleh and Elgammal [22] used the WikiArt database for training and testing in order to classify paintings into different genres. While the first used a very small subset, the later two cases focused on 10 classes from the database (Abstract, Cityscape, Genre, Illustration, Landscape, Nude, Portrait, Religious, Sketch and Study and Still life) leading to ~63 K images.

We have adopted the division from Karayev et al. [18] as working with a complete version of Wikiart. Furthermore, we stress that in our case the images from training and testing are completely different and are randomly chosen.

In order to compare our results to the ones reported by the mentioned articles, we selected the same classes of paintings for training and testing. While in a case [22], the test-to-train ratio is mentioned, in the second, [25], it is not. Under these circumstances the comparison with prior art is, maybe, less accurate.

The results (showed in Table 1) indicates that the proposed method gives similar results with previous [25] with the difference that they use a smaller network (AlexNet) but fine–tuned, while we have used a larger one, but initialized from scratch. Furthermore, we report the average over 5 runs.

6.2   Confusion Matrix

Visual examples of paintings are shown in Fig. 4. The confusion matrix for the best performer on the 26-class experiment is in Fig. 2. We have marked classes that are particular confused. It should be noted that from a human point of view, there is certain confusion between similar genres such as historical–battle–religious, portrait–self portrait, poster–illustration, animal–wildlife etc. Some of these confusable images are, in fact, shown in Fig. 4. Consequently we argue that the top-5 error is also relevant, as in many cases there are multiple genre labels that can be truthfully associated with one image. For the best proposed alternative, ResNet with 34 layers the Top-5 error is 11.85% - corresponding to a **88.15%** accuracy. For the 10-class experiment the top–5 accuracy is **96.75%**.

6.3   Additional Experiments

For the following experiments we will refer solely to the 26 classes test as it the most complete.

As many experiments require a significant amount of time, we restrained the training to 125 epochs. In this case the training takes ~20 h on NVidia GeForce

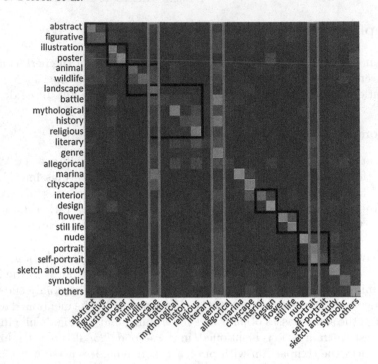

Fig. 2. The confusion matrix for the 26 class with 300 epochs training variant.

980 TI compared to 55 h for the 300 epochs alternative at the expense of 2% accuracy (Fig. 3).

Stochastic Effect. The first test studied the effect of the stochastic nature of deep neural networks. Factors such as the random initialization of all the

Fig. 3. Examples of the genres as illustrated in the database. Please note that other genre labels may be used for each image, thus arguing for the use of top–5 classification.

parameters can influence the results of any considered network. We ran the ResNet 34 several times on 26 classes and the accuracy results had a mean of 59.1% (top–1 accuracy) and a standard deviation of 0.33%. The results underline the fact that even though there is some variation caused by randomness, it does not influence the system significantly.

Influence Artistic Style. Prior art [29] suggested that even the case of the scene, in fact a deep network builds object detectors and it can recognize objects it has seen before. To study this aspect we devise the following experiment. Given the full 79 K images genre database, we selected all the images that are associated with Cubist and Naive Art styles and placed them in testing, resulting in 4,132 images for evaluation and ∼75 K for training. While numerically this is a weaker test than previous, the results are considerably worse (50.82% top–1 accuracy and 82.10% top–5). This is due to the fact that these particular styles are rather different from the rest and the learner had no similar examples in the training database. Also these results argues for *a style oriented domain adaptation.*

(a)   (b)   (c)   (d)

Fig. 4. Illustration of the domain transfer and domain adaptation experiments. Column (b) and (c) marks the image transformed. Columns are: (a) original photograph; (b) the image obtained after the Laplacian transfer [2]; (c) the image obtained after the neural transfer [14]; (d) the reference painting.

6.4   Domain Transfer

The domain transfer experiments consisted of separately augmenting certain classes with examples from the SUN database. Table 2 contains both overall performance for each class augmentation and the change brought to the concerned class (shown in the "Modifier" column). This measure takes into account the value of correct classifications of the regular networks, rather than number of existing samples. The experiment assumes adding each class separately. The

Table 2. The effect of adding extra samples from the SUN database. The "Acc" and "Acc-5" refers to the overall set accuracy when only the first and the top–5 results are taken into account. "Modif" refers to the improvement of the particular class.

Transfer	Original			Laplacian - [2]			Neural - [14]		
Class	Acc [%]	Acc-5 [%]	Modif. [%]	Acc [%]	Acc-5 [%]	Modif. [%]	Acc [%]	Acc-5 [%]	Modif. [%]
Cityscape	58.8	88.37	0.415	58.95	87.59	0.53	n/a	n/a	n/a
Flower painting	58.66	87.23	0.7	59.25	87.89	1.59	57.21	86.9	3.97
Interior	58.33	87.73	17.54	58.30	87.46	3.78	n/a	n/a	n/a
Landscape	59.32	88.43	3.85	58.61	88.13	0.18	n/a	n/a	n/a
Marina	58.35	87.78	15.7	59.50	88.13	1.55	57.22	87.33	1.16

neural style transfer method is very slow requiring 10–30 min to adapt an image. Thus we restrict to augmenting only Marina and Flower Painting classes.

Adding all transferred (adapted) images produced a similar effect, the variation being smaller than stochastic variance (overall accuracy of 59.05%).

As one may notice the overall results are not conclusive, variation being in the stochastic variance. However each of the transfers augmented the classification on the respective class. The most visible numerical effect over the entire database is obtained by adding original interior images; this is the only class where the number of added images is significantly larger than the associated paintings. Also given the transfer, the improvement is associated with styles such academism or realism, that contain very realistic rendering of the original scene, without much abstraction.

The images produced by the Laplacian transfer, while they look more abstract, do not seem "painted"; this domain adaptation does not improve the objective evaluation. The transfer, here, focuses on local contrast and grayscale dynamic range, while CNNs are related to structure. As shown in Fig. 4(b), there is no impression of painting in the images produced, thus it is hardly related with testing examples and is unhelpful when partitioning the data space.

We have found to be somehow surprising that, although the neural style produce images, which seem similar to a painting, its numerical effect is not as dramatic as we expected. However, we believe that the explanation is given by the quantity. The process is lengthy and we only added a small number of images, that are not able to actually fill the data space so that the CNN is able to draw rigorous borders. To see if this is the case we devised two experiments where the number of transferred images is comparable with the one from the standard database.

Neurally Transferring few Images on a Small Database. For these experiments, we have produced, using the neural style transfer algorithm [14], images for three genres: "cityscape" – 262; "flower paintings" – 180 and "marina" – 229. We have considered the case with 26 classes.

For the first experiment, we aimed to see what happens if, for the chosen classes, instead of paintings we provide mostly transferred images. The results are illustrated in Table 3. Initially, we have removed completely any training data for the three classes; obviously there was no correct recognition for these classes.

Than, we have added only the transferred images. In this case there are correct recognitions, even for paintings in abstract style, while quite few, showing that the neural transfer may help and provides relevant data. Afterwards we have added iteratively more images and we noted increasing recognitions. At the end we have removed the transferred images and noted a decrease, again confirming the beneficial effect of domain adaptation.

Another observation related to results is that the impact on the "flower paintings" genre is much reduced when compared to other two, "cityscape" and "marina". A possible explanations is related to the content: for paintings, the "flower paintings" genres refers typically to still flowers in a vase; in contrast for images, "flowers" are from garden, occupying much smaller areas from the image. Such an example is in Fig. 4.

For the second experiment we have reduce the contribution of each class to a number comparable with those transferred (namely to 250). The numerical results are showed in lower part of Table 3. One may see that if the quantity of neural transferred paintings is comparable with original data and the content of the two sets is similar (i.e. for cityscape and marina it is, while for flower is much less), the transfer is again beneficial. Thus, this experiments also shows that the neural style transfer may act as a domain adaptation function.

7 Discussion

In this paper we discuss the CNN capabilities to recognize the scene in a painting. A first contribution is that we clearly showed that machine learning systems

Table 3. Painting recognition accuracy, when classes that can be augmented by neural style transfer [14] have a diminished number of original paintings. The classes of interest are "cityscape", "flower paintings", "marina". Number of images transferred: "cityscape" – 262; "flower paintings" – 180 and "marina" – 229. Paintings used for training in other classes are 57363. For the second experiment, we have considered 250 paintings per class, totalling for 23 classes 5750.

Exp. No.	Paint. in other classes	Paints per interest class	Transferred images	Recognized images			
				Cityscape (From 764)	Flower (From 252)	Marina (From 259)	All classes (From 15708)
1	all	0	0	0	0	0	8024
	all	0	all	37	0	6	8084
	all	50	all	29	17	5	8255
	all	200	all	107	67	24	8364
	all	250	all	149	73	68	8584
	all	250	0	121	48	59	8328
2	5750	0	all	125	24	48	3354
	5750	250	0	219	91	120	3357
	5750	250	all	287	141	145	3561

(deep CNN) are confused by the abstraction level from art. The experiment with abstract art showed that they cannot easily generalize with respect to style.

Furthermore we experimented with domain transfer as an alternative to increase the overall performance and we have found that: (1) sheer numbers of photographs have a beneficial effect by improving the performance over styles with realistic depictions; (2) the CNNs are confused by artistic rendering (i.e. artistic style); (3) the neural transfer style may act as domain transfer adaptation. However significant increase into artistic scene recognition is opposed by the large duration of the neural style transfer. Thus speed-ups of the latter are highly necessary.

Acknowledgments. The work was supported by a grant of the Romanian National Authority for Scientific Research and Innovation, CNCS - UEFISCDI, number PN-II-RU-TE-2014-4-0733.

References

1. Agarwal, S., Karnick, H., Pant, N., Patel, U.: Genre and style based painting classification. In: WACV, pp. 588–594 (2015)
2. Aubry, M., Paris, S., Hasinoff, S.W., Kautz, J., Durand, F.: Fast local Laplacian filters: theory and applications. ACM Trans. Graph. **33**(5), 167 (2014)
3. Aubry, M., Russell, B., Sivic, J.: Painting-to-3D model alignment via discriminative visual elements. ACM Trans. Graph. **33**(2), 14 (2013)
4. Ben-David, S., Blitzer, J., Crammer, K., Kulesza, A., Pereira, F., Wortman Vaughan, J.: A theory of learning from different domains. Mach. Learn. J. **79**(1–2), 1–2 (2010)
5. Bentkowska-Kafel, A., Coddington, J.: Computer vision and image analysis of art. In: SPIE (2010)
6. Cai, H., Wu, Q., Hall, P.: Beyond photo-domain object recognition: Benchmarks for the cross-depiction problem. In: ICCV workshops - TASK, pp. 262–273 (2015)
7. Cichy, R., Khosla, A., Pantazis, D., Torralba, A., Oliva, A.: Comparison of deep neural networks to spatio-temporal cortical dynamics of human visual object recognition reveals hierarchical correspondence. Scientific reports 6 (2016)
8. Condorovici, R.G., Florea, C., Vertan, C.: Painting scene recognition using homogenous shapes. In: Blanc-Talon, J., Kasinski, A., Philips, W., Popescu, D., Scheunders, P. (eds.) ACIVS 2013. LNCS, vol. 8192, pp. 262–273. Springer, Cham (2013). doi:10. 1007/978-3-319-02895-8_24
9. Crowley, E., Zisserman, A.: The state of the art: object retrieval in paintings using discriminative regions. In: BMVC (2014)
10. Crowley, E.J., Zisserman, A.: The art of detection. In: Hua, G., Jégou, H. (eds.) ECCV 2016. LNCS, vol. 9913, pp. 721–737. Springer, Cham (2016). doi:10.1007/978-3-319-46604-0_50
11. DiCarlo, J., Zoccolan, D., Rust, N.: How does the brain solve visual object recognition? Neuron **73**(3), 415–434 (2012)
12. Dollar, P., Appel, R., Belongie, S., Perona, P.: Fast feature pyramids for object detection. T. PAMI **36**(8), 1532–1545 (2014)
13. Donahue, J., Jia, Y., Vinyals, O., Hoffman, J., Zhang, N., Tzeng, E., Darrell, T.: DeCAF: a deep convolutional activation feature for generic visual recognition. In: ICML (2014)

14. Gatys, L., Ecker, A., Bethge, M.: A neural algorithm of artistic style. In: CVPR (2016)
15. Hall, P., Cai, H., Wu, Q., Corradi, T.: Cross-depiction problem: recognition and synthesis of photographs and artwork. Comput. Visual Media 1(2), 91–103 (2015)
16. He, K., Zhang, X., Ren, S., Sun, J.: Deep residual learning for image recognition. In: CVPR (2016)
17. Herranz, L., Jiang, S., Li, X.: Scene recognition with CNNS: objects, scales and dataset bias. In: CVPR (2016)
18. Karayev, S., Trentacoste, M., Han, H., Agarwala, A., Darrell, T., Hertzmann, A., Winnemoeller, H.: Recognizing image style. In: BMVC (2014)
19. Krizhevsky, A., Sutskever, I., Hinton, G.E.: Imagenet classification with deep convolutional neural networks. In: Pereira, F., Burges, C.J.C., Bottou, L., Weinberger, K.Q. (eds.) Advances in Neural Information Processing Systems 25. pp. 1097–1105 (2012)
20. Lu, J., Behbood, V., Hao, P., Zuo, H., Xue, S., Zhang, G.: Transfer learning using computational intelligence: a survey. Knowl.-Based Syst. 80, 14–23 (2015)
21. Saenko, K., Kulis, B., Fritz, M., Darrell, T.: Adapting visual category models to new domains. In: Daniilidis, K., Maragos, P., Paragios, N. (eds.) ECCV 2010. LNCS, vol. 6314, pp. 213–226. Springer, Heidelberg (2010). doi:10.1007/978-3-642-15561-1_16
22. Saleh, B., Elgammal, A.: Large-scale classification of fine-art paintings: learning the right metric on the right feature. In: International Conference on Data Mining Workshops. IEEE (2015)
23. Sewards, T.V.: Neural structures and mechanisms involved in scene recognition: a review and interpretation. Neuropsychologia 49(3), 277–298 (2011)
24. Simonyan, K., Zisserman, A.: Very deep convolutional networks for large-scale image recognition. In: ECCV (2014)
25. Tan, W.R., Chan, C.S., Aguirre, H.E., Tanaka, K.: Ceci n'est pas une pipe: a deep convolutional network for fine-art paintings classification. In: ICIP (2016)
26. Vedaldi, A., Fulkerson, B.: VLFeat: an open and portable library of computer vision algorithms. In: ACM MM, pp. 1469–1472 (2010)
27. Xiao, J., Hays, J., Ehinger, K.A., Oliva, A., Torralba, A.: Sun database: large-scale scene recognition from abbey to zoo. In: CVPR, pp. 3485–3492 (2010)
28. Zeiler, M.D., Fergus, R.: Visualizing and understanding convolutional networks. In: Fleet, D., Pajdla, T., Schiele, B., Tuytelaars, T. (eds.) ECCV 2014. LNCS, vol. 8689, pp. 818–833. Springer, Cham (2014). doi:10.1007/978-3-319-10590-1_53
29. Zhou, B., Khosla, A., Lapedriza, A., Oliva, A., Torralba, A.: Object detectors emerge in deep scene CNNS. In: ICLR (2015)
30. Zhou, B., Lapedriza, A., Xiao, J., Torralba, A., Oliva, A.: Learning deep features for scene recognition using places database. In: NIPS (2014)

Foreign Object Detection in Multispectral X-ray Images of Food Items Using Sparse Discriminant Analysis

Gudmundur Einarsson[1]([✉]), Janus N. Jensen[1], Rasmus R. Paulsen[1],
Hildur Einarsdottir[1], Bjarne K. Ersbøll[1], Anders B. Dahl[1],
and Lars Bager Christensen[2]

[1] DTU Compute, Technical University of Denmark,
Richard Petersens Plads, Building 324, 2800 Kongens Lyngby, Denmark
{guei,jnje,rapa,hildr,bker,abda}@dtu.dk
[2] Teknologisk Institut, Gregersensvej 9, 2630 Taastrup, Denmark
lbc@teknologisk.dk
http://www.compute.dtu.dk/english
http://www.teknologisk.dk

Abstract. Non-invasive food inspection and quality assurance are becoming viable techniques in food production due to the introduction of fast and accessible multispectral X-ray scanners. However, the novel devices produce massive amount of data and there is a need for fast and accurate algorithms for processing it. We apply a sparse classifier for foreign object detection and segmentation in multispectral X-ray. Using sparse methods makes it possible to potentially use fewer variables than traditional methods and thereby reduce acquisition time, data volume and classification speed. We report our results on two datasets with foreign objects, one set with spring rolls and one with minced meat. Our results indicate that it is possible to limit the amount of data stored to 50% of the original size without affecting classification accuracy of materials used for training. The method has attractive computational properties, which allows for fast classification of items in new images.

Keywords: X-ray · Multispectral · Sparse classification · Foreign object detection

1 Introduction

One of the many purposes of X-ray scanning is to provide quality control and assurance in food production industry. The usage of X-rays provides non-destructive means of examining food items and the data can be used to verify that the content is free of anomalies or foreign objects. The usage of multispectral X-ray scanning has been used successfully in detecting explosives [12], and compares well to an X-ray dual-energy sandwich detector [8]. Foreign objects found in food items consist mostly of insects, wood chips, stone pebbles, sand/dust

© Springer International Publishing AG 2017
P. Sharma and F.M. Bianchi (Eds.): SCIA 2017, Part I, LNCS 10269, pp. 350–361, 2017.
DOI: 10.1007/978-3-319-59126-1_29

and plastic. These objects might be present in the raw materials, or accidentally introduced during the manufacturing process [6], where organic materials pose the main challenge for detection. Grating-based imaging techniques [9,10], that measure the attenuation, scattering and refraction of X-ray beams, have shown great promise in detecting organic foreign objects [6]. Although grating based methods are promising, they still have not been scaled to be used in a production line. Multispectral X-ray scanners exist with a conveyor belt setup, where a single acquisition takes around 5 s on the setup we used. Certain foreign objects can be detected in multispectral X-ray images of food items using a sparse classifier, which gives the potential for storing fewer data and making classification and acquisition faster.

According to the Beer-Lambert Law (BLL), the intensity of an X-ray beam decreases as it passes through matter with exponential decay depending on the distance traveled and the medium.

$$I = I_0 e^{-\mu \rho d} \tag{1}$$

I_0 in Eq. 1 corresponds to the initial X-ray intensity, μ and ρ together form the linear absorption coefficient, where μ corresponds to mass absorption and ρ corresponds to density, and finally d corresponds to the distance traveled by the beam. For a given simple material, this equation allows us to do inference on either the thickness or type of material we have, in case either of them is known. In our case, we are interested in food items. Food items are particularly challenging since their shape and material composition can greatly vary. There is considerate specimen to specimen variation along with potentially inhomogeneous materials which makes inference about the material composition difficult.

Instead of trying to model the signal according to the BLL, we apply a data-driven approach, where we train a classifier to recognize whether foreign objects are present by training it on multispectral X-ray image samples of a given food product and foreign objects in the food. The dimensionality of the data is high and poses problems for data storage and processing speed. We thus seek a sparse classifier in order to examine whether decreasing the data dimensionality will still result in reasonable accuracy and a low false discovery rate. We choose sparse linear discriminant analysis (SDA) [4] for this task, since it perfectly fits the requirements. This classification method performs variable selection and dimensionality reduction in the optimization process, which allows us to identify which spectra in the images are relvant for the given classification task. This is achieved via an elastic net regularizer [14]. The elastic net regularizer also allows for the construction of a Tikhonov regularization matrix, that can be further tailored to the specific classification task. Knowing which spectra are relevant for the classification task, allows for compressing the data, to only store the relevant spectra, and it could also give some domain knowledge on which spectra are most different between certain materials. A similar method that could also be considered for this task is sparse partial least squares, [3].

To generate the data for the classifier, we first preprocess it. The main part is normalizing each pixel w.r.t. the maximum intensity, which gives better contrast between different materials. These steps are further explained in the next section. Note that maximum 6 scans (images) were used for generating labels for training, a process that takes around 10 s per image. So the process of generating data for training and training the classifier is fast. This process can further be automated for a given target application.

We will examine to what extent we can detect foreign objects in two types of food materials and report which objects we detect.

The paper's outline consists of a description of the data and acquisition process, where we explain the scanner setup, the properties of the data we obtain and the preprocessing. Next, we describe how the data sets are prepared for the classifier and a description of the classifier. Finally, we present results to evaluate the performance and some discussion.

2   Data and Acquisition

The scanner used for data acquisition is a MULTIX multispectral X-ray scanner with three daisy chained detection modules, providing line scans of $3 \times 128 = 384$ pixels, where the pixel side length is 800 μm. The energy of the photons is measured over 128 energy bins, (also referred to as channels), where the energy for our experiments is set to 90 kV. This spectrometric scanner is made from a combination of a semiconductor crystal and advanced electronics, capable of measuring the energy of every incident X-ray photon. The material signature is acquired in real-time and stored in raw format [2,12]. This scanning technology has been compared to a dual energy sandwich detector, (where two detector modules are used with a single shot exposure), showing better detection for explosive materials and a lower false discovery rate (FDR) [8]. We aim to examine which part of the spectra is best suited for foreign object detection in food and to what extent can we detect foreign objects in organic food items using a sparse classifier.

The MULTIX scanner provides images in a binary format, where pixel intensities are encoded as 16-bit unsigned integers. To make sure that there are no scaling differences between samples, we scale the intensities by looking at an average measurement for a patch of air in the image and find the peak value over all the channels. We scale all values by the inverse of the value at the peak, such that the maximum attenuation corresponds to one, this ensures that no scaling differences are between different images. The different attenuation profiles in an image of spring rolls can be seen in Fig. 1. Foreign objects that are not "inside" a food item give a very different attenuation profile from the ones that are inside the food items. The attenuation profile of the food items naturally varies a lot by the thickness of the item, thus making it more difficult to work with products where the thickness varies much.

After this initial scaling, we remove line artifacts. These artifacts appear as strides in the image where two different detector modules are attached (See Fig. 3). To achieve this we start by creating an average air profile. We use a patch of pixels from a corner of the image where there are no overlapping detector

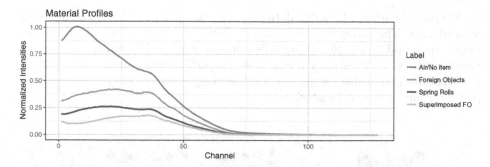

Fig. 1. Intensity profiles for different materials in an image of spring rolls with foreign objects. The green line above the blue one corresponds to foreign objects that are not superimposed on food items, while the bottom most purple line corresponds to foreign objects that are superimposed on top of food items. The data is scaled such that the peak for air/no item is at 1. The profiles are further normalized (not depicted here) such that the maximum value of every pixel is 1. This is a crude normalization for depth/thickness and gives us data that better represents differences between materials. (Color figure online)

modules and we are certain that it contains no items. The mean over the samples gives us an average air profile, similar to the red one seen in Fig. 1, i.e. 128 values that should represent no items, we call that vector $\mu_{\text{Air}} \in \mathbb{R}^{128 \times 1}$. For each column in the scanning direction of the image, we now look at the mean of the first 50 pixels. This gives us another profile which is specific to that particular column, i.e. a local mean profile, which we call $\mu_{\text{local}} \in \mathbb{R}^{128 \times 1}$. Now we need to find the scaling difference between μ_{Air} and μ_{local}, that is the vector $\mathbf{s} \in \mathbb{R}^{128 \times 1}$ which is the solution to the following equation, where on the left-hand side we have elementwise multiplication.

$$\mathbf{s}\mu_{\text{Air}} = \mu_{\text{Air}} \tag{2}$$

The solution to Eq. 2 is simply found via elementwise division of the mean vectors. Now for each pixel in this particular column, we multiply the 128 values with the scaling vector $\mathbf{s}$ elementwise. This process is depicted in Fig. 2, where the strides have been removed in the middle image.

Finally, we do a crude normalization for depth. For every pixel in the image, we find the maximum c_{max} of the 128 channels and multiply each of the 128 values by $1/c_{\text{max}}$, such that all values in each pixel lie between 0 and 1 and the maximum is 1. This should give us data that better represents differences between materials, rather than thickness since different materials have their maximum intensities at different channels, (see Fig. 1, where this scaling has not been done and the maximum intensity appears in different channels). All the preprocessing steps are depicted in Fig. 2.

We test our approach on two datasets, one with spring rolls and another with minced meat. The spring rolls dataset is challenging in a sense that the thickness varies considerably and the spring rolls can also overlap. The minced meat data varies much by thickness since it contains strings/filaments of meat that overlap.

The objects in Table 1 were used for the imaging, where they were superimposed on top of the food material.

Fig. 2. An illustration of the preprocessing steps for the images. Left most image shows channel 10 in the raw data (minced meat). The image in the middle shows channel 10 in image after the removal of the strides/line artifacts. The rightmost image shows channel 10 when each pixel has been scaled by the maximum value in each of its channels. The final normalization step gives more contrast between different materials although the contrast between meat and no item is less in this particular channel. The intensities are scaled linearly from black (lowest) to white (highest) in the shown images.

2.1 Spring Rolls

The spring rolls data set consists of scans of 8 different bags of spring rolls. Each bag was scanned 20 times, then refrozen and scanned again a day later 20 times each. The foreign objects were superimposed on the bags. The foreign objects were also scanned individually 10 times and each bag was also scanned 2 times without any foreign objects. Figure 3 shows four of the image channels in a grayscale. Most of the contrast between the different materials seems to be present in the first channels, which can also be seen in Fig. 1. The different scans provide variation in position and rotation of the food items and the different bags provide shape differences for the dataset. The spring rolls were contained in a plastic bag.

2.2 Minced Meat

A single plastic box containing 1 kg of minced meat was used for all the scans. First, 5 scans were produced with no items, then the meat was scanned 5 times without any foreign objects. Finally, the meat was scanned with 3 sets of foreign objects, 10 times for each set. The types of foreign objects in each of the three sets is described in Table 1 and a sample image from the data can be seen in Fig. 3.

Table 1. Foreign objects used for the scans of minced meat. The items used for the spring rolls are the same excluding the last seven items. Set 3 consists of the same items as used in [6]. PTFE is an acronym for Polytetrafluoroethylene, more commonly referred to as Teflon.

Material	Number of pieces	Size range	Dataset
Quartz balls	5	1–4 mm	Set 1 for minced meat
Aluminium balls	6	2–7 mm	Set 1 for minced meat
Soft bone	5	5 mm long	Set 1 for minced meat
Bone phantoms	5	2–6 mm	Set 1 for minced meat
Polycarbonate balls	6	3.2–8 mm	Set 2 for minced meat
Ceramic balls	6	2–8 mm	Set 2 for minced meat
Glass balls	6	1–5 mm	Set 2 for minced meat
PTFE balls	6	1.6–5 mm	Set 2 for minced meat
Wood	4	2–8 mm	Set 3 for minced meat
Stone pebbles	4	2–8 mm	Set 3 for minced meat
Soft plastic	4	2–8 mm	Set 3 for minced meat
Hard plastic	4	2–8 mm	Set 3 for minced meat
Metal	4	2–8 mm	Set 3 for minced meat
Rubber	4	2–8 mm	Set 3 for minced meat
Glass	4	2–8 mm	Set 3 for minced meat

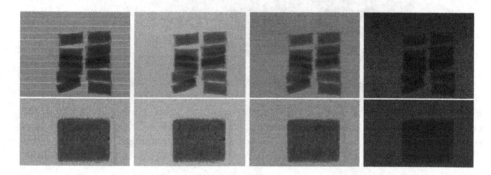

Fig. 3. Raw grayscale images of different channels from a spring roll sample (top row) and minced meat (bottom row) generated with the MULTIX scanner. From left to right are channels 2, 20, 50 and 100. The contrast decreases the higher we go in the channels and the variation in the measurements increases. The foreign objects can be seen as small black dots in the images and are most visible in the second image, better visible in Fig. 4. Strides have been removed in the meat data to show the difference compared to the strides that are present in the spring rolls images shown here. The intensities are scaled linearly from black (lowest) to white (highest). In this image, the line artifacts have not been removed and the individual pixels have not been scaled by the maximum value.

3   Method

For a given scanned food item, we would like to classify which parts of the image contain food and which contain foreign objects. To achieve this we first need to construct a dataset for training a classifier.

For the spring rolls dataset, we manually select four regions from five scans. In each scan, we select a region containing no items, one containing spring rolls and finally two regions with the most visually distinct foreign objects. This selection process is depicted in Fig. 4. To encode the neighborhood information, we treat a single observation of a given pixel as the 5×128 values from itself and the pixels directly above, below and on the left and right. So each observation contains $128 \times 5 = 640$ variables. This should give us more robustness for detection of different materials.

Fig. 4. Selection of data used for training the classifier. Three classes are selected, the enclosed region of the green box represents the no item or air class, the blue region represents the food item and the red boxes represent the foreign objects. For this illustration, the red boxes are a little bit larger than in practice. (Color figure online)

After selecting the regions from five scans we can generate a matrix where rows represent observations and the $p = 640$ variables are represented as columns. Each image yields around 30 pixels of foreign objects, the other classes (spring rolls and air) are randomly subsampled, such that we have equal number of observations in each class, so we end up with a matrix $\mathbf{X}$ of dimension $n \times p = n \times 640$, where the value of n is around 500 to 600 pixels. The labels are represented in an indicator matrix $\mathbf{Y}$, which has an equal number of rows as $\mathbf{X}$, but the number of columns is equal to the number of classes K, in this case, three. If observation i belongs to class j, then $\mathbf{Y}_{ij}$ is 1 and the other values in the same row are zero. We employ a similar methodology to setup the minced meat dataset. After the data is set up we normalize it by subtracting the mean and scaling the variables such that they have unit variance. The same procedure is done for the minced meat data, so we have two different data sets.

The R programming language [11] was used for all the processing of the image data and classification. The package `imager` [1] was used for manually extracting regions from the images. The `imager` package is an R interface to the Cimg C++ library [13].

3.1 Sparse Linear Discriminant Analysis

We apply SDA, [4], to solve the present classification problem. SDA is a statistical learning method [7], which falls under the category of supervised classifiers and is a sparse version of the more basic method linear discriminant analysis (LDA). The method can handle many classes and it can also handle the case when we have more variables than observation, $p \gg n$ problems, with regularization. The underlying problem can be formulated in different ways, but we approach it by sparse optimal scoring.

$$(\theta_k, \beta_k) = \underset{\theta \in \mathbb{R}^K, \beta \in \mathbb{R}^p}{\operatorname{argmin}} \ \|\mathbf{Y}\theta - \mathbf{X}\beta\|^2 + \lambda_1 \beta^T \Omega \beta + \lambda_2 \|\beta\|_1$$
$$\text{s.t.} \frac{1}{n}\theta^T \mathbf{Y}^T \mathbf{Y}\theta = 1, \ \ \theta^T \mathbf{Y}^T \mathbf{Y}\theta_l = 0 \ \forall l < k, \tag{3}$$

In the sparse optimal scoring formulation, (Eq. 3), the $\mathbf{X}$ data matrix and $\mathbf{Y}$ indicator matrix are the same as the ones described in the last section. We seek the discriminant vectors β_i, $i \in \{1, 2, ..., K - 1\}$, which we use to project the data into a lower dimensional space. Classification is performed in this lower dimensional space by classifying an observation as belonging to the class corresponding to the nearest centroid, where the labeled data is used to estimate the centroids. θ serves the purpose of avoiding the masking problem, i.e. such that class centroids are not colinear in the lower dimensional representation, it is not needed for classification of new observations, only in training. The second and third terms in the minimization problem form an elastic net penalty [14], which serves as a regularizer and allows us to solve the problem in the case of more variables than observations. The scaling parameters λ_1 and λ_2 are selected via cross-validation. In our case, the Ω in the first part of the elastic net penalty is a diagonal $p \times p$ matrix, which penalizes the magnitude of the coefficients in the β_i vectors.

The SDA method, (without the elastic net penalty), is a linear map to a lower dimensional representation like principal component analysis (PCA), but in SDA we project the data to a lower dimensional space such that we maximize the variance between classes and minimize the variation within classes for optimal linear separation. There are also sparse versions of PCA [5], which SDA is more similar to. The centroids in the lower dimensional space can be thought of as means of different multivariate normal distributions which all have the same covariance structure, therefore we get linear decision boundaries, like in classical LDA [7]. Other classifiers can also be used on the projected data, an example

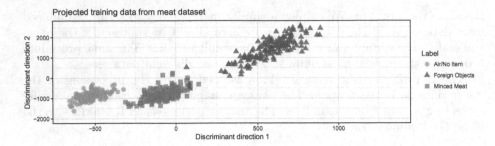

Fig. 5. Visualization of the training data used for the classification of the minced meat data after projection with the discriminant vectors. The classes are almost perfectly separated along the first discriminant direction, while the foreign objects and meat are rather close. The second discriminant vector further separates the meat and foreign objects.

of such projected data is depicted in Fig. 5, it is the training data used for the minced meat dataset.

4 Results

The classification results from the classifier trained on the spring rolls data are summarized in Table 2, where we show the number of detected pixels in images not contained in the training or validation set. The SDA method was trained only on 6 images from dataset 1 and the training data images are not included in the table. The final training set was balanced and consisted of 519 measurements with 640 variables. The training error is 0% with 48% sparsity, i.e. only 48% of the values in the discriminant vectors are non-zero. This means that almost half the variables are irrelevant for this particular classification task. Values corresponding to the same channels in the pixels were non-zero very consistently in the 640 variables corresponding to a pixel and its four neighbors. 10-fold cross-validation was used to tune the sparsity parameter and should be noted that sparsity of 5% only yielded 5% error on the training data, meaning that very few variables are more critical than others.

One thing to note about the results in Table 2 is that consistently fewer foreign object pixels were detected in the scans which only contained foreign objects. That is because the training set only consisted of foreign objects superimposed on the spring rolls. A very low number of false positives were detected in the data set which consisted of only spring rolls and no foreign objects (average 4.21 pixels). In most of the images, only 0, 1, 2 or 3 pixels were detected except for a single outlier were 42 pixels were detected, which inflates the standard deviation.

The classification results for the classifier which was trained on the minced meat dataset are summarized in Table 3, where we show the number of pixels detected in images which were not part of the training or validation set.

Table 2. Results for the spring rolls dataset where we present the number of pixels detected as foreign objects on average in 4 different datasets. Dataset 1 consists of 8 bags of spring rolls with foreign objects superimposed. Dataset 2 consists of the same 8 bags, where the spring rolls have been refrozen and scanned a day later. The other datasets are scans with only the spring rolls or only the foreign objects.

Spring rolls	Dataset 1	Dataset 2	Only FOs	No FOs
Average number of detected pixels	155.87	158.00	112.60	4.21
Standard deviation of detected pixels	16.18	38.01	40.73	11.31
Number of images predicted on	136	149	10	14

The main difference from the spring rolls dataset is that no false positives were discovered in the scans that contained no foreign objects. Otherwise, there is consistent variation between datasets in the number of foreign object pixels discovered. Another difference is that in the cross-validation process the sparsity regularization parameter that was chosen yields 17% sparsity. This means that we could certainly get away with storing fewer data for this approach.

Table 3. Results for the minced meat dataset where we present the number of pixels detected as foreign objects on average in 5 different datasets. The first dataset is 5 scans of nothing, i.e. empty scans. Datasets 1, 2 and 3 consist of 10 scans each, two which were used for training. The difference of the foreign objects in each dataset can be found in Table 1. The last dataset is meat without any foreign objects.

Minced Meat	Nothing	Dataset 1	Dataset 2	Dataset 3	No FOs
Average number of detected pixels	0.00	117.62	190.12	211.50	0.00
Standard deviation of detected pixels	0.00	11.49	9.06	8.90	0.00
Number of images predicted on	5	8	8	8	5

Some example results are presented in Fig. 6. The foreign objects detected were mostly metals, and also some stone pebbles, quartz, and glass. The smallest objects detected are 2–3 mm in diameter.

5 Discussion

We achieve good detection on the type of foreign objects we can detect. The items used for training are the ones that are represented in the detection, we do not manage to generalize to all the scanned foreign objects. This is mainly because of low signal to noise ratio, especially for the objects that have low absorption. The undetected items are also not represented in the training set, if they would be included we would potentially get more false positives, because they are not as distinct as the metals, stone pebbles, and quartz, thus moving the decision boundary closer to the food item. One approach to try to get a more general detector would be to train

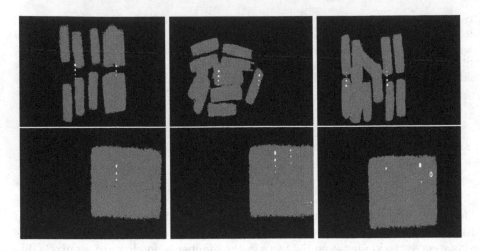

Fig. 6. Example results from the classifier on both datasets. The white color indicates foreign objects. The top row shows the spring rolls and the bottom row shows from left to right an example from dataset 1,2 and 3. The foreign objects detected were stone pebbles, metals, quartz, and glass. These were the foreign objects used for training.

with as many different types of foreign objects as possible, that are detectable, and find the discriminant vectors from SDA or use a semi-supervised approach. Then the new data can be projected similar to Fig. 5 and the air and food item classes can be described there with normal distributions or other ways to encapsulate the two classes. Then everything outside those classes would be classified as foreign objects. Each type of foreign object could also be modeled on its own, then we could encapsulate what is known, and everything outside the known classes would belong to an *unknown* class.

One way to augment the current measurements would be to have some way to estimate the thickness of the materials that are being scanned. That would be a good additional variable for a data-driven approach, or it could be used for normalization. This is already being done in some commercial products using a laser to map the height of the food product.

We can say that for the data sets used we can get away with storing half of the data or less. But this is both application and material dependent. Different applications could yield different foreign objects and different materials have different intensity profiles, meaning that some variables/channels are redundant in some cases and useful in others. This would have to be dealt with for each specific application.

6   Conclusion

We have demonstrated that we can achieve robust detection of certain foreign objects in the data sets used in this work. This was done in a completely data-driven manner by applying a sparse classifier to the normalized data. There is great potential for using an approach similar to the one we present, which could help with storing fewer data and processing the results faster.

Acknowledgements. This work is supported by the Lundbeck foundation, the Technical University of Denmark and the NEXIM research project funded by the Danish Council for Strategic Research (contract no. 11-116226) within the Program Commission on Health, Food and Welfare. We would like to thank the anonymous reviewers for providing valuable comments on the manuscript.

References

1. Barthelme, S.: imager: Image Processing Library Based on 'CImg' (2016). r package version 0.31, https://CRAN.R-project.org/package=imager
2. Brambilla, A., Ouvrier-Buffet, P., Rinkel, J., Gonon, G., Boudou, C., Verger, L.: CdTe linear pixel X-ray detector with enhanced spectrometric performance for high flux X-ray imaging. IEEE Trans. Nuclear Sci. **59**(4), 1552–1558 (2012)
3. Chung, D., Keles, S., et al.: Sparse partial least squares classification for high dimensional data. Stat. Appl. Genet. Mol. Biol. **9**(1), 17 (2010)
4. Clemmensen, L., Hastie, T., Witten, D., Ersbøll, B.: Sparse discriminant analysis. Technometrics **53**(4), 406–413 (2011). Taylor & Francis
5. d'Aspremont, A., El Ghaoui, L., Jordan, M.I., Lanckriet, G.R.: A direct formulation for sparse PCA using semidefinite programming. SIAM Rev. **49**(3), 434–448 (2007)
6. Einarsdóttir, H., Emerson, M.J., Clemmensen, L.H., Scherer, K., Willer, K., Bech, M., Larsen, R., Ersbøll, B.K., Pfeiffer, F.: Novelty detection of foreign objects in food using multi-modal X-ray imaging. Food Control **67**, 39–47 (2016)
7. Friedman, J., Hastie, T., Tibshirani, R.: The elements of statistical learning. Springer series in statistics, vol. 1. Springer, Berlin (2001)
8. Gorecki, A., Brambilla, A., Moulin, V., Gaborieau, E., Radisson, P., Verger, L.: Comparing performances of a CdTe X-ray spectroscopic detector and an X-ray dual-energy sandwich detector. J. Instrumen. **8**(11), P11011 (2013)
9. Pfeiffer, F., Bunk, O., David, C., Bech, M., Le Duc, G., Bravin, A., Cloetens, P.: High-resolution brain tumor visualization using three-dimensional X-ray phase contrast tomography. Phys. Med. Biol. **52**(23), 6923 (2007)
10. Pfeiffer, F., Weitkamp, T., Bunk, O., David, C.: Phase retrieval and differential phase-contrast imaging with low-brilliance X-ray sources. Nat. Phys. **2**(4), 258–261 (2006)
11. R Core Team: R: a language and environment for statistical computing. R Foundation for Statistical Computing, Vienna, Austria (2015). https://www.R-project.org/
12. Rebuffel, V., Rinkel, J., Tabary, J., Verger, L.: New perspectives of X-ray techniques for explosive detection based on CdTe/CdZnTe spectrometric detectors. International Symposium on Digital Industrial Radiology and Computed Tomography-We, vol. 2, pp. 1–8 (2011)
13. Tschumperlé, D.: The CImg library. In: IPOL 2012 Meeting on Image Processing Libraries, p. 4 (2012)
14. Zou, H., Hastie, T.: Regularization and variable selection via the elastic net. J. Roy. Stat. Soc.: Ser. B (Stat. Methodol.) **67**(2), 301–320 (2005)

Sparse Approximation by Matching Pursuit Using Shift-Invariant Dictionary

Karl Skretting$^{(\boxtimes)}$ and Kjersti Engan

University of Stavanger, 4036 Stavanger, Norway
{karl.skretting,kjersti.engan}@uis.no

Abstract. Sparse approximation of signals using often redundant and learned data dependent dictionaries has been successfully used in many applications in signal and image processing the last couple of decades. Finding the optimal sparse approximation is in general an NP complete problem and many suboptimal solutions have been proposed: greedy methods like Matching Pursuit (MP) and relaxation methods like Lasso. Algorithms developed for special dictionary structures can often greatly improve the speed, and sometimes the quality, of sparse approximation.

In this paper, we propose a new variant of MP using a Shift-Invariant Dictionary (SID) where the inherent dictionary structure is maximally exploited. The dictionary representation is simple, yet flexible, and equivalent to a general M channel synthesis FIR filter bank. Adapting the MP algorithm by using the SID structure gives a fast and compact sparse approximation algorithm with computational complexity of order $\mathcal{O}(N \log N)$. In addition, a method to improve the sparse approximation using orthogonal matching pursuit, or any other block-based sparse approximation algorithm, is described. The SID-MP algorithm is tested by implementing it in a compact and fast C code (Matlab mex-file), and excellent performance of the algorithm is demonstrated.

Keywords: Sparse approximation · Matching pursuit · Dictionary · Shift-invariant

1 Introduction

Obtaining a sparse representation of a signal is an important part of many signal processing tasks. A typical way to do this is to use a dictionary of atoms, and to represent the signal as a linear sum of some of the available atoms. A sparse approximation of a signal $\mathbf{x}$ of size $N \times 1$, allowing a (small) error $\mathbf{r}$, can be written as

$$\hat{\mathbf{x}} = D\mathbf{w} = \sum_k w(k)\mathbf{d}_k, \quad \mathbf{x} = \hat{\mathbf{x}} + \mathbf{r} \tag{1}$$

where D is the $N \times K$ dictionary, $\mathbf{d}_k$ of size $N \times 1$ is atom number k, $w(k)$ is coefficient k, and $\mathbf{w}$ is the $K \times 1$ coefficient vector. For a sparse approximation, only a limited number, s, of the coefficients are non-zero.

© Springer International Publishing AG 2017
P. Sharma and F.M. Bianchi (Eds.): SCIA 2017, Part I, LNCS 10269, pp. 362–373, 2017.
DOI: 10.1007/978-3-319-59126-1_30

Common predefined dictionaries are the Discrete Cosine Transform (DCT) basis, wavelet packages and Gabor bases, both orthogonal and overcomplete variants can be used. Advantages of these predefined dictionaries are: They can be implicitly stored, they are, in general, signal independent and thus can successfully be used for most classes of signals, and fast algorithms are available for common operations. The overlapping transforms, i.e. wavelets and filter banks, perform better than the block-based transforms, like DCT, on many common signals. This is because the structure in the overlapping transforms does not need the signal to be divided into (small) blocks and thus they avoid the blocking artifacts.

Learned dictionaries are an alternative to predefined dictionaries. Since they are learned, i.e. adapted to a certain class of signals given by a relevant set of training signals, they will usually give better sparse approximation results for signals belonging to the class. Many methods have been proposed for dictionary learning; K-SVD [2], ODL [14], MOD [8,9] and RLS-DLA [23]. All these methods are block-based and the blocks should be rather small to limit the number of free variables in the dictionary; as seen from Eq. 1 the number of elements in D is NK.

Structured learned dictionaries try to combine the advantages of predefined dictionaries and general (all elements are free) learned dictionaries [17,22]. Imposing a structure on the dictionary allows the atoms to be overlapping and the dictionary could be very large and still have a reasonable number of free variables. In addition, the structure can make dictionary learning computationally more tractable. Several variants of structured dictionaries are possible [1,3,9,22]. An example of a structured dictionary that has been shown to be useful is the Shift-Invariant Dictionary (SID) [12,18,20]. In this paper, we propose a flexible representation for the SID structure.

The sparse approximation step is an important part when both using and learning a dictionary. It is often the computationally most expensive part, thus its speed is very important. Popular approaches are the l_1-norm minimization methods like LARS/Lasso [7], and the greedy algorithms directly using the l_0-quasinorm. For the latter approach Matching Pursuit (MP) [15] and its many variants, like orthogonal matching pursuit [19], are much used. MP algorithms typically has computationally complexity of $\mathcal{O}(N^3)$ and this effectively limits the dictionary size. To overcome this restriction FFT-based algorithms tailored for shift-invariant dictionaries have been developed [10].

FFT-based algorithms work very well for regular dictionaries, i.e. Gabor, harmonic, chirp or DCT based atoms, but for short arbitrary waveforms time-domain algorithms should be expected to perform better. As an example the execution times for Matlab functions `fftfilt()` and `filter()` can be compared for different signal and filter lengths; the time domain approach is faster for filters shorter than 300 and a signal length of 10000 samples. For many learned shift-invariant dictionaries, the atoms will be arbitrary and of limited (short) length, and sparse approximation should benefit from using a well-designed and well-implemented time domain based algorithm. Therefore, this is what we present here.

The matching pursuit variant presented in this paper, denoted as SID-MP, is completely implemented in time domain but still has the same low computationally complexity of order $\mathcal{O}(N \log N)$ as the FFT-based algorithms, it uses a simple and flexible representation for the shift-invariant dictionary, and it allows the atoms to be shifted in steps larger than one. For short arbitrary shape atoms, it should be faster than FFT-based algorithms.

Orthogonal Matching Pursuit (OMP) finds better approximations than MP using the same number of non-zeros [4] and fast algorithms are available for small dictionaries. For large shift-invariant dictionaries (true) OMP is slow, but a low complexity approximate OMP variant has been presented by Mailhé et al. [11]. This variant is FFT-based and achieves results close to true OMP with execution times close to MP. We propose another OMP variant, denoted SID-OMP. Its main idea is to divide the long signal into blocks (segments) of moderate size that are processed by OMP or any other block-oriented sparse approximation algorithm. By doing several iterations, or using partial search [24] instead of OMP, better approximation than (true) OMP can be achieved. The SID structure is used in the proposed method, it has computationally complexity of order $\mathcal{O}(N)$ but the constant factor is large and depends on the segment size.

The organization of this paper is: Sect. 2 presents the representation of the SID structure. The proposed algorithm SID-MP is presented in Sect. 3 and the SID-OMP method in Sect. 4. Finally, in Sect. 5, it is shown that the proposed algorithms are fast and efficient.

2   The Shift-Invariant Dictionary

A Shift-Invariant Dictionary (SID) should be represented independently of the signal size, and it should be easy to extend it to match the (possibly large) signal length N. In this paper, we suggest to use a dictionary structure, denoted

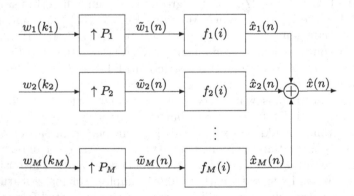

Fig. 1. A general synthesis filter bank of M channels (filters). Filter $\mathbf{f}_m$ has length Q_m and upsampling factor P_m, these values may be different for each filter. $\tilde{w}_m(n)$ is the upsampled coefficients for filter m. If each filter $\mathbf{f}_m$ here is given as the reversed (and shifted) atom $\mathbf{q}_m$ the synthesis equation can be written as in Eq. 5.

as SID structure, equivalent to the general synthesis filter bank in Fig. 1. A matrix representation as in Eq. 1 is equivalent to the M channels filter bank in Fig. 1 when each channel is represented by one submatrix D_m and the dictionary is a concatenation of these:

$$D = [D_1, D_2, \cdots, D_M], \tag{2}$$

One submatrix, D_m, contains one atom (the reversed synthesis filter) and its shifts (below the shift $P_m = 1$ is visualized):

$$D_m = \begin{bmatrix} q_m(1) & & & \\ q_m(2) & q_m(1) & & \\ \vdots & \vdots & \ddots & \\ q_m(Q_m) & q_m(Q_m - 1) & \ddots & q_m(1) \\ & q_m(Q_m) & \ddots & q_m(2) \\ & & \ddots & \vdots \\ & & & q_m(Q_m) \end{bmatrix}. \tag{3}$$

Atom m is denoted as $\mathbf{q}_m$ with elements $q_m(i)$ for $i = 1, 2, \ldots, Q_m$. It has length Q_m and is shifted P_m positions down for each new column in D_m. Here it is assumed that each atom has unit norm and is real. Note that the support for all shifts of an atom is completely within the dictionary column (length N). There is neither circular extension nor mirroring at the ends. This means that the number of columns in submatrix D_m is given as:

$$K_m = \lceil (N - Q_m + 1)/P_m \rceil. \tag{4}$$

$\lceil \cdot \rceil$ denotes the ceiling function. The dictionary D has size $N \times K$ where $K = \sum_{m=1}^{M} K_m$. The overlapping factor, i.e. the ratio K/N (as $N \to \infty$), is $\sum_{m=1}^{M} 1/P_m$. The SID structure is given by the number of atoms M, the atom lengths $\{Q_m\}_{m=1}^{M}$, their shifts $\{P_m\}_{m=1}^{M}$, and their $Q = \sum_{m=1}^{M} Q_m$ values which are collected into one vector $\mathbf{q^T} = [\mathbf{q}_1^T, \mathbf{q}_2^T, \ldots \mathbf{q}_M^T]$. This structure for a shift-invariant dictionary is actually quite flexible, and many dictionaries fit quite well into this structure. It is a subclass of the flexible structure dictionary introduced in [3].

The sparse approximation equation can be divided into blocks as

$$\hat{\mathbf{x}} = D\mathbf{w} = [D_1, \ldots, D_M] \begin{bmatrix} \mathbf{w}_1 \\ \vdots \\ \mathbf{w}_M \end{bmatrix} \tag{5}$$

$$= D_1 \mathbf{w}_1 + \cdots + D_M \mathbf{w}_M = \hat{\mathbf{x}}_1 + \cdots + \hat{\mathbf{x}}_M,$$

where the approximation is split into M terms, each corresponding to a submatrix of the dictionary. It is easily seen that this is an alternative representation of a general synthesis FIR filter bank [5] of M filters as in Fig. 1.

Matching pursuit algorithm:

1    $\mathbf{w} = \mathrm{MP}(D, \mathbf{x}, s)$
2    $\mathbf{w} := \mathbf{0}, \mathbf{r} := \mathbf{x}$
3   while not Finished
4        $\mathbf{c} := D^T \cdot \mathbf{r}$
5   find $k : |c(k)| = \max_j |c(j)|$
6        $w(k) := w(k) + c(k)$
7        $\mathbf{r} := \mathbf{r} - c(k) \cdot \mathbf{d}_k$
8a   Finished := $(\|\mathbf{r}\| < \text{some Limit})$
8b   Finished := $(s \text{ non-zero entries in } \mathbf{w})$
9   end
10   return

Fig. 2. The matching pursuit algorithm. The columns (atoms) of D should have unit norm. Input s is the target number of non-zero coefficients to select.

3   Matching Pursuit Using SID

Matching Pursuit (MP), presented and analyzed in [6,15], is the simplest of the matching pursuit algorithms. An overview of the algorithm is given in Fig. 2. The MP algorithm is quite simple, but it is not very efficient. The loop may be executed many times ($\geq s$), sometimes selecting the same dictionary atom multiple times, and the calculation of the inner products (line 4) is demanding, where approximately NK multiplications and additions are done. As both K and the number of non-zero coefficients to select, s, are (usually) of the order $\mathcal{O}(N)$ this gives the computational complexity of MP as $\mathcal{O}(N^3)$. Using FFT and taking advantage of the dictionary structure, and all shift steps equal to one ($P_m = 1$), shift-invariant MP can be implemented by an algorithm of the order $\mathcal{O}(N \log N)$ [10]. Below, we present a simple time domain algorithm that exploits the SID structure, and allows different shift steps $P_m \geq 1$. It follows the steps of MP as seen in Fig. 2, but has the complexity order of $\mathcal{O}(N \log N)$ or $\mathcal{O}(MQN \log N)$ if the (constant) SID size is included.

The proposed algorithm is done in a straightforward way in time domain, but exploiting the SID structure maximally and thus making the inner loop as fast as possible. An outline of the algorithm is given in Fig. 3 and the similarity to MP in Fig. 2 is obvious. The input is the SID which is given by the number of atoms M, the atom lengths $\{Q_m\}_{m=1}^M$, their shifts $\{P_m\}_{m=1}^M$, and their Q values, the signal $\mathbf{x}$ (length N) and the target number of non-zeros s. In the analysis below the SID is assumed to be constant and independent of signal length N.

As for MP, initialization sets $\mathbf{w}$ to zeros and the residual $\mathbf{r}$ to $\mathbf{x}$, but here the calculation of the inner products, $\mathbf{c}$, is done before the main loop. Since the atoms has fixed support the complexity is $\mathcal{O}(N)$. Line 5 in Fig. 2 has the computational complexity of $\mathcal{O}(N)$ which is not wanted inside a loop. To avoid this a *max-heap*

Matching pursuit using a shift-invariant dictionary:

```
1    w = SID-MP(sid,x, s)
2       w := 0,    r := x
3       c := D^T · r,    InitHeap
4       while not Finished
5           k =: topHeap,
6           w(k) := w(k) + c(k)
7           r := r − c(k) · d_k
8           Update c and Heap
9           Check if Finished
10      end
11      return
```

Fig. 3. The Matching pursuit algorithm using shift-invariant dictionary. Input s is the target number of non-zero coefficients to select. The inner products are stored in a heap, initialized by the InitHeap-function, top extracted by topHeap-function, and in line 8 the inner products and the heap are updated, but only for atoms where the inner product has changed, i.e. the atoms with support overlapping the support of the selected atom.

data structure is used, to build the heap outside the loop is $\mathcal{O}(N)$. Inside the loop max is found in $\mathcal{O}(1)$ and the heap is updated in $\mathcal{O}(\log N)$.

The most demanding task inside the loop is to update the involved inner products and their place in the heap. But, since all atoms have local support the number of inner products to update is limited, depending only of the SID structure and independent of signal length N. The update step (line 8) is thus $\mathcal{O}(\log N)$ which also becomes the order of the loop. The loop is done at least $s = \mathcal{O}(N)$ times. This gives the computational cost for the SID-MP algorithm in the order of $\mathcal{O}(N \log N)$.

An implementation of this algorithm could be both compact and fast and it will be available on the web: http://www.ux.uis.no/~karlsk/dle/.

4   Orthogonal MP Using SID

Orthogonal Matching Pursuit (OMP), somewhere denoted as Order Recursive Matching Pursuit (ORMP), finds better approximations than MP using the same number of non-zeros [4]. A throughout description of OMP/ORMP can be found in [4] or [24] and is not included here. For small (or moderately sized) dictionaries OMP is efficient, and fast implementations using QR-factorization are available: SPAMS from Mairal et al. [13] and OMPbox from Rubinstein et al. [21]. Nevertheless, these effective implementations have time complexity of order $\mathcal{O}((K + s^2)N) = \mathcal{O}(N^3)$ where N and K refer to the size of the dictionary for the signal block.

To do OMP on a large signal using a shift-invariant dictionary is more complicated than to do MP; since orthogonalization is done when a new atom is selected
the effect on the residual, and thus the updated inner products, goes beyond the
support of the selected atom. But the effect decreases as the distance from the
selected atom increases. Using this fact, an approximate OMP algorithm can be
developed in a way similar to the MP algorithm [11]. Another approach is used
in the proposed method. It utilizes the local support of the atoms to divide the
large OMP problem into many moderately sized OMP problems. The segment
size is fixed and independent of N giving the computational complexity as $\mathcal{O}(N)$,
but the constant factor is large $\mathcal{O}(N_{seg}^3)$ where N_{seg} is the signal segment length
used in the OMP problems.

The proposed method, denoted as SID-OMP, has almost the same input as
the SID-MP algorithm in Fig. 3: the shif-invariant dictionary represented as a
SID structure, the long signal $\mathbf{x}$ of length N, and the target number of non-
zero coefficients s *or* an initial coefficient vector $\mathbf{w}$ having the target number s
non-zero elements. The initial coefficient vector $\mathbf{w}$ is found by the SID-MP algorithm in previous section if s is given as input, or $\mathbf{w}$ could be found by another
preferred initialization method and given directly as input. The algorithm then
does some few iterations, and in each iteration improved coefficients are found
but the *number* of non-zero coefficients is unchanged. The coefficient vector $\mathbf{w}$
is returned. A detailed description of one iteration follows:

1. The signal is divided into L consecutive segments, the segment length is
   N_{seg} and the signal may be longer than the total length of the L segments,
   $N = N_b + LN_{seg} + N_e$. The excess portions are N_b samples at the beginning
   and N_e samples at the end of the signal, $0 \leq N_b, N_e < N_{seg}$.
2. The coefficients for atoms with support completely within one segment are
   set to zero, for each segment the number of coefficients set to zero is stored:
   s_i for $i = 1, 2, \ldots, L$. The other coefficients are not changed. The original
   coefficients (and segment residuals) are also stored, and may be restored for
   some of the segments (in step 5 below).
3. The residual is calculated using the remaining coefficients, i.e. only the non-
   zero coefficients belonging to atoms with support that span two segments.
4. The residual segments are used as input for the block-oriented sparse approximation algorithm. If the segment size is chosen correctly, depending on the
   SID (P_m values), the dictionary D_{seg} will be the same for all segments and
   all iterations. For each segment the target number of non-zeros is s_i as found
   in step 2.
5. The coefficients are updated. If the sparse representation for a segment is
   better than what it was the new coefficients are used, if not the coefficients
   are set to the values they had when the iteration started. This way each
   iteration will improve (or keep) the sparse approximation.
6. Go back and do next iteration using a new segmentation of the original signal,
   i.e. new values for N_b and N_e. A few (4–8) iterations are sufficient.

One remark should be made to the algorithm as it is described above. The
locations of the non-zeros coefficients will only change slowly from one iteration

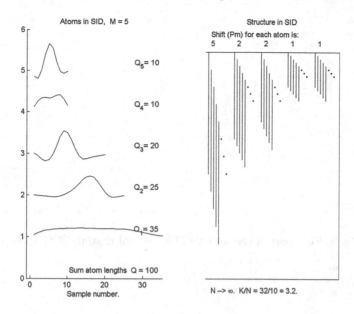

Fig. 4. Example of a shift-invariant dictionary. This structure is used in Table 1 and first row in Table 2.

to the next, and it will take many iterations to move one non-zero coefficient a long distance. This problem with the algorithm can be avoided by selecting a good initial distribution, as SID-MP will give, or by modifying the algorithm: For all segments $(i = 1, 2, \ldots, L)$ the sparse approximation in step 4 could find solutions for $s_i - 1$, s_i and $s_i + 1$ non-zero coefficients and the respective residual norms, $r_{i,-1}, r_{i,0}, r_{i,+1}$, can be calculated. A non-zero coefficient should be moved from segment i to segment j if $(r_{j,0} - r_{j,+1}) > (r_{i,-1} - r_{i,0})$. Repeating this will move non-zeros from one segment to another as long as this operation will reduce the total error. The locations of the non-zeros will now move more quickly between signal parts. We should also note that SID-OMP does not need to use OMP as the sparse approximation algorithm, any block-oriented algorithm will do.

5 Experiments

In this section we present three simple experiments to illustrate how the proposed methods work. All experiments are conducted on an ECG signal from the MIT arrhythmia database [16]. The used signal, MIT100 where a short segment is shown in Fig. 5, is a normal heart rhythm, and the reason for using such an ECG signal is that it will obviously benefit from a SID structure. This signal was also used for dictionary learning. As the purpose of these experiments is to test sparse approximation only, not a particular application, the actual signal or dictionaries used are not that important.

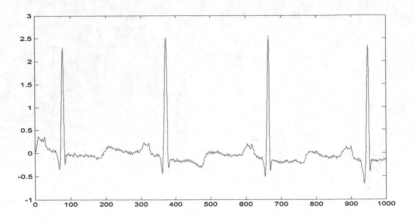

Fig. 5. First part of the used MIT100 normal rhythm ECG signal.

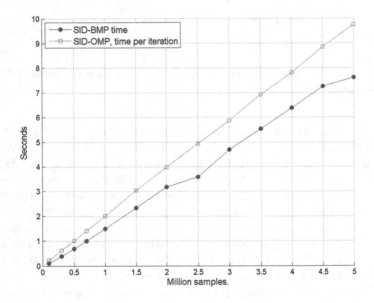

Fig. 6. Running times on an Intel Core i5 3.2 GHz CPU PC for the SID-MP algorithm shown on line marked by disks. The line marked by boxes is running time for one iteration of the SID-OMP algorithm. The segment dictionary D_{seg} has size 360×1170. Note that SID-OMP should do 4–8 iterations.

Experiment 1 is done to show how well, i.e. how fast, the SID-MP and SID-OMP algorithms run as the signal size increases. A SID with structure $M = 5$, $Q_m = \{35, 25, 20, 10, 10\}$ and $P_m = \{5, 2, 2, 1, 1\}$ as illustrated in Fig. 4, is used to make a sparse approximation of the MIT100 signal. The sparsity factor throughout the experiment is kept at 0.08, i.e. the number of non-zero coefficients is $0.08\,N$, where N is the signal length. The MIT100 signal is 250000 samples long, and longer test signals are made simply by repeating it. Figure 6 shows the running time for SID-MP as the signal length increases, it scales (as expected) close to linear order and it is fast. For the longest signal, $N = 5$ million samples and the total number of dictionary atoms is $K = 15.6$ million, the number of non-zero coefficients is 0.4 million, and to find these takes less than 8 s on an Intel Core i5 3.2 GHz CPU PC. Another advantage of the proposed algorithm is that it is small (the compiled mex-file is 23 kB) and it does not depend on any special pre-installed libraries. The running times for one iteration for SID-OMP are also shown in Fig. 6, this line shows the linear order for computational times.

In **Experiment 2** SID-OMP is tested using different segment sizes N_{seg} and different number of iterations. For one segment size OMP is replaced by the computationally more demanding partial search algorithm [24] in the sparse approximation step. All tests use 250000 samples of the MIT100 signal and the sparsity factor is 0.08. Doing eight iterations the Signal to Noise Ratio (SNR) improves for each iteration as shown in Table 1. Most of the improvements is achieved already after two iterations, and using smaller segment sizes keeps on improving SNR for more iterations. After eight iterations, using $N_{seg} = 150$ is the better option. Also, using many small segments is faster that using fewer and larger segments. Thus it is better to use smaller segments than larger, but note that N_{seg} should always be larger than $\max_m Q_m$.

Table 1. This table shows that SID-OMP in Sect. 4 improves the sparse approximation (SA) after SID-MP, SNR after SID-MP was 23.63. The table shows the achieved SNR after each of the iterations in SID-OMP. Time is shown for 8 iterations. The SA step in each iteration was performed by the OMP implementation in SPAMS except for the last line where the computationally more demanding Partial Search (PS) was used for comparison. There are 20000 non-zero coefficients and signal length is 250000. The segment size N_{seg}, and thus the segment dictionary D_{seg}, varies for the different rows.

SA in SID-OMP	D_{seg}	SNR after iteration number (SID-MP: 23.63)								time [s]
		1	2	3	4	5	6	7	8	
OMP	100 × 275	24.56	25.02	25.17	25.26	25.35	25.39	25.41	25.43	1.92
OMP	150 × 435	24.76	25.15	25.25	25.32	**25.39**	**25.42**	**25.44**	**25.44**	2.29
OMP	200 × 595	24.88	25.20	25.27	25.34	25.38	25.40	25.42	25.43	2.73
OMP	250 × 755	24.95	25.23	25.31	**25.35**	25.38	25.40	25.42	25.43	3.32
OMP	300 × 915	25.02	25.25	25.30	25.34	25.37	25.39	25.40	25.41	3.77
OMP	400 × 1235	25.09	25.27	25.31	25.34	25.36	25.37	25.38	25.39	5.23
OMP	500 × 1555	**25.10**	**25.28**	**25.32**	25.34	25.36	25.37	25.38	25.39	7.09
PS	150 × 435	25.36	25.88	26.03	26.12	26.15	26.19	26.21	26.23	67.5

Experiment 3 tests SID-MP and SID-OMP using five different dictionaries, i.e. SID structures. The first three dictionaries are learned for the MIT100 signal, the first is as in Fig. 4. The fourth dictionary is a SID structure with atoms set as low-pass or band-pass filters, $Q_m \in \{128, 64, 32, 16\}$ and $P_m \in \{8, 4, 2, 1\}$, and the fifth dictionary has synthesis atoms from the modified DCT transform, $Q_m = 64$ and $P_m = 1$. The results are shown in Table 2.

Table 2. The achieved SNR for 250000 samples of the MIT100 signal using different
SID structures: dictionary 1–3 are learned, dictionary 4 is a predefined filter bank, and
5 is a modified DCT. For each row, achieved SNR and the computational time for the
two methods are shown.

SID structure				SID-MP		SID-OMP	
No	M	Q	K/N	SNR	time [s]	SNR	time [s]
1	5	100	3.2	23.61	0.3	25.44	2.3
2	6	200	3.3	24.15	0.4	25.74	4.0
3	12	622	7.0	26.20	1.5	27.84	5.9
4	16	960	7.5	21.92	1.8	23.76	5.9
5	32	2048	32.0	18.11	8.8	19.65	16.3

6 Conclusion

The presented SID-MP is a variant of the matching pursuit algorithm that can
be used for a shift-invariant dictionary structure. This structure is flexible as it
encompasses any system that can be expressed as a general synthesis FIR filter
bank of M filters, including the shift-invariant dictionary. The algorithm is fast
and scales well to large signals. We also presented the SID-OMP method that
allows us to use any (fast) block-oriented MP algorithm to further improve the
sparse approximation.

References

1. Aharon, M., Elad, M.: Sparse and redundant modeling of image content using an
   image-signature-dictionary. SIAM J. Imaging Sci. **1**(3), 228–247 (2008)
2. Aharon, M., Elad, M., Bruckstein, A.: K-SVD: an algorithm for designing overcomplete dictionaries for sparse representation. IEEE Trans. Signal Process. **54**(11),
   4311–4322 (2006). http://dx.doi.org/10.1109/TSP.2006.881199
3. Barzideh, F., Skretting, K., Engan, K.: The flexible signature dictionary. In: Proceedings of 23rd European Signal Processing Conference, EUSIPCO-2015, Nice,
   France, September 2015
4. Cotter, S.F., Adler, J., Rao, B.D., Kreutz-Delgado, K.: Forward sequential algorithms for best basis selection. IEE Proc. Vis. Image Signal Process **146**(5), 235–
   244 (1999)
5. Cvetković, Z., Vetterli, M.: Oversampled filter banks. IEEE Trans. Signal Process.
   46(5), 1245–1255 (1998)
6. Davis, G.: Adaptive nonlinear approximations. Ph.D. thesis, New York University,
   September 1994
7. Efron, B., Hastie, T., Johnstone, I., Tibshirani, R.: Least angle regression. Ann.
   Stat. **32**, 407–499 (2004)
8. Engan, K., Aase, S.O., Husøy, J.H.: Method of optimal directions for frame design.
   In: Proceedings of ICASSP 1999, Phoenix, USA, pp. 2443–2446, March 1999

9. Engan, K., Skretting, K., Husøy, J.H.: A family of iterative LS-based dictionary learning algorithms, ILS-DLA, for sparse signal representation. Digit. Signal Proc. **17**, 32–49 (2007)
10. Krstulovic, S., Gribonval, R.: MPTK: matching pursuit made tractable. In: Proceedings ICASSP 2006, Toulouse, France, vol. 3, pp. 496–499, May 2006
11. Mailhé, B., Gribonval, R., Bimbot, F., Vandergheynst, P.: A low complexity orthogonal matching pursuit for sparse signal approximation with shift-invariant dictionaries. In: Proceedings ICASSP 2009, Taipei, Taiwan, pp. 3445–3448, April 2009
12. Mailhé, B., Lesage, S., Gribonval, R., Bimbot, F.: Shift-invariant dictionary learning for sparse representations: extending K-SVD. In: Proceedings of the 16th European Signal Processing Conference (EUSIPCO-2008), Lausanne, Switzerland, August 2008
13. Mairal, J., Bach, F., Ponce, J.: Sparse modeling for image and vision processing. Found. Trends Comput. Graph. Vis. **8**(2–3), 85–283 (2014)
14. Mairal, J., Bach, F., Ponce, J., Sapiro, G.: Online dictionary learning for sparse coding. In: ICML 2009: Proceedings of the 26th Annual International Conference on Machine Learning, pp. 689–696. ACM, New York, June 2009
15. Mallat, S.G., Zhang, Z.: Matching pursuit with time-frequency dictionaries. IEEE Trans. Signal Process. **41**(12), 3397–3415 (1993)
16. Massachusetts Institute of Technology: The MIT-BIH Arrhythmia Database CD-ROM, 2 edn. MIT (1992)
17. Muramatsu, S.: Structured dictionary learning with 2-D non-separable oversampled lapped transform. In: Proceedings ICASSP 2014, Florence, Italy, pp. 2643–2647, May 2014
18. O'Hanlon, K., Plumbley, M.D.: Structure-aware dictionary learning with harmonic atoms. In: Proceedings of the 19th European Signal Processing Conference (EUSIPCO-2011), Barcelona, Spain, August 2011
19. Pati, Y.C., Rezaiifar, R., Krishnaprasad, P.S.: Orthogonal matching pursuit: recursive function approximation with applications to wavelet decomposition. In: Proceedings of Asilomar Conference on Signals Systems and Computers, November 1993
20. Pope, G., Aubel, C., Studer, C.: Learning phase-invariant dictionaries. In: Proceedings ICASSP 2013, Vancouver, Canada, May 2013
21. Rubinstein, R., Zibulevsky, M., Elad, M.: Efficient implementation of the K-SVD algorithm using batch orthogonal matching pursuit. Technical report, CS Technion, Haifa, Israel, April 2008
22. Rubinstein, R., Zibulevsky, M., Elad, M.: Double sparsity: learning sparse dictionaries for sparse signal approximation. IEEE Trans. Signal Process. **58**(3), 1553–1564 (2010)
23. Skretting, K., Engan, K.: Recursive least squares dictionary learning algorithm. IEEE Trans. Signal Process. **58**, 2121–2130 (2010). doi:10.1109/TSP.2010.2040671
24. Skretting, K., Husøy, J.H.: Partial search vector selection for sparse signal representation. In: NORSIG-2003, Bergen, Norway (2003). http://www.ux.uis.no/~karlsk/

Diagnosis of Broiler Livers
by Classifying Image Patches

Anders Jørgensen[1($\boxtimes$)], Jens Fagertun[2], and Thomas B. Moeslund[1]

[1] Media Technology, Aalborg University, Aalborg, Denmark
andjor@create.aau.dk
[2] IHFood A/S, Copenhagen, Denmark

Abstract. The manual health inspection are becoming the bottleneck at poultry processing plants. We present a computer vision method for automatic diagnosis of broiler livers. The non-rigid livers, of varying shape and sizes, are classified in patches by a convolutional neural network, outputting maps with probabilities of the three most common diseases. A Random Forest classifier combines the maps to a single diagnosis. The method classifies 77.6% livers correctly in a problem that is far from trivial.

1 Introduction

To stay competitive with low salary countries, poultry processing plants in the western world, strive to increase their slaughter rates. With line speeds already over 180 bpm (birds per minute), the manual health inspection is now becoming the bottleneck. EU legislation require its member states to perform a post mortem inspection [7] and though no minimum inspection time is specified, it calls for enough time to do a proper inspection. On top of that, local laws in some countries require up to three veterinarians per slaughter line [11].

Post mortem inspection includes both carcass and viscera. If either shows visual signs of a disease both must be discarded. In Sweden and other European countries [11], a mirror is added to the inspection site to allow the inspectors to see the opposite side of the viscera and carcasses. But this also effectively doubles the number of interest points that the inspectors must focus on, making their job twice as hard.

In Denmark, a normal flock has about 1% of the birds discarded due to diseases [5]. Spotting that one bird out of 100 can be a straining task and the inspectors are therefore required to have frequent breaks. This calls for an automated computer vision system to aid the veterinarians and help the processing plants to stay competitive. Such a system could automatically identify unhealthy viscera and discard them. This paper focuses on broiler livers and proposes a method for classifying the three most common liver diseases.

2 Related Works

Computer vision inspection of poultry is a relatively new scientific field, whereas food inspection in general is more mature [10,14]. Recent work by Panagou

P. Sharma and F.M. Bianchi (Eds.): SCIA 2017, Part I, LNCS 10269, pp. 374–385, 2017.
DOI: 10.1007/978-3-319-59126-1_31

et al. [13] displays a non-destructive technique for determining the microbiological quality of beef. Elmasry et al. [6] used hyperspectral near infrared imaging for predicting pH and tenderness and proved it could be an alternative to traditional measuring methods, though it will require some effort to move the system in-line. Feng and Sun [8] developed a method for looking at bacteria by measuring the Total Viable Count in chicken breast fillets with a hyperspectral imaging system.

Others have looked at physical defects, like detecting anomalies on poultry carcases as described in this review by Xiong et al. [17]. One such anomaly can be skin tumours, which Xu et al. [18] detects by employing Principal Component Analysis, Discrete Wavelet Transform and Kernel Discriminant Analysis and combining the classification results. Nakariyakul and Casasent [12] proposed a new Adaptive Branch and Bound algorithm for fast feature selection also when detecting poultry skin tumours. They found their method to be roughly 10 times faster that other Brand and Bound methods.

Speed is an important factor when inspecting poultry at the slaughter lines in a processing plant. Chao et al. [3] presented an in-line system for identifying wholesome and unwholesome chicken. Tests were conducted at a line speed of 140 bpm, yet they concluded that their method should be capable of running at 200 bpm also. Yoon et al. [20] used a hyperspectral line-scan camera for detecting fecal matter and ingesta on poultry carcasses. The system was tested at 140 bpm and 180 bpm, where the latter was deemed the upper limit for the system.

When inspecting viscera the first challenge is to detect and segment the organs. Viscera do not have the same rigidity as the carcass and are often transported in a clamp on a conveyor belt, causing it to swing and rotate as it moves along. Work by Amaral et al. [1] proposed a weighted atlas auto-context algorithm to segment pig viscera in RGB images. Similarly Philipsen et al. [15] proposed a method for segmenting heart, lung and liver in poultry viscera using RGB-D images. The extra dimension proved to give a small performance boost over RGB, but also increased the complexity of the image recording.

3 Data Set

The data set, recorded for this work, consists of images of poultry viscera captured in-line just after evisceration at a poultry processing plant. The RGB images' resolution is 2048×1500.

The focus of this paper is the broiler liver, which is the organ causing the largest number of rejects [5]. In collaboration with a veterinarian working in the field, a total of $1,476$ images have been graded into the categories listed in Table 1. An example image from each class can be seen in Fig. 1. A healthy (**H**) liver has a smooth surface, but can still vary in colour between different shades of red/brown. Cobblestone (**C**) typically affects the entire liver where it looks like dark trenches on the surface. Perihepatitis (**P**) and Necrotic Hepatitis (**N**) can affect just smaller areas of the liver. With **P** the peritoneum gets infected and attaches it self to the liver. **N** appears as small spots which can affect the entire liver or only a small part.

Table 1. Poultry viscera in dataset. "Patch Classifier" shows the number of images used for the patch extraction. "Overall" indicates that the images are used in evaluation of the entire pipeline.

Data set		Data set after split	
Liver diagnosis	# Total	# Patch classifier	# Overall
Healthy (**H**)	542	240	302
Cobblestone Liver (**C**)	458	232	226
Perihepatitis (**P**)	270	125	145
Necrotic Hepatitis (**N**)	206	115	91

The liver is segmented by the method developed by Philipsen et al. [15]. For this work the method has been adapted to RGB images instead of RGB-D images. This yields some decent but imperfect results, as it can be seen in Fig. 1, where the green line indicates the border of the segmentation. Some of the heart might be segmented as liver or sometimes the white peritoneum fails to be segmented with the liver. Improving the segmentation method is not within the scope of this article.

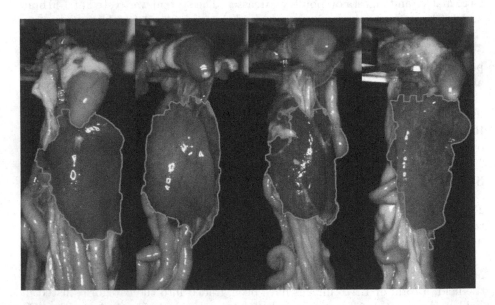

Fig. 1. The four liver diagnoses. From left to right: Healthy liver, Cobblestone liver, Perihepatitis, Necrotic Hepatitis. The green border marks the segmentation found by the segmentation algorithm. The third liver from the left is not fully segmented in the top left part. (Color figure online)

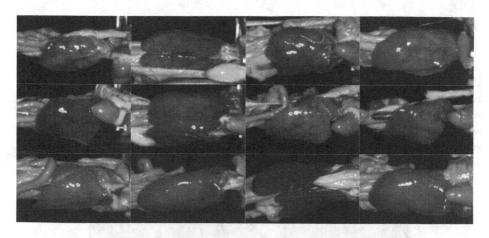

Fig. 2. Variation in the class **C**. Top left is the most severe case, bottom right is the least severe. (Color figure online)

In-Class Variation. The grading is discrete in the sense that a liver belongs to one and only one of the four classes, and there is no severity grading within each class. This is the way the viscera is currently being graded by veterinarians at the processing plant. As a result there is a large in-class variation as it can be seen in Fig. 2, where all livers are diagnosed with **C**. The red/brown tone of the liver can also change widely between broilers. All livers in Fig. 2 could be healthy if judged by colour alone and not texture.

4   Method

Because a disease does not necessarily cover the entire liver it is often the case that different areas of the same liver can be classified differently. The presented method works by finding areas affected by a disease with a patch classifier and based on these results classify the entire liver.

The overall classification consists of the following steps. First a highlight mask is created as described in Sect. 4.1. A convolutional neural network (CNN) trained on image patches constructs a probability map for each class, see Sect. 4.2. The maps are combined with a weight mask, see Sect. 4.3, made from the liver segmentation and the highlight mask. Features are extracted from the weighted maps and fed to a Random Forest classifier giving the overall prediction for the liver, see Sect. 4.4. An overview of the pipeline can be seen in Fig. 3.

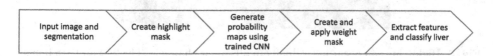

Fig. 3. Overview of the method pipeline.

4.1   Highlight Detection

The images are captured in-line at a poultry processing plant. When a set of viscera passes the camera, led diodes flash at high intensity to generate enough light to illuminate the viscera and enable the camera to use a low exposure time, in order to limit motion blur. As a result this generates some unwanted highlights on the wet glossy surfaces of the liver. Three examples can be seen in Fig. 4. In some images the highlights also appears as speckles. This happens because the evisceration machine in some cases forms small bubbles on the viscera's surface as they are extracted from the carcass.

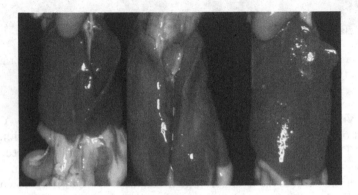

Fig. 4. Examples of livers with highlights. In the image to the right some highlights appear as speckles.

The colour of pixels affected by highlights depends more on the light source than on the object they are portraying. These pixels are therefore considered as noisy and must be removed before further analysis.

The light source consists of red, green and blue LEDs. Because of the spatial distance between the LEDs some highlights occur only as a result of the reflection from one colour e.g. the green. Figure 5 shows an example, and the results of

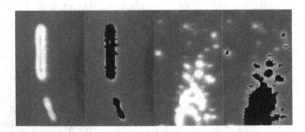

Fig. 5. Highlights can occur individually in all colour channels. The 2nd and 4th image show the results of the highlight removal. (Color figure online)

the highlight removal. The following rules are used to create the highlight mask, which is 0 in highlight areas, otherwise 255:

$$r_{x,y} = g_{x,y} = b_{x,y} = 0 \quad \text{if} \quad g_{x,y} > 175$$
$$r_{x,y} = g_{x,y} = b_{x,y} = 0 \quad \text{if} \quad b_{x,y} \geq r_{x,y} \text{ and } b_{x,y} > 200$$
$$r_{x,y} = g_{x,y} = b_{x,y} = 0 \quad \text{if} \quad g_{x,y} \geq r_{x,y} \text{ and } g_{x,y} > b_{x,y}$$
$$r_{x,y} = g_{x,y} = b_{x,y} = 255 \quad \text{else}$$

Bright green does not naturally occur in the viscera and any high values of green must therefore be caused by a reflection. The blue colour only appears in darker areas, so bright blue pixels are also discarded as highlight. The threshold values have been found through empirical studies.

4.2   Convolutional Neural Network Patch Classifier

It has been chosen to classify the input images in patches. The livers are non-rigid bodies that varies in shape and size and are unevenly affected by the diseases. This makes it difficult to extract the same features for each liver. Dividing the liver into patches allows for classes with less variation than full image livers and each patch will have the same shape and size. Similar approaches have been employed in [9,19], with good results. The patches are used in a CNN which will act as the patch classifier. The CNN will find the best features for discriminating the four classes.

A set of patches have been manually extracted from each class, to train and evaluate the CNN. An extra set of patches, miscellaneous (**M**), have also been extracted. This class contains areas that cannot be represented by the other classes. This may be other organs like the heart or intestines, that mistakenly have been included in the segmentation. It can also be post mortem bruises or damages to the liver that obscure the underlying class. All patches measures 75×75 pixels. This resolution was found to be large enough to cover the texture patterns of all diseases, and small enough to be easy placeable within the livers. Table 2 shows some examples of the extracted patches.

To avoid data cross contamination, a set of images have been randomly selected for each class from the full data set, as shown in Table 1. These images will only be used for training and evaluation of the CNN. Table 3 shows the count of extracted patches. The number of images used for extraction varies based on the size and severity of the disease. For **C**, **P** and **N**, it has not been possible to extract the same amount of patches as for **H**. Patches for the class **M** have been extracted from images from the other four classes.

The CNN is implemented as a small sequential network with Keras [4] and Theano [16]. A representation of the network, as it is implemented in Keras, can be seen in Fig. 6. A softmax activation layer is used to generate the output, which is a vector with normalised probabilities for each class. The code is available at https://bitbucket.org/andjor/2017_broiler_livers.

All training and validation patches are augmented with 36 rotations in increments of $10°$. The validation set is used to pick the epoch with the best model.

Table 2. Six examples from each extracted class for the patch classifier.

Table 3. Patches extracted from the full images. Test patches are extracted from the Test images. M patches have been extracted from images from all other classes. The number in the parentheses is with augmentation.

Diag	Full images		Extracted patches		
	Train/val	Test	Train	Val	Test
H	180	60	306 (11016)	75 (2700)	100
C	152	80	235 (8460)	75 (2700)	100
P	89	36	205 (7380)	75 (2700)	100
N	77	38	200 (7200)	75 (2700)	100
M	-	-	323 (11628)	75 (2700)	100

Looking at the validation loss in Fig. 7, it can be seen that the smallest validation loss happened at epoch 24.

Generating Probability Maps. The CNN is used to predict patches sampled from the input images. The bounding box from the segmentation is used as region of interest in the input image. To further reduce the calculation time, the CNN is only applied to every n'th pixel in both the x and y direction, effectively down-sampling the generated probability maps by a factor of n compared to the original. A map will be generated for each class, where each element holds the probability that the corresponding region belongs to that class.

Fig. 6. Structure of the CNN, as it is implemented in Keras [4]. More information about the layers is present in the Keras documentation.

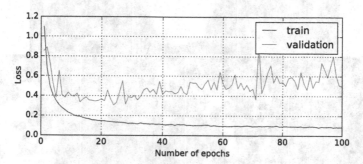

Fig. 7. Training and validation loss for the CNN. A snapshot of the model is saved at the smallest validation loss, which is epoch 24.

4.3   Creating the Weight Mask

As the patch classifier classifies pixels based on a surrounding square region it is important to justify how trustworthy that region is. Some pixels in the region can be outside the liver mask or the region can contain a high number of highlight pixels. To take this into account a weight mask is created by anding the liver segmentation mask and the highlight mask. The weights are then calculated by moving a sliding window, 75×75 pixels, across the joint mask with the same step size used in the patch classifier. For each step the weight, α, of the region is calculated as

$$\alpha = \begin{cases} 1 - \frac{b}{b_{\max}} & \text{if } b < b_{\max} \\ 0 & \text{else} \end{cases} \qquad (1)$$

$$b_{\max} = 0.1 \cdot w \cdot h$$

where b is the number of black pixels (highlight or not liver) in the patch and w and h is the width and height of the patch. If 10% or more of the patch is black, the weight is set to zero.

An example of the generated probability maps, with the weight mask applied, can be seen in Fig. 8.

4.4   Random Forest Liver Classifier

The classification of the liver is based on the weighted probability maps. As it can be seen in Fig. 8, will a liver often respond in all four classes and the **M**

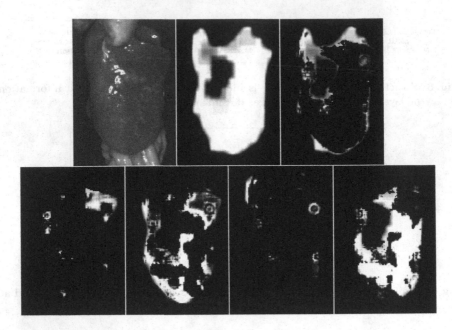

Fig. 8. Probability maps for an input image diagnosed with **N**. Input image top left. Weight mask is top middle. Top right is the probability map for **M**. In the bottom from the left to the right: **H**, **C**, **P** and **N**.

class. This can be because of misclassified patches or simply because the disease only affects a small part of the liver.

For a data driven approach, features are extracted from each probability map. These are features like the count of non zero probabilities in the map, the sum of all probabilities or the ratio of non zero probabilities compared to the map size. A total of 8 features are extracted per probability map, 40 features per liver. All are fed to a Random Forest Classifier [2], responsible for the liver classification. Random Forest is chosen as it works well with non normalised features. The number of trees is 100 and the minimum split size is 1.

5   Results

The performance of the CNN patch classifier, calculated on the test patches, can be seen in Table 4. Each class contains 100 sample patches. The total true positive rate is 95.4%.

For the results of the entire method the step size n is set to 5, which means the algorithm will moving 5 pixels when generating the probability maps and the weight mask. This was chosen as a good balance between speed up and resolution of the probability maps. The patches are still highly correlated as only 6.$\bar{6}$% pixels will be "new data" when moving the classifier one step in the horizontal direction.

Table 4. Classification matrix for the CNN. 100 sample patches in each class.

		Prediction					True positive
		H	C	P	N	M	
True	H	98	2	0	0	0	98.0%
	C	5	91	2	2	0	91.0%
	P	0	1	98	0	1	98.0%
	N	1	0	0	98	1	98.0%
	M	1	4	3	0	92	92.0%

The images in the "Overall" column in Table 1 was used to evaluate the performance over the entire pipeline in Fig. 3. To measure both accuracy and precision, the classification rates are measured over 10 five fold runs, using stratified sampling. The average true positive rate is 77.6%. Table 5 shows the confusion matrix average and one standard deviation for the 10 five fold runs.

Table 5. Classification matrix for the overall system. Average ± one standard deviation.

		Prediction				True Positive
		H	C	P	N	
True	H	242.5 ± 3.47	27.2 ± 1.93	32.3 ± 2.71	0.0 ± 0.00	80.3%
	C	37.1 ± 3.57	176.3 ± 4.37	10.1 ± 1.66	2.5 ± 0.71	78.0%
	P	34.1 ± 3.28	7.9 ± 0.74	103.0 ± 3.06	0.0 ± 0.00	71.0%
	N	10.1 ± 0.57	8.6 ± 0.52	1.0 ± 0.47	71.3 ± 0.48	78.4%

The processing time per liver image is 3–5 s depending on the size of the liver. The tests were performed on a laptop with Intel i7-4720HQ CPU, 16 GB ram and a Nvidia 960m GPU.

6 Conclusion

The patch classifier correctly classifies 95.4% of the test patches, see Table 4. **C** has the most misclassifications, many fall in **H**, which is consistent with the fact that **C** slowly evolves from **H**, as shown in Fig. 2. Many early stage **C** livers will be confused with **H** livers. **M** has the second most misclassifications, primarily spread between **C** and **P**. **C** might, in severe cases, look like bruises, and **P** can be all white much like the intestines.

A good performance of the patch classifier does not directly translate to a good performance of the overall system. The patches were hand picked and it is more important that the patches fully represent the variation in their classes than getting a high performance on very specialised classes. It is the authors' beliefs that **H** is well represented, but **C**, **P** and **N** could make use of more samples. It was not possible to extract more patches for this work, without leaving too few images for the overall performance test. For future works it might help to use other augmentation techniques on the training data.

The overall system had an average classification error of 22.4%. Some of these errors are no doubt a result of the patch classifier, not classifying patches correctly. In cases where diseases like **P** and **N** only affects a small part of the liver, it makes the probability maps very susceptible to misclassifications. This makes it difficult for the Random Forest classifier to set an appropriate threshold. Some of this could be remedied with better features for the probability maps.

Fig. 9. All livers are diagnosed with **P**, yet they look very different.

It is not known how much influence the imperfect segmentation have on the classification results. Manually annotating all livers would be a slow process and one could argue that the current segmentation is closer to what it would be in a real scenario. Yet to isolate the true performance of this method, a perfect segmentation is needed.

Some errors are due to the large variation in the classes. The livers in Fig. 9 are all diagnosed with **P**, but they look very different. For future work it might be beneficial to categorise the diagnoses into smaller subclasses. This could be helped by having a more fine-grained grading that includes the severity of the diagnoses. With this, better features for the probability maps and more training data it should be possible to improve the performance of this method.

References

1. Amaral, T., Kyriazakis, I., Mckenna, S.J., Ploetz, T.: Weighted atlas auto-context with application to multiple organ segmentation. In: Proceedings of WACV (2016)
2. Breiman, L.: Random forests. Mach. Learn. **45**(1), 5–32 (2001)
3. Chao, K., Yang, C.C., Kim, M.S.: Line-scan spectral imaging system for online poultry carcass inspection. J. Food Process Eng. **34**(2011), 125–143 (2011)
4. Chollet, F.: Keras (2015). https://github.com/fchollet/keras
5. Claudi-magnussen, C., Daugaard, H.: Automatiserede hjælpeværktøjer til kødkontrol på kyllingeslagterierne. Technical report, Danish Meat Research Institute (2011)

6. Elmasry, G., Sun, D.W., Allen, P.: Near-infrared hyperspectral imaging for predicting colour, pH and tenderness of fresh beef. J. Food Eng. **110**(1), 127–140 (2012)
7. EU: REGULATION (EC) No 854/2004: laying down specific rules for the organisation of official controls on products of animal origin intended for human consumption (2004)
8. Feng, Y.Z., Sun, D.W.: Determination of total viable count (TVC) in chicken breast fillets by near-infrared hyperspectral imaging and spectroscopic transforms. Talanta **105**, 244–249 (2013)
9. Hou, L., Samaras, D., Kurc, T., Gao, Y.: Patch-based convolutional neural network for whole slide tissue image classification, p. 7 (2015)
10. Huang, H., Liu, L., Ngadi, M.O.: Recent developments in hyperspectral imaging for assessment of food quality and safety. Sensors **14**(4), 7248–7276 (2014). (Basel, Switzerland)
11. Löhren, U.: Overview on current practices of poultry slaughtering and poultry meat inspection, pp. 1–58 (2012)
12. Nakariyakul, S., Casasent, D.P.: Fast feature selection algorithm for poultry skin tumor detection in hyperspectral data. J. Food Eng. **94**(3), 358–365 (2009)
13. Panagou, E.Z., Papadopoulou, O., Carstensen, J.M., Nychas, G.J.E.: Potential of multispectral imaging technology for rapid and non-destructive determination of the microbiological quality of beef filets during aerobic storage. Int. J. Food Microbiol. **174**, 1–11 (2014)
14. Park, B., Lu, R. (eds.): Hyperspectral Imaging Technology in Food and Agriculture. Food Engineering Series. Springer, New York (2015)
15. Philipsen, M.P., Jørgensen, A., Escalera, S., Moeslund, T.B.: RGB-D segmentation of poultry entrails. In: Perales, F.J.J., Kittler, J. (eds.) AMDO 2016. LNCS, vol. 9756, pp. 168–174. Springer, Cham (2016). doi:10.1007/978-3-319-41778-3_17
16. Theano Development Team: Theano: A Python framework for fast computation of mathematical expressions. arXiv e-prints abs/1605.02688 (2016)
17. Xiong, Z., Xie, A., Sun, D.W., Zeng, X.A., Liu, D.: Applications of hyperspectral imaging in chicken meat safety and quality detection and evaluation: a review (2014)
18. Xu, C., Kim, I., Kim, M.S.: Poultry skin tumor detection in hyperspectral reflectance images by combining classifiers. In: Kamel, M., Campilho, A. (eds.) ICIAR 2007. LNCS, vol. 4633, pp. 1289–1296. Springer, Heidelberg (2007). doi:10.1007/978-3-540-74260-9_114
19. Xu, Y., Jia, Z., Ai, Y., Zhang, F., Lai, M., I-Chao Chang, E.: Deep Convolutional activation features for large scale brain tumor histopathology image classification and segmentation, pp. 947–951 (2015)
20. Yoon, S.C., Park, B., Lawrence, K.C., Windham, W.R., Heitschmidt, G.W.: Line-scan hyperspectral imaging system for real-time inspection of poultry carcasses with fecal material and ingesta. Comput. Electron. Agricult. **79**(2), 159–168 (2011)

Historical Document Binarization Combining Semantic Labeling and Graph Cuts

Kalyan Ram Ayyalasomayajula(✉) and Anders Brun

Department of Information Technology, Centre for Image Analysis,
Uppsala University, Uppsala, Sweden
{kalyan.ram,anders.brun}@it.uu.se

Abstract. Most data mining applications on collections of historical documents require binarization of the digitized images as a preprocessing step. Historical documents are often subjected to degradations such as parchment aging, smudges and bleed through from the other side. The text is sometimes printed, but more often handwritten. Mathematical modeling of appearance of the text, background and all kinds of degradations, is challenging. In the current work we try to tackle binarization as pixel classification problem. We first apply semantic segmentation, using fully convolutional neural networks. In order to improve the sharpness of the result, we then apply a graph cut algorithm. The labels from the semantic segmentation are used as approximate estimates of the text and background, with the probability map of background used for pruning the edges in the graph cut. The results obtained show significant improvement over the state of the art approach.

Keywords: Binarization · Semantic labeling · Deep learning · Graph cut · Zero shot learning

1 Introduction

In historical document image analysis, binarized images record each pixel as background (parchment/paper) or foreground (text/ink) by preserving most of the relevant visual information in the image. A high-quality binarization significantly simplifies further tasks to be performed on the document image, such as word spotting and transcription. The challenges commonly faced in this area are similar to the problem of uneven illumination for thresholding algorithms. However, historical documents may in addition have other artefacts such as; bleed through; fading or paling of the ink in some areas; smudges, stains and blots covering the text; text on textured background and handwritten documents with heavy-feeble pen strokes for cursive or calligraphic effects to name a few. In general, this makes the task of historical document binarization very challenging as shown in Fig. 1. The task is often subjective and garners interest in the field, which has led to the document image binarization content (DIBCO) [12] for automatic methods with minimum parameter setting.

© Springer International Publishing AG 2017
P. Sharma and F.M. Bianchi (Eds.): SCIA 2017, Part I, LNCS 10269, pp. 386–396, 2017.
DOI: 10.1007/978-3-319-59126-1_32

The field of research into binarization in general has led to many important methods. A classical approach is the thresholding approach from Otsu [11], which tries to maximize the gray level separation between foreground (FG) and back ground (BG) classes. In an ideal scenario it often suffices to fall back on such a global threshold method. However, local intensity variation and other artifacts introduced in image capturing procedure have led to a more *locally adaptive* techniques, such as the methods from Niblack [10], Sauvola and Pietikainen [13]. The techniques discussed in all these methods are generic and applicable to any image in general, however winning entries of DIBCO in the past have developed methods intended to improve binarization through modeling properties of FG/BG specific to documents images. Lu et al. [8] have for instance modeled background using polynomial smoothing followed by local thresholding on detected text strokes, Bar-Yousef et al. [2] iteratively grow FG and BG within a 7 × 7 window.

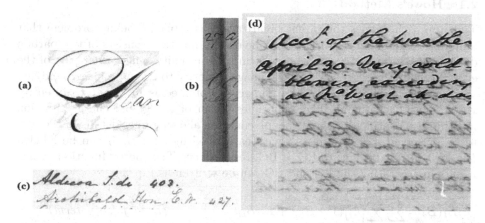

Fig. 1. Examples of typical image degradations are show in the figure (a) smudging of text (b) degradation of paper from aging (c) unevenness of ink in writing (d) bleed through from the ink on the other side of the document

The basic algorithm proposed in this paper draws motivations from other ideas that employed use of Markov random fields (MRF) for binarization such as Howe [4], Mishra et al. [9]. These methods take both the global as well as the local aspects of the image into consideration in order achieve pixel labeling. The former uses the image Laplacian to obtain invariance to BG intensity and builds a graph from the image using Canny edges of the image. A graph-cut is performed with suitable *source-sink* nodes, which are seeds points for FG-BG pixels respectively. Although the fundamental idea governing Howe's method have been explored previously as separate methods, combining them into an energy function proved particularly effective. Further improvement of Howe's approach was proposed by Ayyalasomayajula and Brun [1] by efficient detection of seeds for the source-sink nodes and effective edge pruning by exploiting the inherent topology by defining a binarization space.

The proposed method improves upon our previous work in Ayyalasomaya-jula and Brun [1]. The method builds upon three ideas, basic idea is borrowed from the Howe's approach [5] of defining the binarization as labeling problem. The next step is to incorporate the benefits of defining good seed points for foreground and background regions and idea of edge pruning to delineate them as discussed in our previous work [1]. We use the labeled outputs and BG class probability map from a fully convolution neural network as FG-BG seed points and improving edge estimates respectively. Further understanding of the key contributions in current work would require an overview of Howe's and our previous methods. The following section captures the essence of these methods drawing attentions to the aspects that have been improved.

2 Motivation

2.1 Howe's Method

This approach defines the target binarization as a pixel labeling problem that minimizes a global energy function. The energy function consists of two parts a data-fidelity term and a smoothness term for continuity. The former part of the energy relies on the Laplacian of the image intensity to estimate foreground/ink and background/parchment. The Laplacian obtained is invariant to intensity variation in background. The smoothness term of the global energy function incorporates continuities along the ink contours allowing finer details in the text to be preserved. For an image I with each pixel indexed (i, j) can be labeled $B_{ij} \in \{0,1\}$ if it belongs to FG or BG, respectively. The energy function, which results in separating the FG and BG can be written as Eq. (1)

$$E_I(B) = \sum_{i=0}^{m} \sum_{j=0}^{n} [L_{ij}^0(1 - B_{ij}) + L_{ij}^1 B_{ij}] \qquad \text{(Data fidelity term)}$$

$$+ \sum_{i=0}^{m} \sum_{j=0}^{n} C_{ij}^h(B_{ij} \neq B_{i+1,j}) \qquad \text{(Horizontal continuity term)} \quad (1)$$

$$+ \sum_{i=0}^{m} \sum_{j=0}^{n} C_{ij}^v(B_{ij} \neq B_{i,j+1}) \qquad \text{(Vertical continuity term)}$$

where $L_{ij}^0, L_{ij}^1, C_{ij}^h, C_{ij}^v$ are costs associated with labeling a pixel as belonging to FG, BG, smoothness along horizontal and vertical directions respectively at a pixel indexed (i,j). Labeling costs are governed by Eq. (2)

$$L_{ij}^0 = \nabla^2 I_{ij}$$
$$L_{ij}^1 = \begin{cases} -\nabla^2 I_{ij}, & I_{ij} \leq \mu_{ij}^r + 2\sigma_{ij}^r \\ \phi, & I_{ij} > \mu_{ij}^r + 2\sigma_{ij}^r \end{cases} \qquad (2)$$

where $\nabla^2 I$, $\nabla^2 I_{ij}$, are the Laplacian of image I and Laplacian value at the pixel (i,j) respectively. The $\nabla^2 I_{ij}$ values are enough to label the pixel as belonging to

FG or BG however this could lead to extremely large values at image borders due
to BG class size far exceeding FG class size in typical handwritten document. In
order to avoid such strong label associations a conservative strategy is applied
to BG labeling as indicated in the modified form of L_{ij}^1 in Eq. (2), where pixels
with intensities more than two standard deviations $(2\sigma_{ij}^r)$ brighter than the local
mean μ_{ij}^r, as computed over nearby pixels weighted by a Gaussian of radius r
are assigned a constant cost ϕ. This facilitates the possibility of label switching
with some penalty but not restricting it all together.

Smoothness cost is governed by Eq. (3)

$$C_{ij}^h = \begin{cases} 0, & E_{ij} \wedge (I_{ij} < I_{i+1,j}) \\ 0, & E_{i+1,j} \wedge (I_{ij} \geq I_{i+1,j}) \\ c, & otherwise \end{cases}$$

$$C_{ij}^v = \begin{cases} 0, & E_{ij} \wedge (I_{ij} < I_{i,j+1}) \\ 0, & E_{i+1,j} \wedge (I_{ij} \geq I_{i,j+1}) \\ c, & otherwise \end{cases} \quad (3)$$

which encourages label consistency on either sides of the edges. As per DIBCO
requirement Eq. (3) includes edges in the FG, which is a reasonable choice to
make. From Eqs. (2, 3) it can be inferred that there are five parameters; Guassian
radius r, label switching cost ϕ, discontinuity cost c, and thresholds (t_{hi}, t_{lo})
governing Canny edges E_{ij}, that can be associated with the energy function in
Eq. (1). Of all the parameters mentioned c and t_{hi} are critical for performance. It
is worth noting that both these parameters are associated with edge information
and smoothness term of the energy function.

2.2 Topological Clustering Approach

This method is similar in spirit to Lelore and Bouchara [6] where image pixels
are divided into three classes: ink, background and unknown. The FG and BG
cluster help define the unknown pixels as belonging to one of the two classes.
In this approach an ordered triplet (I, dI_h, dI_v) corresponding to pixel intensity,
gradient in horizontal and vertical direction respectively are used to represent
every pixel in a three-dimensional space defined as the *binarization space* denoted
by $\mathscr{S}$ in Eq. (4). It can be noted that the region $\mathscr{S}$ is bounded as shown below

$$\mathscr{S} = \begin{cases} 0 \leq I \leq 255, \\ -255 \leq dI_h \leq 255, \\ -255 \leq dI_v \leq 255. \end{cases} \quad (4)$$

This theoretical framework helps to define a *topology* [1] within this space
with a natural way of defining hierarchical clusters with neighborhood constraints. Figure 2 shows the space $\mathscr{S}$ for a typical image from the DIBCO dataset
with points in red and blue representing the FG and BG pixels, respectively.

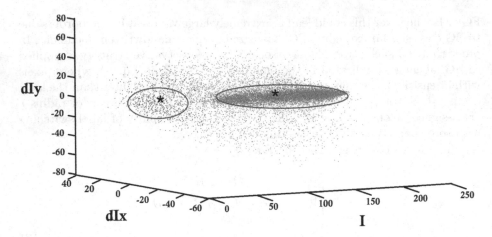

Fig. 2. Distribution of FG and BG pixels as points in binarization space in red and blue respectively, for the file '*2009_H02.bmp*' from the DIBCO dataset. (Color figure online)

The FG and BG clusters are encircled with ellipses for a *core-object* denoted in * for a given ε-reachable neighborhood N_ε. For all practical purposes with a certain abuse of notation we can think of *core-object* as choice of cluster center with ε denoting the maximum distance separating a pixel from core point in $\mathscr{S}$. The advantage in this approach is that it allows to iterative refine the cluster hierarchically based on N_ε with respect to a core object. Upon using the core-objects as estimates for source-sink, the information about the cluster associated with core objects help in including only the relevant edges for the graph cut to make a better segmentation of FG from BG thus improving the result.

2.3   Proposed Method

From Howe's method and the clustering approach, it can be inferred that results of graph-cut can benefit a lot from an approach that can produce better edge estimates. To meet this end we propose an approach to assist the graph-cut through source-sink and edge estimates from a *Fully Convolutional Neural Network* (FCNN). The idea is built from the fact that FCNNs have been very effective in producing state-of-the-art results in semantic segmentation [14]. The binarization results can be posed as a semantic segmentation problem with two class instances. Figure 3 shows the architecture of a FCNN as a *single-stream* and is taken from [14]. It learns to combine coarse, high layer information with fine, low layer information. Pooling and prediction layers are shown as gray grids, reveal relative spatial coarseness, while intermediate convolutional layers are shown in blue as vertical lines.

As the output from a strided convolution layer or a pooling layer is a down sampled version of the input to the layer the resolution at the final layer will be lesser than the input image. To match the output labels for each pixel at the

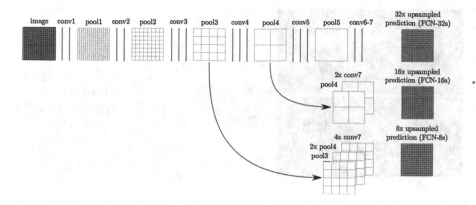

Fig. 3. Architecture learns to combine coarse, high layer information with fine, low layer information. Pooling and prediction layers are shown as grids that reveal relative spatial coarseness, while intermediate layers are shown as vertical lines. FCN-32s, FCN-16s, FCN-8s represent the prediction stride corresponding to pixels in a single step. (Color figure online)

same resolution as the input, upsampling layers are incorporated. Depending on the number of upsampling layers used, the FCNN in [14] can have three variations of stream-nets. Referring to Fig. 3 the first row of an FCN-32s: depicts a single-stream net, it upsamples stride 32 predictions back to pixels in a single step. Second row FCN-16s: combines predictions from both the final layer and the pool4 layer, at stride 16, it allows the network to predict finer details, while retaining high-level semantic information. Third row FCN-8s: combines additional predictions from pool3, at stride 8, to provide further precision. Combining the low level features with high level semantic information is achieved through skip connections that allow for a better gradient propagation while training. A trained network can be used to predict the probability of a pixel belonging to FG/BG class. A final label can be obtained from a soft-max layer with each pixel labeled based on the maximum probability of it belonging to a class.

All the observations made so far have been incorporated into our proposed method in the following steps:

– The labels obtained from FCNN output act as very good source and sink estimates.
– The network performs very well in estimating the background, but gives a very conservative estimate of the foreground labels. This is due to the class imbalance between FG and BG pixels, as the FCN is optimized for *overall accuracy* [14] on class label prediction.
– The probability map of the BG class thresholded on mean BG class probability is good indication of FG/BG separation.

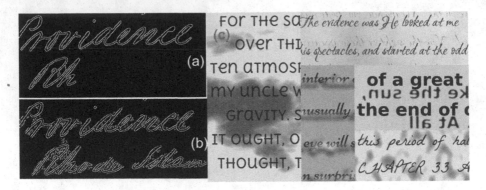

Fig. 4. (a) shows edge map from Canny threshold picked through optimization as described in [5], (b) shows the pruned edge map from the class probabilities, (c) shows few samples from synthetic text data

- A graph is constructed by including edges within this thresholded probability mask give a good estimate of the FG spread as shown by comparing Fig. 4(a) and (b).
- These source, sink and edge map estimates serve as optimal inputs for a graph-cut algorithm to be applicable.

3   Experiments

3.1   DIBCO Data Processing

The experiments were conducted on the *DIBCO* [12] datasets for binarization consisting of 76 images. The dataset was divided into training and validation sets with 70–30 split. The ground truth and input images were converted into 500×300 pixels of cropped images with 100 pixels horizontally and vertically overlap. This creates create more data for training through augmentation of cropped overlapping input images as mentioned previously. The cropped images of 500×300 pixels allow the trained model to fit into memory in Convolutional Architecture for Fast Feature Embedding framework [15] (commonly known as CAFFE) on a on an NVIDIA Titan GTX. The model was then initialized with weights from a model pre-trained on PASCAL-VOC dataset available at [16].

3.2   Training

The FCN-8 architecture was used as it has a better receptive field which translates to accurate pixel labeling for binarization. In order to train the network on binarization data, the weights for layers till FC7 are loaded from the pre-trained model and layers beyond FC7 are trained on the DIBCO dataset using the ground truth labels. The training was continued for 150,000 iteration till an accuracy of 75% for mean Intersection over Union (mIoU) was obtained for predicted vs. ground truth segmented regions.

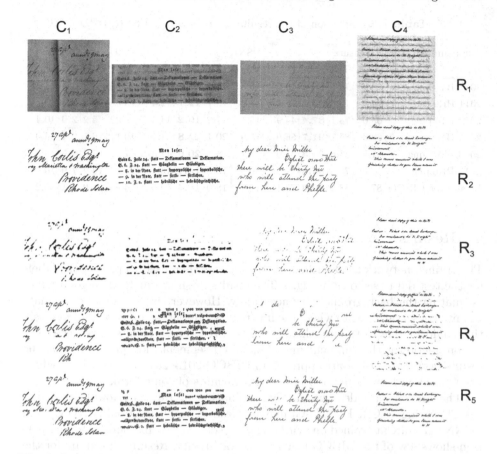

Fig. 5. Image Rows R_1, R_2, R_3, R_4 and R_5 shows the DIBCO images, ground truth, FCNN, Howe and FCNN+GraphCut outputs respectively for images shown in columns $C_1 - C_4$.

3.3 Synthetic Data Processing

In order check for any bias that could arise due to over-fitting the network to data, the experiments were repeated by training the network on synthetic data. Documents resembling historical handwritten and printed material were generated synthetically. Various filters were applied to resemble background textures and degradations in the parchment. The text was generated using handwriting and machine printed fonts from the Google^{TM} fonts [17]. Figure 4(c) shows few cropped images from the synthetic dataset. The results from binarization on DIBCO dataset using the network trained on the synthetic dataset are presented in Table 1 in FC(S) column and can be compared with a network trained on DIBCO dataset in FC column.

Table 1. Comparison of the results for F-Measure, PSNR, DRD

File name	FMeasure				PSNR				DRD			
	Howe	TCl	FC	FC(S)	Howe	TCl	FC	FC(S)	Howe	TCl	FC	FC(S)
2011_HW6.png	78.9	76.1	**86.4**	43.9	15.6	14.4	**17.2**	12.7	5.50	8.6	**3.4**	11.7
2011_PR2.png	74.9	79.9	**80.1**	78.4	11.4	12.7	**12.7**	12.4	13.4	9.6	**9.3**	10.1
2012_H13.png	63.0	78.1	**87.4**	44.9	15.5	17.2	**19.2**	14.5	8.02	5.2	**2.9**	10.4
2013_HW5.png	69.3	85.7	**91.7**	86.9	15.9	20.1	**22.8**	21.1	19.1	6.1	**2.4**	4.0
2013_HW7.png	51.4	51.6	**74.3**	34.0	18.1	18.1	**20.0**	17.3	7.28	7.3	**4.5**	8.8
2013_PR6.png	70.0	72.8	**84.8**	79.5	10.3	10.9	**14.1**	13.1	21.9	18.9	**7.9**	10.1
Mean over DIBCO	87.7	88.3	**89.8**	75.1	17.8	18.0	**18.4**	16.0	4.2	4.0	**3.1**	7.2

4   Results

The trained network as described previously was then used to predict the labels for the test data as shown in Fig. 5. The results as shown in R_3 do a good job of estimating the back-ground very accurately. However the fore-ground estimate is quite conservative as to where the ink/text is present, as mentioned previously this could be due to the imbalance in class sizes. These results have been used in conjunction with a min-cut max-flow based segmentation [3] to improve the results over state-of-the-art approach in DIBCO 2013–2016 binarizations; which were based on Howe's method [4,5] or its variants of some form.

The Table 1 provides quantitative evaluation of Howe's (Howe), Topological Clustering (TCl), FCNN+GraphCut trained on DIBCO dataset (FC) and FCNN+GraphCut trained on synthetic data set (FC(S)) methods. The comparison shows few of the DIBCO metrics [12] for brevity. Results shown are for the files which have shown more then 10% gain on any one of the metrics in absolute scale. A high score on F-Measure, Peak Signal to noise ratio (PSNR) and a low Distance Reciprocal Distortion Metric (DRD) [7] is desirable for an algorithm, these metrics are defined in the appendix for ease of reference.

The F-Measure is a well understood metric for measuring classification accuracy where there is 2.1% gain in mean absolute percentage over the entire DIBCO dataset with 76 files, which is significant improvement. PSNR metric is useful when comparing with the topological clustering with FCNs as it can serve as a measure to associate unlabeled pixels in binarization space $\mathscr{S}$ to the exact classes where there is a 3.3% gain in mean relative scale. Both these metrics are generic, however DRD is more tailored for binarization of document images. There is a significant boost of 27.1% in results over mean relative scale for DRD metric. Figure 5 compares the results qualitatively for the methods to provide a visual cue into the gain achieved by various methods.

4.1   Conclusions and Future Work

The current experimental framework shows relevance of FCNNs in segmentation based tasks such as binarization in documents to obtain state-of-the-art results.

The possibilities as stated below are speculations into intended directions to pursue our future research into this field. The current approach can be extended to other tasks such as layout analysis and de-noising of historical documents. The method can benefit greatly from training on synthetic data generated by mimicking various degradations through filter operation thus enabling zero shot learning. Though unable to achieve state-of-the-art results on synthetic data the preliminary results show potential for improvement and benefit along this direction. The graph-cut can be integrated as a loss layer into the network to provide end-to-end binarization. Also training the network on a modified loss layer based on DIBCO metrics may lead to improved results.

Acknowledgment. This project is a part of q2b, From quill to bytes, an initiative sponsored by the Swedish Research Council (Vetenskapsrådet D.Nr 2012-5743) and Riksbankens Jubileumsfond (R.Nr NHS14-2068:1) and Uppsala university. The authors would like to thank Fredrik Wahlberg and Tomas Wilkinson of Dept. of Information Tech., Uppsala University and also the anonymous reviewers for their constructive criticism in improving the manuscript.

Appendix

F-Measure

$$F - Measure = \frac{2 \times Recall \times Precision}{Recall + Precision} \tag{5}$$

where, $Recall = \frac{TP}{TP+FN}$, $Precision = \frac{TP}{TP+FP}$, TP, FP, FN denote the True Positive, False Positive and False Negative values, respectively.

Peak Signal to Noise Ratio

$$PSNR = 10\log\left(\frac{C^2}{MSE}\right) \tag{6}$$

where, $MSE = \frac{\sum_M^{i=1}\sum_N^{j=1}\left[I(i,j)-I'(i,j)\right]^2}{MN}$. PSNR is a measure of how close is one image to another. Higher the value of PSNR, more is the similarity between binarized image and the ground truth. Note that C equals to the difference between foreground and background, M, N are the width and height of the image respectively.

Distance Reciprocal Distortion Metric

The DRD Metric serves as a measure of visual distortion in a binary document images [7]. It correlates with the human visual perception and measures the distortion for all the $\mathbb{S}$ flipped pixels as follows:

$$DRD = \frac{\sum_{k=1}^{S} DRD_k}{NUBN} \tag{7}$$

where $NUBN$ is the number of non-uniform (gray pixels) 8×8 blocks in the ground truth (GT) image, and DRD_k is the distortion of the k-th flipped pixel that is calculated using a 5×5 normalized weight matrix W_{Nm} as defined in [7]. DRD_k equals to the weighted sum of the pixels in the 5×5 block of the GT that differ from the centered k-th flipped pixel at (x, y) in the binarization result image $\mathbb{B}$ as defined below:

$$DRD_k = \sum_{i=-2}^{2} \sum_{j=-2}^{2} |GT_k(i, j) - B_k(i, j)| \times W_{Nm}(i, j) \qquad (8)$$

References

1. Ayyalasomayajula, K.R., Brun, A.: Document binarization using topological clustering guided Laplacian energy segmentation. In: Proceedings of ICFHR, pp. 523–528 (2014)
2. Bar-Yosef, I., Beckman, I., Kedem, K., Dinstein, I.: Binarization, character extraction and writer identification of historical Hebrew calligraphy documents. Int. J. Doc. Anal. Recogn. 9(2), 89–99 (2007)
3. Boykov, Y., Kolmogorov, V.: An experimental comparison of min-cut/max-flow algorithms for energy minimization in vision. PAMI 26(9), 1124–1137 (2004)
4. Howe, N.: A Laplacian energy for document binarization. In: International Conference on Document Analysis and Recognition, pp. 6–10 (2011)
5. Howe, N.R.: Document binarization with automatic parameter tuning. Int. J. Doc. Anal. Recognit. 16(3), 247–258 (2012). doi:10.1007/s10032-012-0192-x
6. Lelore, T., Bouchara, F.: Super-resolved binarization of text based on FAIR algorithm. In: International Conference on Document Analysis and Recognition, pp. 839–843 (2011)
7. Lu, H., Kot, A.C., Shi, Y.Q.: Distance-reciprocal distortion measure for binary document images. IEEE Signal Process. Lett. 11(2), 228–231 (2004)
8. Lu, S., Su, B., Tan, C.L.: Document image binarization using background estimation and stroke edges. Int. J. Doc. Anal. Recogn. 13(4), 303–314 (2010)
9. Mishra, A., Alahari, K., Jawahar, C.V.: An MRF model for binarization of natural scene text. In: International Conference on Document Analysis and Recognition (2011)
10. Niblack, W.: An Introduction to Digital Image Processing. Prentice-Hall, Englewood Cliffs (1986)
11. Otsu, N.: A threshold selection method from gray level histograms. IEEE Trans. Syst. Man Cybern. 9, 62–66 (1979)
12. Pratikakis, I., Gatos, B., Ntirogiannis, K.: ICDAR: document image binarization contest (DIBCO 2011). In: International Conference on Document Analysis and Recognition, pp. 1506–1510 (2011)
13. Sauvola, N., Pietikainen, M.: Adaptive document image binarization. Pattern Recogn. 33(2), 225–236 (2000)
14. Shelhamer, E., Long, J., Darrell, T.: Fully convolutional networks for semantic segmentation (2016). arXiv:1605.06211
15. Yangqing, J., Evan, S., Jeff, D., Sergey, K., Jonathan, L., Ross, G., Sergio, G., Trevor, D.: Caffe: convolutional architecture for fast feature embedding, arXiv preprint (2014). arXiv:1408.5093
16. http://dl.caffe.berkeleyvision.org/fcn8s-atonce-pascal.caffemodel
17. https://github.com/google/fonts

Convolutional Neural Networks for Segmentation and Object Detection of Human Semen

Malte S. Nissen[1,2,3,4](✉), Oswin Krause[1], Kristian Almstrup[2,3],
Søren Kjærulff[4], Torben T. Nielsen[4], and Mads Nielsen[1]

[1] Department of Computer Science, University of Copenhagen,
Copenhagen, Denmark
nissen@di.ku.dk
[2] Department of Growth and Reproduction, Rigshospitalet,
University of Copenhagen, Copenhagen, Denmark
[3] International Center for Research and Research Training in Endocrine
Disruption of Male Reproduction and Child Health (EDMaRC), Rigshospitalet,
University of Copenhagen, Copenhagen, Denmark
[4] ChemoMetec A/S, Allerød, Denmark

Abstract. We compare a set of convolutional neural network (CNN)
architectures for the task of segmenting and detecting human sperm cells
in an image taken from a semen sample. In contrast to previous work,
samples are not stained or washed to allow for full sperm quality analysis,
making analysis harder due to clutter. Our results indicate that training
on full images is superior to training on patches when class-skew is properly handled. Full image training including up-sampling during training
proves to be beneficial in deep CNNs for pixel wise accuracy and detection performance. Predicted sperm cells are found by using connected
components on the CNN predictions. We investigate optimization of a
threshold parameter on the size of detected components. Our best network achieves 93.87% precision and 91.89% recall on our test dataset
after thresholding outperforming a classical image analysis approach.

Keywords: Deep learning · Segmentation · Convolutional neural
networks · Human sperm · Fertility examination

1 Introduction

Sperm Quality Analysis (SQA) involves measuring concentration, morphology,
and motility [13] of sperm cells. For the application to animal sperm cells, there
exist a number of commercial Computer-Aided Sperm Analysis (CASA) systems,
such as the Hamilton-Thorne *IVOS-II* and *CEROS-II*[1] and the *Sperm Class
Analyzer*[2].

---

[1] http://www.hamiltonthorne.com/.
[2] http://www.micropticsl.com/products/sperm-class-analyzer-casa-system/.

© Springer International Publishing AG 2017
P. Sharma and F.M. Bianchi (Eds.): SCIA 2017, Part I, LNCS 10269, pp. 397–406, 2017.
DOI: 10.1007/978-3-319-59126-1_33

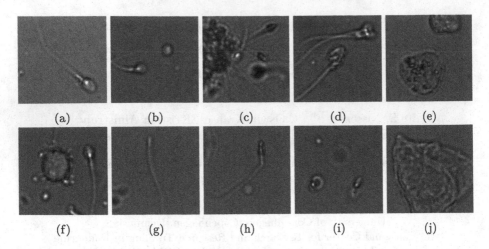

Fig. 1. Examples of debris, variations, and morphological abnormalities: normal sperm cell (a, b), aggregated cells out of focus (c), agglutinated cells (d), round cells (e, f), headless sperm (g), sperm head seen from the side or morphologically abnormal (h, i), circular tails (i), and other types of artifacts and debris (b, f, j).

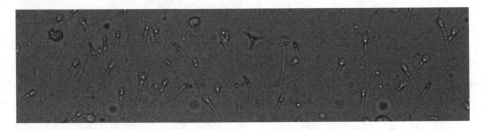

Fig. 2. 1200 × 300 pixel cut-out of image from the dataset

Human semen samples have a significantly lower quality of sperm cells compared to most animals [7], which increases the accuracy demand on the analysis. Moreover, human semen is often cluttered with debris and cells other than normal mature sperms. Figure 1 shows examples of typical debris, variations, and morphological abnormalities of human sperm samples. Figure 2 shows a section of a typical image.

In practice, staining and smearing are often used for preparation of samples to highlight specific properties of the cells [1–4, 10], but the sample needs to be in its natural form for motility estimation. This article focuses on the first step of SQA, image segmentation and detection of non-stained human sperm cells as analyzed by Ghasemian et al. [4] and Hidayahtullah and Zuhdi [6]. These algorithms apply classical image analysis techniques to solve the problem. To our knowledge no deep learning techniques have been applied yet.

Our approach focuses on deep convolutional neural networks (CNN) to segment the sperm cells in the image. There are three main challenges in this

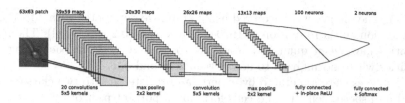

Fig. 3. Illustration of the 2-conv CNN

approach: Firstly, every pooling layer in a CNN reduces resolution by at least 50%; after three layers of pooling, every pixel of the result encodes the information of an 8×8 area of the original image. Secondly, CNNs are often trained on image patches, however there is a huge class imbalance between background and sperm pixels, where sperm pixels are significantly harder to detect. Lastly, we need to cluster the segmentations to objects. Imperfect predictions of the networks often lead to spurious detections, which need to be removed. One way to do this is to use thresholding on the size of clusters, leading to an arbitrary threshold parameter. This parameter needs to be chosen carefully.

We investigate possible solutions to these challenges. While using max-pooling layers is possible without reducing resolution [5], an exponential amount of time in the number of pooling layers is required. This makes it infeasible in practice as the results have to be computed quickly enough to allow video analysis. We follow Long et al. [9] and investigate up-sampling on the output of the CNN during training and testing. Ronneberger et al. [11] proposed a more complex architecture, which we disregard since predictions would be too slow for our application. Further, we compare training on image patches with training on the full images, where class-labels are re-weighted to correct the class-skew.

For comparison we implemented the sperm head detection method proposed by Ghasemian et al. [4]. This method has a similar threshold parameter as our method which has to be adapted for a fair comparison. For this, we propose a way to adapt the thresholding parameters using the product of precision and recall on the final detections.

The paper is organized as follows: Sect. 2 describes the dataset and the CNN architectures used. Experiments are described in Sect. 3. Results are given in Sect. 4 and discussed in Sect. 5. Finally, we conclude in Sect. 6.

2   Method

Dataset. We have constructed a dataset of 765 grayscale images of 35 independent sperm samples. The 35 samples were individually diluted using a solution of Bicarbonate-Formalin (as devised by WHO [13]) to get an appropriate amount of cells in each image (between 2 and 290 sperm cells) and to fixate them. Fixation facilitates sedimentation of the cells to the bottom of the counting chamber, ensuring that all cells are roughly in the same focal plane. In order to have cells both in and out of focus, reflecting the optical variation, Z-stacks of images were acquired. The images were acquired using an image cytometer with $20 \times$ optical magnification and a resolution of 1920×1440 pixels ($0.2 \mu m/$pixel). The image intensities have been quantized from 14- to 8-bit images. In each image the intensities where normalized to lie between zero and one.

The images were annotated by experts and registered into two classes: background and sperm cells. Round cells form an important part of the background and were therefore also annotated. The tip of the head and the neck point was registered for each sperm cell while the circumference was annotated for each round cell. Pixel-segmentation ground truths are generated by creating an ellipse at the center of each sperm cell head with radius $r_1 = \frac{1}{4} l_{cell}$ and $r_2 = \frac{2}{3} r_1$ where l_{cell} is the length of the cell head.

We split the samples into 70% train and 30% test data based on stratified sampling on the average number of sperm cells in the full images of each sample. This ensures that images from the same sample are part of the same split as they contain correlated data. Hence, one sample being part of testing data is never represented in the training data.

From the training dataset we generated an additional dataset of extracted patches from the images using the annotated classes. This patch dataset contains 63×63 pixel patches which are labelled by their ground truth in the center pixel. The size of the patches is chosen to allow the entire head, which is typically 25 pixels long, and a small part of the tail to be included. From each image, we extract up to 3,000 patches, split into 40% sperm cells, 40% background and 20% round cells. The numbers were chosen to cover the variety of debris in the background class (round cells contribute a lot to the variability of the background). Random rotation and flipping is applied before extracting each patch. Table 1 shows statistics for the resulting datasets. Note that the dataset contains a total of 38,708 sperm cells of which 23,997 are included in the train set and 14,711 are included in the test set.

Table 1. Data statistics

Statistic	Train	Test	Total
Images	540	225	765
Sperm cells	23,997	14,711	38,708
Patches	1,424,341	601,290	2,025,631

Networks. We define seven networks to test against each other. The first network is called *2-conv*. It is defined for input patches and illustrated in Fig. 3. It is a standard CNN with two convolutional, ReLU and max-pooling layers followed by two fully connected layers separated by another ReLu layer and including 50% dropout during training. The network *3-conv* is obtained by adding an additional set of convolution, ReLU, and max-pooling layers. The networks are defined with receptive fields of size 63×63 using 20 filters in each of their convolution layers and 100 filters in their fully convolutional layer.

Table 2. Experiment results m_{IU}, threshold, and m_{pred} for all eight methods

Method	m_{IU}	Threshold	m_{pred} (s)
2-conv	0.6658	200	0.145
2-conv-full	0.7080	200	0.143
2-conv-full-up	0.6805	250	0.143
3-conv	0.6556	200	0.119
3-conv-full	0.6497	150	0.119
3-conv-full-up	0.6661	300	**0.116**
3-conv-full-up-inc	**0.7387**	150	0.364
Baseline [4]	0.5679	400	-

For prediction on the full images, the fully connected layers are substituted with fully convolutional layers as described by Long et al. [9] to allow for faster computation. As each max-pooling layer divides the spatial resolution of the output by a factor of 2 in each dimension, we further perform bilinear upscaling of the network output probabilities to obtain a pixel-wise segmentation.

To compare whether training on full images is beneficial compared to patch-based training, we define the architectures *2-conv-full* and *3-conv-full*, which have the same structure as *2-conv* and *3-conv* in the prediction phase and are trained on full images with the final up-sampling removed. Finally, the architectures *2-conv-full-up* and *3-conv-full-up* also incorporate the bilinear up-sampling into the training process. The networks trained on full images use a receptive field of size 64×64 and the same number of filters[3]. We further add a network *3-conv-full-up-inc* with the same receptive field size but with 64, 128, and 256 filters in the convolution layers and 1024 filters in the fully convolutional layer. We omit the network *2-conv-full-up-inc* due to limitations in the framework used.

When testing the networks, we perform post-processing of the full size output probabilities in two steps: Firstly, we choose the most probable class as output for each pixel. Secondly, we cluster pixel-wise segmentation to objects by computing the 8-neighbourhood connected components and removing components smaller than a threshold t. The value of this threshold is found in Sect. 4.

3 Experiments

The 2-conv and 3-conv architectures have been trained on the patch dataset and tested on the full image dataset, whereas all other networks have been trained and tested on the full image dataset. The outputs of 2-conv-full and 3-conv-full are smaller than the label masks of the full images. We therefore downsample the label masks by factors 4 and 8 respectively. This is done by taking every 4th or 8th pixel corresponding to the center of the receptive field of the output.

---

[3] The difference comes from the fact that it is easier to define a center-pixel in 63×63 receptive fields.

All networks are trained by optimizing the cross-entropy between the predicted and ground truth label. To compensate for the class skew in the full images during training we re-weight the classes according to their distribution. The weight w_i of class i is defined as $w_i = \frac{1}{n_i \sum_j \frac{1}{n_j}}$ where n_i is the number of pixels belonging to class i. Omitting the re-weighting led to far inferior results classifying everything as background.

The architectures have been trained for 200 epochs using the Adam solver [12] with mini-batches of 256 patches or 1 full image (1920 · 1440 "samples"). For training we chose learning rate $\alpha = 0.001$, moment 1 $\beta_1 = 0.9$, moment 2 $\beta_2 = 0.999$, and $\epsilon = 10^{-8}$. We implemented the networks using Caffe [8], and the experiments have been carried out using a single Titan X GPU.

The baseline method [4] consists of three major steps: Noise reduction, object region detection, and sperm head localization. The method assumes that all sufficiently large object regions are sperm cells and therefore filters out all object regions smaller than a chosen threshold. This threshold is crucial for the performance of the algorithm and needs to be chosen carefully.

On an object level we are interested in finding each sperm cell. For this purpose we use the two measures precision $= \frac{TP}{TP+FP}$ and recall $= \frac{TP}{TP+FN}$, where TP is the number of true positives, FP is the number of false positives and FN is the number of false negatives. A predicted sperm cell is categorized as TP if it covers more than half the area of a ground truth sperm cell. Each predicted cell can only count as one positive, i.e. a predicted cell covering more than half the area of two sperm cells counts as one true positive and one false negative. We evaluate precision and recall for multiple thresholds on the training data to get a precision-recall (PR) curve for every method. We choose the threshold value that maximizes the product between precision and recall.

Mean intersection over union (mean IU) m_{IU} is used to quantify the pixelwise segmentation performance as described by Long et al. [9]:

$$m_{IU} = \frac{1}{2} \sum_i \left(\frac{p_{ii}}{\sum_j (p_{ij} + p_{ji}) - p_{ii}} \right)$$

where p_{ij} is the number of occurences of class i predicted as class j. We have chosen this measure since it is invariant to the aforementioned class skew.

Finally, fast computations is one of the requirements for automatic SQA. We therefore record the execution time of computing a prediction and object removal on all 765 full images and compute the mean execution time m_{pred} per image. Our baseline method implementation is not as optimized as our networks and therefore we omit the results.

4   Results

Results of the mean IU m_{IU}, thresholds found by maximizing the product between precision and recall on training data, and mean execution time m_{pred} for each method are given in Table 2.

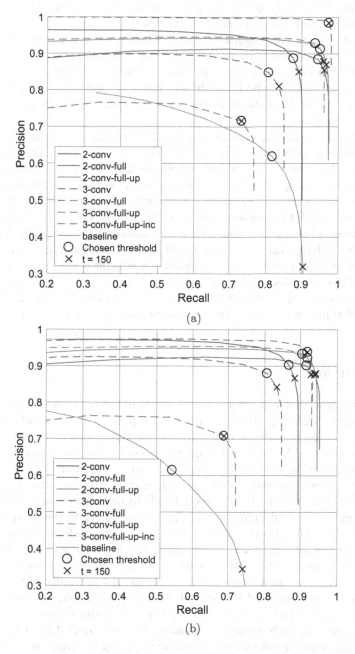

Fig. 4. Precision-recall graphs for (a) train and (b) test for all networks. 2-conv networks are plotted using fully-drawn lines and 3-conv networks using dashed lines. Same colours of lines indicate same parameters. The black circles indicate the point on each graph that corresponds to the threshold t reported in Table 2, while the black crosses indicate the points for $t = 150$.

Generally when considering m_{IU}, the networks trained on full images with up-sampling perform better than the networks trained on patches. All networks perform better than the baseline method. Training on full images without up-sampling leads to better results for 2-conv-full but worse for 3-conv-full. The 2-conv networks perform better than their 3-conv equivalents. The network *3-conv-full-up-inc* performs best, but it also has a considerably higher execution time $m_{pred} = 0.364$ than the other methods spanning the range of 0.116–0.145 seconds per image.

The results for the object detection are given in Fig. 4. The figure shows the precision-recall graphs for (a) train and (b) test for all methods. The graphs have ends due to the smallest and largest thresholds considered (0–1, 000). We plot 2-conv networks using fully-drawn lines and 3-conv networks using dashed lines. Same colours of lines indicate same parameters. The black circles indicate the point on each graph that corresponds to the threshold reported in Table 2, while the black crosses indicate the points for a threshold of 150. The baseline performs considerably different on the train and test set even though there is no training involved apart from the choice of threshold. It performs considerably worse than our networks except 3-conv-full. The best method is 3-conv-full-up-inc having 93.87% precision and 91.89% recall on the test set using threshold 150. While some overfitting can be seen between training and test, it still outperforms the other methods.

5   Discussion

Our results show that using neural networks is beneficial compared to the classical approach. The large difference in baseline performance indicate that there is a large variation between samples. We believe that we have captured the variation of a sperm cell in our train and test sets, but we have not captured all possible combinations of cells in an entire image. Given our limited number of individual samples, there are some cell concentration differences. The baseline performance difference is likely caused by these cell concentration differences. Our networks are not affected by these differences except to the degree expected from overfitting. All networks except 3-conv-full-up-inc perform almost the same on train and test data whereas 3-conv-full-up-inc is showing clear signs of overfitting. This indicates that our networks are sufficiently complex to cover the variation of the data and that even larger networks are unlikely to generalize better. As we have not used the test set for model selection, we can expect the performance on the test set to be close to the true performance.

Up-sampling has different effects on mean IU and object detection. For mean IU detecting object boundaries is important. As up-sampling is equivalent to blurring it is not beneficial for mean IU when the model is already able to accurately describe the shape of the objects. This can be seen in the difference in its effect on networks with two and three max-pooling layers. We hypothesize that training using up-sampling gives us true predictions with cluster areas closer to the true size of sperm cells. This makes it easier to distinguish sperm cells from a specific type of debris (Fig. 1b and i) easily mistaken for the head of a sperm cell but having a slightly smaller area.

When omitting up-sampling, there is no general tendency when comparing patch-based and full-image training. For 2-conv networks, full-image training seems to profit from the increased variation in the data while patch-based training profits from the weighting of round cells in the background. This can be seen by the differences in precision and recall for the two methods in Fig. 4.

When we compare the PR-curves, we see that the choice of a fixed threshold can be misleading. It turns out that the ranking of the networks can change depending on the choice of it. However, the chosen thresholds on the training set lead to consistent rankings on the test set in our case. Introducing the threshold and optimizing it leads to far superior results for all networks compared to choosing an arbitrary value. The obtained precision and recall seems reasonable for the purpose of identifying sperm cells in a semen sample, however it needs clinical testing for verification of its performance in practice.

6   Conclusion

In this paper, we have used deep convolutional neural networks for the task of sperm cell segmentation and object detection. In this task, we are constrained by the computation time as well as the accuracy demands, which make it harder to train networks with many pooling layers. To mitigate both problems we explored the use of full image training and up-sampling of the network outputs in order to increase performance. We specifically investigated thresholding on the size of detected components. Choosing the product of precision and recall leads to a robust estimate of threshold parameter. For deeper networks, up-sampling appears necessary to achieve good segmentation and object detection performance. The same does not necessarily hold for more shallow networks.

Our method outperformed a classical image analysis method which can be considered state-of-the-art. Overall the system sensitivity and precision are sufficiently high to be valuable for human sperm analysis systems.

Acknowledgements. This work is partly funded by the Innovation Fund Denmark (IFD) under File No. 4135-00169B. We would like to thank Department of Growth and Reproduction, Rigshospitalet, Denmark, for helping with annotation of our data.

References

1. Bijar, A., Pe, A., Mikaeili, M., et al.: Fully automatic identification and discrimination of sperm's parts in microscopic images of stained human semen smear. Scientific Research Publishing (2012)
2. Carrillo, H., Villarreal, J., Sotaquira, M., Goelkel, M.A., Gutierrez, R.: A computer aided tool for the assessment of human sperm morphology. In: Proceedings of the 7th IEEE International Conference on Bioinformatics and Bioengineering, pp. 1152–1157 (2007)

3. Chang, V., Saavedra, J.M., Castañeda, V., Sarabia, L., Hitschfeld, N., Härtel, S.: Gold-standard and improved framework for sperm head segmentation. Comput. Methods Progr. Biomed. **117**(2), 225–237 (2014)
4. Ghasemian, F., Mirroshandel, S.A., Monji-Azad, S., Azarnia, M., Zahiri, Z.: An efficient method for automatic morphological abnormality detection from human sperm images. Comput. Methods Progr. Biomed. **122**(3), 409–420 (2015)
5. Giusti, A., Cireşan, D.C., Masci, J., Gambardella, L.M., Schmidhuber, J.: Fast image scanning with deep max-pooling convolutional neural networks. arXiv preprint (2013). arXiv:1302.1700
6. Hidayatullah, P., Zuhdi, M.: Automatic sperms counting using adaptive local threshold and ellipse detection. In: International Conference on Information Technology Systems and Innovation (ICITSI), pp. 56–61 (2014)
7. van der Horst, G., Mortimer, S.T., Mortimer, D.: The future of computer-aided sperm analysis. Asian J. Androl. **17**, 4 (2015)
8. Jia, Y., Shelhamer, E., Donahue, J., Karayev, S., Long, J., Girshick, R., Guadarrama, S., Darrell, T.: Caffe: convolutional architecture for fast feature embedding. arXiv preprint (2014). arXiv:1408.5093
9. Long, J., Shelhamer, E., Darrell, T.: Fully convolutional networks for semantic segmentation. In: Proceedings of the IEEE Conference on Computer Vision and Pattern Recognition, pp. 3431–3440 (2015)
10. Medina-Rodríguez, R., Guzmán-Masías, L., Alatrista-Salas, H., Beltrán-Castañón, C.: Sperm cells segmentation in micrographic images through lambertian reflectance model. In: Azzopardi, G., Petkov, N. (eds.) CAIP 2015. LNCS, vol. 9257, pp. 664–674. Springer, Cham (2015). doi:10.1007/978-3-319-23117-4_57
11. Ronneberger, O., Fischer, P., Brox, T.: U-Net: convolutional networks for biomedical image segmentation. In: Navab, N., Hornegger, J., Wells, W.M., Frangi, A.F. (eds.) MICCAI 2015. LNCS, vol. 9351, pp. 234–241. Springer, Cham (2015). doi:10.1007/978-3-319-24574-4_28
12. Kingma, D.P., Ba, J.: Adam: a method for stochastic optimization. arXiv preprint (2014). arXiv:1412.6980
13. World Health Organization and others: WHO laboratory manual for the examination and processing of human semen. World Health Organization, Geneva (2010)

Convolutional Neural Networks for False Positive Reduction of Automatically Detected Cilia in Low Magnification TEM Images

Anindya Gupta[1]([⊠]), Amit Suveer[2], Joakim Lindblad[2,3], Anca Dragomir[4],
Ida-Maria Sintorn[2,5], and Nataša Sladoje[2,3]

[1] T.J. Seebeck Department of Electronics,
Tallinn University of Technology, Tallin, Estonia
anindya.gupta@ttu.ee
[2] Department of IT, Centre for Image Analysis,
Uppsala University, Uppsala, Sweden
{amit.suveer,joakim.lindblad,
ida.sintorn,natasa.sladoje}@it.uu.se
[3] Mathematical Institute,
Serbian Academy of Sciences and Arts, Belgrade, Serbia
[4] Department of Surgical Pathology,
Uppsala University Hospital, Uppsala, Sweden
anca.dragomir@igp.uu.se
[5] Vironova AB, Stockholm, Sweden

Abstract. Automated detection of cilia in low magnification transmission electron microscopy images is a central task in the quest to relieve the pathologists in the manual, time consuming and subjective diagnostic procedure. However, automation of the process, specifically in low magnification, is challenging due to the similar characteristics of non-cilia candidates. In this paper, a convolutional neural network classifier is proposed to further reduce the false positives detected by a previously presented template matching method. Adding the proposed convolutional neural network increases the area under Precision-Recall curve from 0.42 to 0.71, and significantly reduces the number of false positive objects.

Keywords: Convolutional neural network · Primary Ciliary Dyskinesia · Template maching · Transmission electron microscopy

1 Introduction

Primary Ciliary Dyskinesia (PCD) is a rare genetic disorder resulting in dysfunctional cilia - the hairlike structures protruding from certain cells. Dysfunctionality of cilia can result in severe chronic respiratory infection, and infertility in both genders. To diagnose the disorder, pathologists examine the morphological appearance of cilia (∼220–250 nm) using transmission electron microscopy

© Springer International Publishing AG 2017
P. Sharma and F.M. Bianchi (Eds.): SCIA 2017, Part I, LNCS 10269, pp. 407–418, 2017.
DOI: 10.1007/978-3-319-59126-1_34

(TEM). Qualitative analysis of cilia in the TEM images is still largely subjective and manual diagnosis is laborious, monotonous, and hugely time consuming (diagnosis takes ca. two hours per sample). An expert pathologist has to zoom in and out at locations of cilia which possibly exhibit structural information necessary for correct diagnosis. Navigation through the huge search space, together with change of magnification, is very demanding. Hence, there is an inevitable requisite for the automation of the cilia detection and diagnosis process. However, it is not feasible to acquire images which cover the whole sample at a magnification that allows structural analysis; such an acquisition would take tens of hours. Furthermore, objects of interest are rare, very small, and not spreading over more than a couple of percents of the total sample. Locating these regions of interest at low magnification, and acquiring high magnification images only at selected locations, would therefore be highly beneficial.

Automated detection of cilia structures (of a quality sufficient for diagnosis) at low magnification is a challenging task due to (1) their similar characteristics with the large number of non-cilia structures, and (2) variance in the size, shape and appearance of the individual cilia structures. The task becomes more complicated also due to noise and the non-homogeneous background at low magnification, see Fig. 1.

Lately, availability of large amounts of data and strong computational power have rapidly increased the popularity of machine learning approaches (deep learning). Convolutional neural networks (CNN) [10] have outperformed the state-of-the-art in many computer vision applications [8]. Similarly, the applicability of

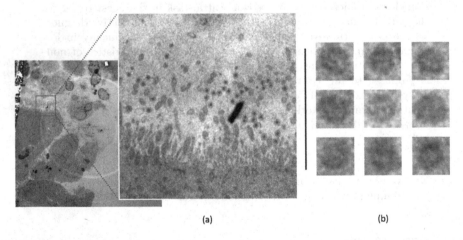

(a)   (b)

Fig. 1. (a) Low magnification TEM image of 4096×4096 pixels utilized for training purpose with the magnified view of 350×350 pixel bounding box (marked in red) with indicated ground truth marked by an expert pathologist. Here, cilia candidates marked with blue dots are of the suitable quality. (b) Some examples of patches extracted by previously reported method [15], the first and second rows contain true positives (TP) whereas patches in the third row are false positives (FP). Note the high similarity between the classes, this makes the problem a serious challenge. (Color figure online)

CNN is also investigated in the medical image analysis field [1,11]. In particular, their capability to learn discriminative features while trained in a supervised fashion makes them useful for automated detection of structures in, e.g., electron microscopy images. For instance, Ciresan *et al.* [5] reported a CNN model to segment the neuronal membranes in electron microscopy images; in [19], a CNN with autoencoder for automated detection of nuclei in high magnification (HM) microscopy images was employed.

Previously, a template matching (TM) method to detect cilia candidates in low magnification TEM images was proposed [15]. Considering that we aim at locating regions highly populated by good quality cilia, for further HM image acquisition and analysis, it is crucial that the identification of such regions is not misled by a large number of false positives (FP). In the current work, we aim at improving the performance by incorporating a dedicated CNN model in the cilia detection scheme with the special focus on reducing the number of FP. A performance benchmark for the proposed model is presented, and independent validation on an additional image is performed.

2 Image Data

Two low magnification (LM) TEM images from different patients, each with ca. 200 cilia structures, are used for training and independent validation purposes. Both images are acquired with a FEI Tecnai G2 F20 TEM and a bottom mounted FEI Eagle 4K × 4K HR CCD camera, resulting in 16-bit gray scale TIFF images of size 4096 × 4096 pixels.

For each LM image field, a set of mid magnification (MM) images are acquired, where the ground truth, i.e., true cilia candidates of promising quality for diagnosing at HM (not dealt with in this paper), are manually marked by an expert pathologist (author AD). Some examples of extracted patches of marked cilia candidates are shown in Fig. 1(b). The field of view (FOV) for a MM (2900×) image is 15.2 μm and for a LM (690×) image, it is 60.6 μm.

3 Method

The overall detection workflow consists of two stages: (1) Template matching as described in [15], and (2) further FP reduction using a 2-D CNN model, which is the core of this paper.

3.1 Initial Candidate Detection

Template matching based on normalized cross-correlation (NCC) and a customized synthetic template is used to detect the initial cilia candidates. The cross correlation image is thresholded at a suitable threshold, followed by area filtering and position filtering, meaning that only the best hit in a local region is kept as a candidate [15].

3.2 Data Partitioning and Augmentation

For each candidate position, we extracted patches of 23×23 pixels centered at a given position $p = (x, y)$. The patch size was chosen in order to contain a cilia object (~19–20 pixels diameter), and some local background around the cilia instances (~3 pixels) to include sufficient context information.

A training set of cilia, as well as non-cilia candidates, was extracted from the training image based on ground truth markings made by our expert pathologist (author AD), in MM images covering the same area of the sample. All true cilia (a total of 136) regardless of their match score, i.e., their NCC values, were chosen. A set of 272 non-cilia candidates was extracted from different NCC levels in order to represent non-cilia objects with high similarity to good cilia (136 randomly chosen non-cilia objects with NCC values ≥ 0.5) as well as non-cilia objects more different from true cilia (136 randomly chosen objects with NCC threshold values between 0.2 and 0.5).

While training a CNN model, an imbalanced dataset can mislead the optimization algorithm to converge to a local minimum, wherein the predictions can be skewed towards the candidates of the majority class, resulting in an over-fitted model. To avoid overfitting, candidates from both classes (i.e. cilia and non-cilia) are augmented. Augmentation on test data has shown a considerable improvement in terms of robustness of the system, as it, if designed properly for the problem at hand [3].

Prior to the augmentation step, the candidates are randomly divided into training, validation and test sets. The training set consists of 82 cilia and 164 non-cilia candidates whereas the validation and test sets, each consists of 27 cilia and 54 non-cilia candidates. The candidates are augmented using affine transformations (rotation, scaling and shear) and bilinear interpolation. Horizontal flipping is applied to the cilia candidates to balance the sets. A fully automated script is created to perform the combination of seven random angular rotations (0–360°), six random scalings within ±10% range and five random shearings within 5% range in both x- and y- directions, resulting in 1050 augmented variations for each candidate. The augmentation scheme is applied separately for each subset to ensure independency of the training set from the validation and test sets.

3.3 2-D CNN Configuration

The architecture of the proposed CNN model is initially derived from the LeNet architecture [9]. The motivation behind this choice is its efficiency, as well as lower computational cost compared to the architectures such as Alexnet [8] and VGGnet [13]. These models have extended the functionality of LeNet into a much larger neural network with often better performance but at a cost of a massive increase in number of parameters and computational time. Training of such large networks is still difficult due to the lack of powerful ways to regularize the models and large feature sizes in many layers [16]. Hence, we decided to empirically modify the LeNet architecture to fit our application.

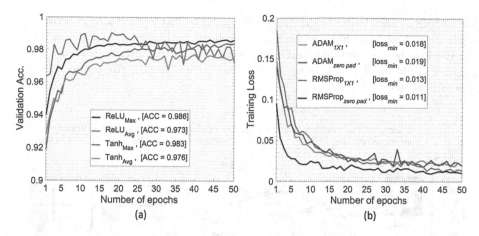

Fig. 2. Performance curves of different configuration: (a) validation accuracy for different activation functions and pooling layer combinations; (b) training loss for different optimizers with zero-padding and kernel of 1×1.

In our modified architecture, the default activation function i.e., hyperbolic tangent (tanh) [18] is replaced with Rectified linear units (ReLU) [12]. In comparison to the tanh, the constant gradient of ReLUs results in faster learning and also reduces the problem of vanishing gradient. We also implemented the maxpooling layer instead of average pooling as subsampling layer [8]. A comparative performance of both activation functions with different subsampling layers are shown in Fig. 2(a). The figure shows the accuracy for each configuration at different number of epochs. It is noticeable that the performance is better when ReLU was configured with maxpooling layer, resulting in higher accuracy after 50 epochs.

We also compared the usability of zero-padding and 1×1 convolution filters (as suggested in [16]) for two different optimizers, Adam [7] and RMSProp [17]. A kernel of size 1×1 in the first convolutional layer reduced the number of parameters (difference of 1 120 parameters compared to the zero-padding), thus keeping the computations reasonable. Comparatively, in either configuration, RMSProp with zero-padding resulted in a better training loss, as shown in Fig. 2(b). We thus, selected the configuration with minimum training loss. Moreover, several parameters (number of layers, kernel size, training algorithm, and number of neurons in the dense layer) were also experimentally determined.

In the proposed CNN classifier, the input patches are initially padded with a three pixels thick frame of zeros in order to keep the spatial sizes of the patches constant after the convolutional layers, as well as to keep the border information up to the last convolutional layer. Next, two consecutive convolutional layers and subsampling layers are used in the network. The first convolutional layer consists of 32 kernels of size $6 \times 6 \times 1$. The second convolutional layer consists of 48 kernels of size $5 \times 5 \times 32$. The subsampling layer is set as the maximum values in non-overlapping windows of size 2×2 (stride of 2). This reduces the size of the output of each convolutional layer by half. The last layer is a fully connected layer with 20 neurons followed by a softmax layer for binary classification. ReLU are used in the convolutional and dense layers, where the activation y for a given input x is obtained as $y = max\ (0,\ x)$. The architecture of the proposed CNN model is shown in Fig. 3.

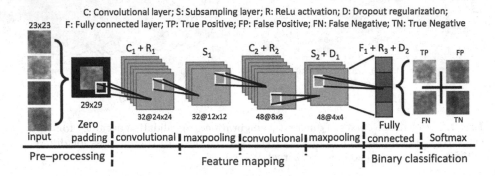

Fig. 3. An overview of the proposed CNN model.

3.4   Network training

The training of the classifier was performed in a 5-fold cross-validation scheme. For each fold, the candidates were randomly split into five blocks to ensure that each set was utilized as test set once. The distribution of candidates in each fold was kept as shown in Table 1.

Table 1. The number of cilia and non-cilia candidates in the different sets. Candidates marked in bold are finally utilized for building the model.

Set	Training	Validation	Test
Cilia	82	27	27
Aug (cilia)	172 364	56 754	56 754
Non-cilia	164	54	54
Aug (non-cilia)	172 364	56 754	56 754
Final set	**344 728**	**113 508**	**113 508**

On the given training dataset, RMSProp [17] is used to efficiently optimize the weights of the CNN. RMSProp is an adaptive optimization algorithm, which normalizes the gradients by utilizing the magnitude of recent gradients. The weights are initialized using normalized initialization as proposed in [6] and updated in a mini-batch scheme of 128 candidates. The biases were initialized with zero and learning rate was set to 0.001. A dropout of 0.5 is implemented as regularization, on the output of the last convolutional layer and the dense layer to avoid overfitting [14]. Softmax loss (cross-entropy error loss) is utilized to measure the error loss. The CNN model is implemented using theano backend in Keras [4]. The average training time is approximately 48 s/epoch on a GPU GeForce GTX 680.

4   Experimental Results and Discussion

The performance of the proposed CNN model was evaluated in terms of *Precision, Recall, Area under the Precision-Recall curve (AUC)*, and *F-score*, defined as:

$$Precision = \frac{TP}{TP + FP}, \qquad Recall = \frac{TP}{TP + FN},$$

$$F\text{-}score = 2 \times \frac{Precision \times Recall}{Precision + Recall}, \qquad AUC = \int_0^1 P(r)dr.$$

The *AUC* is the average of precision $P(r)$ over the interval $(0 \leq r \leq 1)$, and $P(r)$ is a function of recall r. Additionally, for different NCC threshold levels, the Free-response Receiver Operating Characteristic (FROC) curve [2] was utilized to measure the sensitivities at a specific number of false positives per image. The FROC curve is an extension of the receiver operating characteristic (ROC) curve, which can be effective when multiple candidates are present in a single image. It plots the Recall (Sensitivity) against the average number of false positives per images. FROC is more sensitive at detecting small differences between performances and has higher statistical discriminative power [2].

4.1   Quantitative results

Figures 4(a) and (b) show the precision-recall curves corresponding to cilia detection for the CNN classifier applied after thresholding the template matching at different NCC levels (0.2, 0.3, 0.4, and 0.5), as well as the detection when using only template matching (which includes NCC thresholding at 0.546), as proposed in [15], for the training and test image, respectively. In the figures, the AUC is also stated. The results show that adding a CNN classifier significantly improves the AUC to 0.82 and 0.71 compared to the AUC of 0.48 and 0.42, for both the training and test image, respectively, at an NCC threshold level of 0.5.

The FROC curve for the proposed CNN applied to the training and test images when the template matching result was thresholded at different NCC levels (0.2, 0.3, 0.4, and 0.5) is shown in Fig. 5(a)–(b). This corresponds to the sensitivity of the classifier against total number of FP per image.

A classification confusion matrix is also shown in Table 2. The matrix shows the performance of the classifier for both the training and test image, in terms of TP (true positive), FP (false positive), FN (false negative), and TN (true negative), at equal error rate. At an NCC threshold level 0.5, the template matching method detected 212 (73 cilia and 139 non-cilia) candidates as potential cilia candidates. Amongst these, in the Table 2(A), the proposed CNN classifier correctly

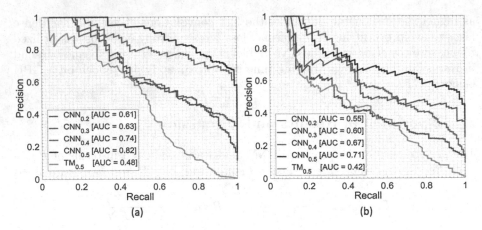

Fig. 4. Precision-recall curves of the CNN classifier at different NCC threshold levels shown together with the AUC for the template matching approach(TM) [15] for (a) training, (b) test images

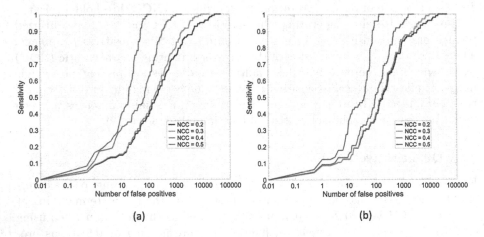

Fig. 5. FROC curves of the CNN classifier for (a) training image (b) test image at different NCC threshold levels. The number of FP are shown on a logarithmic scale.

classified 47 (TP) out of 73 (TP+FN) cilia candidates whereas from the set of 139 (FP+TN) non-cilia candidates, 26 non-cilia candidates (FP) were wrongly classified as cilia candidates by our proposed CNN classifier. We observe, in the training image, at equal error rate (Table 2(A)), the classifier also performed well when tested with the candidates extracted at an NCC threshold level of 0.4, but it eventually underperformed for the test image. The achieved results led us to finally conclude that the proposed CNN model yields a stable performance if it is incorporated with the candidates extracted at an NCC threshold level of 0.5. This observation is supported by the F-Score curves, shown in Fig. 6. Comparatively for

Table 2. Classification matrix of the CNN classifier at different NCC threshold levels for: (A) training image and (B) test image; at equal error rate.

A: Training image (Equal error rate)								
		0.2		**0.3**		**0.4**		**0.5**
TP	FP	51	85	50	80	64	51	47	26
FN	TN	85	48 004	80	18 035	51	1 113	26	113

B: Test image (Equal error rate)								
		0.2		**0.3**		**0.4**		**0.5**
TP	FP	38	67	37	66	37	60	37	36
FN	TN	67	45 926	66	18 348	60	2 658	36	188

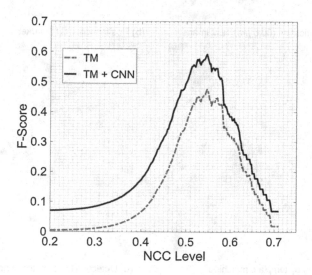

Fig. 6. F-score curves, for the test image, showing the improvement in overall performance by adding a CNN classifier with template matching approach(TM) [15] at different NCC threshold levels

the test image, at an NCC level of 0.546 (as suggested in [15]), the proposed CNN model increases the overall F-Score from 0.47 to 0.59.

4.2 False positive reduction results

Detection results of the proposed CNN model on a ROI of 650 × 650 for the test LM TEM image, at an NCC level of 0.5, are shown in Fig. 7(c)–(d). Figure 7(c) shows the detection results of the initial candidate detection step (template matching method, [15]) whereas Fig. 7(d) shows the improved results achieved by incorporating the proposed CNN model as an FP reduction step. In these images, the blue circles, red crossed circles, and green squares represent the candidates that have been correctly detected (TP), the candidates that have been

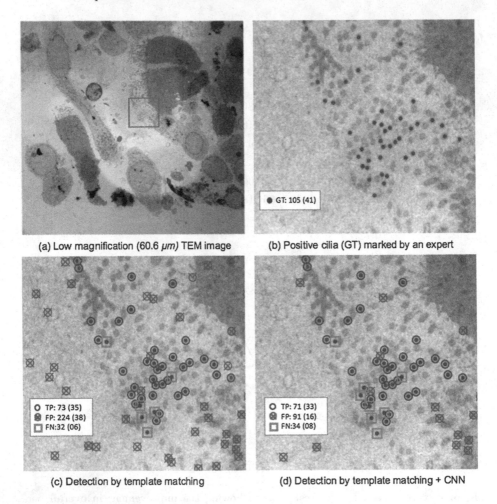

(a) Low magnification (60.6 *μm*) TEM image   (b) Positive cilia (GT) marked by an expert

GT: 105 (41)

○ TP: 73 (35)
⊠ FP: 224 (38)
□ FN:32 (06)

○ TP: 71 (33)
⊠ FP: 91 (16)
□ FN:34 (08)

(c) Detection by template matching   (d) Detection by template matching + CNN

Fig. 7. Illustration of cilia detection results. (a) The 4096×4096 test image, (b) a 650×650 example subregion of the test image, (c) same subregion after initial template matching method, and (d) after proposed CNN classifier. The numbers are given for the whole image and for the ROI is in parenthesis. Here, blue circles, red crossed circles, and green squares represent the TP, FP, and FN, respectively. (Color figure online)

erroneously detected as cilia (FP), and the cilia that were missed with respect to the manually ascertained ground truth delineations and initial detection step (FN), respectively. These results show the potential of our CNN model for cilia detection in low magnification TEM images.

Examples of classified candidate image patches in the test image are shown in Fig. 8. The images marked in the first row are the TP and FP candidates from both methods (i.e., TM and CNN). In the second row, TP candidates detected by TM but erroneously classified as FN by CNN; and FP candidates detected by TM, which are successively classified as TN by proposed classifier.

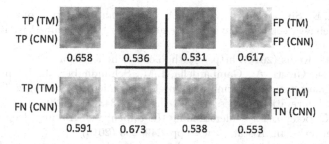

Fig. 8. Examples of candidates (with their corresponding NCC values) detected or missed by the proposed CNN model in the test image at an NCC level of 0.5. The first row shows TP's and FP's of both methods. The second row shows TP and FP candidates which are missed and successively classified by the CNN method, respectively.

5 Conclusion

In this paper, a CNN classifier is presented as a false positive reduction step for automated detection of cilia candidates in low magnification TEM images. The results suggest that adding a CNN classifier as a FP reduction step certainly improves the performance and results in an increased F-Score from 0.47 to 0.59. It was also investigated whether utilizing a CNN classifier as an additional refinement step would allow for using a lower NCC threshold in order to not discard true cilia objects in the template matching step. This was however, not found to be practically suitable as lowering the NCC threshold increases the number of candidates to analyze tremendously while only rather few additional true candidates are detected. It will be interesting in the future to develop and investigate a CNN model for the whole automated cilia detection problem, without relying on a first template matching step. This is currently not possible as it requires more training and test data.

Acknowledgments. The work is supported by Skype IT Academy Stipend Program, EU Institutional grant IUT19-11 of Estonian Research Council and the Swedish Innovation Agency's MedTech4Health program grant no. 2016-02329. J. Lindblad and N. Sladoje are supported by the Ministry of Education, Science, and Technological Development of the Republic of Serbia through projects ON174008 and III44006.

References

1. Brosch, T., Yoo, Y., Li, D.K.B., Traboulsee, A., Tam, R.: Modeling the variability in brain morphology and lesion distribution in multiple sclerosis by deep learning. In: Golland, P., Hata, N., Barillot, C., Hornegger, J., Howe, R. (eds.) MICCAI 2014. LNCS, vol. 8674, pp. 462–469. Springer, Cham (2014). doi:10.1007/978-3-319-10470-6_58
2. Chakraborty, D.: A status report on free-response analysis. Radiat. Prot. dosimetry **139**, 20–25 (2010)

3. Chatfield, K., Simonyan, K., Vedaldi, A., Zisserman, A.: Return of the devil in the details: delving deep into convolutional nets. In: British Machine Vision Conference (BMVC) (2014)

4. Chollet, F.: Keras (2015). https://github.com/fchollet/keras

5. Ciresan, D., Giusti, A., Gambardella, L.M., Schmidhuber, J.: Deep neural networks segment neuronal membranes in electron microscopy images. In: Advances in Neural Information Processing Systems, pp. 2843–2851 (2012)

6. Glorot, X., Bengio, Y.: Understanding the difficulty of training deep feedforward neural networks. In: Aistats, vol. 9, pp. 249–256 (2010)

7. Kingma, D., Ba, J.: Adam: a method for stochastic optimization. In: Proceedings of the 3rd International Conference on Learning Representations (ICLR) (2015)

8. Krizhevsky, A., Sutskever, I., Hinton, G.E.: Imagenet classification with deep convolutional neural networks. In: Advances in Neural Information Processing Systems, pp. 1097–1105 (2012)

9. LeCun, Y., Bottou, L., Bengio, Y., Haffner, P.: Gradient-based learning applied to document recognition. Proc. IEEE **86**(11), 2278–2324 (1998)

10. LeCun, Y., Hinton, G.: Deep learning. Nature **521**(7553), 436–444 (2015)

11. Li, R., Zhang, W., Suk, H.-I., Wang, L., Li, J., Shen, D., Ji, S.: Deep learning based imaging data completion for improved brain disease diagnosis. In: Golland, P., Hata, N., Barillot, C., Hornegger, J., Howe, R. (eds.) MICCAI 2014. LNCS, vol. 8675, pp. 305–312. Springer, Cham (2014). doi:10.1007/978-3-319-10443-0_39

12. Nair, V., Hinton, G.E.: Rectified linear units improve restricted Boltzmann machines. In: 27th International Conference on Machine Learning, pp. 807–814 (2010)

13. Simonyan, K., Zisserman, A.: Very deep convolutional networks for large-scale image recognition. In: Proceedings of the 3rd International Conference on Learning Representations (ICLR) (2015)

14. Srivastava, N., Hinton, G.E., Krizhevsky, A., Sutskever, I., Salakhutdinov, R.: Dropout: a simple way to prevent neural networks from overfitting. J. Mach. Learn. Res. **15**(1), 1929–1958 (2014)

15. Suveer, A., Sladoje, N., Lindblad, J., Dragomir, A., Sintorn, I.M.: Automated detection of cilia in low magnification transmission electron microscopy images using template matching. In: 13th IEEE International Symposium on Biomedical Imaging (ISBI), pp. 386–390. IEEE (2016)

16. Szegedy, C., Liu, W., Sermanet, P., Reed, S., Anguelov, D., Erhan, D., Vanhoucke, V., Rabinovich, A.: Going deeper with convolutions. In: Proceedings of the IEEE Conference on Computer Vision and Pattern Recognition, pp. 1–9 (2015)

17. Tieleman, T., Hinton, G.: Lecture 6.5-RmsProp: divide the gradient by a running average of its recent magnitude. COURSERA: Neural Networks for ML (2012)

18. Vogl, T.P., Rigler, A., Zink, W., Alkon, D.: Accelerating the convergence of the back-propagation method. Biol. Cybern. **59**(4–5), 257–263 (1988)

19. Xu, J., Xiang, L., Liu, Q., Gilmore, H., Wu, J., Tang, J., Madabhushi, A.: Stacked saparse autoencoder (SSAE) for nuclei detection on breast cancer histopathology images. IEEE Trans. Med. Imag. **35**(1), 119–130 (2016)

Deep Kernelized Autoencoders

Michael Kampffmeyer[1]($\boxtimes$), Sigurd Løkse[1], Filippo M. Bianchi[1] (iD),
Robert Jenssen[1], and Lorenzo Livi[2]

[1] Machine Learning Group,
UiT–The Arctic University of Norway, Tromsø, Norway
michael.c.kampffmeyer@uit.no
http://site.uit.no/ml/
[2] Department of Computer Science,
University of Exeter, Exeter, UK

Abstract. In this paper we introduce the deep kernelized autoencoder, a neural network model that allows an explicit approximation of (i) the mapping from an input space to an arbitrary, user-specified kernel space and (ii) the back-projection from such a kernel space to input space. The proposed method is based on traditional autoencoders and is trained through a new unsupervised loss function. During training, we optimize both the reconstruction accuracy of input samples and the alignment between a kernel matrix given as prior and the inner products of the hidden representations computed by the autoencoder. Kernel alignment provides control over the hidden representation learned by the autoencoder. Experiments have been performed to evaluate both reconstruction and kernel alignment performance. Additionally, we applied our method to emulate kPCA on a denoising task obtaining promising results.

Keywords: Autoencoders · Kernel methods · Deep learning · Representation learning

1 Introduction

Autoencoders (AEs) are a class of neural networks that gained increasing interest in recent years [18,23,25]. AEs are used for unsupervised learning of *effective* hidden representations of input data [3,11]. These representations should capture the information contained in the input data, while providing meaningful features for tasks such as clustering and classification [2]. However, what an *effective* representation consists of is highly dependent on the target task.

In standard AEs, representations are derived by training the network to reconstruct inputs through either a bottleneck layer, thereby forcing the network to learn how to compress input information, or through an over-complete representation. In the latter, regularization methods are employed to, e.g., enforce sparse representations, make representations robust to noise, or penalize sensitivity of the representation to small changes in the input [2]. However, regularization provides limited control over the nature of the hidden representation.

© Springer International Publishing AG 2017
P. Sharma and F.M. Bianchi (Eds.): SCIA 2017, Part I, LNCS 10269, pp. 419–430, 2017.
DOI: 10.1007/978-3-319-59126-1_35

In this paper, we hypothesize that an *effective* hidden representation should capture the relations among inputs, which are encoded in form of a kernel matrix. Such a matrix is used as a prior to be reproduced by inner products of the hidden representations learned by the AE. Hence, in addition to minimizing the reconstruction loss, we also minimize the normalized Frobenius distance between the prior kernel matrix and the inner product matrix of the hidden representations. We note that this process resembles the kernel alignment procedure [26].

The proposed model, called *deep kernelized autoencoder*, is related to recent attempts to bridge the performance gap between kernel methods and neural networks [5, 27]. Specifically, it is connected to works on interpreting neural networks from a kernel perspective [21] and the Information Theoretic-Learning Auto-Encoder [23], which imposes a prior distribution over the hidden representation in a variational autoencoder [18].

In addition to providing control over the hidden representation, our method also has several benefits that compensate for important drawbacks of traditional kernel methods. During training, we learn an explicit approximate mapping function from the input to a kernel space, as well as the associated back-mapping to the input space, through an end-to-end learning procedure. Once the mapping is learned, it can be used to relate operations performed in the approximated kernel space, for example linear methods (as is the case of kernel methods), to the input space. In the case of linear methods, this is equivalent to performing non-linear operations on the non-transformed data. Mini-batch training is used in our proposed method in order to lower the computational complexity inherent to traditional kernel methods and, especially, spectral methods [4, 15, 24]. Additionally, our method applies to arbitrary kernel functions, even the ones computed through ensemble methods. To stress this fact, we consider in our experiments the probabilistic cluster kernel, a kernel function that is robust with regards to hyperparameter choices and has been shown to often outperform counterparts such as the RBF kernel [14].

2   Background

2.1   Autoencoders and Stacked Autoencoders

AEs simultaneously learn two functions. The first one, *encoder*, provides a mapping from an input domain, $\mathcal{X}$, to a code domain, $\mathcal{C}$, i.e., the hidden representation. The second function, *decoder*, maps from $\mathcal{C}$ back to $\mathcal{X}$. For a single hidden layer AE, the encoding function $E(\cdot; \mathbf{W}_E)$ and the decoding function $D(\cdot; \mathbf{W}_D)$ are defined as

$$\begin{aligned} \mathbf{h} &= E(\mathbf{x}; \mathbf{W}_E) = \sigma(\mathbf{W}_E \mathbf{x} + \mathbf{b}_E) \\ \tilde{\mathbf{x}} &= D(\mathbf{h}; \mathbf{W}_D) = \sigma(\mathbf{W}_D \mathbf{h} + \mathbf{b}_D), \end{aligned} \tag{1}$$

where $\sigma(\cdot)$ denotes a suitable transfer function (e.g., a sigmoid applied component-wise), $\mathbf{x}$, $\mathbf{h}$, and $\tilde{\mathbf{x}}$ denote, respectively, a sample from the input space, its hidden representation, and its reconstruction; finally, $\mathbf{W}_E$ and $\mathbf{W}_D$ are the weights and $\mathbf{b}_E$ and $\mathbf{b}_D$ the bias of the encoder and decoder, respectively. For the sake of readability, we implicitly incorporate $\mathbf{b}_E, \mathbf{b}_D$ in the notation. Accordingly, we can rewrite

$$\tilde{\mathbf{x}} = D(E(\mathbf{x}; \mathbf{W}_E); \mathbf{W}_D). \tag{2}$$

In order to minimize the discrepancy between the original data and its reconstruction, the parameters in Eq. 1 are typically learned by minimizing, usually through stochastic gradient descent (SGD), a reconstruction loss

$$L_r(\mathbf{x}, \tilde{\mathbf{x}}) = \|\mathbf{x} - \tilde{\mathbf{x}}\|_2^2. \tag{3}$$

Differently from Eq. 1, a stacked autoencoder (sAE) consists of several hidden layers [11]. Deep architectures are capable of learning complex representations by transforming input data through multiple layers of nonlinear processing [2]. The optimization of the weights is harder in this case and pretraining is beneficial, as it is often easier to learn intermediate representations, instead of training the whole architecture end-to-end [3]. A very important application of pretrained sAE is the initialization of layers in deep neural networks [25]. Pretraining is performed in different phases, each of which consists of training a single AE. After the first AE has been trained, its encoding function $E(\cdot; \mathbf{W}_E^{(1)})$ is applied to the input and the resulting representation is used to train the next AE in the stacked architecture. Each layer, being trained independently, aims at capturing more abstract features by trying to reconstruct the representation in the previous layer. Once all individual AEs are trained, they are unfolded yielding a pretrained sAE. For a two-layer sAE, the encoding function consists of $E(E(\mathbf{x}; \mathbf{W}_E^{(1)}); \mathbf{W}_E^{(2)})$, while the decoder reads $D(D(\mathbf{h}; \mathbf{W}_D^{(2)}); \mathbf{W}_D^{(1)})$. The final sAE architecture can then be fine-tuned end-to-end by back-propagating the gradient of the reconstruction error.

2.2 A Brief Introduction to Relevant Kernel Methods

Kernel methods process data in a kernel space $\mathcal{K}$ associated with an input space $\mathcal{X}$ through an implicit (non-linear) mapping $\phi : \mathcal{X} \rightarrow \mathcal{K}$. There, data are more likely to become separable by linear methods [6], which produces results that are otherwise only obtainable by nonlinear operations in the input space. Explicit computation of the mapping $\phi(\cdot)$ and its inverse $\phi^{-1}(\cdot)$ is, in practice, not required. In fact, operations in the kernel space are expressed through inner products (kernel trick), which are computed as Mercer kernel functions in input space: $\kappa(\mathbf{x}_i, \mathbf{x}_j) = \langle \phi(\mathbf{x}_i), \phi(\mathbf{x}_j) \rangle$.

As a major drawback, kernel methods scale poorly with the number of data points n: traditionally, memory requirements of these methods scale with $\mathcal{O}(n^2)$ and computation with $\mathcal{O}(n^2 \times d)$, where d is the dimension [8]. For example, kernel principal component analysis (kPCA) [24], a common dimensionality reduction technique that projects data into the subspace that preserves the maximal amount of variance in kernel space, requires to compute the eigendecomposition of a kernel matrix $\mathbf{K} \in \mathbb{R}^{n \times n}$, with $K_{ij} = \kappa(x_i, x_j), x_i, x_j \in \mathcal{X}$, yielding a computational complexity $\mathcal{O}(n^3)$ and memory requirements that scale as $\mathcal{O}(n^2)$. For this reason, kPCA is not applicable to large-scale problems. The availability of efficient (approximate) mapping functions, however, would reduce the complexity, thereby enabling these methods to be applicable on larger datasets [5]. Furthermore, by providing an approximation for $\phi^{-1}(\cdot)$, it would be possible to directly control and visualize data represented in $\mathcal{K}$. Finding an explicit inverse mapping from $\mathcal{K}$ is a central problem in several applications, such as image denoising performed with kPCA, also known as the pre-image problem [1,13].

2.3   Probabilistic Cluster Kernel

The Probabilistic Cluster Kernel (PCK) [14] adapts to inherent structures in the data and it does not depend on any critical user-specified hyperparameters, like the width in Gaussian kernels. The PCK is trained by fitting multiple Gaussian Mixture Models (GMMs) to input data and then combining these models into a single kernel. In particular, GMMs are trained for a variety of mixture components $g = 2, 3, \ldots, G$, each with different randomized initial conditions $q = 1, 2, \ldots, Q$. Let $\boldsymbol{\pi}_i(q, g)$ denote the *posterior distribution* for data point $\mathbf{x}_i$ under a GMM with g mixture components and initial condition q. The PCK is then defined as

$$\kappa_{\mathrm{PCK}}(\mathbf{x}_i, \mathbf{x}_j) = \frac{1}{Z} \sum_{q=1}^{Q} \sum_{g=2}^{G} \boldsymbol{\pi}_i^T(q, g) \boldsymbol{\pi}_j(q, g), \tag{4}$$

where Z is a normalizing constant.

Intuitively, the posterior distribution under a mixture model contains probabilities that a given data point belongs to a certain mixture component in the model. Thus, the inner products in Eq. 4 are large if data pairs often belong to the same mixture component. By averaging these inner products over a range of G values, the kernel function has a large value only if these data points are similar on both global scale (small G) and local scale (large G).

3   Deep Kernelized Autoencoders

In this section, we describe our contribution, which is a method combining AEs with kernel methods: the deep kernelized AE (dkAE). A dkAE is trained by minimizing the following loss function

$$L = (1 - \lambda)L_r(\mathbf{x}, \tilde{\mathbf{x}}) + \lambda L_c(\mathbf{C}, \mathbf{P}), \tag{5}$$

where $L_r(\cdot, \cdot)$ is the reconstruction loss in Eq. 3. λ is a hyperparameter ranging in $[0, 1]$, which weights the importance of the two objectives in Eq. 5. For $\lambda = 0$, the loss function simplifies to the traditional AE loss in Eq. 2. $L_c(\cdot, \cdot)$ is the code loss, a distance measure between two matrices, $\mathbf{P} \in \mathbb{R}^{n \times n}$, the kernel matrix given as prior, and $\mathbf{C} \in \mathbb{R}^{n \times n}$, the inner product matrix of codes associated to

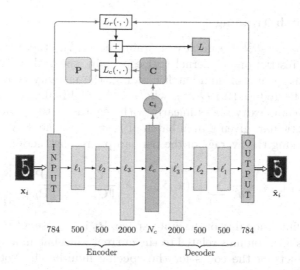

Fig. 1. Schematic illustration of dkAE architecture. Loss function L depends on two terms. First, $L_r(\cdot, \cdot)$, is the reconstruction error between true input $\mathbf{x}_i$ and output of dkAE, $\tilde{\mathbf{x}}_i$. Second term, $L_c(\cdot, \cdot)$, is the distance measure between matrices $\mathbf{C}$ (computed as inner products of codes $\{\mathbf{c}_i\}_{i=1}^n$) and the target prior kernel matrix $\mathbf{P}$. For mini-batch training matrix $\mathbf{C}$ is computed over the codes of the data in the mini-batch and that distance is compared to the submatrix of $\mathbf{P}$ related to the current mini-batch.

input data. The objective of $L_c(\cdot, \cdot)$ is to enforce the similarity between $\mathbf{C}$ and the prior $\mathbf{P}$. A depiction of the training procedure is reported in Fig. 1.

We implement $L_c(\cdot, \cdot)$ as the normalized Frobenius distance between $\mathbf{C}$ and $\mathbf{P}$. Each matrix element C_{ij} in $\mathbf{C}$ is given by $C_{ij} = E(\mathbf{x}_i) \cdot E(\mathbf{x}_j)$ and the code loss is computed as

$$L_c(\mathbf{C}, \mathbf{P}) = \left\| \frac{\mathbf{C}}{\|\mathbf{C}\|_F} - \frac{\mathbf{P}}{\|\mathbf{P}\|_F} \right\|_F. \tag{6}$$

Minimizing the normalized Frobenius distance between the kernel matrices is equivalent to maximizing the traditional kernel alignment cost, since

$$\left\| \frac{\mathbf{C}}{\|\mathbf{C}\|_F} - \frac{\mathbf{P}}{\|\mathbf{P}\|_F} \right\|_F = \sqrt{2 - 2A(\mathbf{C}, \mathbf{P})}, \tag{7}$$

where $A(\mathbf{C}, \mathbf{P}) = \frac{\langle \mathbf{C}, \mathbf{P} \rangle_F}{\|\mathbf{C}\|_F \|\mathbf{P}\|_F}$ is exactly the kernel alignment cost function [7,26]. Note that the distance in Eq. 7 can be implemented also with more advanced differentiable measures of (dis)similarity between PSD matrices, such as divergence and mutual information [9,19]. However, these options are not explored in this paper and are left for future research.

In this paper, the prior kernel matrix $\mathbf{P}$ is computed by means of the PCK algorithm introduced in Sect. 2.3, such that $\mathbf{P} = \mathbf{K}_{\mathrm{PCK}}$. However, our approach is general and *any* kernel matrix can be used as prior in Eq. 6.

3.1   Mini-Batch Training

We use mini batches of k samples to train the dkAE, thereby avoiding the computational restrictions of kernel and especially spectral methods outlined in Sect. 2.2. Making use of mini-batch training, the memory complexity of the algorithm can be reduced to $O(k^2)$, where $k \ll n$. Finally, we note that the computational complexity scales linearly with regards to the parameters in the network. In particular, given a mini batch of k samples, the dkAE loss function is defined by taking the average of the per-sample reconstruction cost

$$L_{\text{batch}} = \frac{1-\lambda}{kd} \sum_{i=1}^{k} L_r(\mathbf{x}_i, \tilde{\mathbf{x}}_i) + \lambda \left\| \frac{\mathbf{C}_k}{\|\mathbf{C}_k\|_F} - \frac{\mathbf{P}_k}{\|\mathbf{P}_k\|_F} \right\|_F, \quad (8)$$

where d is the dimensionality of the input space, $\mathbf{P}_k$ is a subset of $\mathbf{P}$ that contains only the k rows and columns related to the current mini-batch, and $\mathbf{C}_k$ contains the inner products of the codes for the specific mini-batch. Note that $\mathbf{C}_k$ is re-computed in each mini batch.

3.2   Operations in Code Space

Linear operations in code space can be performed as shown in Fig. 2. The encoding scheme of the proposed dkAE explicitly approximates the function $\phi(\cdot)$ that maps an input $\mathbf{x}_i$ onto the kernel space. In particular, in a dkAE the feature vector $\phi(\mathbf{x}_i)$ is approximated by the code $\mathbf{c}_i$. Following the underlying idea of kernel methods and inspired by Cover's theorem [6], which states that a high dimensional embedding is more likely to be linearly separable, linear operations can be performed on the code. A linear operation on $\mathbf{c}_i$ produces a result in the code space, $\mathbf{z}_i$, relative to the input $\mathbf{x}_i$. Codes are mapped back to the input space by means of a decoder, which in our case approximates the inverse mapping $\phi(\cdot)^{-1}$ from the kernel space back to the input domain. Unlike other kernel methods where this explicit mapping is not defined, this fact permits visualization and interpretation of the results in the original space.

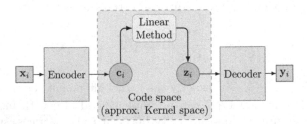

Fig. 2. The encoder maps input $\mathbf{x}_i$ to $\mathbf{c}_i$, which lies in code space. In dkAEs, the code domain approximates the space associated to the prior kernel $\mathbf{P}$. A linear method receives input $\mathbf{c}_i$ and produces output $\mathbf{z}_i$. The decoder maps $\mathbf{z}_i$ back to input space. The result $\mathbf{y}_i$ can be seen as the output of a non-linear operation on $\mathbf{x}_i$ in input space.

4 Experiments and Results

In this section, we evaluate the effectiveness of dkAEs on different benchmarks. In the first experiment we evaluate the effect on the two terms of the objective function (Eq. 8) when varying the hyperparameters λ (in Eq. 5) and the size of the code layer. In a second experiment, we study the reconstruction and the kernel alignment. Further we compare dkAEs approximation accuracy of the prior kernel matrix to kPCA as the number of principle components increases. Finally, we present an application of our method for image denoising, where we apply PCA in the dkAE code space $\mathcal{C}$ to remove noise.

For these experiments, we consider the MNIST dataset, consisting of 60000 images of handwritten digits. However, we use a subset of 20000 samples due to the computational restrictions imposed by the PCK, which we use to illustrate dkAEs ability to learn arbitrary kernels, even if they originate from an ensemble procedure. We train the PCK by fitting the GMMs on a subset of 200 training samples, with the parameters $Q = G = 30$. Once trained, the GMM models are applied on the remaining data to calculate the kernel matrix. We use 70%, 15% and 15% of the data for training, validation, and testing, respectively.

4.1 Implementation

The network architecture used in the experiments is $d - 500 - 500 - 2000 - N_c$ (see Fig. 1), which has been demonstrated to perform well on several datasets, including MNIST, for both supervised and unsupervised tasks [12,20]. Here, N_c refers to the dimensionality of the code layer. Training was performed using the sAE pretraining approach outlined in Sect. 2.1. To avoid learning the identify mapping on each individual layer, we applied a common [16] regularization technique where the encoder and decoder weights are tied, i.e., $W_E = W_D^T$. This is done during pretraining and fine-tuning. Unlike in traditional sAEs, to account for the kernel alignment objective, the code layer is optimized according to Eq. 5 *also* during pretraining.

Size of mini-batches for training was chosen to be $k = 200$ randomly, independently sampled data points; in our experiments, an epoch consists of processing $(n/k)^2$ batches. Pretraining is performed for 30 epochs per layer and the final architecture is fine-tuned for 100 epochs using gradient descent based on Adam [17]. The dkAE weights are randomly initialized according to Glorot et al. [10].

4.2 Influence of Hyperparameter λ and Size N_c of Code Layer

In this experiment, we evaluate the influence of the two main hyperparameters that determine the behaviour of our architecture. Note that the experiments shown in this section are performed by training the dkAE on the training set and evaluating the performance on the validation set. We evaluate both the out-of-sample reconstruction L_r and L_c. Figure 3(a) illustrates the effect of λ for a fixed value $N_c = 2000$ of neurons in the code layer. It can be observed

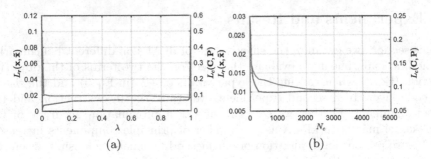

Fig. 3. (a): Tradeoff when choosing λ. High λ values result in low L_c, but high reconstruction cost, and vice-versa. (b): Both L_c and reconstruction costs decrease when code dimensionality N_c increases.

that the reconstruction loss L_r increases as more and more focus is put on minimizing L_c (obtained by increasing λ). This quantifies empirically the trade-off in optimizing the reconstruction performance and the kernel alignment at the same time. Similarly, it can be observed that L_c decreases when increasing λ. By inspecting the results, specifically the near constant losses for λ in range [0.1,0.9] the method appears robust to changes in hyperparameter λ.

Analyzing the effect of varying N_c given a fixed $\lambda = 0.1$ (Fig. 3(b)), we observe that both losses decrease as N_c increases. This could suggest that an even larger architecture, characterized by more layers and more neurons w.r.t. the architecture adopted, might work well, as the dkAE does not seem to overfit, due also to the regularization effect provided by the kernel alignment.

4.3   Reconstruction and Kernel Alignment

According to the previous results, in the following experiments we set $\lambda = 0.1$ and $N_c = 2000$. Figure 4 illustrates the results in Sect. 4.2 qualitatively by displaying a set of original images from our test set and their reconstruction for the chosen λ value and a non-optimal one. Similarly, the prior kernel (sorted by class in the figure, to ease the visualization) and the dkAEs approximated kernel matrices, relative to test data, are displayed for two different λ values. Notice that, to illustrate the difference with a traditional sAE, one of the two λ values is set to zero. It can be clearly seen that, for $\lambda = 0.1$, both the reconstruction and the kernel matrix, resemble the original closely, which agrees with the plots in Fig. 3(a).

Inspecting the kernels obtained in Fig. 4, we compare the distance between the kernel matrices, $\mathbf{C}$ and $\mathbf{P}$, and the ideal kernel matrix, obtained by considering supervised information. We build the ideal kernel matrix $\mathbf{K}_I$, where $K_I(i,j) = 1$ if elements i and j belong to same class, otherwise $K_I(i,j) = 0$. Table 1 illustrates that the kernel approximation produced by dkAE outperforms a traditional sAE with regards to kernel alignment with the ideal kernel. Additionally it can be seen that the kernel approximation actually improves slightly

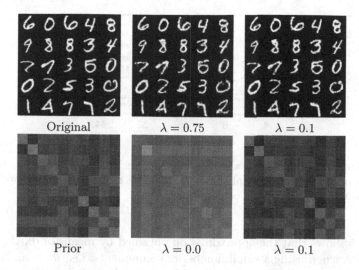

Fig. 4. Illustrating the reconstruction and kernel alignment trade-off for different λ values. We note that the reconstruction for a small λ is generally better (see also Fig. 3(a)), but that small λ yields high L_c.

Table 1. Computing L_c with respect to an ideal kernel matrix $\mathbf{K}_I$ for our test dataset (10 classes) and comparing relative improvement for the three kernels in Fig. 4. Prior kernel $\mathbf{P}$, a traditional sAE ($\lambda = 0$) $\mathbf{K}_{AE}$, and dkAEs $\mathbf{C}$.

Kernel	$L_c(\cdot, \mathbf{K}_I)$	Improvement [%] vs.		
		$\mathbf{P}$	$\mathbf{K}_{AE}$	$\mathbf{C}$
$\mathbf{P}$	1.0132	0	12.7	−0.2
$\mathbf{K}_{AE}$	1.1417	−11.3	0	−11.4
$\mathbf{C}$	1.0115	0.2	12.9	0

on the kernel prior, which we hypothesise is due to the regularization that is imposed by the reconstruction objective.

4.4 Approximation of Kernel Matrix Given as Prior

In order to quantify the kernel alignment performance, we compare dkAE to the approximation provided by kPCA when varying the number of principal components. For this test, we take the kernel matrix $\mathbf{P}$ of the training set and compute its eigendecomposition. We then select an increasing number of components m (with $m \geq 1$ components associated with the largest eigenvalues) to project the input data: $\mathbf{Z}_m = \mathbf{E}_m \mathbf{\Lambda}_m^{1/2}, d = 2, ..., N$. The approximation of the original kernel matrix (prior) is then given as $\mathbf{K}_m = \mathbf{Z}_m \mathbf{Z}_m^T$. We compute the distance between $\mathbf{K}_m$ and $\mathbf{P}$ following Eq. 7 and compare it to dissimilarity between $\mathbf{P}$ and $\mathbf{C}$. For evaluating the out-of-sample performance, we use the Nyström approximation for kPCA [24] and compare it to the dkAE kernel approximation on the test set.

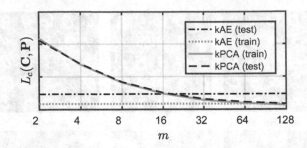

Fig. 5. Comparing dkAEs approximation of the kernel matrix to kPCA for an increasing number of components. The plot shows that dkAE reconstruction is more accurate for low number (i.e., $m < 16$) of components.

Figure 5 shows that the approximation obtained by means of dkAEs outperforms kPCA when using a small number of components, i.e., $m < 16$. Note that it is common in spectral methods to chose a number of components equal to the number of classes in the dataset [22] in which case, for the 10 classes in the MNIST dataset, dkAE would outperform kPCA. As the number of selected components increases, the approximation provided by kPCA will perform better. However, as shown in the previous experiment (Sect. 4.3), this does not mean that the approximation performs better with regards to the ideal kernel. In fact, in that experiment the kernel approximation by dkAE actually performed at least as well as the prior kernel (kPCA with all components taken into account).

4.5 Linear Operations in Code Space

Here we hint at the potential of performing operations in code space as described in Sect. 3.2. We try to emulate kPCA by performing PCA in our learned kernel space and evaluate the performance on the task of denoising. Denoising is a task that requires both a mapping to the kernel space, as well as a back-projection. For traditional kernel methods no explicit back-projection exists, but approximate solutions to this so called pre-image problem have been proposed [1,13]. We chose the method proposed by Bakir et al. [1], where they use kernel ridge regression, such that a different kernel (in our case an RBF) can be used for the back-mapping. As it was a challenging to find a good σ for the RBF kernel that captures all numbers in the MNIST dataset, we performed this test on the 5 and 6 class only. The regularization parameter and the σ required for the back-projection where found via grid search, where the best regularization parameter (according to MSE reconstruction) was found to be 0.5 and σ as the median of the euclidean distances between the projected feature vectors.

Both models are fitted on the training set and Gaussian noise is added to the test set. For both methods 32 principle components are used. Table 2 illustrates that dkAE+PCA outperforms kPCAs reconstruction with regards to mean squared error. However, as this is not necessarily a good measure for

Table 2. Mean squared error for reconstruction.

Noise std.	kPCA	dkAE+PCA
0.25	0.0427	0.0358

Fig. 6. Original images (left), the reconstruction with kPCA (center) and with dkAE+PCA (right).

denoising [1], we also visualize the results in Fig. 6. It can be seen that dkAE yields sharper images in the denoising task.

5   Conclusions

In this paper, we proposed a novel model for autoencoders, based on the definition of a particular unsupervised loss function. The proposed model enables us to learn an approximate embedding from an input space to an arbitrary kernel space as well as the projection from the kernel space back to input space through an end-to-end trained model. It is worth noting that, with our method, we are able to approximate arbitrary kernel functions by inner products in the code layer, which allows us to control the representation learned by the autoencoder. In addition, it enables us to emulate well-known kernel methods such as kPCA and scales well with the number of data points.

A more rigorous analysis of the learned kernel space embedding, as well as applications of the code space representation for clustering and/or classification tasks, are left as future works.

Acknowledgments. We gratefully acknowledge the support of NVIDIA Corporation with the donation of the GPU used for this research. This work was partially funded by the Norwegian Research Council FRIPRO grant no. 239844 on developing the *Next Generation Learning Machines*.

References

1. Bakir, G.H., Weston, J., Schölkopf, B.: Learning to find pre-images. In: Advances in Neural Information Processing Systems, pp. 449–456 (2004)
2. Bengio, Y., Courville, A., Vincent, P.: Representation learning: a review and new perspectives. IEEE Trans. Pattern Anal. Mach. Intell. **35**(8), 1798–1828 (2013)
3. Bengio, Y.: Learning deep architectures for ai. Found. Trends Mach. Learn. **2**(1), 1–127 (2009)
4. Boser, B.E., Guyon, I.M., Vapnik, V.N.: A training algorithm for optimal margin classifiers. In: Proceedings of the Fifth Annual Workshop on Computational Learning Theory, pp. 144–152 (1992)
5. Cho, Y., Saul, L.K.: Kernel methods for deep learning. In: Advances in Neural Information Processing Systems 22, pp. 342–350 (2009)

6. Cover, T.M., Thomas, J.A.: Elements of Information Theory. Wiley, New York (1991)
7. Cristianini, N., Elisseeff, A., Shawe-Taylor, J., Kandola, J.: On kernel-target alignment. In: Advances in Neural Information Processing Systems (2001)
8. Dai, B., Xie, B., He, N., Liang, Y., Raj, A., Balcan, M.F.F., Song, L.: Scalable kernel methods via doubly stochastic gradients. In: Advances in Neural Information Processing Systems, pp. 3041–3049 (2014)
9. Giraldo, L.G.S., Rao, M., Principe, J.C.: Measures of entropy from data using infinitely divisible kernels. IEEE Trans. Inf. Theory **61**(1), 535–548 (2015)
10. Glorot, X., Bengio, Y.: Understanding the difficulty of training deep feedforward neural networks. In: Proceedings of the International Conference on Artificial Intelligence and Statistics (AISTATS10) (2010)
11. Hinton, G.E., Salakhutdinov, R.R.: Reducing the dimensionality of data with neural networks. Science **313**(5786), 504–507 (2006)
12. Hinton, G.E., Osindero, S., Teh, Y.W.: A fast learning algorithm for deep belief nets. Neural Comput. **18**(7), 1527–1554 (2006)
13. Honeine, P., Richard, C.: A closed-form solution for the pre-image problem in kernel-based machines. J. Sig. Process. Syst. **65**(3), 289–299 (2011)
14. Izquierdo-Verdiguier, E., Jenssen, R., Gómez-Chova, L., Camps-Valls, G.: Spectral clustering with the probabilistic cluster kernel. Neurocomputing **149**, 1299–1304 (2015)
15. Jenssen, R.: Kernel entropy component analysis. IEEE Trans. Pattern Anal. Mach. Intell. **32**(5), 847–860 (2010)
16. Kamyshanska, H., Memisevic, R.: The potential energy of an autoencoder. IEEE Trans. Pattern Anal. Mach. Intell. **37**(6), 1261–1273 (2015)
17. Kingma, D., Ba, J.: Adam: A method for stochastic optimization. arXiv preprint arXiv:1412.6980 (2014)
18. Kingma, D.P., Welling, M.: Auto-encoding variational bayes. arXiv preprint arXiv:1312.6114 (2013)
19. Kulis, B., Sustik, M.A., Dhillon, I.S.: Low-rank kernel learning with Bregman matrix divergences. J. Mach. Learn. Res. **10**, 341–376 (2009)
20. Maaten, L.: Learning a parametric embedding by preserving local structure. In: International Conference on Artificial Intelligence and Statistics, pp. 384–391 (2009)
21. Montavon, G., Braun, M.L., Müller, K.R.: Kernel analysis of deep networks. J. Mach. Learn. Res. **12**, 2563–2581 (2011)
22. Ng, A.Y., Jordan, M.I., Weiss, Y., et al.: On spectral clustering: analysis and an algorithm. In: Advances in Neural Information Processing Systems, pp. 849–856 (2001)
23. Santana, E., Emigh, M., Principe, J.C.: Information theoretic-learning auto-encoder. arXiv preprint arXiv:1603.06653 (2016)
24. Schölkopf, B., Smola, A., Müller, K.R.: Nonlinear component analysis as a kernel eigenvalue problem. Neural Comput. **10**(5), 1299–1319 (1998)
25. Vincent, P., Larochelle, H., Lajoie, I., Bengio, Y., Manzagol, P.A.: Stacked denoising autoencoders: learning useful representations in a deep network with a local denoising criterion. J. Mach. Learn. Res. **11**, 3371–3408 (2010)
26. Wang, T., Zhao, D., Tian, S.: An overview of kernel alignment and its applications. Artif. Intell. Rev. **43**(2), 179–192 (2015)
27. Wilson, A.G., Hu, Z., Salakhutdinov, R., Xing, E.P.: Deep kernel learning. In: Proceedings of the 19th International Conference on Artificial Intelligence and Statistics, pp. 370–378 (2016)

Spectral Clustering Using *PCKID* – A Probabilistic Cluster Kernel for Incomplete Data

Sigurd Løkse[1(✉)], Filippo M. Bianchi[1] ⓘ, Arnt-Børre Salberg[2],
and Robert Jenssen[1,2]

[1] Machine Learning Group, UiT – The Arctic University of Norway,
Tromsø, Norway
{sigurd.lokse,filippo.m.bianchi}@uit.no
[2] Norwegian Computing Center, Oslo, Norway
http://site.uit.no/ml

Abstract. In this paper, we propose *PCKID*, a novel, robust, kernel function for spectral clustering, specifically designed to handle incomplete data. By combining posterior distributions of Gaussian Mixture Models for incomplete data on different scales, we are able to learn a kernel for incomplete data that does not depend on any critical hyperparameters, unlike the commonly used RBF kernel. To evaluate our method, we perform experiments on two real datasets. *PCKID* outperforms the baseline methods for all fractions of missing values and in some cases outperforms the baseline methods with up to 25% points.

Keywords: Missing data · Robustness · Kernel methods · Spectral clustering

1 Introduction

Clustering is of utmost importance in the field of machine learning, with a vast literature and many practical applications [7]. Over the past decades, a huge variety of methods have been proposed. These range from simple linear methods like k-means [21], to more recent advanced methods, like spectral clustering [4,14,15,22,23]. Spectral clustering is a family of highly performing clustering algorithms, currently considered state of the art. In spectral clustering, the eigenvectors and eigenvalues (spectrum) of some similarity matrix are exploited to generate a beneficial representation of the data, such that a simple method like k-means could be utilized to generate a partitioning, even with non-linearly separable data.

Analyzing incomplete datasets (with missing features) is a big challenge within clustering methods and data analysis in general, since encountering incomplete data is common in real applications. For instance, an entry in the dataset may not be recorded if a sensor is failing or a field in a questionnaire is left unanswered. Both supervised and unsupervised methods have been proposed to deal with incomplete data. In the supervised setting, we have e.g. a max–margin

© Springer International Publishing AG 2017
P. Sharma and F.M. Bianchi (Eds.): SCIA 2017, Part I, LNCS 10269, pp. 431–442, 2017.
DOI: 10.1007/978-3-319-59126-1_36

framework, where geometric interpretations of the margin is used to account for missing data [1], an approach based on training one SVM per missingness pattern [17] and the "best" Bayesian classifier [12] approach. In the unsupervised setting, there are mixture model formulations accounting for missing features, including both non–Bayesian approaches [5,10] and Bayesian approaches [11]. In general, a common approach is to apply imputation techniques [3] to estimate the missing values and then proceeding with the analysis on the imputed, complete, data set. None of these approaches come without challenges since the best choice of imputation technique is often very dependent on the data, and moreover difficult to evaluate.

In this paper, we propose as a new approach to integrate in a synergistic manner recent advances in spectral clustering and kernel methods with existing probabilistic methods for dealing with incomplete data. In particular, we exploit the Probabilistic Cluster Kernel (PCK) framework [6], which combines posterior distributions of Gaussian Mixture Models (GMMs) on different scales to learn a robust kernel function, capturing similarities on both a global and local scale. This kernel function is robust with regards to hyperparameter choices, since instead of assuming some structure in the data, the ensemble of GMMs adapt to the data manifold. We hypothesize that by integrating GMMs specifically designed to handle incomplete data [10] into the PCK framework for spectral clustering, we will be able to cluster incomplete data sets in a more robust manner compared to existing approaches. The proposed approach for building the kernel matrix to be used for spectral clustering in our framework, is denoted the *Probabilistic Cluster Kernel for Incomplete Data (PCKID)*.

2   Background Theory

2.1   Missing Data Mechanisms

Let $\mathbf{x} = \{x_i\}$ denote a data vector and let $\mathbf{x}^o$ and $\mathbf{x}^m$ denote the observed- and missing features of $\mathbf{x}$. Define $\mathbf{r} = \{r_i\}$, where $r_i = 1$ if $x_i \in \mathbf{x}^m$ and zero otherwise to be the *missing indicator* for $\mathbf{x}$. In order to train a model that accounts for values in the dataset that are not observed, one has to rely on assumptions that describe how missing data occurs. In this section, we describe the three main missing data mechanisms that characterize the structure of $\mathbf{r}$ [17].

Missing Completely at Random (MCAR). Features are said to be *missing completely at random* (MCAR) if the features are missing independently from both the observed values $\mathbf{x}^o$ and the missing values $\mathbf{x}^m$. That is,

$$P(\mathbf{r}|\mathbf{x}) = P(\mathbf{r}).$$

This is the missingness assumption on the data that leads to the simplest analysis. However, this assumption is rarely satisfied in practice.

Missing at Random (MAR). If the *features* are missing independently of their *values*, the features are said to be *missing at random* (MAR). Then the missingness of the features are only dependent of the *observed* values, such that

$$P(\mathbf{r}|\mathbf{x}) = P(\mathbf{r}|\mathbf{x}^{\mathrm{o}}).$$

This missing data mechanism is often assumed when working with missing data, since many real world missing data are generated by this mechanism. For instance, a blood test of a patient might be missing if it is only taken given some other test (observed value) exceeds a certain value.

Not Missing at Random (NMAR). If the missingness of a feature is dependent on their values, it is said to be not missing at random (NMAR), that is

$$P(\mathbf{r}|\mathbf{x}) = P(\mathbf{r}|\mathbf{x}^{\mathrm{m}}).$$

For instance, NMAR occurs when a sensor measurement is discarded because it goes beyond the maximum value that the sensor can handle.

2.2   Gaussian Mixture Models for Incomplete Data

In this section, we briefly summarize how to implement Gaussian Mixture Models (GMM) when the data have missing features. This model will be exploited as the foundation for *PCKID* to learn a robust kernel function. For details, we address the interested reader to [10].

A GMM is used to model the probability density function (PDF) for given dataset. In a GMM, a data point $\mathbf{x}_i$ is assumed to be sampled from a multivariate Gaussian distribution $\mathcal{N}_k(\mathbf{x}_i|\boldsymbol{\mu}_k, \boldsymbol{\Sigma}_k)$ with probability π_k and $k \in [1, K]$, where K corresponds to the number of mixture components. Accordingly, the PDF of the data is modeled by a *mixture* of Gaussians, such that

$$f(\mathbf{x}) = \sum_{k=1}^{K} \pi_k \mathcal{N}(\mathbf{x}|\boldsymbol{\mu}_k, \boldsymbol{\Sigma}_k). \tag{1}$$

The maximum likelihood estimates for the parameters in this model can be approximated through the Expectation Maximization (EM) algorithm.

When the data have missing features, we assume that the elements in a data vector $\mathbf{x}_i$ can be partitioned into two components; one observed part $\mathbf{x}_i^{\mathrm{o}}$ and one missing part $\mathbf{x}_i^{\mathrm{m}}$ as explained in Sect. 2.1. Then, one can construct a binary matrix $\mathbf{O}_i$ by removing the rows from the identity matrix corresponding to the missing elements $\mathbf{x}_i^{\mathrm{m}}$, such that $\mathbf{x}_i^{\mathrm{o}} = \mathbf{O}_i \mathbf{x}_i$. Given the mean vector $\boldsymbol{\mu}_k$ and the covariance matrix $\boldsymbol{\Sigma}_k$ for mixture component k, the mean and covariance matrix for the *observed* part of missingness pattern i is given by

$$\boldsymbol{\mu}_{k,i}^{\mathrm{o}} = \mathbf{O}_i \boldsymbol{\mu}_k$$
$$\boldsymbol{\Sigma}_{k,i}^{\mathrm{o}} = \mathbf{O}_i \boldsymbol{\Sigma}_k \mathbf{O}_i^{T}.$$

Algorithm 1. EM algorithm for incomplete data GMM

1: Initialize $\hat{\boldsymbol{\mu}}_k^{(0)}$, $\hat{\boldsymbol{\Sigma}}_k^{(0)}$, $\hat{\pi}_k^{(0)}$ and $\hat{\gamma}_{i,k}^{(0)}$ for $k \in [1, K]$ and $i \in [1, N]$.
2: **while** not converged **do**
3:     **E-Step:** Compute

$$\hat{\gamma}_{k,i}^{(\ell)} = \frac{\hat{\pi}_k^{(\ell)} \mathcal{N}\left(\mathbf{x}_i^o | \hat{\boldsymbol{\mu}}_{k,i}^{o(\ell)}, \hat{\boldsymbol{\Sigma}}_{k,i}^{o(\ell)}\right)}{\sum_{j=1}^K \hat{\pi}_j^{(\ell)} \mathcal{N}\left(\mathbf{x}_i^o | \hat{\boldsymbol{\mu}}_{j,i}^{o(\ell)}, \hat{\boldsymbol{\Sigma}}_{j,i}^{o(\ell)}\right)}$$

$$\hat{\mathbf{Y}}_{k,i}^{(\ell)} = \hat{\boldsymbol{\mu}}_k^{(\ell)} + \hat{\boldsymbol{\Sigma}}_k^{(\ell)} \hat{\mathbf{S}}_{k,i}^{o(\ell)}\left(\mathbf{x}_i - \hat{\boldsymbol{\mu}}_k^{(\ell)}\right)$$

4:     **M-Step:** Compute the next model parameters, given by

$$\hat{\pi}_k^{(\ell+1)} = \frac{1}{N} \sum_{i=1}^N \hat{\gamma}_{k,i}^{(\ell)}$$

$$\hat{\boldsymbol{\mu}}_k^{(\ell+1)} = \frac{\sum_{i=1}^N \hat{\gamma}_{k,i}^{(\ell)} \hat{\mathbf{Y}}_{k,i}^{(\ell)}}{\sum_{i=1}^N \hat{\gamma}_{k,i}^{(\ell)}}$$

$$\hat{\boldsymbol{\Sigma}}_k^{(\ell+1)} = \frac{\sum_{i=1}^N \hat{\boldsymbol{\Omega}}_{k,i}^{(\ell)}}{\sum_{i=1}^N \hat{\gamma}_{k,i}^{(\ell)}},$$

where

$$\boldsymbol{\Omega}_{k,i}^{(\ell)} = \hat{\gamma}_{k,i}^{(\ell)}\left(\left(\hat{\mathbf{Y}}_{k,i}^{(\ell)} - \hat{\boldsymbol{\mu}}_k^{(\ell+1)}\right)\left(\hat{\mathbf{Y}}_{k,i}^{(\ell)} - \hat{\boldsymbol{\mu}}_k^{(\ell+1)}\right)^T\right.$$

$$\left. + \left(\mathbf{I} - \hat{\boldsymbol{\Sigma}}_k^{(\ell)} \hat{\mathbf{S}}_{k,i}^{o(\ell)}\right) \hat{\boldsymbol{\Sigma}}_k^{(\ell)}\right).$$

5: **end while**

By defining

$$\mathbf{S}_{k,i}^o = \mathbf{O}_i^T {\boldsymbol{\Sigma}_{k,i}^o}^{-1} \mathbf{O}_i,$$

one can show that, under the MAR assumption, the EM procedure outlined in Algorithm 1 will find the parameters that maximizes the likelihood function [10].

Note that, even though the notation in this paper allows for a unique missingness pattern for each data point $\mathbf{x}_i$, one missingness pattern is usually shared between several data points. Thus, to improve efficiency when implementing Algorithm 1, one should sort the data points by missingness pattern such that parameters that are common across data points are calculated only once [10].

Diagonal Covariance Structure Assumption. In some cases, when the dimensionality of the data is large compared to the number of data points, in combination with many missingness patterns, one could consider assuming a diagonal covariance structure for the GMM for computational efficiency and numerical stability when inverting covariance matrices. This will of course limit the models to not encode correlations between dimensions, but for some tasks it provides a good approximation that is a viable compromise when limited computational resources are available. In this case, covariance matrices are encoded in d-dimensional vectors, which simplify the operations in Algorithm 1.

Let $\hat{\boldsymbol{\sigma}}_k$ be the vector of variances for mixture component k and let $\hat{\mathbf{s}}_{k,i}$ be a vector with elements $\hat{s}_{k,i}(\ell) = \frac{1}{\sigma_k(\ell)}$ if element ℓ of data point $\mathbf{x}_i$ is observed and

$\widehat{s}_{k,i}(\ell) = 0$ otherwise. Define

$$\widehat{\mathbf{y}}_{k,i} = \widehat{\boldsymbol{\mu}}_k + \widehat{\boldsymbol{\sigma}}_k \odot \widehat{\mathbf{s}}_{k,i} \odot (\mathbf{x}_i - \widehat{\boldsymbol{\mu}}_k), \tag{2}$$

and

$$\boldsymbol{\omega}_{k,i} = \widehat{\gamma}_{k,i} \left((\widehat{\mathbf{y}}_{k,i} - \widehat{\boldsymbol{\mu}}_k) \odot (\widehat{\mathbf{y}}_{k,i} - \widehat{\boldsymbol{\mu}}_k) + \widehat{\boldsymbol{\sigma}}_k - \widehat{\boldsymbol{\sigma}}_k \odot \widehat{\mathbf{s}}_{k,i} \odot \widehat{\boldsymbol{\sigma}}_k \right) \tag{3}$$

where $\odot$ denotes the Hadamard (element wise) product. Estimating the parameters with an assumption of diagonal covariance structure is then a matter of exchanging $\widehat{\mathbf{Y}}_{k,i}$ and $\boldsymbol{\Omega}_{k,i}$ with $\widehat{\mathbf{y}}_{k,i}$ and $\boldsymbol{\omega}_{k,i}$ respectively in Algorithm 1.

2.3 Spectral Clustering Using Kernel PCA

Spectral clustering is a family of clustering algorithms, where the spectrum, i.e. the eigenvalues and eigenvectors, of some similarity matrix is exploited for clustering of data separated by non-linear structures [4,14,15,22,23]. Most spectral clustering algorithms employ a two-stage approach, with (i) a non-linear feature generation step using the spectrum and (ii) clustering by k-means on top of the generated features [14,19]. Some have employed a strategy where the final clustering step is replaced by spectral rotations [15,20] or by replacing both steps with kernel k-means [2], which is difficult to initialize. In this work, we employ the two-stage approach where we use kernel PCA [18] to generate k-dimensional feature vectors, for then to cluster these using k-means.

Kernel PCA. Kernel PCA implicitly performs PCA in some reproducing kernel Hilbert space $\mathcal{H}$ given a positive semidefinite kernel function $\kappa \colon \mathcal{X} \times \mathcal{X} \to \mathbb{R}$, which computes inner products in $\mathcal{H}$. If we define a *kernel matrix*, $\mathbf{K}$, whose elements are the inner products $\kappa(\mathbf{x}_i, \mathbf{x}_j) = \langle \phi(\mathbf{x}_i), \phi(\mathbf{x}_j) \rangle_{\mathcal{H}}$, this matrix is positive semidefinite, and may be decomposed as $\mathbf{K} = \mathbf{E}\boldsymbol{\Lambda}\mathbf{E}^T$, where $\mathbf{E}$ is a matrix with the eigenvectors as columns and $\boldsymbol{\Lambda}$ is the diagonal eigenvalue matrix of $\mathbf{K}$. Then it can be shown that the k-dimensional projections onto the principal components in $\mathcal{H}$ is given by

$$\mathbf{Z} = \mathbf{E}_k \boldsymbol{\Lambda}_k^{\frac{1}{2}}, \tag{4}$$

where $\boldsymbol{\Lambda}_k$ consists of the k largest eigenvalues of $\mathbf{K}$ and $\mathbf{E}_k$ consists of the corresponding eigenvectors.

The traditional choice of kernel function is an RBF kernel, defined as

$$\kappa(\mathbf{x}_i, \mathbf{x}_j) = e^{-\frac{1}{2\sigma^2} \|\mathbf{x}_i - \mathbf{x}_j\|^2}, \tag{5}$$

where the σ parameter defines the width of the kernel.

3   *PCKID* – A Probabilistic Cluster Kernel for Incomplete Data

In this paper, we propose a novel procedure to construct a kernel matrix based on models learned from data with missing features, which we refer to as *PCKID*.

In particular, we propose to learn similarities between data points in an unsupervised fashion by fitting GMMs to the data with different initial conditions $q \in [1, Q]$ and a range of mixture components, $g \in [2, G]$ and combine the results using the posterior probabilities for the data points. That is, we define the kernel function as

$$\kappa_{PCKID}(\mathbf{x}_i, \mathbf{x}_j) = \frac{1}{Z} \sum_{q=1}^{Q} \sum_{g=2}^{G} \boldsymbol{\gamma}_i^T(q, g) \boldsymbol{\gamma}_j(q, g), \tag{6}$$

where $\boldsymbol{\gamma}_i(q, g)$ is the posterior distribution for data point $\mathbf{x}_i$ under the model with initial condition q and g mixture components and Z is a normalizing constant. By using Algorithm 1 to train the models, we are able to learn the kernel function from the inherent structures of the data, even when dealing with missing features. In this work, we use this kernel for spectral clustering.

The *PCKID* is able to capture similarities on both a local and a global scale. When a GMM is trained with many mixture components, each mixture component covers a small, *local* region in feature space. On the contrary, when the GMM is trained with a small number of mixture components, each mixture component covers a large, *global* region in feature space. Thus, if two data points are similar under models on all scales, they are likely to be similar, and will have a large value in the *PCKID*. This procedure of fitting models to the data on different scales, ensures robustness with respect to parameters, as long as Q and G are set sufficiently large. Thus, we are able to construct a kernel function that is robust with regards to parameter choice. This way of constructing a robust kernel is similar to the methodology used in ensemble clustering and recent work in spectral clustering [6]. However, such recent methods are not able to explicitly handle missing data.

According to the ensemble learning methodology [13,24], we build a powerful learner by combining multiple weak learners. Therefore, one does not need to run the EM algorithm until convergence, but instead perform just a few iterations[1]. This also has the positive side-effect of encouraging diversity, providing efficiency and preventing overfitting. To further enforce diversity, it is beneficial to use subsampling techniques to train different models on different subsets of the data and evaluate the complete kernel on the full dataset.

3.1   Initialization

For each mixture model that is trained, one needs to provide an initialization. Since we are fitting large models to data that in practice does not necessarily fit these models, the initialization needs to be reasonable in order to avoid computational issues when inverting covariance matrices. An initialization procedure that has been validated empirically for the *PCKID* is

1. Use mean imputation to impute missing values.
2. Draw K random data points from the input data and use them as initial cluster centers.

---

[1] For instance, 10 iterations.

3. Run *one* k-means iteration to get initial cluster assignments and means.
4. Calculate the empirical covariance matrix from each cluster and calculate empirical prior probabilities for the mixture model based on the cluster assignments.

Data with imputed values is only used to be able to calculate initial means and covariances. *When training the model, data without imputed values is used.*

4   Experiments

4.1   Experiment Setup

PCKID **Parameters.** In order to illustrate that *PCKID* does not need any parameter tuning, the parameters are set to $Q = G = 30$ for all experiments. In order to increase diversity, each model in the ensemble is trained on a random subset of 50% of the whole dataset. The kernel is evaluated on the full dataset, once the models are trained. Each GMM is trained for 10 iterations with a diagonal covariance structure assumption.

Baseline Methods. For the baseline methods, missing data is handled with imputation techniques, in particular, (i) zero imputation, (ii) mean imputation (iii) median imputation and (iv) most frequent value imputation. To produce a clustering result, each of these imputation techniques is coupled with (i) k-means on the data and (ii) spectral clustering using an RBF kernel, where the kernel function is calculated by (5).

Since no hyperparameters need to be tuned in in *PCKID*, the kernel width σ of the RBF is calculated with a rule of thumb. In particular, σ is set to 20% of the median pairwise distances in the dataset, as suggested in [8]. This is in agreement with unsupervised approaches, where labels are not known and cross validation on hyperparameters is not possible.

Performance Metric. In order to assess the performance of *PCKID*, its supervised clustering accuracy is compared with all baseline models. The supervised clustering accuracy is computed by

$$ACC = \max_{\mathcal{M}} \frac{\sum_{i=1}^{n} \delta\{y_i = \mathcal{M}(\widehat{y}_i)\}}{n}, \tag{7}$$

where y_i is the ground truth label, $\widehat{y}_i$ is the cluster label assigned to data point i and $\mathcal{M}(\cdot)$ is the label mapping function that maximizes the matching of the labels. This is computed using the Hungarian algorithm [9].

Clustering Setup. Spectral clustering with k clusters is performed by mapping the data to a k dimensional empirical kernel space and clustering them with k-means as described in Sect. 2.3. For all methods, k-means is run 100 times. The final clustering is chosen by evaluating the k-means cost function and choosing the partitioning with the lowest cost. The number of clusters, k, is assumed known.

4.2   MNIST 5 vs. 6

In this experiment, subsets containing 1000 of the MNIST 5 and 6 images are clustered. The subsets consist of a balanced sample, i.e. there are approximately the same amount of images from each class. The images are unraveled to 784 dimensional vectors, which are used as the input to the algorithms. Missing data is generated by randomly choosing a proportion p_m of the images and removing one of the four quadrants in the image according to the MAR mechanism. These missingness patterns are illustrated in Fig. 1(a). In each test, we consider different probabilities of having missing quadrants, i.e. $p_m \in \{0.0, 0.1, 0.2, \ldots, 0.9\}$, Each method is run 30 times for each value of p_m, with a unique random subset of the data for each run. Since there are dimensions in the dataset where there is no variation between images, they are removed before training the GMMs. These are dimensions without information, and causes problems when inverting the covariance matrices. The number of dimensions with variance varies across the runs, since the subset from the dataset and the missingness is randomly sampled for each run. The number of dimensions with variance is approximately 500.

Figure 1(b) shows a plot of the mean clustering accuracy over the 30 runs versus the missingness proportion p_m. The proposed method outperforms the baseline methods for all p_m. Although the clustering accuracy declines slightly when the p_m increases, the results are quite stable.

Figure 2(a)–(b) shows two dimensional representations using kernel PCA on *PCKID* with $p_m = 0$ and $p_m = 0.9$, respectively. The shape of the markers indicate ground truth class, while the color indicate the clustering result. It is interesting to see that although the plot with no missing data has a smoother structure, the overall topology seems to be very similar when $p_m = 0.9$.

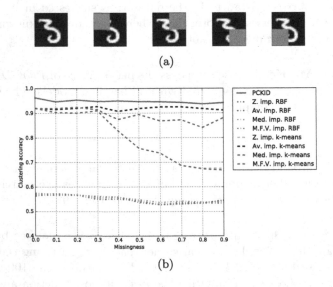

Fig. 1. (a): Example of missingness patterns. Gray pixels are considered missing. (b): Mean clustering accuracy as a function of the percentage of images with missing values.

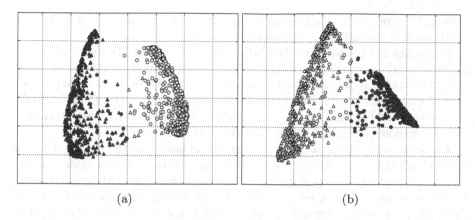

$$(a) \qquad\qquad\qquad\qquad (b)$$

Fig. 2. Example of embedding and clustering in kernel space with (a): No missingness, (b): 90% missingness. The marker indicates the true label, while the color indicates the clustering results. (Color figure online)

The two-classes seem to be less separable in the plot with more missing data, which is not surprising, given the numerical clustering results in Fig. 1(b).

When considering the approach of k-means directly on data with imputed values, we see that none of the imputation techniques perform as well as *PCKID*, although in this case mean imputation works reasonably well. To explain performance improvements as p_m increases, it is possible that the missingness patterns chosen for this experiment introduce some noise that provides a form of regularization that is beneficial to certain imputation techniques, or maybe the balance in the dataset is helping the mean of the observed values to not introduce bias towards one class. With median–, zero– and most frequent value imputation, the clustering accuracy starts to decline around $p_m = 0.3$, with zero imputation and most frequent value imputation following almost exactly the same path. This is likely due to the nature of the data, where many of the dimensions actually contains zeros in most of the images. The most frequent value in most dimensions will then be zero.

Spectral clustering using an RBF kernel completely fails in this experiment, which is probably due to a sub-optimal kernel width. However, this illustrates the difficulty with an unsupervised problem, where no prior information is given, making cross-validation virtually impossible without expertise knowledge on the data.

4.3 Land Cover Clustering

In this experiment, we cluster pixels in high resolution land cover images contaminated with clouds, also used for classification in [16,17]. The data consists of three Landsat ETM+ images covering Hardangervidda in southern Norway, in addition to elevation and slope information. With 6 bands in each image, the total dimensionality of the data is 20. In this dataset, a value is considered missing if a pixel in an image is contaminated by either clouds or snow/ice. For details on how the dataset is constructed, see [16].

The pixels in the image are labeled as one of 7 classes: (1) *water*, (2) *ridge*, (3) *leeside*, (4) *snowbed*, (5) *mire*, (6) *forest* and (7) *rock*. In this experiment, we exclude the water class, since it is easy to separate from the other classes in the Norwegian mountain vegetation. To investigate how the *PCKID* handle the different combination of classes, we restrict the analysis to pairwise classes. Each dimension is standardized on the observed data.

The average clustering accuracy for each combination of the chosen classes is reported in Table 1. The average is computed over 30 runs of each algorithm. We see that *PCKID* seems to perform better for most class pairs. Although it might struggle with some classes, most notably class 2. For the class pair 3–5, *PCKID* wins with a clustering accuracy of 0.563, which is not much better than random chance in a two-class problem. It is however worth to note that the classes labels are set according the vegetation at the actual location, which is not necessarily the group structure reflected in the data. The class combinations where *PCKID* really outperforms the other methods seems to be when class 7 (rocks) is present in the data, where we improve performance by up to 25% points with regards to the baseline methods.

Table 1. Average clustering accuracy over 30 runs for different combinations of classes in the Hardangervidda dataset. The best results are marked in bold. The baseline methods are: ZI (zero imputation), AI (average imputation), MI (median imputation) and MFVI (most frequent value imputation), combined with either k-means or spectral clustering using an RBF kernel.

Classes	PCKID	Spectral clustering, RBF				k-means			
		ZI	AI	MI	MFVI	ZI	AI	MI	MFVI
2–3	0.580	0.610	0.610	0.624	**0.627**	0.601	0.601	0.601	0.605
2–4	0.536	0.663	0.663	0.663	**0.674**	0.591	0.591	0.590	0.597
2–5	0.661	0.589	0.589	0.598	0.605	**0.671**	**0.671**	0.663	0.652
2–6	**0.712**	0.578	0.578	0.571	0.594	0.672	0.672	0.664	0.639
2–7	**0.868**	0.519	0.519	0.516	0.501	0.854	0.854	0.858	0.862
3–4	0.698	0.505	0.505	0.505	0.511	0.697	0.697	0.711	**0.722**
3–5	**0.563**	0.521	0.521	0.511	0.516	0.534	0.534	0.540	0.540
3–6	**0.620**	0.565	0.565	0.562	0.564	0.521	0.521	0.519	0.523
3–7	**0.933**	0.501	0.501	0.726	0.522	0.577	0.577	0.599	0.603
4–5	0.764	0.517	0.517	0.512	0.510	0.839	0.839	0.847	**0.848**
4–6	**0.897**	0.517	0.517	0.547	0.547	**0.897**	**0.897**	0.894	0.880
4–7	**0.931**	0.550	0.550	0.547	0.534	0.687	0.687	0.687	0.718
5–6	**0.740**	0.623	0.623	0.644	0.672	0.554	0.554	0.602	0.606
5–7	**0.956**	0.687	0.687	0.667	0.698	0.706	0.706	0.706	0.706
6–7	**0.970**	0.767	0.767	0.752	0.696	0.759	0.759	0.759	0.670

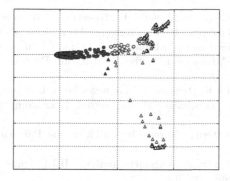

Fig. 3. Example of mapping for the *forest–rock* class pair. Colors indicate clustering, while the shape of the marker indicates the ground truth label.

Figure 3 shows an example of a mapping for the *forest–rock* class pair, where it seems like the *rock* class, as defined by the ground truth, actually consists of two separate structures in the KPCA embedding using *PCKID*. This demonstrates the power of *PCKID*s ability to adapt to the inherent structures in the data.

5 Conclusion

In this paper, we have proposed *PCKID*, a novel kernel function for spectral clustering, designed to (i) explicitly handle incomplete data and (ii) be robust with regards to parameter choice. By combining posterior distributions of Gaussian Mixture Models for incomplete data on different scales, *PCKID* is able to learn similarities on the data manifold, yielding a kernel function without any *critical* hyperparameters to tune. Experiments have demonstrated the strength of our method, by improved clustering accuracy compared to baseline methods, while keeping parameters fixed for all experiments.

Acknowledgments. This work was partially funded by the Norwegian Research Council FRIPRO grant no. 239844 on developing the *Next Generation Learning Machines*.

References

1. Chechik, G., Heitz, G., Elidan, G., Abbeel, P., Koller, D.: Max-margin classification of data with absent features. JMLR **9**, 1–21 (2008)
2. Dhillon, I.S., Guan, Y., Kulis, B.: Kernel k-means: spectral clustering and normalized cuts. In: Proceedings of the Tenth ACM SIGKDD International Conference on Knowledge Discovery and Data Mining, pp. 551–556. ACM (2004)
3. Dixon, J.K.: Pattern recognition with partly missing data. IEEE Trans. Syst. Man Cybern. **9**(10), 617–621 (1979)

4. Filippone, M., Camastra, F., Masulli, F., Rovetta, S.: A survey of kernel and spectral methods for clustering. Pattern Recogn. **41**(1), 176–190 (2008)
5. Ghahramani, Z., Jordan, M.I.: Supervised learning from incomplete data via an EM approach. In: Advances in Neural Information Processing Systems, pp. 120–120 (1994)
6. Izquierdo-Verdiguier, E., Jenssen, R., Gómez-Chova, L., Camps-Valls, G.: Spectral clustering with the probabilistic cluster kernel. Neurocomputing **149**, 1299–1304 (2015)
7. Jain, A.K.: Data clustering: 50 years beyond k-means. Pattern Recogn. Lett. **31**(8), 651–666 (2010)
8. Jenssen, R.: Kernel entropy component analysis. IEEE Trans. Pattern Anal. Mach. Intell. **32**(5), 847–860 (2010)
9. Kuhn, H.W.: The hungarian method for the assignment problem. Naval Res. logistics Q. **2**(1–2), 83–97 (1955)
10. Lin, T.I., Lee, J.C., Ho, H.J.: On fast supervised learning for normal mixture models with missing information. Pattern Recogn. **39**(6), 1177–1187 (2006)
11. Marlin, B.M.: Missing data problems in machine learning. Ph.D. thesis, University of Toronto (2008)
12. Mojirsheibani, M., Montazeri, Z.: Statistical classification with missing covariates. J. Roy. Stat. Soci. Ser. B (Statistical Methodology) **69**(5), 839–857 (2007)
13. Monti, S., Tamayo, P., Mesirov, J., Golub, T.: Consensus clustering: a resampling-based method for class discovery and visualization of gene expression microarray data. Mach. Learn. **52**(1–2), 91–118 (2003)
14. Ng, A.Y., Jordan, M.I., Weiss, Y., et al.: On spectral clustering: analysis and an algorithm. In: Advances in Neural Information Processing Systems, pp. 849–856 (2001)
15. Nie, F., Zeng, Z., Tsang, I.W., Xu, D., Zhang, C.: Spectral embedded clustering: a framework for in-sample and out-of-sample spectral clustering. IEEE Trans. Neural Netw. **22**(11), 1796–1808 (2011)
16. Salberg, A.B.: Land cover classification of cloud-contaminated multitemporal high-resolution images. IEEE Trans. Geosci. Remote Sens. **49**(1), 377–387 (2011)
17. Salberg, A.B., Jenssen, R.: Land-cover classification of partly missing data using support vector machines. Int. J. Remote Sens. **33**(14), 4471–4481 (2012)
18. Schölkopf, B., Smola, A., Müller, K.-R.: Kernel principal component analysis. In: Gerstner, W., Germond, A., Hasler, M., Nicoud, J.-D. (eds.) ICANN 1997. LNCS, vol. 1327, pp. 583–588. Springer, Heidelberg (1997). doi:10.1007/BFb0020217
19. Shi, J., Malik, J.: Normalized cuts and image segmentation. IEEE Trans. Pattern Anal. Mach. Intell. **22**(8), 888–905 (2000)
20. Stella, X.Y., Shi, J.: Multiclass spectral clustering. In: ICCV, pp. 313–319 (2003)
21. Theodoridis, S., Koutroumbas, K.: Pattern Recognition, 4th edn. Academic Press, 4th edn. (2008)
22. Von Luxburg, U.: A tutorial on spectral clustering. Stat. Comput. **17**(4), 395–416 (2007)
23. Yang, Y., Xu, D., Nie, F., Yan, S., Zhuang, Y.: Image clustering using local discriminant models and global integration. IEEE Trans. Image Process. **19**(10), 2761–2773 (2010)
24. Zimek, A., Gaudet, M., Campello, R.J., Sander, J.: Subsampling for efficient and effective unsupervised outlier detection ensembles. In: Proceedings of the 19th ACM SIGKDD International Conference on Knowledge Discovery and Data Mining, pp. 428–436. ACM (2013)

Automatic Emulation by Adaptive Relevance Vector Machines

Luca Martino$^{(\boxtimes)}$, Jorge Vicent, and Gustau Camps-Valls

Image Processing Laboratory (IPL), Universitat de València, Valencia, Spain
`luca.martino@uv.es`
`http://isp.uv.es`

Abstract. This paper introduces an automatic methodology to construct emulators for costly radiative transfer models (RTMs). The proposed method is sequential and adaptive, and it is based on the notion of the acquisition function by which instead of optimizing the unknown RTM underlying function we propose to achieve accurate approximations. The proposed methodology combines the interpolation capabilities of a modified Relevance Vector Machine (RVM) with the accurate design of an acquisition function that favors sampling in low density regions and flatness of the interpolation function. The proposed Relevance Vector Machine Automatic Emulator (RAE) is illustrated in toy examples and for the construction of an optimal look-up-table for atmospheric correction based on MODTRAN5.

Keywords: Radiative transfer model · Relevance Vector Machines · Emulation · Self-learning · Look-up table · Interpolation · MODTRAN

1 Introduction

In many fields of Science, Engineering and Technology, the mathematical and physical models are implemented in computer programs known as *simulators* [25]. Simulators are increasingly popular nowadays in social sciences, social network modeling and (electronic) commerce as well. These simulators aim to model and reproduce the complex real-world phenomena accurately. On the downside, simulators typically require large computational cost and memory resources, as well as the introduction of complicated *ad hoc* rules in the programs. Despite their good performance and reputation in the related fields, these shortcomings impede its wide practical adoption. In the last decades, machine learning has played a key role in the field by proposing surrogate models, known commonly as *emulators*, which try to reproduce (learn) the complex input-output mapping built by simulators from empirical data. This is typically done by running the simulator for a number of input factors and situations, thus yielding

The research was funded by the European Research Council (ERC) under the ERC-CoG-2014 SEDAL project (grant agreement 647423), and the Spanish Ministry of Economy and Competitiveness (MINECO) through the project TIN2015-64210-R.

© Springer International Publishing AG 2017
P. Sharma and F.M. Bianchi (Eds.): SCIA 2017, Part I, LNCS 10269, pp. 443–454, 2017.
DOI: 10.1007/978-3-319-59126-1_37

a training dataset, which is then used to fit regression models able to perform out-of-sample predictions.

Statistical emulators were actually developed in the 1980s for general purposes [9, 21, 25]. Emulators typically rely on Gaussian Processes and neural networks because of their flexibility, accuracy, and computational efficiency. Once the emulator model is developed, one can run *approximate* simulations very efficiently because the testing time for machine learning models is commonly very low (often linear or sub-linear). This actually allows one to analyze the system for which the simulator was built with far more instantiations and situations, as well as to perform *sensitivity analysis* (that is, analyze the relative relevance of the drivers) in a more robust manner. We find interesting applications of Gaussian process emulators for input feature selection and sensitivity analysis, and uncertainty quantification of the outputs given the uncertainty of the inputs [5, 19].

In this paper, we will focus on the perhaps most active field nowadays building emulators, that of Earth Science. In Earth observation and climate science one typically has access to physical models encoding the variable relations. These physical models are either called process-based models in global change ecology, radiative transfer models (RTMs) in remote sensing, or climate models in detection-and-attribution schemes for climate science applications. RTMs simulate the system as $\mathbf{y} = f(\mathbf{x}, \boldsymbol{\omega})$, where $\mathbf{y}$ is a measurement obtained by the satellite (e.g. radiance); the vector $\mathbf{x}$ represents the state of the biophysical variables on the Earth; $\boldsymbol{\omega}$ contains a set of controllable conditions (e.g. wavelengths, viewing direction, time, Sun position, and polarization); and $f(\cdot)$ is a function which relates $\mathbf{y}$ with $\mathbf{x}$. Such a function f is typically considered to be nonlinear, smooth and continuous. The goal in inverse modeling is to obtain an accurate model $g(\cdot) \approx f^{-1}(\cdot)$, parametrized by $\boldsymbol{\theta}$, which approximates the biophysical variables $\mathbf{x}$ given the data $\mathbf{y}$ received by the satellite, i.e. $\hat{\mathbf{x}} = g(\mathbf{y}, \boldsymbol{\theta})$. In emulation mode, however, we are interested in approximating the RTM well, that is, obtain a machine learning model $\widehat{f}$ that approximates the RTM code f at a fraction of time and accurately. Obtaining such a model is a game changer since one can do model inversion, sensitivity analysis and parameter retrieval much more efficiently than with the original simulator.

Here we are concerned about using the emulator to replace RTMs and then perform model inversion. Such inversion typically requires large multidimensional look-up tables (LUTs), which are precomputed for their later interpolation [11]. However, the computation of these LUTs still imposes a large computation burden, requiring techniques of parallelization and execution in computer grids. In order to further reduce this computation burden, a possible strategy is to select an optimal subset of *anchor or landmark points* in order to reduce the error of the interpolation of LUTs. Compact and informative LUTs give raise, in turn, to interesting possibilities for emulating RTMs [24]. In this work, we address the problem of optimal selection of the points to be included in the LUT and the construction of the emulator simultaneously.

The field has received attention from (apparently unrelated) fields in statistical signal processing and machine learning. The problem has been cast as *exper-*

imental optimal design [6,17,20,26] of interpolators of arbitrary functions f. To reduce the number of direct runs of the system (evaluations of f), a possible approach is to construct an approximation of f starting with a set of support points. This approximation is then sequentially improved incorporating novel points given a suitable statistical rule. This topic is also related to different research areas: optimal nonuniform sampling, quantization and interpolation of continuous signals [8,16], the so-called Bayesian Optimization (BO) problem [12,18,27], and active learning [2,7,10,28]. Finally, an interesting alternative approach is based on *adaptive gridding*, where the aim is to construct a partitioning of $\mathcal{X}$ into cells of equal size, where the cell edges have different lengths depending on their spatial direction. This was the approached followed in [4]. In order to find these lengths, the proposed method uses a modification of the Relevance Vector Machines (RVMs) [3,22].

In this paper we introduce a simpler and more general approach. The proposed method is a sequential, adaptive and automatic construction of the emulator based on the notion of the *acquisition function*, similarly to the BO approach [12,18]. Unlike in BO, our goal is not the optimization of the unknown underlying function f but its accurate *approximation* $\widehat{f}$. Given a set of initial points, the emulator is built automatically with the online addition of new nodes maximizing the acquisition function at each iteration. Theoretically, the acquisition function should incorporate (a) geometric information of the unknown function f, and (b) information about the distribution of the current nodes. Indeed, areas of high variability of $f(\mathbf{x})$ require the addition of more points as well as areas with a small concentration of nodes require the introduction of new inputs. Thus, the experimental design problem is converted into a sequential optimization problem where the function to be optimized involves geometric and spatial information (regardless of the dimensionality of the input space).

The rest of the paper is outlined as follows. Next Sect. 2 describes the general ideas behind the automatic emulation. In Sect. 3, we introduce the proposed automatic emulator scheme, and revises each of the building blocks (regression and acquisition functions). In Sect. 4, we give experimental evidence of performance of the proposal in several numerical examples, one of them involving the optimization of the nodes for the construction of an optimal LUT for atmospheric correction involving a computationally demanding physical model. We conclude in Sect. 5 with some remarks and outline of the further work.

2   Automatic Emulation

This section introduces the proposed scheme for automatic emulation. We start by fixing the notation and presenting the processing scheme. Then we will detail all the ingredients that allows to design an adaptive acquisition function in conjunction with the RVM model.

2.1   Emulation Scheme

Let us consider a D-dimensional input space $\mathcal{X}$, i.e., $\mathbf{x} = [x_1, \ldots, x_D]^{\mathsf{T}} \in \mathcal{X} \subset \mathbb{R}^D$ and, for the sake of simplicity, we assume that $\mathcal{X}$ is bounded. Let us denote the system to be emulated as $f(\mathbf{x}) : \mathcal{X} \mapsto \mathbb{R}$, e.g. a complicated transfer equations modeled with an expensive RTM. Given an input matrix $\mathbf{X}_t = [\mathbf{x}_1, \cdots, \mathbf{x}_{m_t}]$ of dimension $D \times m_t$, we have a vector of outputs $\mathbf{y}_t = [y_1, \ldots, y_{m_t}]^{\mathsf{T}}$, where $y_t = f(\mathbf{x}_t)$, where the index $t \in \mathbb{N}^+$ denotes the t-th iteration of the algorithm. Essentially, at each iteration t one performs an *interpolation* or a *regression*, providing $\widehat{f}_t(\mathbf{x}|\mathbf{X}_t, \mathbf{y}_t)$, followed by an *optimization* step that updates the acquisition function, $A_t(\mathbf{x})$, updates the set $\mathbf{X}_{t+1} = [\mathbf{X}_t, \mathbf{x}_{m_t+1}]$ adding a new node, set $m_{t+1} = m_t + 1$ and $t \leftarrow t + 1$. The procedure is repeated until a suitable stopping condition is satisfied. We assume scalar outputs in order to simplify the description of the technique, yet the algorithm can be easily extended to multi-output settings. The algorithm is outlined in Algorithm 1 and Fig. 1 shows a graphical representation of a generic automatic emulator. In this work, $f(\mathbf{x})$ represents RTM but, more generally, it could be any costly function.

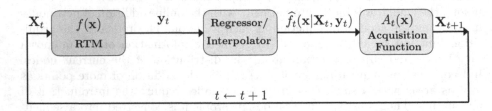

Fig. 1. Graphical representation of an automatic emulator. In this work, the function $f(\mathbf{x})$ represents RTM but, more generally, it can be any function costly to be evaluated.

Algorithm 1. Automatic Emulation.

1: **Regression:** Apply an regression procedure, following some pre-established method, providing $\widehat{f}_t(\mathbf{x}|\mathbf{X}_t, \mathbf{y}_t)$, given the current matrix of input/support points $\mathbf{X}_t = [\mathbf{x}_1, \ldots, \mathbf{x}_{m_t}]^{\mathsf{T}}$, and $\mathbf{y}_t = [y_1, \ldots, y_{m_t}]^{\mathsf{T}}$.

2: **Optimization:** Update the acquisition function obtaining $A_t(\mathbf{x})$ and set

$$\mathbf{x}_{m_t+1} = \max_{\mathbf{x} \in \mathcal{X}} A_t(\mathbf{x}).$$

Moreover, include the new point in the set of nodes, i.e., $\mathbf{X}_{t+1} = [\mathbf{X}_t; \mathbf{x}_{m_t+1}]$, and set $m_{t+1} = m_t + 1$.

3: **Check Stop Condition:** For instance, whether the number of desired points was reached, or if $\|\widehat{f}_t(\mathbf{x}) - \widehat{f}_{t-1}(\mathbf{x})\| \leq \epsilon$, then stop. Otherwise, set $t \leftarrow t + 1$ and come back to step 1.

2.2   Conceptual Definition of the Acquisition Function

Let us start describing the general properties that a generic acquisition function $A_t(\mathbf{x})$ should satisfy. Conceptually, an acquisition function can be defined as the product of two functions, a *geometry term* $G_t(\mathbf{x})$ and a *diversity term* $D_t(\mathbf{x})$:

$$A_t(\mathbf{x}) = G_t(\mathbf{x})D_t(\mathbf{x}), \tag{1}$$

where $G_t(\mathbf{x}) : \mathcal{X} \mapsto \mathbb{R}^+$, $D_t(\mathbf{x}) : \mathcal{X} \mapsto \mathbb{R}^+$ and hence $A_t(\mathbf{x}) : \mathcal{X} \mapsto \mathbb{R}^+$ (i.e., $A(\mathbf{x}) \geq 0$). The first function $G_t(\mathbf{x})$ represents some suitable geometrical information of the hidden function f. The second function $D_t(\mathbf{x})$ depends on the distribution of the points in the current vector $\mathbf{X}_t$. More specifically, $D_t(\mathbf{x})$ will have a greater probability mass around empty areas within $\mathcal{X}$, whereas $D_t(\mathbf{x})$ will be approximately zero close to the support points and exactly zero at the support points, i.e., $D_t(\mathbf{x}_i) = 0$, for $i = 1, \ldots, m_t$ and $\forall t \in \mathbb{N}$. As a consequence, a suitable acquisition function satisfies the following condition,

$$A_t(\mathbf{x}_i) = 0, \quad \text{for } i = 1, \ldots, m_t, \text{ and } \forall t \in \mathbb{N}. \tag{2}$$

3   RVM Automatic Emulator (RAE)

In this section, we specify the implementation of an automatic emulator based on a variant of the RVM method, called RAE. Thus, we introduce the Adaptive RVM (A-RVM) (employed as regressor in RAE) and then we describe a suitable construction of an acquisition function $A_t(\mathbf{x})$, taking into account important information provided by the hyperparameters of A-RVM previously optimized.

3.1   Adaptive Relevance Vector Machine (A-RVM)

Let us consider the standard regression model

$$y = f(\mathbf{x}) + \epsilon,$$

where $\epsilon \sim \mathcal{N}(0, \sigma_e^2)$ and we observe N data pairs, $\{\mathbf{x}_n, y_n\}_{n=1}^N$. We also denote $\mathbf{X} = [\mathbf{x}_1, \ldots, \mathbf{x}_N]$, $\mathbf{y} = [y_1, \ldots, y_N]^\mathsf{T}$, and define the following N basis functions

$$\phi_n(\mathbf{x}|\mathbf{x}_n) = \exp\left(-\frac{\|\mathbf{x} - \mathbf{x}_n\|^2}{2\delta_n^2}\right) : \mathcal{X} \times \mathcal{X} \to \mathbb{R} \tag{3}$$

Hence, we have N functions centered in the data inputs $\mathbf{x}_n$'s, each one with a different scale parameter δ_n^2. Moreover, let us define the $N \times N$ matrix $[\boldsymbol{\Phi}]_{i,j} := \phi_j(\mathbf{x}_i|\mathbf{x}_j)$, and the $N \times 1$ vector $\boldsymbol{\phi}(\mathbf{x}, \mathbf{X}) = [\phi_1(\mathbf{x}|\mathbf{x}_1), \ldots, \phi_N(\mathbf{x}|\mathbf{x}_N)]^\mathsf{T}$. We assume that the hidden function f can be expressed as $f(\mathbf{x}) = \boldsymbol{\phi}(\mathbf{x}, \mathbf{X})^\mathsf{T}\mathbf{w}$, where $\mathbf{w}$ is an unknown vector. Furthermore, we consider a Gaussian prior over the $N \times 1$ weight vector $\mathbf{w}$, i.e., $\mathbf{w} \sim \mathcal{N}(\mathbf{0}, \boldsymbol{\Sigma}_p)$. The Minimum Mean Square Error (MMSE) solution is

$$\widehat{\mathbf{w}} = \frac{1}{\sigma_e^2}\left(\frac{1}{\sigma_e^2}\boldsymbol{\Phi}\boldsymbol{\Phi}^\mathsf{T} + \boldsymbol{\Sigma}_p^{-1}\right)^{-1}\boldsymbol{\Phi}\mathbf{y} = \boldsymbol{\Sigma}_p\boldsymbol{\Phi}\left(\boldsymbol{\Phi}^\mathsf{T}\boldsymbol{\Sigma}_p\boldsymbol{\Phi} + \sigma_e^2\mathbf{I}_N\right)^{-1}\mathbf{y}. \tag{4}$$

Hence, since $\widehat{f}(\mathbf{x}) = \phi(\mathbf{x}, \mathbf{X})^\mathsf{T}\widehat{\mathbf{w}}$, then

$$\widehat{f}(\mathbf{x}) = \phi(\mathbf{x}, \mathbf{X})^\mathsf{T}\boldsymbol{\Sigma}_p\boldsymbol{\Phi}\left(\boldsymbol{\Phi}^\mathsf{T}\boldsymbol{\Sigma}_p\boldsymbol{\Phi} + \sigma_e^2\mathbf{I}_N\right)^{-1}\mathbf{y} = \mathbf{k}^\mathsf{T}\left(\mathbf{K} + \sigma_e^2\mathbf{I}_N\right)^{-1}\mathbf{y}, \qquad (5)$$

where we have set $\mathbf{K} = \boldsymbol{\Phi}^\mathsf{T}\boldsymbol{\Sigma}_p\boldsymbol{\Phi}$ and $\mathbf{k}^\mathsf{T} = \phi(\mathbf{x}, \mathbf{X})^\mathsf{T}\boldsymbol{\Sigma}_p\boldsymbol{\Phi}$.

On the adaptive lengthscales. We remark again that the number of basis function ϕ_n is exactly N as the number of data and $N + 1$ hyperparameters, $\boldsymbol{\theta} = [\delta_1, \ldots, \delta_N, \sigma_e^2]^\mathsf{T}$, N scale parameters δ_n (one per each function ϕ_n) and the variance of the measurement noise σ_e^2.

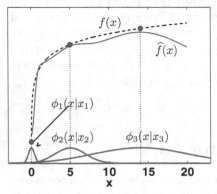

Clearly, the use of different δ_n's fosters the flexibility of the regression method. Moreover, and more important for our purpose, each parameter δ_n contains *geometric information* about the hidden function $f(\mathbf{x})$. Indeed, the parameters δ_n's of the functions ϕ_n's located in regions with a greater variation of $f(\mathbf{x})$ are smaller than the parameters δ_n's of the functions ϕ_n's located in regions where $f(\mathbf{x})$ is flatter. Roughly speaking, a great value of δ_n means that ϕ_n is located in an area where f is virtually flat, whereas a small value of δ_n is obtained when ϕ_n is located in a region where f has an high derivative, for instance. This consideration is very useful in order to build a suitable acquisition function. An illustrative example of the adaptive property is shown in Fig. 2.

Fig. 2. Graphical representation of adaptive RVM regression model. The regressor $\widehat{f}(x) = \sum_{n=1}^3 w_n\phi_n(x|x_n)$ is a linear combination of $N = 3$ basis functions, each one with a different scale δ_n, $n = 1, 2, 3$.

On hyperparameter tuning. The tuning of the hyperparameters is performed maximizing the log-marginal likelihood,

$$J(\boldsymbol{\theta}) = \log[p(\mathbf{y}|\boldsymbol{\theta})] = -\frac{1}{2}\mathbf{y}^\mathsf{T}(\mathbf{K} + \sigma_e^2\mathbf{I}_N)^{-1}\mathbf{y} - \frac{1}{2}\log\left[\det(\mathbf{K} + \sigma_e^2\mathbf{I}_N)\right] + c,$$

where c is a constant and $\mathbf{K} = \boldsymbol{\Phi}^\mathsf{T}\boldsymbol{\Sigma}_p\boldsymbol{\Phi}$. We consider a simulated annealing approach for finding the global maximum of $J(\boldsymbol{\theta})$ [13,15,23]. More sophisticated optimization methods could exploit information about the gradient of $J(\boldsymbol{\theta})$.

3.2   Proposed Acquisition Function

As we have observed before, the N learnt parameters δ_n's of M-RVM contain geometric information about the hidden function $f(\mathbf{x})$. Thus, in order to create a suitable acquisition function we need to take into account also the distribution of the other inputs in the current iteration of the algorithm. A possibility is to define the acquisition function as

$$A(\mathbf{x}) = \prod_{i=1}^{m_t} V_i(\mathbf{x}|\mathbf{x}_i), \tag{6}$$

where $V_i(\mathbf{x}|\mathbf{x}_i) = (\mathbf{x}-\mathbf{x}_i)^2 \exp(-\frac{(\mathbf{x}-\mathbf{x}_i)^2}{\ell \delta_i})$, being $\ell > 0$ a parameter chosen by the user. The function $V_i(\mathbf{x}|\mathbf{x}_i)$ has the analytical structure of a Nakagami density [14]. The Nakagami density has been widely studied from an analytical point of view. Thus, different features of $V_i(\mathbf{x}|\mathbf{x}_i)$ are known analytically and it can be easily normalized.

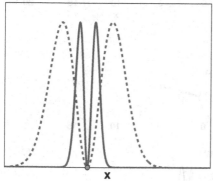

Fig. 3. Example of two functions $V_i(\mathbf{x}|\mathbf{x}_i)$ for the same $\mathbf{x}_i$ (circle) with a high (solid) and low (dashed) δ_n.

Note that each $V_i(\mathbf{x}|\mathbf{x}_i)$ depends on the parameter δ_i and $V_i(\mathbf{x}_i|\mathbf{x}_i) = 0$ (Fig. 3). Hence, $A(\mathbf{x}_i) = 0$ for all $i = 1,\dots,m_t$. The function $V_i(\mathbf{x}|\mathbf{x}_i)$ is symmetric with respect to $\mathbf{x}_i$ and it is bimodal with modes located at

$$\mathbf{x}^* = \mathbf{x}_i \pm \sqrt{\ell \delta_i}. \tag{7}$$

When δ_i is small, the two modes are closer to each other and closer to $\mathbf{x}_i$, whereas when δ_i is high the modes are far away from $\mathbf{x}_i$. Figure 4 shows a sequence of approximations $\{\widehat{f}_t(x)\}_{t=0}^3$ of $f(x)$, and the corresponding acquisition functions at each iteration.

4   Experimental Results

4.1   Synthetic Example

In order to test RAE, first we consider a toy example where we can compare the achieved approximation $\widehat{f}_t$ with the underlying function f which is unknown in the real-world applications. In this way, we can exactly check the true accuracy of the obtained approximation using different schemes. For the sake of simplicity, we consider the function

$$f(x) = \log(x), \tag{8}$$

with $x \in \mathcal{X} = (0, 20]$ hence $D = 1$ (see Fig. 5). Even in this simple scenario, the procedure used for selecting new points is relevant as confirmed the results provided below. We start with $m_0 = 3$ support points, $\mathbf{X}_0 = [0.01, 10, 20]$. We add sequentially and automatically other 10 points using different techniques: **(a)** completely random choice, uniformly within $\mathcal{X} = [0, 20]$, **(b)** a deterministic procedure filling the greatest distance between two consecutive points, adding the middle point of this interval, **(c)** Latin Hypercube (LHC) sampling [17], **(d)** RAE with $\ell = 6$ in the functions $V_i(\mathbf{x}|\mathbf{x}_i)$. In all cases, we use the M-RVM regression scheme (for simplicity, we also set $\sigma_e^2 = 0.1$). The optimization of the remaining hyperparameters is obtained

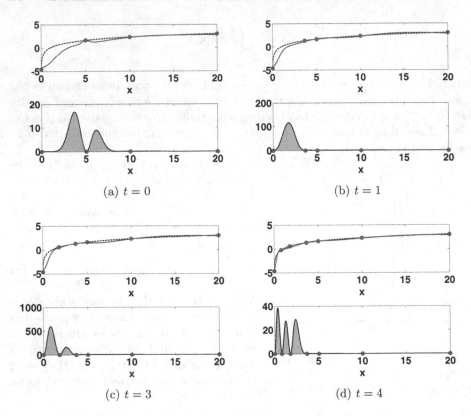

Fig. 4. A sequence of approximations $\widehat{f}_0(x)$, $\widehat{f}_1(x)$, $\widehat{f}_2(x)$ and $\widehat{f}_3(x)$ (top) of the function $f(x) = \log(x)$ (dashed line) obtained by RAE starting with 3 points and the corresponding acquisition functions $A_0(x)$, $A_1(x)$, $A_2(x)$, $A_3(x)$ and $A_4(x)$ (bottom). The nodes are depicted by circles. In this example, we have set $\ell = 6$.

by a parallel simulated annealing approach [13,15]. Note the deterministic procedure always adds sequentially the following points, obtaining $\mathbf{X}_{10} = [\mathbf{X}_0, 15, 5.005, 12.5, 17.5, 7.5025, 11.25, 13.75, 16.25, 18.75]$ (recalling that $\mathbf{X}_0 = [0.01, 10, 20]$ for all methods). At each run, the results can slightly vary even for the deterministic procedure due to the optimization of the hyperparameters (we use a simulated annealing approach that is a stochastic optimization technique [13]). We average all the results over 500 independent runs.

We compute the L_2 distance between $\widehat{f}_t(x)$ and $f(x)$ at each iteration, obtained by the different method We show the evolution of the averaged L_2 distance versus the number of support points m_t (that is $m_t = t + m_0$) in Fig. 5 (specifically, we show the median curves obtained over the 500 runs). We can observe that the RAE scheme outperforms the other methods, providing the smallest distances between f and $\widehat{f}_t$. The proposed technique gives a clear advantage in the first iterations: with only $m_t = 7$ points provides an error of $\approx 10^{-1}$, whereas the other methods require more than 12 points for obtain-

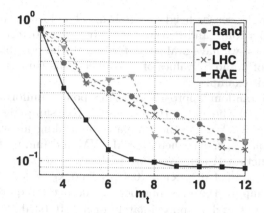

Fig. 5. The L_2 distance (in log-scale) between $f(x)$ and $\widehat{f_t}(x)$ versus the number of the number of support points m_t, that is $m_t = t + 3$ in this example. The results of RAE are shown with squares, the LHC curve is shown with x-marks, the curve of the deterministic procedure is depicted with triangles and the results of the completely random approach is illustrated by circles.

ing the same error value. After adding a certain number of nodes (enough for covering the entire domain $\mathcal{X}$), clearly the results become similar. The difficulty of the automatic problem is clearly represented by the curve of the deterministic procedure: the attempt of filling as fast as possible all the domain $\mathcal{X}$ trying to maximizing the distance among the nodes is not the best strategy, in general. In some specific scenarios, after the addition of one point, the performance can even get worse (it can vary depending on the regression scheme employed). This happens with the deterministic procedure and the addition of the 6-th and 7-th node (see Fig. 5). Figure 4 depicts the function $f(x)$ in Eq. (8), the approximation $\widehat{f_t}(x)$, and the acquisition function $A_t(x)$ obtained in four consecutive iterations of RAE, considering a specific run of the algorithm.

4.2   Emulation of Costly Radiative Transfer Codes

Our second example focuses on the optimization of selected points for a MODTRAN5-based LUT. MODTRAN5 is considered as *de facto* standard atmospheric RTM for atmospheric correction applications [1]. This RTM solves the radiative transfer equation in the atmosphere considering the effect of scattering and absorption by gasses and aerosols for a flexible configuration of viewing and illumination conditions and surface reflectance. In our test application, and for the sake of simplicity, we have considered $D = 2$ with the Aerosol Optical Thickness at 550 nm (τ) and ground elevation (h) as key input parameters. The underlying function $f(\mathbf{x})$ consists therefore on the execution of MODTRAN5 at given values of τ and h at the single output wavelength of 760 nm (i.e. bottom of the O_2-A band). As the parameters a_i, b_i and c in the toy example above, other MODTRAN5 input parameters are set to standard atmospheric and geometric conditions (e.g. mid-latitude summer atmosphere, rural aerosol, nadir view,

$55°$ solar zenith angle, sensor height at $100\,\mathrm{km}$). The input parameter space is bounded to 0.05–0.4 for τ and 0–$3\,\mathrm{km}$ for h. In order to test the accuracy of the different schemes, we have evaluated $f(x)$ at all the possible 1750 combinations of 35 values of τ and 50 values of h. Namely, this thin grid represents the ground-truth in this example.

We test (a) a random approach choosing points uniformly within $\mathcal{X} = [0.05, 0.4] \times [0, 3]$, (b) the Latin Hypercube (LHC) sampling (see, e.g., [17]) and (c) RAE ($\ell = 3$). We use the simulated annealing algorithm [13,15] for both, optimizing the hyper-parameters of M-RVM and finding the maximum of the acquisition function $A_t(\mathbf{x})$. We start with $m_0 = 5$ points $\mathbf{x}_1 = [0.05, 0]^\top$, $\mathbf{x}_2 = [0.05, 3]^\top$, $\mathbf{x}_3 = [0.4, 0]^\top$, $\mathbf{x}_4 = [0.4, 3]^\top$ and $\mathbf{x}_5 = [0.2, 1.5]^\top$ for all the techniques. We compute the final number of nodes m_t required to obtain an ℓ_2 distance between f and $\widehat{f}$ approximately of $\eta \in \{0.03, 0.2\}$, with the different methods. The results, averaged over 10^3 runs, are shown in Table 1. RAE requires the addition of ≈ 3 new points to obtain a distance ≈ 0.2 and ≈ 6 new points to obtain a distance ≈ 0.03.

Table 1. Averaged number of nodes needed for obtaining a L_2 error $\approx \eta$.

L_2 error η	Random	Latin Hypercube	RAE
0.2	19.25	11.03	7.58
0.03	28.43	16.69	11.19

5   Conclusions

We introduced an automatic method to construct surrogate models, also known as emulators, and optimal look-up-tables for costly physical models (as RTMs). We proposed an iterative scheme combining the interpolation capabilities of RVMs with the design of an acquisition function that fosters a suitable choice of the nodes and flatness of the interpolation function. We illustrated the good capabilities of the method in a synthetic example and a real example involving atmospheric correction based on the computationally expensive MODTRAN5 model. Future work is tied to the development of multi-output schemes and testing in other costly RTMs.

References

1. Berk, A., Anderson, G., Acharya, P., Bernstein, L., Muratov, L., Lee, J., Fox, M., Adler-Golden, S., Chetwynd, J., Hoke, M., Lockwood, R., Gardner, J., Cooley, T., Borel, C., Lewis, P., Shettle, E.: MODTRAN5: 2006 update. The International Society for Optical Engineering (2006)
2. Beygelzimer, A., Dasgupta, S., Langford, J.: Importance-weighted active learning. In: International Conference on Machine Learning (ICML), pp. 49–56 (2009)
3. Bishop, C.M.: Pattern recognition. Mach. Learn. 128, 1–58 (2006)

4. Busby, D.: Hierarchical adaptive experimental design for Gaussian process emulators. Reliab. Eng. Syst. Saf. **94**, 1183–1193 (2009)
5. Camps-Valls, G., Verrelst, J., Muñoz Marí, J., Laparra, V., Mateo-Jiménez, F., Gomez-Dans, J.: A survey on Gaussian processes for earth observation data analysis. IEEE Geosci. Remote Sens. Mag. **4**(2), 58–78 (2016)
6. Chaloner, K., Verdinelli, I.: Bayesian experimental design: a review. Stat. Sci. **10**(3), 237–304 (1995)
7. Cohn, D., Ghahramani, Z., Jordan, M.: Active learning with statistical models. J. Artif. Intell. Res. **4**, 129–145 (1996)
8. Cover, T.M., Thomas, J.A.: Elements of Information Theory. Wiley-Interscience, New York (1991)
9. Currin, C., Mitchell, T., Morris, M., Ylvisaker, D.: A Bayesian approach to the design and analysis of computer experiments, September 1988
10. Dasgupta, S.: Analysis of a greedy active learning strategy. In: Advances in Neural Information Processing Systems (NIPS) **16**(3), pp. 337–344 (2004)
11. Guanter, L., Richter, R., Kaufmann, H.: On the application of the MODTRAN4 atmospheric radiative transfer code to optical remote sensing. Int. J. Remote Sens. **30**(6), 1407–1424 (2009)
12. Gutmann, M.U., Corander, J.: Bayesian optimization for likelihood-free inference of simulator-based statistical models. J. Mach. Learn. Res. **16**, 4256–4302 (2015)
13. Kirkpatrick, S.K., Gelatt, C.D., Vecchi, M.P.: Optimization by simulated annealing. Science **220**(4598), 671–680 (1983)
14. Luengo, D., Martino, L.: Almost rejectionless sampling from Nakagami-m distributions (m $\geq$ 1). IET Electron. Letters **48**(24), 1559–1561 (2012)
15. Martino, L., Elvira, V., Luengo, D., Corander, J., Louzada, F.: Orthogonal parallel MCMC methods for sampling and optimization. Digit. Signal Proc. **58**, 64–84 (2016)
16. Marvasti, F.: Nonuniform Sampling: Theory and Practice. Kluwer Academic Publishers, New York (2001)
17. McKay, M.D., Beckman, R.J., Conover, W.J.: A comparison of three methods for selecting values of input variables in the analysis of output from a computer code. Technometrics **21**(2), 239–245 (1979)
18. Mockus, J.: Bayesian Approach to Global Optimization. Kluwer Academic Publishers, Dordrecht (1989)
19. Oakley, J.E., O'Hagan, A.: Probabilistic sensitivity analysis of complex models: a Bayesian approach. J. Roy. Stat. Soc. **66B**, 751–769 (2004)
20. O'Brien, T.E., Funk, G.M.: A gentle introduction to optimal design for regression models. Am. Stat. **57**(4), 265–267 (2003)
21. O'Hagan, A.: Curve fitting and optimal design for predictions. J. Roy. Stat. Soc. **40B**, 1–42 (1978)
22. Rasmussen, C.E., Williams, C.K.I.: Gaussian Processes for Machine Learning. The MIT Press, Cambridge (2005)
23. Read, J., Martino, L., Luengo, D.: Efficient Monte Carlo optimization for multi-label classifier chains. In: IEEE International Conference on Acoustics, Speech, and Signal Processing (ICASSP), pp. 1–5 (2013)
24. Rivera, J., Verrelst, J., Gómez-Dans, J., Muñoz Marí, J., Moreno, J., Camps-Valls, G.: An emulator toolbox to approximate radiative transfer models with statistical learning. Remote Sens. **7**(7), 9347–9370 (2015)
25. Sacks, J., Welch, W.J., Mitchell, T.J., Wynn, H.P.: Design and analysis of computer experiments. Stat. Sci. **4**, 409–423 (1989)

26. da Silva Ferreira, G., Gamerman, D.: Optimal design in geostatistics under preferential sampling. Bayesian Anal. **10**(3), 711–735 (2015)
27. Snoek, J., Larochelle, H., Adams, R.P.: Practical Bayesian optimization of machine learning algorithms. Neural Information Processing Systems (NIPS), pp. 1–9 (2012). arXiv:1206.2944 (2012)
28. Verrelst, J., Dethier, S., Rivera, J., Muñoz-Marí, J., Camps-Valls, G., Moreno, J.: Active learning methods for efficient hybrid biophysical variable retrieval. IEEE Geosci. Remote Sens. Lett. **13**(7), 1012–1016 (2016)

Image Processing and Applications

Deep Learning for Polar Bear Detection

Scott Sorensen[1(✉)], Wayne Treible[1], Leighanne Hsu[1], Xiaolong Wang[6],
Andrew R. Mahoney[2], Daniel P. Zitterbart[3,4,5], and Chandra Kambhamettu[1]

[1] University of Delaware, Newark, DE, USA
sorensen@udel.edu
[2] University of Alaska Fairbanks, Fairbanks, AK, USA
[3] Alfred Wegener Institute for Polar and Marine Research, Bremerhaven, Germany
[4] Applied Ocean Physics and Engineering,
Woods Hole Oceanographic Institution, Woods Hole, MA, USA
[5] Biophysics Group, Erlangen University, Erlangen, Germany
[6] Samsung Research America, Mountain View, CA, USA

Abstract. Marine mammals in the Arctic are threatened by a changing climate and increasing human activity in the region. International
laws protect these animals, however detecting and identifying them is
not always easy. We have developed a multimodal approach using an
omnidirectional thermal camera system, and an optical band stereo system operating in parallel. Using a unified framework for transfer learning
with convolutional neural networks in both modalities we have trained
a system to detect and classify mammals as well as habitat indicators
in the images from both camera systems. Our experiments show that
mammal habitat can be identified reliably using these techniques, and
our analysis provides a framework for real world use cases.

Keywords: Deep learning · Thermal imaging · Marine mammals

1 Introduction

As climate changes across the globe, the Arctic faces uncertainty, as do the
animals that call it home. Increased human activity in the form of shipping,
sightseeing, and oil development means that arctic animals are becoming more
exposed to people and machinery. International laws and regulations like United
Nations Convention on the Law of the Sea, and the International Convention
for the Regulation of Whaling have put in place some protection on marine
mammals. However as development increases, automatic methods for identifying
these animals are increasingly needed. We have developed a system for detecting
habitat of polar bears using a two camera system operating in different modalities
with different fields of view and operating goals. The two camera systems are
shown in Fig. 1.

In summer 2012 the Polar Sea Ice Topography REconstruction System, or
PSITRES, and the FIRST-Navy IR system were both deployed aboard the RV

S. Sorensen—Now employed at Vision Systems Incorporated, Providence RI, USA.

P. Sharma and F.M. Bianchi (Eds.): SCIA 2017, Part I, LNCS 10269, pp. 457–467, 2017.
DOI: 10.1007/978-3-319-59126-1_38

Fig. 1. The PSITRES and FIRST-Navy camera systems aboard the RV Polarstern

Polarstern during the ARKXXVII/3 research cruise. The cruise spent over two months at sea throughout a large area of the central Arctic ocean and surrounding coastal seas. These two camera systems continuously recorded ice conditions around the ship, and have directly observed many animals as well as evidence of their presence.

1.1 Camera System

The two camera systems have heterogeneous imaging capabilities, and vary drastically in capabilities. They have non overlapping field of views, different resolutions, and they operate in different portions of the spectrum. Both systems are shown in Fig. 1.

PSITRES. The Polar Sea Ice Topography REconstruction System, or PSITRES is an optical band 3D camera system consisting of a stereo pair of 5 megapixel machine vision cameras on a 2 m baseline and an optional 10 megapixel center camera with a wider field of view. The stereo cameras are synchronized by a hardware trigger and captured images at a rate of 1 frame every 3 s for the duration of the ARKXXVII/3 cruise. The entire system is mounted to the flying deck of the ship off the port side and observes a patch of ice and water as shown in Fig. 2.

The FIRST-Navy IR System. The FIRST-Navy IR system is an omnidirectional gimbal stabilized long wave infrared (LWIR) camera system that records images at 7200 × 576 resolution at 5 frames per second as shown in Fig. 3. The system collects thermal images with a 360° horizontal and 18° vertical field of view. The system is mounted on the ship's crows nest.

1.2 Problem Statement

In this work we present a method for detecting polar bears in images collected by both the PSITRES and FIRST-Navy IR camera systems. While the images

Fig. 2. Sample PSITRES Imagery with polar bear prints

Fig. 3. Sample FIRST-Navy IR imagery

collected by these systems are dissimilar, we have developed a common approach to transfer learning, that allows us to use the same training scheme for both image types.

2   Related Work

Camera systems have been used to detect animals and marine mammals in particular for some time. Works utilizing thermal cameras for identifying the denning sites of polar bears dating back to the 1970s [4]. Much of the work has been done using aerial imagery [2,4]. These techniques have predominantly been manual or kept people in the loop, including the use of Infrared Binoculars [3]. Fully automatic detection of whales has been studied from a variety of thermal imaging platforms [5] including the FIRST-Navy IR system used in this work [14].

Machine learning has been used for classifying images that may contain animals with many recent classification and detection datasets targeting common animal types. The ImageNet Large Scale Visual Recognition Challenge dataset [10] contains 1000 different classes including polar bears, and birds. A number of works have targeted this challenge using convolution neural networks [7,11].

2.1 Camera Systems

The PSITRES and FIRST-Navy system have both been subjects of a variety of computer vision and image analysis research including stereo and Structure From Motion reconstruction techniques [8,9], whale blow detection [14], and image retrieval [6]. More recently the two camera systems were integrated into a common coordinate system and Virtual Reality framework [12]. This work aligned the two camera systems into a unified coordinate system as shown in Fig. 4. This work facilitates metric distance measurements in both images for measuring distance from the ship, which will play a role in our experiments in Sect. 4.4.

Fig. 4. The 3D alignment and reprojection developed in [12].

3 Methods

In this section we will discuss our techniques for preprocessing images, and our framework for transfer learning. The two modalities of image vary significantly from each other, and therefore detection is treated differently. In practice these differences manifest themselves predominantly in how we treat preprocessing and labeling the data.

3.1 IR Preprocessing

The IR images themselves are captured from a sensor mounted on the crow's nest of the ship, but are not at the highest point, meaning that a portion of the crow's nest and radar mast are present in every image. The stabilization used on the sensor means that these components move relative to sensor and are not fixed in the images. To combat this we have masked off a region in the images larger than the area corresponding to ship regions. This mask covers the entire crows nest and radar mast, which are relatively stable, as well as a large area around the railings and other components that move more relative to the sensor. We only process regions outside of this masked area throughout the remainder of our technique.

The IR images are high resolution and only contain small salient regions with animals. To reduce the computational load of detecting animals in these images we leverage the fact that the animals in these images are warm blooded and stand out against the cold environment in thermal images. To do this we employ a simple intensity threshold ($I_\tau = 150$) to eliminate image regions with nothing that could be a warm blooded animal. This threshold was selected because it excludes the vast majority of unimportant image regions while still remaining maximally inclusive to animals in the scenes.

After thresholding based on intensity we are left with a number of variable sized image patches containing animals or other warm components in the scene. These other scene components consist melt ponds and other regions of ice and water that are warmer than the surrounding scene due to solar heating. Some of these regions can be quite large (on the order of hundreds of meters). We place an additional constraint that these regions are of an appropriate size by putting a threshold on patch size of ($W_\tau = 200$ and $H_\tau = 100$). This exceeds the projected size of even the largest bear appearing closest to the ship in our data, but still includes small ambient regions.

We use the resulting small patches for classification using the transfer learning scheme discussed in Sect. 3.3. To generate a training set we have developed a GUI that was used to label more than 10,000 patches as containing either bears, birds, seals or ambient components. Sample patches for each category are shown

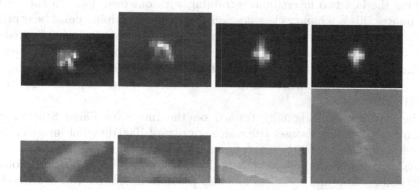

Fig. 5. Patches containing bears (top left two images), birds (top right two images) and ambient components (bottom row)

in Fig. 5. These patches are then stored with labels to use for training and testing of the machine learning approaches.

3.2   PSITRES Data Preparation

The PSITRES system has a much smaller viewing volume than the FIRST-Navy system and views only an area adjacent to the ship. As a result the animals who are wary of a large, noisy ship do not enter the field of view of the cameras. There are however indicators of habitat that do enter the camera system's field of view. Blood, scat, and other indications of animal presence are all left on the ice, but by far the most common that are readily apparent in PSITRES images are footprints. Not all ice conditions cause footprints to appear, but footprints are an indication that bears are present in the region (Fig. 6).

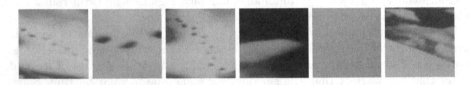

Fig. 6. Positive samples with patches (left three images), and negative samples without patches (right three images).

3.3   Transfer Learning Scheme

We have formulated our problem of detection differently in both modalities of image, however the differences are mostly manifest in the preprocessing steps, and the treatment of the results. Training is done using the same scheme of transfer learning for classification. We have fine-tuned a pre-trained network by replacing the last two layers, and retraining with our own data. In the visible band images this is a binary classification of patches containing polar bear prints or patches without polar bear prints. In LWIR the problem is complicated by other animals, and we have used a 3 label scheme with bears, birds, and ambient components. We initialize networks for both modalities of images using the InceptionNet [13] implementation in Google's Tensor Flow deep learning framework [1].

The network was originally trained on the ImageNet Large Scale Visual Recognition Challenge dataset [10] which consists of 1000 different image classes. This network consists of 22 layers, composed of convolution, pooling, and softmax operations. We have formulated our problem as a binary classification problem in the visible band and a 3 class labeling problem in LWIR. To accommodate the large change in number of labels we have modified the network using a new softmax layer with the corresponding number of outputs in the classification domain.

4   Experiments and Analysis

We have developed experiments to both validate our classification/detection scheme as well as to validate some aspects unique to our problem. Validating our approach to detection is done using traditional means, but since this framework has been developed for the application of polar bear detection from a vessel in polar regions, we aim to quantify how well detection works in this context.

4.1   Cross Validation

To validate our classification and detection framework we have used ten fold cross validation. This means we train 10 models with non overlapping testing sets spanning our data. We evaluate accuracy on the testing set for each fold and average the results. We have conducted a few different experiments within this framework to evaluate both LWIR and visible band classification.

The main criteria for evaluation is accuracy on the testing set averaged across each fold. In the following subsection we discuss different patch sizes for PSITRES imagery, but in general we found a trend of larger patch sizes resulting in higher accuracy, so we will report visible band results on the largest patch sizes of 160×160. Table 1 shows results for both image modalities.

Table 1. Performance results in LWIR and visible band

	10 fold accuracy
LWIR	97.46%
Visible	90.67%

4.2   Patch Size

Since we extract patches around individual prints (which can contain other prints), we have experimented with different patch sizes to evaluate how this affects accuracy. We have run ten fold cross validation on 8 different datsets of varying patch sizes. The patch sizes are of 20×20 to 160×160 in increments of 20 pixels. Figure 7 shows results for each patch size including the average across all ten folds as well as the accuracy of the top performing fold. This shows a trend of larger patch size and increasing accuracy, which is likely due to image patches containing a greater number of individual prints.

4.3   Supplementary Validation

The LWIR data features many consecutive frames of the same individual bears. While the bears move and this results in a larger training set, many of the images are homogeneous. Even with 10 fold cross validation there is a high chance that for a given testing image the model was trained on a highly similar image. To

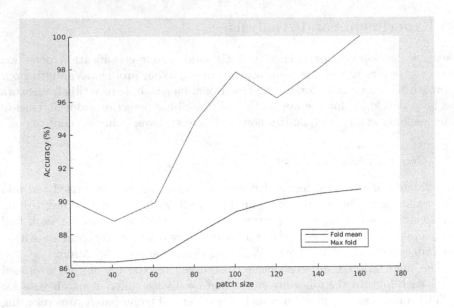

Fig. 7. Results for different patch sizes

ensure that the model is robust we have isolated a secondary test set of images consisting of an individual bear that was not part of the training, testing or validation set. After training we apply each model to the isolated image patches and compare the resulting label to the ground truth.

Each model from the training folds achieved a perfect accuracy of 100% on all 11 image patches with the isolated bear. Furthermore the minimum reported confidence (softmax output) from any classification was 0.577, which is a convincing majority for a 3 class labeling problem.

4.4   Use Case

Since this work aims to detect polar bears from a ship in ice covered waters one of the most practical pieces of information for users is how far away the bears are. On research vessels such as the RV Polarstern, scientists carry out ice stations, where they work on the ice. Bears in the vicinity of people working on the ice is dangerous and protocol dictates evacuation. In a more general setting, giving marine mammals such as polar bears a wide breadth is not only important, but a legal requirement. We aim to quantify the conditions under which bears can be detected, and put these in terms of real units.

To evaluate performance in these terms we have conducted an experiment to evaluate the maximum range for detection of bears in LWIR. To do this we haves used the reprojection scheme developed by [12] and measured the Euclidean

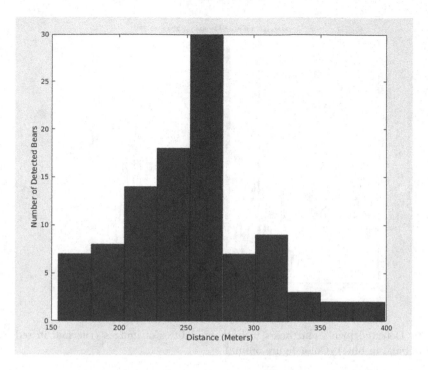

Fig. 8. The distribution of distance from the sensor to the 100 smallest detected bears.

distance from the sensor to the detected bears in our dataset. We have taken the smallest 100 detected regions and reprojected the center-point of the patch using spherical projection and ray tracing. The distribution of these distances is shown in Fig. 8 as a histogram. This shows that a bear can be detected by our scheme at up to almost half a kilometer.

4.5   Habitat Identification

We have run the outlined approach for the LWIR and PSITRES data from 2012-09-17 which had the highest concentration of polar bears from the cruise. Figure 9 shows results for both modalities for the morning, showing detected bears in the thermal modality as vertical red lines. The detected paw prints are shown in blue as the fraction of patches in the image with a moving average 1D filter applied to the noisy time series data. This shows a generally high concentration of prints over the course of the morning, with a few bears spotted in clusters, which is accurate to the real experience on the ship.

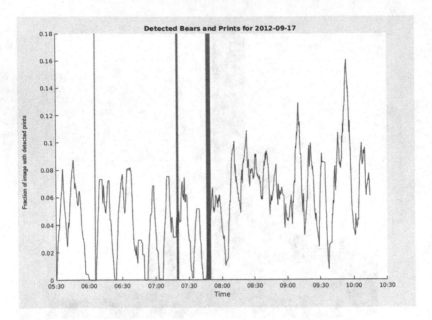

Fig. 9. Detected prints and bears in both modalities of images (thermal in red, and visible band in blue) (Color figure online)

5   Conclusion

We have presented a framework for detection of polar bears and polar bear footprints from images acquired by heterogeneous camera systems that were deployed aboard an icebreaker. The two systems, an omnidirectional thermal camera and a visible band stereo system, offer complementary information that can be utilized by those operating in the Arctic Ocean. We have developed a transfer learning scheme using convolutional neural networks which is applied to patches from both image modalities, casting the problem as a multi-label classification task. We have evaluated this approach with experiments that have been designed to validate this approach and overcome problems present in each set of images. The system can detect polar bears at up to a half kilometer distance and can detect paw prints which indicate the presence of these animals.

Acknowledgments. We would like to thank the crew and scientific party of the Polarstern ARK-XXVII/3 research cruise. This work is funded by NSF CDI Type I grant 1124664.

References

1. Abadi, M., Agarwal, A., Barham, P., Brevdo, E., Chen, Z., Citro, C., Corrado, G.S., Davis, A., Dean, J., Devin, M., Ghemawat, S., Goodfellow, I., Harp, A., Irving, G., Isard, M., Jia, Y., Jozefowicz, R., Kaiser, L., Kudlur, M., Levenberg, J., Mané, D., Monga, R., Moore, S., Murray, D., Olah, C., Schuster, M., Shlens, J., Steiner, B., Sutskever, I., Talwar, K., Tucker, P., Vanhoucke, V., Vasudevan, V., Viégas, F., Vinyals, O., Warden, P., Wattenberg, M., Wicke, M., Yu, Y., Zheng, X.: TensorFlow: Large-scale machine learning on heterogeneous systems, 2015. Software available from tensorflow.org
2. Amstrup, S.C., York, G., McDonald, T.L., Nielson, R., Simac, K.: Detecting denning polar bears with forward-looking infrared (flir) imagery. Bioscience **54**(4), 337–344 (2004)
3. Baldacci, A., Carron, M., Portunato, N.: Infrared detection of marine mammals. Technical report, Nato Undersea Research Centre, December 2005
4. Brooks, J.W.: Infra-red scanning for polar bear. Bears: Their Biol. Manage. **2**, 138–141 (1972)
5. Graber, J.: Land-based infrared imagery for marine mammal detection. Ph.D. thesis, University of Washington (2011)
6. Lu, G., Sorensen, S., Kambhamettu, C.: Fast ice image retrieval based on a multilayer system. In: IS&T/SPIE Electronic Imaging, pp. 90300Q–90300Q. International Society for Optics and Photonics (2014)
7. Ouyang, W., Wang, X., Zeng, X., Qiu, S., Luo, P., Tian, Y., Li, H., Yang, S., Wang, Z., Loy, C.-C., et al.: Deepid-net: deformable deep convolutional neural networks for object detection. In: Proceedings of the IEEE Conference on Computer Vision and Pattern Recognition, pp. 2403–2412 (2015)
8. Rohith, M.V., Rhein, S., Lu, G., Sorensen, S., Mahoney, A.R., Eicken, H., Ray, G.C., Kambhamettu, C.: Iterative reconstruction of large scenes using heterogeneous feature tracking. In: CVPR Workshops, pp. 407–412. IEEE Computer Society (2013)
9. Rohith, M.V., Sorensen, S., Rhein, S., Kambhamettu, C.: Shape from stereo and shading by gradient constrained interpolation. In: ICIP, pp. 2232–2236. IEEE (2013)
10. Russakovsky, O., Deng, J., Su, H., Krause, J., Satheesh, S., Ma, S., Huang, Z., Karpathy, A., Khosla, A., Bernstein, M., Berg, A.C., Fei-Fei, L.: ImageNet large scale visual recognition challenge. Int. J. Comput. Vis. (IJCV) **115**(3), 211–252 (2015)
11. Simonyan, K., Zisserman, A.: Very deep convolutional networks for large-scale image recognition. CoRR, abs/1409.1556 (2014)
12. Sorensen, S., Kolagunda, A., Mahoney, A.R., Zitterbart, D.P., Kambhamettu, C.: A virtual reality framework for multimodal imagery for vessels in polar regions. In: Proceedings of the International Conference on Multimedia Modeling, January 2017
13. Szegedy, C., Liu, W., Jia, Y., Sermanet, P., Reed, S.E., Anguelov, D., Erhan, D., Vanhoucke, V., Rabinovich, A.: Going deeper with convolutions. CoRR, abs/1409.4842 (2014)
14. Zitterbart, D.P., Kindermann, L., Burkhardt, E., Boebel, O.: Automatic round-the-clock detection of whales for mitigation from underwater noise impacts. PLoS ONE **8**(8), e71217 (2013)

Crowd Counting Based on MMCNN in Still Images

Tao Wang[1], Guohui Li[1(✉)], Jun Lei[1], Shuohao Li[1], and Shukui Xu[2]

[1] College of Information System and Management,
National University of Defense Technology, Changsha, China
{wangtao1993,guohli,leijun1987,lishuohao}@nudt.edu.cn
[2] The 28th Research Institute of China Electronic Technology
Group Corporation, Nanjing, China
xskgfkd@163.com

Abstract. Accurately estimate the crowd count from a still image with arbitrary perspective and arbitrary crowd density is one of the difficulties of crowd analysis in surveillance videos. Conventional methods are scene-specific and subject to occlusions. In this paper, we propose a Multi-task Multi-column Convolutional Neural Network (MMCNN) architecture for crowd counting and crowd density estimation in still images of surveillance scenes. The MMCNN architecture is an end-to-end system which is robust for images with different perspective and different crowd density. By promoting MCNN with 3×3 filter, the MMCNN could utilize local spatial features from each column. Furthermore, the ground truth density map is generated based on Perspective-Adaptive Gaussian kernels which can better represent the heads of pedestrians. Finally, we use an iterative switching process in our deep crowd model to alternatively optimize the crowd density map estimation task and crowd counting task. We conduct experiments on the WorldExpo'10 dataset and our method achieves better results.

Keywords: Convolutional neural networks · Crowd counting · Crowd density estimation · MMCNN

1 Introduction

Counting crowd pedestrians in surveillance videos attracts a lot of attention in public safety. It is especially significant in major cities where there are public rallies and sports events. In the new year eve of 2015, 35 people died of stampede in Shanghai, China. Unfortunately, there are many more similar disasters take place around the world. Accurately estimating crowds from images or videos is a highly valued problem of computer vision. In practical applications, crowd counting and density estimation are challenging, because of severe occlusions, diverse crowd distributions and scene perspective distortions.

G. Li—Thanks to my tutor for giving me a lot of support and help.

P. Sharma and F.M. Bianchi (Eds.): SCIA 2017, Part I, LNCS 10269, pp. 468–479, 2017.
DOI: 10.1007/978-3-319-59126-1_39

The most intuitive method is detecting and tracking all the people or foreground segmentation, but these methods are indispensable in practical crowd scenes. Many excellent methods [1–4] were constructed based on regression. These methods firstly represent the crowd scenes into feature space, then learn a mapping from the feature space to crowd counts. However, these works used hand-crafted features which is designed by experienced researchers/engineers. And the these features are scene-specific. Cong Zhang *et al.* [5] proposed a framework for crowd counting based on deep convolutional neural networks, their method uses CNN to extract deep features instead of hand-crafted features. A good method of crowd counting can also be extended to other domains, such as counting cells or bacteria from microscopic images, animal crowd estimates in wild scenes, or estimating the number of cars at traffic jams, etc. [5] constructs a new training set (sampled from source domain) which follows the distribution of target domain to adapt the model to the new scenario. But this work is not an end-to-end system, that is, we must crop the image to a same size set in advance and splice them together when counting the crowd. Yingying Zhang *et al.* [6] proposed a Multi-column Convolutional Neural Network to predict crowd scenes' density maps. But their network is very difficult to obtain optima due to its multi-column structure and large-number-pixels regression task.

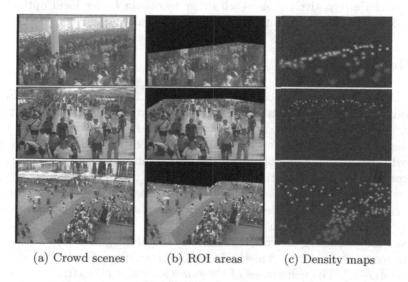

(a) Crowd scenes   (b) ROI areas   (c) Density maps

Fig. 1. (a) Crowd scenes there are many occlusions. All the images are selected from the WorldExop'10 dataset [5,37]. (b) ROI areas of crowd scenes because there are much noise. The area out of ROI is set to zero. (c) Density maps generated by using Perspective-adaptive kernels. The size and shape of each kernel is similar to the heads of pedestrians. These density maps are labels of MMCNN when we training crowd density estimation task.

Contribution of This Paper. Inspired by state-of-the-art method [6], we propose a Multi-task Multi-column Convolutional Neural Network for crowd density estimation and counting. We especially put the focus on density estimation and the target is generating a density map close to the reality. By this density map, we can conveniently find the high density areas and warning in advance. Contributions of this paper are summarized as follows:

(1) We proposed Perspective-Adaptive method to generate head-shaped and head-sized Gaussian Kernels. The Gaussian kernels used in [5] represent the pedestrian contour (head and body) but it does not work well enough when the occlusion is serious. [6] used round Gaussian kernels to describe pedestrians' heads. However, it is not accurate enough when there is perspective distortion.
(2) Based on MCNN [6], we improve the method of merging the feature maps from different columns with 3×3 filter. In this way, the network could consider local spatial information of each feature map.
(3) We proposed a Multi-task Multi-column Convolutional Neural Network (MMCNN) based on MCNN [6]. We use an iterative switching process in our deep crowd model to alternatively optimize the density map estimation task and the count estimation task. In this way, the two different but related task can alternatively assist each other to obtain better local optima and make up for the shortcomings of regression framework.

2   Related Work

Many algorithms have proposed in the literature for crowd counting. Earlier methods [7] adopt a detection-style framework to estimate the number of pedestrians. [8–10] have used a similar detection-based framework for pedestrian counting. In detection-base crowd counting methods, people typically assume a crowd is composed of individual entities which can be detected by some given detectors [11–14]. The limitation of such methods is that occlusion among people in a very dense crowd significantly affects the performance of the detector.

In counting crowds in videos, people have proposed to cluster trajectories of tracked visual features. Such as [15,16]. But these tracking-based methods do not work for estimating crowds from individual still images.

The most extensively used method for crowd counting is feature-based regression, see [17–22]. The main steps of these methods are: (1) extracting the foreground; (2) extracting various features from the foreground, such as area of crowd mask [17,18,20,23], edge count [16–18,24], for texture features [17,25]; (3) utilizing a regression function to estimate the crowd count. Linear [23] or piece-wise linear [24] functions are relatively simple models and get decent performance. Other more advanced methods are ridge regression (RR) [17], Gaussian process regression (GPR) [16], and neural network (NN) [25].

There have also been some works focusing on crowd counting from still images. [26] has proposed to leverage multiple sources of information to compute an estimate of the number of individuals present in an extremely dense crowd. In that work, a dataset of fifty crowd images containing 64 K annotated humans (UCF_CC_50) is introduced. [27] has followed the work and estimated counts by fusing information of multiple sources, namely, interest points (SIFT), Fourier analysis, wavelet decomposition, GLCM features, and low confidence head detections. [28] has utilized the features extracted from a pre-trained CNN to train a support vector machine (SVM) that subsequently generates counts for still images.

Many works introduce deep learning into various surveillance applications, such as person re-identification [29], pedestrian detection [1,30,31], tracking [32], crowd behavior analysis [33] and crowd segmentation [34]. Their success benefits from discriminative power of deep models. Zhang *et al.* [5] has proposed a CNN based method to count crowd in different scenes. They first pre-train a network for certain scenes. When a test image from a new scene is given, they choose similar training data to fine-tune the pre-trained network based on the perspective information and similarity in density map. Their method demonstrates good performance on most existing datasets. But this type of methods need to crop one image into patches of similar sizes, then estimate the total number of each patch, and merge all the patches into one image. In this way, there are inevitably overlaps between patches. And this method is not an end-to-end framework, it is not suitable for practical application such as real-time public safety monitoring. Yingying Zhang *et al.* [6] recently proposed a Multi-column Convolutional Neural Network. They use three columns of CNNs corresponding to filters with receptive fields of different sizes so that the features learned by each column CNNs could deal with perspective effect and different image resolution. But they did not take perspective distortion of heads into consideration.

3   Method

3.1   Perspective-Adaptive Kernels

Many works followed [5] and defined the density map regression ground truth as a sum of Gaussian kernels centered on the location of objects. This kind of density map is suitable for characterizing the density distribution of circle-like objects such as cells and bacteria. [5] uses human-shaped Gaussian kernels to generate density map. But due to severe occlusions (Fig. 1(a)), heads are the main cues to judge whether there exists a pedestrian in a practical surveillance scene. [6] uses circle-like Gaussian kernels to represent the heads of pedestrians in the image. It works when the scene is full of pedestrians but actually the heads in the image are not circle-like because of perspective distortion. Precisely, the shapes of heads are more similar to ellipses than circles. We proposed a method of generating density maps based on Perspective-Adaptive Kernels:

Firstly, we generate each scene's perspective map M by linear regression [5] or geometry-adaptive method [6]. Most of the time, the datasets have provided the perspective map, such as the WorldExpo'10 dataset. The value of each element in the perspective map $M(p)$ denotes the number of pixels in the image representing one square meter at that location in the actual scene. The value of M contains the perspective information. For example, if the same person is standing at two point with different distances from the camera, it will look shorter when he is in the place far away from the camera, the head will be smaller, and the $M(p)$ value will be smaller too, vice versa. After we obtain the perspective map and the center positions of pedestrian head p_i in the region of interest (ROI), we create the crowd density map as:

$$D_i(p) = \sum_{P \subseteq P_i} \frac{1}{\|Z\|} (N_h(p; P_h, \Sigma)) \tag{1}$$

where g is the gradient of perspective map, C_1 and C_2 are empirical coefficients, N_h represents Gaussian kernel and $\Sigma = (\sigma_x, \sigma_y)$, P_h represents the position of head. Assume that perspective distortion only exists in y-axis and the heads in the image is distorted only in y-axis. To exactly represent the pedestrian head, we set the variance $\sigma_x = C_1 M(p)$, $\sigma_y = (1 + gC_2)\sigma_x$.

3.2   Normalized Crowd Density Map

Our MMCNN model has two switchable learning objectives, the main task is to estimate the crowd density map of the input image. Because density map represents the distribution of pedestrians and has abundant spatial information. To ensure that the integration of all density values in a density map equals to the total crowd number in the original image, the whole distribution is normalized by Z. The created density maps are showed in Fig. 1(c). To obtain density maps precisely, we conduct a lot of experiments and find the best empirical coefficients, $C_1 = 0.12$, $C_2 = 0.5$. At this time, the size of Gaussian kernel is close to pedestrian head size in the image and will not appear too small. We only consider the pedestrians in ROI and truncate the Gaussian distribution outside ROI.

3.3   Multi-task Multi-column CNN Model

An overview of our crowd MMCNN model is shown in Fig. 2. Our MMCNN model uses the base framework of MCNN [6]. The input image is original still image from surveillance video. In our model, we use the filters of different sizes to perceptive heads of different scales and there is no need to crop the image into a pre-set size like [5]. Motivated by the good performance Fully Convolutional Networks [36], we promote the MCNN [6] model by using 3×3 filters instead of 1×1 filters to merge the *Concated feature map* into one-channel feature map. Suppose that the shape of *Concated feature map* volume is $42 \times 144 \times 180$, if we use 1×1 filters to merge all the channels, the filter only consider how to compute the weighted average of 42 channels. It is difficult to consider the local relationship between neighboring pixels. But if we use 3×3 filters, the network could consider contextual information (compute weighted average of $42 \times 3 \times 3$ neighboring values) and has a good ability to fit actual scenes.

Our framework contains three parallel CNNs whose filters are with local receptive fields of different sizes. We use the same network structures for all columns like [6] but changed 1 × 1 kernel and the number of filters to slightly increase model's complexity. For each column, the structure is *Conv-ReLU-Pooling-Conv-ReLU-Pooling-Conv-ReLU-Conv*, and we concat all the feature maps from different columns and merge the feature map volume by 1 3 × 3 × 42 filter. Then, we set the neuron corresponding to the area out of ROI to zero based on the ROI mask (images with ROIs are showed in Fig. 1(b)). The ROI mask is optional, because some datasets do not provide ROI information or we do not need to consider ROI in practical problems. And both two pooling layer use 2 × 2 Max pooling kernel.

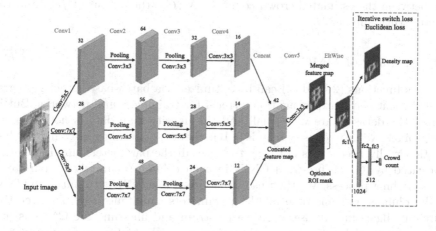

Fig. 2. The structure of the proposed MMCNN model for crowd counting and density map estimation (inherit the structure of MCNN [6]). The text in blue denote manipulations of deep neural network (*Conv_i* means *ith* convolutional layer and *Eltwise* means dot production of two matrix). The cubes in blue denote different feature map volumes after each manipulations. The number above each blue cube is the number of output channels. (Color figure online)

We introduce an iterative switching process in our framework to alternatively optimize the density map estimation task and the count estimation task. The main task is estimating the crowd density map of the input image. Density map prediction needs spatial information and crowd distribution while global count regression only focuses on predicting a numerical value, therefore, using global count as the secondary learning objective can improve the accuracy of our main task. Because of two pooling layers in the CNN model, the output density map is down-sampled to 1/4 size. Since the density map contains rich and abundant local and detailed information, the CNN model can obtain a better representation of crowd image. The total count regression of the input image is treated as the secondary task and the regression head is directly connected after

$Conv5$. Then Euclidean distance is used to measure the difference between the estimated density map (global count) and ground truth. The two loss functions are defined as:

$$L_D(\Theta) = \frac{1}{N} \sum_i^N \|F_d(X_i; \Theta) - D_i\|^2, \tag{2}$$

$$L_C(\Theta) = \frac{1}{N} \sum_i^N \|F_c(X_i; \Theta) - C_i\|^2, \tag{3}$$

where Θ is the set of parameters of the CNN model and N is the number of training samples. L_D is the loss between estimated density map $F_d(X_i; \Theta)$ (the output of $Conv5$) and the ground truth density map D_i. Similarly, L_C is the loss between the estimated crowd count $F_c(X_i; \Theta)$ (the output of $fc3$) and the ground truth number C_i.

3.4   Training of MMCNN

The loss functions (2) and (3) can be optimized via batch-based stochastic gradient descent and backpropagation, typical for training neural networks. But in reality, the datasets are very small and the density map data is not balanced (most pixels' value of density map is 0, but the non-zero values are what the loss function really need), it is not easy to learn all the parameters simultaneously. Motivated by [6], we pre-train CNNs in each column separately by our iterative switching process. We then use these pre-trained CNNs to initialize CNNs in all columns and fine-tune all the parameters simultaneously. The iterative switching algorithm we use when we pre-train and fine-tune the CNNs is just like the learning objective switch algorithm in [5]. We firstly use density map to supervise MMCNN's training process due to its abundant spatial information. And then we use global count as training label to promote the precision of our MMCNN model. In this way, our MMCNN model could easily get the optima.

4   Experiment

We evaluate our MMCNN model on The WorldExpo'10 dataset-the largest and most comprehensive dataset in crowd counting field and compare our model to many other methods in the literature. Implementation of the network and its training are based on $Caffe$ developed by [35].

Evaluation Metric. By following the convention of existing works [5,6] for crowd counting, we evaluate different methods with the absolute error (MAE), which is defined as follow:

$$MAE = \frac{1}{N} \sum_1^N |z_i - p(z_i)|, \tag{4}$$

where N is the number of test images, z_i is the actual number of crowd in the ith image, and $p(z_i)$ is the estimated number of crowd in the ith image. In some ways, MAE indicates the accuracy of the estimates.

The WorldExpo'10 Dataset. WorldExpo'10 crowd counting dataset was firstly introduced by [5, 37]. This dataset is the largest dataset focusing on crowd counting. It contains 1132 annotated video sequences which are captured by 108 surveillance cameras, all from Shanghai World Expo 2010. The authors of [5] provided a total of 199,923 annotated pedestrians at the centers of their heads in 3980 frames. 3380 frames are used in training data. Testing dataset includes five different video sequences, and each video sequence contains 120 labeled frames. Five different regions of interest (ROI) are provided for the test scenes. Some images are shown in Fig. 1(a).

In this dataset, the perspective maps are given. We apply our Perspective-Adaptive Kernels to obtain density maps. After many times of experiments, we set $C_1 = 0.12$, $C_2 = 0.9$ (details in Sect. 3.1). Because of the noise in the area out of ROI, we only consider the ROI regions. So we use *dot production* to set the neuron corresponding to the area out of ROI to zero. We generate the ROI mask matrix corresponding to each scene. Each mask matrix has the same resolution of density map and the value is either 1 or 0. For evaluation, we use the same evaluation metric (MAE) suggested by [5]. The original input image has 720×560 pixels, we generate density maps have a resolution of 144×180 because our MMCNN model has two 2×2 Max pooling layers. The sum of each density map equals to the total count of pedestrian in the image. The input of our model is original image and RIO mask, output is predicted density map or regressed global count, the whole process is illustrated in Fig. 2.

We use 1/10 of training data as validation data to evaluate the performance and help to optimize hyper parameters. We first split the MMCNN model into three single-column multi-task CNNs, and train each column with two learning tasks iteratively. First of all, we train each column with density map and obtain the optima. Then, we append *fully connected layers* at the end of each CNN and train each CNN with global count, we firstly fix the parameters of convolutional layers until the learning curve come into plateau period. Then, we update all the parameters. In this way, we can obtain the optimal model of each column. We then use these pre-trained single-column CNNs to initialize the columns of MMCNN model and update all the parameters simultaneously in the same way of pre-training each column. For comparing 3×3 kernels and 1×1 kernels, we conducted the experiment with 3×3 kernels in the same method above-mentioned. The result is showed in Fig. 3. The figure tells that 3×3 kernels worked, but multi-task training policy played a greater effect.

In the test phase, the multi-column CNN model is fine-tuned with the first 60 labeled frames for every test scene, and the remaining frames are used as the test data. The results of different methods in the five test video sequences can be seen in Table 1. In scene 2 and scene 5, the density of the crowd changes drastically and there are a lot of noise. Such as scene5, many pedestrians are

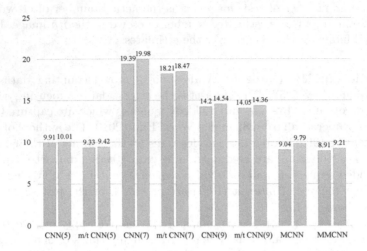

Fig. 3. Comparing single-task CNNs with mulit-task CNNs. Blue bars denote test MAE when using 3×3 kernels and orange bars denote test MAE when using 1×1 kernels. $CNN(i)$ means a single column CNN of MMCNN with single task (crowd density map estimation). $m/t\ CNN(i)$ means single column CNN with multi-task training. $MCNN$ means single-task multi-column CNN. The figure tells that multi-task learning and 3×3 kernels could optimize the performance of our CNN model. And the experiment is conducted on the WorldExpo'10 dataset. (Color figure online)

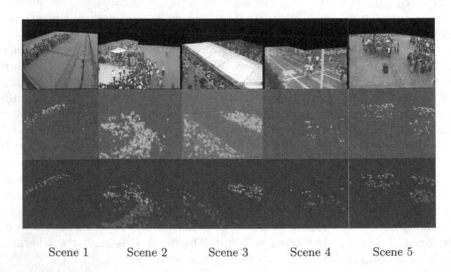

Scene 1   Scene 2   Scene 3   Scene 4   Scene 5

Fig. 4. Test results in each test scene of the WorldExpo'10 dataset. Each column has three pictures which are examples selected from a same scene, the first row are input images with ROIs, the second row are predicted density maps of MMCNN, the third row are the ground truths.

taking umbrella and wearing a hat. In these two scenes, [5] works better. But our model also achieves better performance than Fine-tuned Crowd CNN model [5] in terms of average MAE. Compared to MCNN [6], our model performs better in scene 2 and scene 5 because our 3×3 filter in $Conv5$ could integrate local spatial information in a 3×3 receptive fields and our model has a similar average MAE with MCNN [6] (Fig. 4).

Table 1. Mean absolute errors of the WorldExpo'10 crowd counting dataset

Method	Scene1	Scene2	Scene3	Scene4	Scene5	Average
Zhang et al. [5]	9.8	**14.1**	14.3	22.2	**3.7**	12.9
MCNN [6]	**3.4**	20.6	12.9	13.0	8.1	11.6
MMCNN	**3.4**	18.5	**9.6**	**8.7**	7.9	**9.6**

5   Conclusion

In this paper, we have proposed a Multi-task Multi-column Convolutional Neural Network which can accurately estimate crowd global count and crowd density map in a still image. We evaluate the performance of our model in WorldExpo'10 dataset which is the most representative dataset in crowd counting field. In this dataset, MMCNN outperforms the state-of-art crowd counting methods [6] in each scene and also performs better than Fine-tuned Crowd CNN model [5] in terms of average MAE. Compared to [5], the result reflects that our head-shaped Gaussian kernel is not robust enough in such scene where there are many umbrellas and hats.

Acknowledgement. This work was supported by NSFC (No.71673293).

References

1. Zeng, X., Ouyang, W., Wang, M., Wang, X.: Deep learning of scene-specific classifier for pedestrian detection. In: Fleet, D., Pajdla, T., Schiele, B., Tuytelaars, T. (eds.) ECCV 2014. LNCS, vol. 8691, pp. 472–487. Springer, Cham (2014). doi:10.1007/978-3-319-10578-9_31
2. Chen, K., Gong, S., Xiang, T., Mary, Q., Loy, C.C.: Cumulative attribute space for age and crowd density estimation. In: CVPR (2013)
3. Chen, K., Loy, C.C., Gong, S., Xiang, T.: Feature mining for localised crowd counting. In: BMVC (2012)
4. Loy, C.C., Gong, S., Xiang, T.: From semi-supervised to transfer counting of crowds. In: ICCV (2013)
5. Zhang, C., Li, H., Wang, X., Yang, X.: Cross-scene crowd counting via deep convolutional neural networks. In: CVPR (2015)
6. Zhang, Y., Zhou, D., Chen, S., Gao, S., Ma, Y.: Single-image crowd counting via multi-column convolutional neural network. In: CVPR (2016)

7. Viola, P., Jones, M.J., Snow, D.: Detecting pedestrians using patterns of motion and appearance. Int. J. Comput. Vis. **63**(2), 153–161 (2005)
8. Lin, Z., Davis, L.S.: Shape-based human detection and segmentation via hierarchical part-template matching. Pattern Anal. Mach. Intell. **32**(4), 604–618 (2010)
9. M. Wang and X. Wang. Automatic adaptation of a generic pedestrian detector to a specific traffic scene. In: CVPR, pp. 3401–3408. IEEE (2011)
10. Wu, B., Nevatia, R.: Detection of multiple, partially occluded humans in a single image by Bayesian combination of edgelet part detectors. In: ICCV, vol. 1, pp. 90–97. IEEE (2005)
11. Idrees, H., Soomro, K., Shah, M.: Detecting humans in dense crowds using locally-consistent scale prior and global occlusion reasoning. Pattern Anal. Mach. Intell. **37**(10), 1986–1998 (2005)
12. Zhao, T., Nevatia, R., Wu, B.: Segmentation and tracking of multiple humans in crowded environments. Pattern Anal. Mach. Intell. **30**(7), 1198–1211 (2008)
13. Li, M., Zhang, Z., Huang, K., Tan, T.: Estimating the number of people in crowded scenes by mid based foreground segmentation and head-shoulder detection. In: ICPR, pp. 1–4. IEEE (2008)
14. Ge, W., Collins, R.T.: Marked point processes for crowd counting. In: CVPR, pp. 2913–2920. IEEE (2009)
15. Brostow, G.J., Cipolla, R.: Unsupervised Bayesian detection of independent motion in crowds. In: CVPR, vol. 1, pp. 594–601. IEEE (2006)
16. Rabaud, V., Belongie, S.: Counting crowded moving objects. In: CVPR, vol. 1, pp. 705–711. IEEE (2006)
17. Chan, A.B., Liang, Z.-S.J., Vasconcelos, N.: Privacy preserving crowd monitoring: counting people without people models or tracking. In: CVPR, pp. 1–7. IEEE (2008)
18. Chen, K., Loy, C.C., Gong, S., Xiang, T.: Feature mining for localised crowd counting. In: BMVC, vol. 1, p. 3 (2012)
19. Chan, A.B., Vasconcelos, N.: Bayesian poisson regression for crowd counting. In: ICCV, pp. 545–551. IEEE (2009)
20. Ryan, D., Denman, S., Fookes, C., Sridharan, S.: Crowd counting using multiple local features. In: Digital Image Computing: Techniques and Applications, pp. 81–88. IEEE (2009)
21. Kong, D., Gray, D., Tao, H.: Counting pedestrians in crowds using viewpoint invariant training. In: BMVC, Citeseer (2005)
22. Liu, B., Vasconcelos, N.: Bayesian model adaptation for crowd counts. In: ICCV (2015)
23. Paragios, N., Ramesh, V.: A MRF-based approach for real-time subway monitoring. In: CVPR, vol. 1, p. I-1034. IEEE (2001)
24. Regazzoni, C.S., Tesei, A.: Distributed data fusion for real-time crowding estimation. Signal Process. **53**(1), 47–63 (1996)
25. Marana, A., Costa, L.d.F., Lotufo, R., Velastin, S.: On the efficacy of texture analysis for crowd monitoring. In: International Symposium on Computer Graphics, Image Processing, and Vision, pp. 354–361. IEEE (1998)
26. Idrees, H., Saleemi, I., Seibert, C., Shah, M.: Multi-source multi-scale counting in extremely dense crowd images. In: CVPR, pp. 2547–2554. IEEE (2013)
27. Bansal, A., Venkatesh, K.: People counting in high density crowds from still images. arXiv preprint arXiv: 1507.084452015
28. Tota, K., Idrees, H.: Counting in dense crowds using deep features
29. Li, W., Zhao, R., Xiao, T., Wang, X.: DeepReID: deep filter pairing neural network for person re-identification. In: CVPR (2014)

30. Zeng, X., Ouyang, W., Wang, X.: Multi-stage contextual deep learning for pedestrian detection. In: ICCV (2013)
31. Ouyang, W., Wang, X.: Joint deep learning for pedestrian detection. In: ICCV (2013)
32. Wang, N., Yeung, D.-Y.: Learning a deep compact image representation for visual tracking. In: NIPS (2013)
33. Jing, S., Kai, K., Chang, L.C., Xiaogang, W.: Deeply learned attributes for crowd scene understanding. In: CVPR (2015)
34. Kai, K., Xiaogang, W.: Fully convolutional neural networks for crowd segmentation. arXiv preprint arXiv:1411.4464 (2014)
35. Jia, Y., Shelhamer, E., Donahue, J., Karayev, S., Long, J., Girshick, R., Guadarrama, S., Darrell, T.: Caffe: convolutional architecture for fast feature embedding. arXiv preprint arXiv:1408.5093 (2014)
36. Long, J., Shelhamer, E., Darrell, T.: Fully convolutional models for semantic segmentation. In: CVPR (2015)
37. Zhang, C., Zhang, K., Li, H., Wang, X., Yang, X.: Data-driven crowd understanding: a baseline for a large-scale crowd dataset. IEEE Trans. Multimedia 18(6), 1048–1061 (2016)

Generation and Authoring of Augmented Reality Terrains Through Real-Time Analysis of Map Images

Theodore Panagiotopoulos, Gerasimos Arvanitis$^{(\boxtimes)}$, Konstantinos Moustakas, and Nikos Fakotakis

Electrical and Computer Engineering,
University of Patras, Patras, Greece
theopanag7@gmail.com,
{arvanitis,moustakas}@ece.upatras.gr, fakotaki@upatras.gr

Abstract. In this paper we present a novel method for real time 3D terrain creation and augmented reality rendering based on contour map images. Initially, terrain information is extracted from the contour images and a flat Delaunay triangulated terrain is generated. Then elevation information is added and remedying is performed so as to maintain smooth local surface representation. Then augmented reality rendering is performed in real time using the 2D contour map as a marker to manipulate the 3D terrain. The proposed framework also demonstrates potential editing applications like manual design of auxiliary information on the contour map, like roads, that can be automatically converted into 3D information and rendered on the 3D terrain, thus resulting in a more immersive and intuitive design experience.

Keywords: Augmented reality · Contour lines · Topographic maps

1 Introduction

The main augmented reality (AR) paradigm is related to the superimposition of digital-virtual information on top of the real environment, usually in real time. The virtual objects are traditionally rendered using devices like mobile phones or specialized augmented reality glasses. This powerful new technology brings out the components of the digital world (graphics) into a person's perceived real world by rendering them in a suitable device. Despite the significant advantages, there are still some limitations that are yet to be overcome. A challenging issue is the real time scene acquisition and scene generation.

Cartography is a traditional scientific field with numerous applications, like navigation systems and other contemporary applications. A topographic map is a map characterized by large-scale detail and quantitative representation of relief, usually using contour lines. The traditional way of reading contour lines in such a map is not an easy and intuitive task especially for people without experience.

© Springer International Publishing AG 2017
P. Sharma and F.M. Bianchi (Eds.): SCIA 2017, Part I, LNCS 10269, pp. 480–491, 2017.
DOI: 10.1007/978-3-319-59126-1_40

The proposed framework introduces a computationally efficient image-based approach for real time topographic map sensing and reconstruction. A 3D terrain is thus created and rendered into an augmented reality environment. The 3D representation of topographic maps is more intuitive and engaging for the users. Moreover, it can help users, like civil and surveying engineers or students, to better understand the topography of contour lines, thus increasing perception and immersion.

1.1 Related Work

A method for extraction of contour lines from paper-based topographic maps is presented in [1]. In [2] authors propose an automatic approach to reconstruct gaps in contour lines. This parameterless reconstruction scheme is based on the extrapolation of the gradient orientation field from the available pieces of thinned contours. In [3] an application is presented using camera and HMD for presenting augmented maps. In [4] an application is presented that allows the generation of virtual terrains interactively, using augmented reality markers. This application also allows the user to navigate in the generated virtual environment. [5] describes an educational Augmented Reality application for mobile phones that supports landscape architecture students who learn how to read and analyze contour maps. AR Sandbox [6] is an application that allows users to create topography models by shaping real sand, which is then augmented in real time by an elevation color map, topographic contour lines, and simulated water. The system teaches geographic, geologic, and hydrologic concepts such as how to read a topography map and the meaning of contour lines. In [7] authors developed a browsing tool for visualizing information about geographic surfaces using map-based augmented reality.

In most cases, the terrain has already been created and then is superimposed in the real environment as dictated by the position and orientation of the marker. On contrary the proposed scheme makes dual use of a contour map. First of all it extracts the 3D terrain information dynamically and secondly uses the map itself as a marker to superimpose the 3D terrain.

1.2 Contribution

The contributions of our work are outlined below:

- Fast image processing and rendering for converting the information of a printed map into the corresponding 3D terrain.
- The created 3D objects are rendered into user's mobile device using as marker the topographic map itself.
- An example of dynamic interaction between users and the application is show-cased, allowing the users to design roads on the 2D maps that are automatically converted and superimposed on the 3D terrain.

The paper is organized as follows: Sect. 2 presents a brief overview of the proposed system. In Sect. 3 we describe the algorithms and the basic steps that we follow for the image processing analysis. In Sect. 4 we analyze the way that the 3D objects are created. Moreover, we show how the 3D objects are displayed on user's device and we also present the way for user-application interaction by drawing roads in real time. In Sect. 5 we provide experimental evaluation of the proposed framework, while in Sect. 6 conclusions, limitations and future work is discussed.

2   Framework Overview

The proposed system accomplishes two main operations, the creation of 3D models and the 3D object rendering. For the 3D model creation we use image processing techniques that help us to classify the contour polygons of the topographic map and based on this classification we create the 3D object by performing iterative Delaunnay triangulation. The 3D model creation is the most computational consuming process, thus we use a server for the processing instead of a smartphone. In Fig. 1 we show the architecture scheme of our method. Details and specifics on the building blocks of the proposed framework are presented in the next sections.

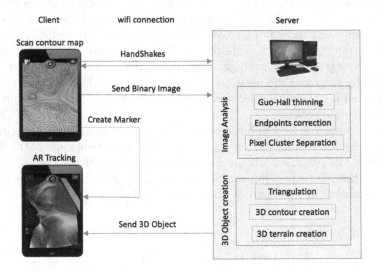

Fig. 1. Architecture scheme of our proposed method

3   Sensing and Image Processing

The sensing and image processing module initially converts the map image in binary form, then performs skeletonization that is followed by denoising so as to finally allow for smooth extraction of the contour lines.

Binary Image. For the creation of binary image we apply an adaptive threshold filter as described in Eq. (1).

$$I_b(i) = \begin{cases} 1 \ if \ A(i) < \frac{\sum_{n=1}^{|N_i|} A(N_i(n)) - \delta}{|N_i|} \\ 0 \qquad \qquad otherwise \end{cases} \quad \forall \, i = 1, L \qquad (1)$$

where I_b represents the binary image, $A(i)$ is the intensity of i pixel, $|N_i|$ is the number of i's neighbors, $N_i(n)$ is the n neighbor of i pixel, L is image length and δ is a small positive value. The adaptive threshold filter has satisfying results even in cases with bad light conditions, as we can see in Fig. 2.

(a)   (b)

Fig. 2. (a) Bad light condition for a grayscale topographic map, (b) binary image created by adaptive threshold filter

Skeletonization and Edges Correction. The binary image is skeletonized using the Guo-Hall thinning algorithm [8] so all lines have the same thickness and we can continue with the classification. An effective contour map model requires connected lines (polygons). However, there are cases where polygons are not continuous. We handle this issue by using the following very simple three step algorithm: (a) neighbors pixels are classified into the same cluster, (b) clusters with low number of pixels are considered as noise and are removed, (c) lines with common points are classified into the same Pixel Cluster, even if they are different polygons (we deal with this limitation in the next step). There are two types of pixels, those that have two or more neighbors and those that have only one (edges) Fig. 3(a). Our goal is to connect all edges with others of the same Pixel Cluster until no edges exist. The parameters that we take into account for applying an efficient connection are: The distance d between two edges and the angle of edge's vectors $\theta_w = 180° - \hat{\theta}_{w_1 w_2}$. We assume that if the extended directions of two edges are connected then it is more likely that they are connected with each other. In the general case we connect edges that maximize the following equation:

$$S = \frac{1}{d \cdot \theta_w} \qquad (2)$$

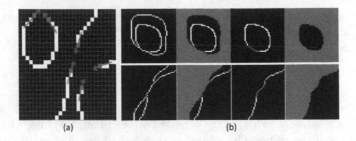

Fig. 3. (a) Example of pixels with only one neighbors (edges) (b) steps for Pixel Cluster classification

Pixel Cluster Separation. As we mentioned earlier, there are cases where two or more different polygons have common pixels as the result of being incorrectly classified into the same Pixel Cluster. Figure 3(b) shows the steps that we follow in order to handle these situations. We start the separation from the outwards polygons. We classify all the pixels into the same class and remove them. Inevitably, lines with missing pixels are created but we complete them and continue with the same procedure until all polygons are separated and classified. At the end we have managed to ensure that each one of the Pixel Clusters is a continuous line that can easily be recognized as a different poly-line.

4   Geometry Processing and Rendering

A 3D model can be represented as an indexed face set $M = (V, F)$, being composed of vertices (V) and the indexed faces (F). Each vertex can be represented as a point in space $v_i = (x_i, y_i, z_i) \ \forall \ i = 1, k$. In this case we create a vector of vertices $\mathbf{v} = [\ \mathbf{x}, \ \mathbf{y}, \ \mathbf{z} \ ]$ in a 3D coordinate space such as $\mathbf{x}, \mathbf{y}, \mathbf{z} \in \Re^{k \times 1}$ and $\mathbf{v} \in \Re^{k \times 3}$. This means that we have a set of k points such that $V = \{v_1, \ v_2, \ ... \ v_k\}$. Additionally, each face is represented as a set of 3 vertices $f_i = [v_{i1}, \ v_{i2}, \ v_{i3}] \ \forall \ i = 1, m$ where $m > k$, so we have m faces $F = \{f_1, \ f_2, \ ... \ f_m\}$.

3D Terrain Model Creation. The Pixel Clusters classification step helps us to recognize the different areas with different heights as shown in Fig. 4(a). We colorized each Pixel Cluster with different color in order to be separated from each other. The sea level (outer level) area is represented always by black color Fig. 4(b). We define each area's relative height H in comparison with the sea level area. We start by setting the black area with value H = 0 and subsequently, for each other area with common border the H value increases by one. We continue the procedure using the same logic and at the end all areas will have been defined.

The previous procedure is a vital step, however, additional information is required for the 3D terrain creation. Indeed, we need to assign each pixel i of the image (polygons and internal pixels) with a height value $h_i \ \forall \ i = 1, L$. For

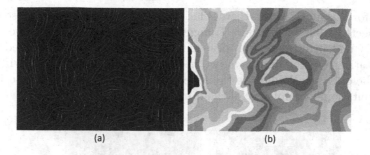

(a)   (b)

Fig. 4. (a) Separated pixel clusters, (b) separated cluster areas (Color figure online)

finding this value, we use linear interpolation and we estimate the weighted average of the distance d between pixel and its relative polygons. More specifically, assuming that pixel i lies between polygons $c1$ and $c2$, with relatives heights H_{c1} and H_{c2} correspondingly, then its height value is estimated according to:

$$h_i = \frac{d_{c1} \cdot H_{c1} + d_{c2} \cdot H_{c2}}{d_{c1} + d_{c2}} \tag{3}$$

where $d_c = \sqrt{(x_i - x_c)^2 + (y_i - y_c)^2}$. However, in cases where the area does not contain other polygons (peaks or valleys) we need to create an auxiliary edge-line that represents a higher or lower height. After this adjustment we can use the linear interpolation as previously. In order to create this auxiliary edge-line we skeletonize the area using the Guo-Hall thinning algorithm. Depending on the case, the edge-line takes value $+0.5$ if the area is a peak or -0.5 if the area is a valley. In Fig. 5 we can see an example of an auxiliary edge-line creation for an area without other enclosed polygons.

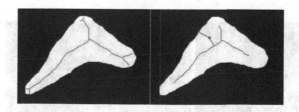

Fig. 5. Polygon's skeletonization

At the end we assign each pixel to a corresponding color based on its height value. The colormap that we use is presented in Fig. 6(c).

Triangulation. A 3D object is represented by a number of vertices that create triangle faces when they are connected with each other. We have knowledge about the position of the vertices based on the position of the pixels and the height information. However, there is lack of knowledge about how these vertices are connected with each other. Triangulation is the solution but we need to go further and investigate the ideal number of points that is needed for the representation. On the one hand, we need a lot of points for preserving detail but on the other we should use the fewest possible points so as not to increase the computational complexity of the framework.

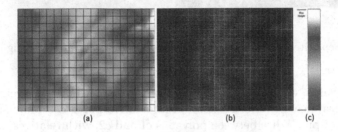

Fig. 6. (a) An example of a grid triangulation where each square is divided into two triangles, (b) a satisfying grid, (c) colormap (Color figure online)

In the context of the proposed framework we used grid triangulation. Figure 6(a) shows a simple example of a grid triangulation and Fig. 6(b) shows a satisfying grid triangulation that can be used in our case. The only thing that we need to consider is the need for more accuracy in situations where a steep slope exists, but using this type of triangulation this requirement is satisfied.

3D Contour Model Creation. Before the triangulation a process of sampling is necessary. In Fig. 7(a) we can see how the sampling algorithm is applied to a polygon, while in Fig. 7(b) we can see a sampling example.

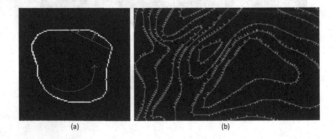

Fig. 7. Example of sampling operation for contour polygons

The sampling algorithm starts by randomly choosing a pixel p_r of a polygon and a straight line is created $l_{p_r p_{r+1}}$ by connecting it with its first counterclockwise neighbor pixel p_{r+1}. The Algorithm 1 describes how it is used in the general case.

```
polygon = 1;
while polygon < polygon.length do
    λ = 1;
    while [Unsearching pixel exists] do
        for i ← 0 to λ do
            estimate l_{p_r p_{r+λ}}
            if max(p_{r+i} ⊥ l_{p_r p_{r+λ}}) > O then
                r = r + λ - 1;                    // set a new initial pixel
                λ = 1;
            else
                λ = λ + 1;      // continue with the next counterclockwise
                    neighbor pixel
            end
        end
    end
    polygon = polygon + 1;
end
```

Algorithm 1. Sampling each polygon

We can change the sampling results by changing the threshold O. The experiments show that sampling by using a very low value of O we can decrease the number of vertices ~88% without significantly decreasing detail. For the triangulation we use the Ear Clipping algorithm as it is described in [9].

4.1 Rendering and Augmented Reality Interaction

At every augmented reality application a special marker is needed that represents the specific area in which the virtual object will appear. This marker must be easily recognizable, meaning that it must has high frequency features like edges. There are two different solutions known as AR marker-based (with target) and AR markerless tracking (without target) [10]. In our case the contour map itself can be used as a marker. For the 3D object presentation [11] we use a transformation $\mathbf{T} = \mathbf{v} \; x \; \mathbf{Q} \; x \; \mathbf{P}$ in order to estimate the points $\mathbf{T} = \{u, m\}$ that are displayed in the screen. where $\mathbf{Q}$ is the Model View Matrix and $\mathbf{P}$ is the Projection Matrix.

4.2 Intuitive, Interactive Authoring

The developed application provides tools that allow users add information with two different ways: (a) by selecting points (b) by drawing lines. In the first case we just have to create a graph and subsequently to present it in the 3D model. In the second case we need to follow the same approach but an image processing step takes initially place.

Create Graph. For the creation of the graph we need to assign each of the selected pixel to the ideal representative faces. Firstly, each one of the selected pixel is converted to a 3D ray. Subsequently, we apply a ray casting algorithm and we choose the triangle faces with the shortest distance.

Image Processing for Drawn Line. Firstly, we separate the colored road from the rest lines. We transform the RGB image to HSV, making the color separation easier. This step gives us a binary image with very good appearance but there are some missing areas Fig. 8 (color recognition). To cover the imperfections we apply a dilation-erosion filter that achieves the creation of a continuous line (road). Next, we apply the Guo-Hall thinning algorithm in order to skeletonize the road. The last step is the creation of a graph from the skeletonized road. For this purpose some extra steps are required: (a) **Extra thinning.** Although removing only ~1 − 2% of pixels, this procedure is vital for our purpose (b) **Crossroads Finding.** We assume that if any pixel has 3 neighbors then it is crossroad (c) **Remove small areas.** If the length of lines between two endpoints is shorter than a threshold then it is removed. (d) **Road sampling.** The same sampling procedure is used as in polygon sampling case, (e) **Graph creation.** After that we have the necessary points for the graph's creation.

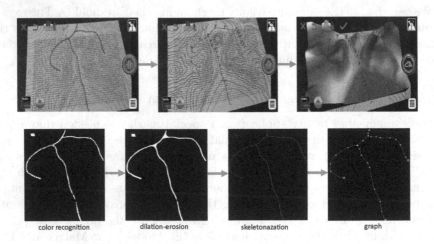

Fig. 8. Steps for real time 3D road creation and rendering in the 3D models by a drawn line. (Color figure online)

5   Experiments and Results

5.1   Timings and Complexity Analysis

Having a stable wifi connection, the time for sending (binary image) and receiving (two 3D models) data between smartphone and server is very fast ~1 s. More time consuming operations, like image processing techniques and 3D model creation, take place in the server exclusively and take 2–5 s to be completed.

The most computational consuming step is the skeletonization algorithm (Thinning) that can correspond to the 20% of the total time. A typical smartphone is capable to handle the rendering of 3D models with approximately 20.000 faces, in augmented reality at interactive rates.

5.2 Examples

We have examined several scenarios using different conditions in each case. Below we present the results of three different experiments representing the most common cases appeared in real scenarios. In the first experiment a significant amount of noise is present, due to bad illumination conditions, Fig. 9. In the second experiment, we have different light conditions, we use a little different color of ink and the paper of the map is a little tearing, as shown in Fig. 10. The third experiment has better light conditions however the contour lines are very closed each other, as illustrated in Fig. 11. The experimental results show that the proposed approach is capable to efficiently manage different case studies overcoming problems caused by bad illumination, different color of ink or illegible lines.

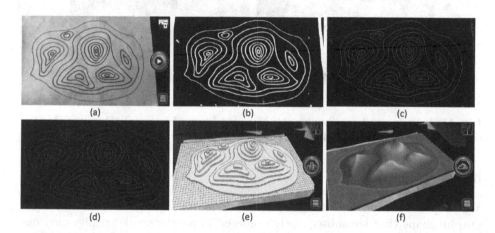

Fig. 9. Experiment 1. (a) Topographic Map, (b) Binary Image, (c) Skeletonized map, (d) Contour Clusters, (e) 3D contour Model, (f) 3D terrain Model

5.3 Hardware Setup and Libraries

We need to manage both client (android smartphone) and server (computer) operations. For the server, the code for the image processing and for 3D objects creation is written in C++. For the client, we use java scripts for the connection with the server and for 3D object presentation and display. The external libraries that are used, are: Vuforia (Augmented Reality), OpenCV (Computer Vision) for image processing, PolyPartition (ear-clipping) for triangulation.

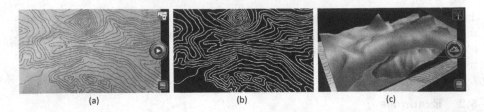

Fig. 10. Experiment 2. (a) Topographic Map, (b) Binary Image, (c) 3D terrain Model (Color figure online)

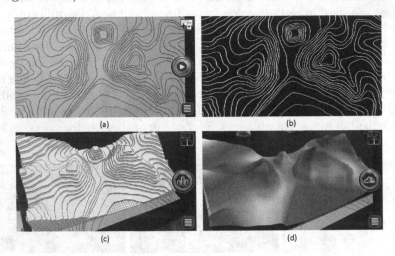

Fig. 11. Experiment 3. (a) Topographic Map, (b) Binary Image, (c) 3D contour Model, (d) 3D terrain Model (Color figure online)

6   Conclusions

We proposed a framework for the reconstruction of 3D terrains based on topographic maps that are subsequently rendered in an augmented reality environment. Preliminary application scenarios demonstrate the potential of the approach for the intuitive and interactive investigation/editing of contour maps.

6.1   Limitations

Despite the fact that the proposed method operates reliably in most of the cases, there are some limitations that may affect the final results, such as: (a) It needs a good resolution camera. (b) We use topographic maps without extra symbols or colors just only simple contour lines.

6.2   Future Work

In a future edition of our work we will try to overcome some of the limitations that we mentioned before and additional to add more new features. Some futures extensions of our work will be: (i) Digit recognition of the contour map, for an automated height estimation. (ii) Simplified contour maps by removing extra symbols. (iii) Adding more user interaction tools in order to support manual creation and handling 3D object structures (e.g. bridges).

References

1. Xin, D., Zhou, X., Zheng, H.: Contour line extraction from paper-based topographic maps. J. Inf. Comput. Sci. **1**(5), 275–283 (2006). ISSN 1746-7659, England, UK
2. Pouderoux, J., Spinello, S.: Global contour lines reconstruction in topographic maps, pp. 779–783, January 2007
3. Bobrich, J., Otto, S.: Augmented maps. In: Symposium on Geospatial Theory, Processing and Applications, Ottawa (2002)
4. Dembogurski, R., Sad, D.O., de Souza Filho, J.L., Silva, R., Vieira, M.B., Dembogurski, B.: Interactive virtual terrain generation using augmented reality markers. SBC J. 3D Interact. Syst. **3**(3), 29–36 (2012)
5. Schroth, O., Zhang, C.: Augmented landform an educational augmented reality tool for landscape architecture students. In: Peer Reviewed Proceedings of Digital Landscape Architecture, ETH Zurich (2014)
6. Augmented Reality Sandbox. http://idav.ucdavis.edu/~okreylos/ResDev/SARndbox/
7. Asai, K., Kondo, T., Kobayashi, H., Mizuki, A.: A geographic surface browsing tool using map-based augmented reality. In: 2008 International Conference Visualisation, London, pp. 93–98 (2008)
8. Guo, Z., Hall, R.W.: Fast fully parallel thinning algorithms. J. CVGIP Image Understand. **55**(3), 317–328 (1992)
9. Eberly, D.: Triangulation by ear clipping. In: Geometric Tools, LLC, November 2002
10. Wagner, D.: Handheld Augmented Reality, Graz, Austria (2007)
11. Ginsburg, D.: OpenGL ES 3.0 Programming Guide, 2nd edn. Addison-Wesley Professional, New York

Solution of Pure Scattering Radiation Transport Equation (RTE) Using Finite Difference Method (FDM)

Hassan A. Khawaja[✉]

Department of Engineering and Safety (IIS-IVT),
UiT The Arctic University of Norway, 9037 Tromsø, Norway
hassan.a.khawaja@uit.no

Abstract. Radiative transfer is the physical phenomenon of energy transfer in the form of electromagnetic radiation. The propagation of radiation through a medium is affected by absorption, emission, and scattering. Radiative Transfer Equation (RTE) have been applied in a many subjects including optics, astrophysics, atmospheric science, remote sensing, etc. Analytic solutions for RTE exist for simple cases, but, for more realistic media with complex multiple scattering effects, numerical methods are required. In the RTE, six different independent variables define the radiance at any spatial and temporal point. By making appropriate assumptions about the behavior of photons in a scattering medium, the number of independent variables can be reduced. These assumptions lead to the diffusion theory (or diffusion equation) for photon transport. In this work, the diffusive form of RTE is discretized, using a Forward-Time Central-Space (FTCS) Finite Difference Method (FDM). The results reveal the radiance penetration according to Beer-Lambert law.

Keywords: Radiation Transport Modelling (RTM) · Radiation Transport Equation (RTE) · Finite Difference Method (FDM) · Forward-Time Central-Space (FTCS)

1 Introduction

Radiative transfer is the physical phenomenon of energy transfer in the form of electromagnetic radiation. Radiative transfer has applications in a wide variety of subjects including optics, astrophysics, oceanography, atmospheric science, remote sensing, infra-red imaging, etc. [3,11,16].

The propagation of radiation through a medium is affected by absorption, emission, and scattering processes [3]. Absorption is the process by which the energy of the photons is transferred to particles such as electrons present in its transmission medium. The transfer of momentum raises the localized kinetic energy of the electrons and hence the temperature. A common daily life example

© Springer International Publishing AG 2017
P. Sharma and F.M. Bianchi (Eds.): SCIA 2017, Part I, LNCS 10269, pp. 492–501, 2017.
DOI: 10.1007/978-3-319-59126-1_41

of absorption is that on a sunny day we can feel the warmth of the sun, even if the surrounding temperature is quite low.

Emission is the process through which radiation transfers in the form of waves/particles. Emissions may originate from a point source, such as a bulb filament or a spark, or from a surface such as an ionized tube, neon bulb, etc.

Scattering is the deviation of the radiation waves/particles from their original path. Scattering occurs as a result of particle-particle collisions. It can be categorized as specular, such as in the case of mirror or diffused reflections [2].

Radiative transfer can be expressed mathematically in the form of radiation transport equation [4]. The equation poses a challenge for those wishing to obtain a definite solution considering its partial differential nature. Therefore, various studies have employed different methodologies. For example, some researchers have used the Monte-Carlo method to solve the radiation transport models [6]. Similarly, other numerical techniques such as the discrete-ordinate method have also been employed to find the solution [8].

This paper focuses on solving the pure scattering radiation transport equation using the finite difference method. This methodology has previously been used to solve the heat equation and simulate an infra-red signature [9,10,14].

2   Radiation Transport Equation

The equation of radiative transfer [13,17] is given mathematically as shown in Eq. (1),

$$\frac{\partial I_v\left(\mathbf{r}, \hat{\mathbf{n}}, t\right)}{c\, \partial t} + \hat{\Omega} \cdot \nabla I_v\left(\mathbf{r}, \hat{\mathbf{n}}, t\right) + \left(k_{v,s} + k_{v,a}\right) I_v\left(\mathbf{r}, \hat{\mathbf{n}}, t\right)$$
$$= j_v\left(\mathbf{r}, t\right) + \frac{1}{4\pi} k_{v,s} \int_{\Omega} I_v\left(\mathbf{r}, \hat{\mathbf{n}}, t\right) d\Omega \quad (1)$$

where I_v the is spectral radiance of electromagnetic waves, c is the speed of light, $\hat{\Omega}$ is the vectorial position of a solid angle (polar angle θ and azimuthal angle φ), Ω is a solid angle, $k_{v,s}$ is the scattering opacity of the medium, $k_{v,a}$ is the absorption opacity of the medium, j_v is the emission coefficient of the medium and t is the time variable.

The energy contents of electromagnetic waves can be calculated as shown in Eq. (2),

$$dE_v = I_v\left(\mathbf{r}, \hat{\mathbf{n}}, t\right) \cos(\theta)\, dv\, dA\, d\Omega\, dt \quad (2)$$

where E_v is the radiation energy, θ is an angle that the unit vector $\hat{\mathbf{n}}$ makes with the normal of elemental area dA, positioned at $\mathbf{r}$, and v is the frequency. This is illustrated in Fig. 1.

Similarly, radiance intensity and energy flux can be written as shown in Eqs. (3) and (4):

$$\phi_v\left(\mathbf{r}, t\right) = \int_{4\pi} I_v\left(\mathbf{r}, \hat{\mathbf{n}}, t\right) d\Omega \quad (3)$$

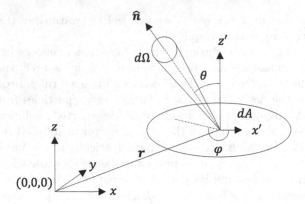

Fig. 1. Radiance energy originating from differential area dA at vectorial position $\mathbf{r}$ within a solid angle $d\Omega$.

$$\psi_v\left(\mathbf{r}, t\right) = \int_{4\pi} \hat{\mathbf{n}}\, I_v\left(\mathbf{r}, \hat{\mathbf{n}}, t\right)\, d\Omega \qquad (4)$$

where ϕ_v and ψ_v are radiance intensity and energy flux respectively. Each of these terms has units of W/m^2.

Analytical solutions to the radiative transfer equation (RTE) exist for simple cases, but, for more realistic mediums and complex multiple scattering effects, computer simulations are required.

3   Methodology

There are six independent variables defining the radiance at any spatial and temporal point in the radiation transport equation. These variables are the x, y, z coordinates from a reference presented in the form of vector $\mathbf{r}$, two dimensional vectorial position of a solid angle $\hat{\Omega}$ (polar angle θ and azimuthal angle φ) and time variable t. By making appropriate assumptions about the behavior of the photons in the scattering medium, the number of independent variables can be reduced. These assumptions lead to the diffusion theory (and diffusion equation) for photon transport. Two assumptions, which permit the application of diffusion theory are:

1. Relative to scattering events, there are very few absorption events. Likewise, after numerous scattering events, few absorption events will occur, and the radiance will become nearly isotropic. This assumption is sometimes called directional broadening.
2. In a primarily scattering medium, the time for substantial current density change is much longer than the time to traverse one transport mean free path. Thus, over one transport mean free path, the fractional change in current density is much less than unity. This property is sometimes called temporal broadening.

It should be noted that both of these assumptions require a high-albedo (pre-dominantly scattering) medium. The diffusion approximation is limited to systems where reduced scattering coefficients are much larger than their absorption coefficients and have a minimum layer thickness of the order of a few transport mean free paths.

From the diffusion approximation, we can write Eq. (5),

$$I_v(\mathbf{r}, \hat{n}, t) = \frac{1}{4\pi}\phi_v(\mathbf{r}, t) + \frac{3}{4\pi}\psi_v(\mathbf{r}, t).\hat{n} \tag{5}$$

by substituting Eq. (5) in Eq. (1), we get Eq. (6),

$$\frac{\partial\phi_v(\mathbf{r}, t)}{c\,\partial t} + k_{v,a}\phi_v(\mathbf{r}, t) - \nabla.\psi_v(\mathbf{r}, t) = j_v(\mathbf{r}, t) \tag{6}$$

by applying Fick's Law [7], we get Eq. (7),

$$\psi_v(\mathbf{r}, t) = -\frac{\nabla\phi_v(\mathbf{r}, t)}{3\left((1 - g)\,k_{v,s} + k_{v,a}\right)} = -D\nabla\phi_v(\mathbf{r}, t) \tag{7}$$

where g is the anisotropy of the medium and D is the diffusion coefficient.

By substituting Eq. (7) in Eq. (6), we get Eq. (8),

$$\frac{\partial\phi_v(\mathbf{r}, t)}{c\,\partial t} + k_{v,a}\phi_v(\mathbf{r}, t) - D\nabla^2\phi_v(\mathbf{r}, t) = j_v(\mathbf{r}, t) \tag{8}$$

Assuming hypothetically that there are zero absorption and zero emission (pure scattering medium), we can simplify Eq. (8) to Eq. (9),

$$\frac{\partial\phi_v(\mathbf{r}, t)}{c\,\partial t} = D\nabla^2\phi_v(\mathbf{r}, t) = D\left(\frac{\partial^2\phi_v(\mathbf{r}, t)}{\partial x^2} + \frac{\partial^2\phi_v(\mathbf{r}, t)}{\partial y^2} + \frac{\partial^2\phi_v(\mathbf{r}, t)}{\partial z^2}\right) \tag{9}$$

where x, y, z are the space dimensions. We can reduce the dimensions to a hypothetical two-dimensional space. This will further simplify the equation as shown in Eq. (10),

$$\frac{\partial\phi_v(\mathbf{r}, t)}{c\,\partial t} = D\left(\frac{\partial^2\phi_v(\mathbf{r}, t)}{\partial x^2} + \frac{\partial^2\phi_v(\mathbf{r}, t)}{\partial y^2}\right) \tag{10}$$

In order to do so, we have to discretize the equation. In this work, we will discretize the equation using a Forward-Time Central-Space (FTCS) Finite Difference Method (FDM) [1,10,12,14]. This results in Eq. (11),

$$
\begin{aligned}
\partial\phi_v(\mathbf{r}, t)_{i,j}^{t+1} = {}& \partial\phi_v(\mathbf{r}, t)_{i,j}^{t} \\
&+ c\,D\frac{\left(\partial\phi_v(\mathbf{r}, t)_{i+1,j}^{t} - 2\partial\phi_v(\mathbf{r}, t)_{i,j}^{t} + \partial\phi_v(\mathbf{r}, t)_{i-1,j}^{t}\right)}{(\Delta x)^2}\Delta t \\
&+ c\,D\frac{\left(\partial\phi_v(\mathbf{r}, t)_{i,j+1}^{t} - 2\partial\phi_v(\mathbf{r}, t)_{i,j}^{t} + \partial\phi_v(\mathbf{r}, t)_{i,j-1}^{t}\right)}{(\Delta y)^2}\Delta t
\end{aligned} \tag{11}
$$

where subscripts i, j are integers representing the computational points in the two-dimensional space domain and the superscript represents the transient state.

A finite difference method (FDM) is a numerical method for solving differential equations such as that in Eq. (10). This method approximates the differentials with differences by discretizing the dependent variable (radiance intensity) in the independent variable domains (space and time). Each discretized value of the dependent variable is referred to as a nodal value.

Equation (11) is solved in a two-dimensional spatial domain, as shown in Fig. 2. The two-dimensional space is discretized in equally spaced quadrilaterals. Each quadrilateral is referenced in two-dimensional space via indices i and j. Indices i and j refer to positions on the horizontal and vertical axes, respectively. The n and m refer to the maximum value of the i and j indices, respectively.

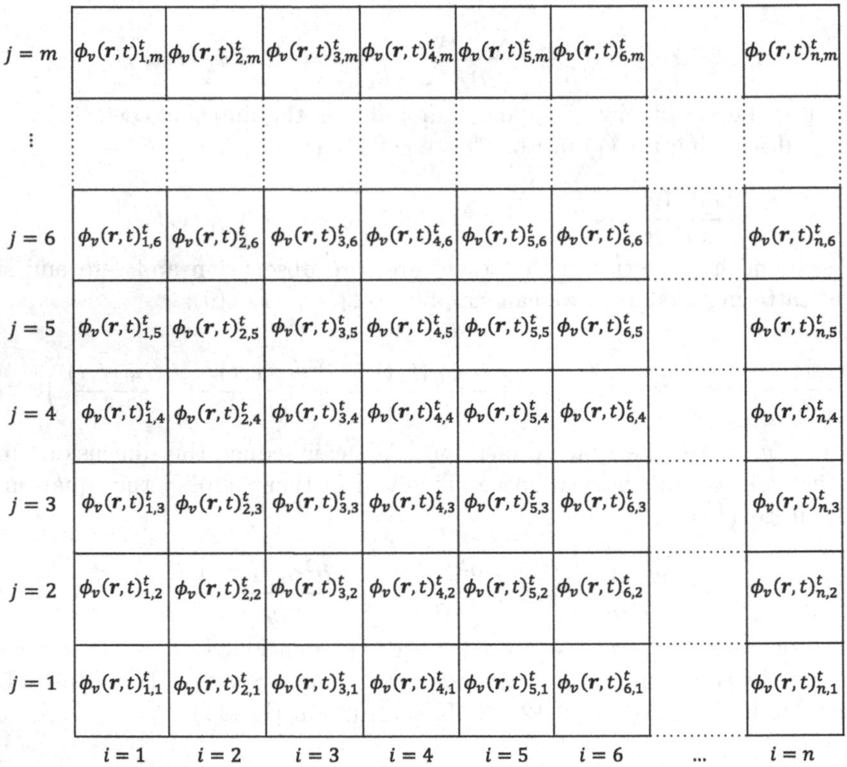

$j = m$	$\phi_v(r,t)^t_{1,m}$	$\phi_v(r,t)^t_{2,m}$	$\phi_v(r,t)^t_{3,m}$	$\phi_v(r,t)^t_{4,m}$	$\phi_v(r,t)^t_{5,m}$	$\phi_v(r,t)^t_{6,m}$		$\phi_v(r,t)^t_{n,m}$
$\vdots$								
$j = 6$	$\phi_v(r,t)^t_{1,6}$	$\phi_v(r,t)^t_{2,6}$	$\phi_v(r,t)^t_{3,6}$	$\phi_v(r,t)^t_{4,6}$	$\phi_v(r,t)^t_{5,6}$	$\phi_v(r,t)^t_{6,6}$		$\phi_v(r,t)^t_{n,6}$
$j = 5$	$\phi_v(r,t)^t_{1,5}$	$\phi_v(r,t)^t_{2,5}$	$\phi_v(r,t)^t_{3,5}$	$\phi_v(r,t)^t_{4,5}$	$\phi_v(r,t)^t_{5,5}$	$\phi_v(r,t)^t_{6,5}$		$\phi_v(r,t)^t_{n,5}$
$j = 4$	$\phi_v(r,t)^t_{1,4}$	$\phi_v(r,t)^t_{2,4}$	$\phi_v(r,t)^t_{3,4}$	$\phi_v(r,t)^t_{4,4}$	$\phi_v(r,t)^t_{5,4}$	$\phi_v(r,t)^t_{6,4}$		$\phi_v(r,t)^t_{n,4}$
$j = 3$	$\phi_v(r,t)^t_{1,3}$	$\phi_v(r,t)^t_{2,3}$	$\phi_v(r,t)^t_{3,3}$	$\phi_v(r,t)^t_{4,3}$	$\phi_v(r,t)^t_{5,3}$	$\phi_v(r,t)^t_{6,3}$		$\phi_v(r,t)^t_{n,3}$
$j = 2$	$\phi_v(r,t)^t_{1,2}$	$\phi_v(r,t)^t_{2,2}$	$\phi_v(r,t)^t_{3,2}$	$\phi_v(r,t)^t_{4,2}$	$\phi_v(r,t)^t_{5,2}$	$\phi_v(r,t)^t_{6,2}$		$\phi_v(r,t)^t_{n,2}$
$j = 1$	$\phi_v(r,t)^t_{1,1}$	$\phi_v(r,t)^t_{2,1}$	$\phi_v(r,t)^t_{3,1}$	$\phi_v(r,t)^t_{4,1}$	$\phi_v(r,t)^t_{5,1}$	$\phi_v(r,t)^t_{6,1}$		$\phi_v(r,t)^t_{n,1}$
	$i = 1$	$i = 2$	$i = 3$	$i = 4$	$i = 5$	$i = 6$	...	$i = n$

Fig. 2. Radiance intensities in the two-dimensional discretized domain. Indices i and j refer to the nodal position of radiance intensity.

For the stability and accuracy of the FDM, it is vital to choose the correct time step value. In this work, the Courant-Friedrichs-Lewy (CFL) condition [5,12] is used to decide the time step size. The CFL condition is given in Eq. (12),

$$2Dc\Delta t \ \leq \ min\left((\Delta x)^2 , (\Delta y)^2\right) \tag{12}$$

where D is the diffusion coefficient (m^2/s), c is the speed of light (m/s), Δt is the time step size (s), and Δx and Δy are the differences in the spatial positions of the nodes (m).

In addition, initial and boundary conditions are required. In this case, boundary conditions were specified such that they resembles to an infinite space. The source radiance was introduced on a small part of the boundary. The flow chart of the method of solution is given in Fig. 3, while the values of the constants are presented in Table 1.

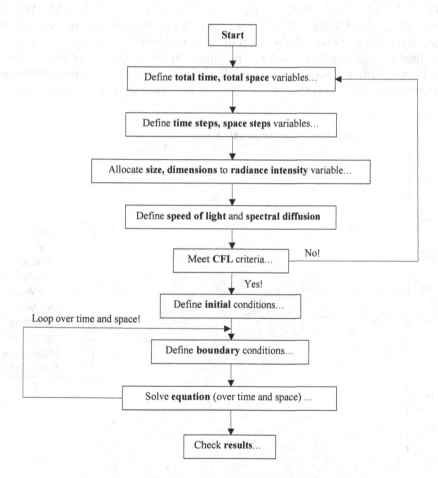

Fig. 3. Flow chart of the method of solution.

Table 1. Values of constants

Constant	Value	Units
D (diffusion coefficient)	0.282×10^{-4}	m^2/s
$\Delta x \times \Delta y$ (size of nodal quadrilateral)	1×1	m
c (speed of light)	3.0×10^8	m/s

4   Results and Discussion

Ten cases are set with inlet radiance intensities of $10\,W/m^2$, $50\,W/m^2$, $100\,W/m^2$, $500\,W/m^2$, $1000\,W/m^2$, $5000\,W/m^2$, $10000\,W/m^2$, $50000\,W/m^2$, $100000\,W/m^2$, and $500000\,W/m^2$. The main reason for this diverse range is to understand the response of the medium and to identify its limiting behavior. It is vital to highlight here that the medium's spectral diffusivity plays a vital role. Given study is limited to only highly diffusive mediums. Figure 4 shows the radiance intensity penetration with different inlet radiance intensities.

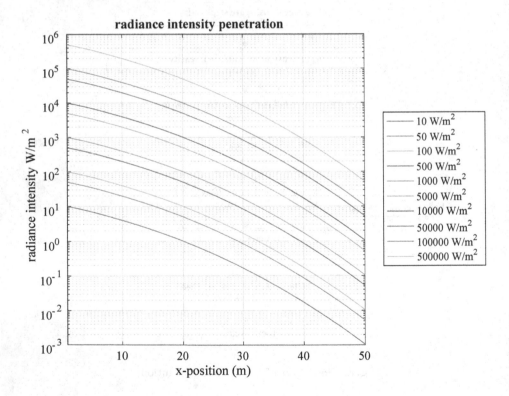

Fig. 4. Radiance intensity penetration

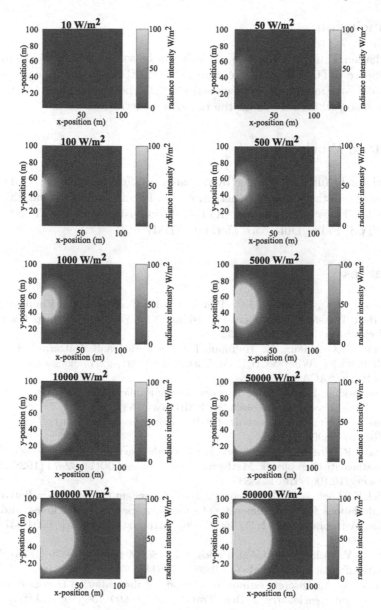

Fig. 5. Contour plots with various radiance intensities.

Contour plots with various inlet radiance intensities are shown in Fig. 5. Lighten regions show the radiance intensity of $100\,\text{W/m}^2$ and above in a two-dimensional space. As expected, the results clearly demonstrate that the radiance penetration has increased with the increased value of the inlet source radiance. This behavior is also in accordance with the Beer-Lambert Law [15].

5 Conclusion

The Radiation Transport Equation (RTE) can be solved using a Forward-Time Central-Space (FTCS) Finite Difference Method (FDM) in special cases such as a highly diffusive mediums (as discussed above). The results show the radiance intensity space, which reflects on the radiance penetration.

6 Future Work

This work can be further extended by varying the diffusion coefficient and observing its impact on the radiance penetration. It will help to find the validity limits of Radiation Transport Equation (RTE) solution using Forward-Time Central-Space (FTCS) Finite Difference Method (FDM).

References

1. Anderson, D.A., Tannehill, J.C., Pletcher, R.H.: Computational Fluid Mechanics and Heat Transfer. Hemisphere Publishing, New York (1984). http://www.osti.gov/scitech/servlets/purl/5012735
2. Bohren, C., Clothiaux, E., Huffman, D.: Absorption and Scattering of Light by Small Particles. Wiley-VCH, New York (2009). https://books.google.no/books?id=peHRPwAACAAJ
3. Chandrasekhar, S.: Radiative Transfer. Dover Publications, New York (1960). https://books.google.no/books?id=CK3HDRwCT5YC
4. Collins, G.W.: The fundamentals of stellar astrophysics (2003). http://ads.harvard.edu/books/1989fsa..book/
5. Courant, R., Friedrichs, K., Lewy, H.: Über die partiellen differenzengleichungen der mathematischen physik. Mathematische Annalen 100(1), 32–74 (1928). http://dx.doi.org/10.1007/BF01448839
6. Daniel, K., Thomas, R.C., Nugent, P.: Time-dependent monte carlo radiative transfer calculations for three-dimensional supernova spectra, light curves, and polarization. Astrophys. J. 651(1), 366 (2006). http://stacks.iop.org/0004-637X/651/i=1/a=366
7. Fick, A.: V. on liquid diffusion. Philos. Mag. Ser. 4 10(63), 30–39 (1855). http://www.tandfonline.com/doi/abs/10.1080/14786445508641925
8. Fiveland, W.A.: Discrete-ordinates solutions of the radiative transport equation for rectangular enclosures. J. Heat Transf. 106(4), 699–706 (1984). http://dx.doi.org/10.1115/1.3246741
9. Khawaja, H.: Applicability extent of 2-d heat equation for numerical analysis of a multiphysics problem. AIP Conference Proceedings 1798(1), 020075 (2017). http://aip.scitation.org/doi/abs/10.1063/1.4972667
10. Khawaja, H.A., Rashid, T., Eiksund, O., Broadal, E., Edvardsen, K.: Multiphysics simulation of infrared signature of an ice cube. Int. J. Multiphys. 10(3), 291–302 (2016). http://dx.doi.org/10.21152/1750-9548.10.3.291
11. Lenoble, J.: Radiative Transfer in Scattering and Absorbing Atmospheres: Standard Computational Procedures. A. Deepak, Hampton Virginia (1985). https://books.google.no/books?id=bZsRAQAAIAAJ

12. Patankar, S.: Numerical Heat Transfer and Fluid Flow. Taylor and Francis, London (1980). http://books.google.co.uk/books?id=5JMYZMX3OVcC

13. Platt, U., Pfeilsticker, K., Vollmer, M.: Radiation and Optics in the Atmosphere, pp. 1165–1203. Springer, New York (2007). http://dx.doi.org/10.1007/978-0-387-30420-5_19

14. Rashid, T., Khawaja, H.A., Edvardsen, K.: Determination of thermal properties of fresh water and sea water ice using multiphysics analysis. Int. J. Multiphys. **10**(3), 277–291 (2016). http://dx.doi.org/10.21152/1750-9548.10.3.277

15. Svanberg, S.: Atomic and Molecular Spectroscopy: Basic Aspects and Practical Applications. Heidelberg (2003). https://books.google.no/books?id=4uLdXjnKGKwC

16. Thomas, G., Stamnes, K.: Radiative Transfer in the Atmosphere and Ocean. Cambridge University Press, Cambridge (2002). https://books.google.no/books?id=DxR2nEp.0CUIC

17. Wang, L., Wu, H.: Biomedical Optics: Principles and Imaging. Wiley, Hoboken, New Jersey (2012). https://books.google.no/books?id=EJeQ0hAB76gC

Optimised Anisotropic Poisson Denoising

Georg Radow[1]([✉]), Michael Breuß[1], Laurent Hoeltgen[1], and Thomas Fischer[2]

[1] Chair for Applied Mathematics, Brandenburg Technical University,
Platz der Deutschen Einheit 1, 03046 Cottbus, Germany
{radow,breuss,hoeltgen}@b-tu.de
[2] Central Analytical Laboratory, Brandenburg Technical University,
Konrad-Wachsmann-Allee 6, 03046 Cottbus, Germany
thomas.fischer@b-tu.de

Abstract. The aim of this paper is to deal with Poisson noise in images arising in electron microscopy. We consider here especially images featuring sharp edges and many relatively large smooth regions together with smaller strongly anisotropic structures. To deal with the denoising task, we propose a variational method combining a data fidelity term that takes into account the Poisson noise model with an anisotropic regulariser in the spirit of anisotropic diffusion. In order to explore the flexibility of the variational approach also an extension using an additional total variation regulariser is studied. The arising optimisation problems can be tackled by efficient recent algorithms. Our experimental results confirm the high quality obtained by our approach.

Keywords: Poisson denoising · Variational methods · Electron microscopy

1 Introduction

The aim of this paper is to deal with noise in certain types of images arising in electron microscopy. A typical example for some of our applications of interest is depicted in Fig. 1. The displayed test image shows gold sputtered Polytetrafluoroethylene (PTFE) particles that were recorded using the secondary electron detector of a ZEISS DSM 962 scanning electron microscope.

Let us briefly discuss the properties of images we deal with here as exemplified via Fig. 1. Beginning with image content, we observe that there are many relatively large smooth (and round) regions with strong edges at particle outlines and particle overlaps. However, there are also relatively thin, elongated structures. Turning to the image disturbances, the image is characterised by random electron discharges appearing as bright horizontally striped artefacts. To circumvent sample destruction by electron impact at the image acquisition process, we had to choose a high scanning speed, which further resulted in a grainy texture.

Our Modelling Approach. Variational methods have been very successful for image denoising and restoration tasks [4]. Their main conceptual advantages are

© Springer International Publishing AG 2017
P. Sharma and F.M. Bianchi (Eds.): SCIA 2017, Part I, LNCS 10269, pp. 502–514, 2017.
DOI: 10.1007/978-3-319-59126-1_42

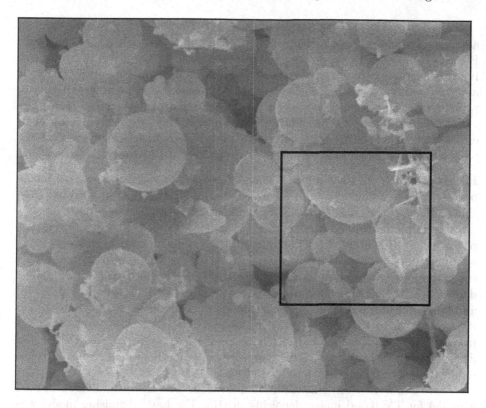

Fig. 1. Noisy image from an electron microscope as an example for a typical application in our lab, image size is $M \times N = 2048 \times 1664$. For image acquisition, we used an accelerating voltage of $20\,\mathrm{kV}$, a working distance of $6\,\mathrm{mm}$ and a magnification of 5000 with a ZEISS DSM 962 scanning electron microscope. The marked image region is later on employed for demonstrating filtering results.

their flexibility with respect to the possible model components and the inherent robustness due to the use of suitable regularisers. Moreover, the components of a variational method can often be easily interpreted and fine tuned. As the images in our applications may show various types of structures of potential interest, we opt to employ a variational approach because of its flexibility. As a further benefit we may devise a model for enhancing specific image structures.

Let us now turn to the variational model. The task is to consider the terms in a corresponding energy functional $E(u)$ that shall be minimised for finding the optimal solution u^*. The usual design incorporates terms for data fidelity and regularisation, respectively, so that on a coarse level of abstraction the variational model reads as:

$$E(u) := R(u) + \tau S(u; u^p). \tag{1}$$

Thereby, u^p denotes the given degraded input image, and $\tau > 0$ is a parameter defining the balance between data term $S(u; u^p)$ and regulariser $R(u)$.

Let us first consider the data term $S(u; u^p)$. We may safely assume that an adequate model for noise in electron microscopy images u^p is Poisson noise [18]. For the regulariser $R(u)$ we need to take into account the types of prominent image structures. We choose here to employ an anisotropic regulariser that is able to preserve the strong edges and the thin elongated structures that are prominent in our input images. Furthermore, exploring the flexibility of the variational approach, we also add a total variation (TV) regulariser in order to enhance edges and achieve visually more plateau-like structures corresponding to individual particles in our application.

Related Work. As the subject of this paper touches various aspects of image processing, we can only attempt to cover here some of the most important works that influenced the considerations for the construction of our approach.

Several methods and approaches for Poisson denoising (PD) have been proposed in previous literature. Often a variance stabilizing transformation (VST) is applied to transform Poisson noisy data into data with Gaussian noise. One such transformation was proposed by Anscombe in [1]. Methods to optimise the denoising with this transformation were presented by Mäkitalo and Foi in [13] and by Azzari and Foi in [2]. An extension of the Anscombe transformation and its application to PD was proposed by Zhang *et al.* [22]. Also, some methods from the field of machine learning have been considered, see *e.g.* Giryes and Elad [9] and Salmon *et al.* [17]. Poisson denoising methods based on multiscale models were studied by Lefkimmiatis *et al.* [11].

There also have been some variational methods for PD. Rudin *et al.* proposed a model for TV based image denoising in [16]. The basic denoising model was customised by Le *et al.* in [10] for PD. As this model is technically relevant for our approach, let us give some more details. As the data term naturally includes the formulation for Poisson noise, the data term in [10] resembles the one we will formulate, as is also the case in [7,8]. However, we will introduce a slight technical modification enabling better algorithmical treatment. In contrast to our work, Le *et al.* consider only TV regularisation. As another major difference we make use of a modern, non-trivial numerical optimisation approach for the numerical realisation, whereas Le *et al.* consider a simple finite difference method solving the Euler-Lagrange equation for their functional. Figueiredo *et al.* employ for the regulariser besides TV also frame-based models. The arising optimisation problems are tackled with the alternating direction method of multipliers. The recent variational model of Feng *et al.* follows the field of experts methodology for devising the regulariser [7].

In summary, we introduce several alternatives in modeling and numerics compared to the works [7,8,10]. Especially, we consider a novel anisotropic model for the problem and make use of efficient numerical tools.

Turning to the use of anisotropic diffusion, Weickert presented an extensive framework in [20]. The anisotropic regulariser we employ here has been proposed before in the context of diffusion [20]. In [21] accurate numerics, such as a family of specific finite difference schemes is given. We also take into account these recent, dedicated approximations within our numerical framework.

2   Poisson Noise

Suppose we have a Poisson noisy image $u^p \in \mathbb{N}_0^{MN}$ with $\mathbb{N}_0 := \mathbb{N} \cup \{0\}$, that is the result of a random experiment with the probability distribution

$$\mathbf{P}(U = u^p | u^0) := \prod_{(i,j) \in I} P(U_{i,j} = u_{i,j}^p | u_{i,j}^0), \tag{2}$$

where for every $(i, j) \in I := \{1, \ldots, M\} \times \{1, \ldots, N\}$ the parameter $u_{i,j}^0 \geq 0$ is supplied by the original image u^0 and P is the discrete Poisson distribution

$$P(X = x | y) := \begin{cases} \dfrac{(y)^x}{x!} \exp(-y), & y > 0, \\ \delta_{\{0\}}(x), & y = 0. \end{cases} \tag{3}$$

Here, $\delta_{\mathscr{S}}(z)$ is the Kronecker delta function that equals 1 if $z \in \mathscr{S}$ and 0 otherwise.

For any $u \in \mathbb{R}^{MN}$ we define sets of indices $I^+(u) := \{(i, j) \in I : u_{i,j} > 0\}$, $I^0(u) := \{(i, j) \in I : u_{i,j} = 0\}$ and $I^-(u) := \{(i, j) \in I : u_{i,j} < 0\}$. While $\mathbf{P}$ in (2) is defined only for nonnegative u^0, the likelihood of any $u \in \mathbb{R}^{MN}$ to be the image that led to u^p can be computed by

$$\mathscr{P}(u | u^p) := \delta_{\{\emptyset\}}(I^-(u)) \prod_{(i,j) \in I \setminus I^-(u)} P(U_{i,j} = u_{i,j}^p | u_{i,j}). \tag{4}$$

With the probability distribution (3) we immediately see that $\mathscr{P}(u | u^p) > 0$ if and only if $I^-(u) = \emptyset$ and $u_{i,j} > 0$ for all $i \in I^+(u^p)$. Because of $\lim_{x \to 0} x \log x = 0$, the log-likelihood is given by

$$\mathscr{L}(u | u^p) := \log(\mathscr{P}(u | u^p)) = \sum_{(i,j) \in I^+(u)} \left(u_{i,j}^p \log u_{i,j} - \log u_{i,j}^p! - u_{i,j} \right)$$
$$- \gamma_{\{\emptyset\}}(I^-(u)) - \sum_{(i,j) \in I^0(u)} \gamma_{\{0\}}(u_{i,j}^p), \tag{5}$$

where $\gamma_{\mathscr{S}}(z)$ is the indicator function that equals 0 if $z \in \mathscr{S}$ and ∞ otherwise. For a given u^p, maximizing $\mathscr{L}(u | u^p)$ is equivalent to minimizing

$$\sum_{(i,j) \in I^+(u)} \left(u_{i,j} - u_{i,j}^p \log u_{i,j} \right) + \gamma_{\{\emptyset\}}(I^-(u)) + \sum_{(i,j) \in I^0(u)} \gamma_{\{0\}}(u_{i,j}^p). \tag{6}$$

Minimizing a data fidelity in form of (6) is an approach for PD, which was used e.g. in [7,8,10].

For the noise model sometimes also a mixed Poisson distribution is used, e.g. in [12]. We assume that the image u^p has only Poisson noise, for a detailed overview of noise sources in scanning electron microscopy see [18].

3   Anisotropic Diffusion

As we rely in this paper on the variational formulation of anisotropic diffusion in the sense of Weickert [20] let us briefly recall some details.

The anisotropic diffusion flow within the image plane during a given time can be described by the partial differential equation (PDE) $\partial_t u = \text{div}\,(D\nabla u)$. Thereby D denotes the diffusion tensor with $D = v_1 v_1^\top + \lambda_2 v_2 v_2^\top$, where the vectors $v_1 \parallel \nabla u^\sigma$ and $v_2 \perp \nabla u^\sigma$ are of unit length and $\lambda_2 = g(\|\nabla u^\sigma\|^2)$, see also [20]. The function $g(\cdot)$ denotes the diffusivity function. By u^σ we denote the convolution of u with a Gaussian of standard deviation σ. For the spatial derivatives in ∇u^σ we use here central differences.

In [21] a discretisation of the term $\text{div}\,(D\nabla u)$ is obtained by minimizing

$$R(u) := \frac{1}{2} \int_\Omega \nabla^\top u D \nabla u \, dx dy \,, \tag{7}$$

where D is in this notation the time-invariant diffusion tensor

$$D = \begin{pmatrix} a(x,y) & b(x,y) \\ b(x,y) & c(x,y) \end{pmatrix}. \tag{8}$$

Following [21] we remark that (7) can be discretised and written as

$$R(u) := -\frac{1}{2} u^\top A u \quad \text{and} \quad \nabla R(u) = -Au \tag{9}$$

with some negative semidefinite matrix A. An explicit time discretisation can be computed by iterating $u^{(l+1)} = u^{(l)} + \alpha A(u^{(l)})u^{(l)}$, where α is a suitable step size and $A(u^{(l)})$ is the matrix A, which stems from the diffusion tensor using the iterate $u^{(l)}$, see [21] for more details. Similar to that, we employ a *lagged diffusivity* approach and evaluate A always at the given iterate in our optimisation method.

4   Our Variational Approach

As indicated we aim to find a local minimiser $u^* \in \mathbb{R}^{MN}$ of

$$E(u) := R(u) + \tau S(u; u^p), \tag{10}$$

see (1). By E we want to combine anisotroptic diffusion with data fidelity $S(\,\cdot\,;u^p)$ customised for a certain noise model. Among other models that we also test here, we especially consider the log-likelihood data term as developed before in (5). In the total, we used for $S := S(u; u^p)$ the following functions:

$$S_2 := \frac{1}{2}\|u - u^p\|_2^2 \,, \qquad S_1 := \|u - u^p\|_1 \,, \tag{11}$$

$$S_\ell := \sum_{(i,j)\in I^+(u)} \left(u_{i,j} - u_{i,j}^p \log u_{i,j}\right) + \gamma_{\{\emptyset\}}(I^-(u)) + \sum_{(i,j)\in I^0(u)} \gamma_{\{0\}}(u_j^p). \tag{12}$$

For the diffusivity $g := g(s^2)$, that contributes to R through the eigenvalue λ_2, some possibilities that we consider are

$$g_{\mathrm{PM}} := \frac{1}{1 + s^2/\lambda^2}, \qquad\qquad g_{\mathrm{ePM}} := \exp\left(-s^2/(2\lambda^2)\right), \qquad\qquad (13)$$

$$g_{\mathrm{Ch}} := \frac{1}{\sqrt{1 + s^2/\lambda^2}}, \qquad g_{\mathrm{W}} := \begin{cases} 1, & s^2 = 0, \\ 1 - \exp\left(\frac{-3.31488}{(s/\lambda)^8}\right), & s^2 > 0. \end{cases} \qquad (14)$$

For the Perona-Malik diffusivity g_{PM} and the exponential Perona-Malik diffusivity g_{ePM} see [15], for the Charbonnier diffusivity g_{Ch} see [5] and for the Weickert diffusivity g_{W} see [20]. Each of the diffusivity functions depend on the contrast parameter λ. While the impact of λ on the diffusion process is different in each diffusivity, edges with a contrast below λ are smoothed out more than those with a contrast above λ, generally speaking.

5   A Numerical Solution Strategy

Our variational model $E(u) := R(u) + \tau S(u; u^p)$ requires the minimisation of a non-convex and non-smooth cost function. A common choice to handle the non-convexity is to embed our method into a lagged diffusivity fixed-point iteration, thereby fixing the diffusivities at the previous iteration and thus, considering

$$R\left(u; \hat{u}^{(l)}\right) := -\frac{1}{2} u^{\mathsf{T}} A\left(\hat{u}^{(l)}\right) u. \qquad\qquad (15)$$

Now, the minimisation of $R(\,\cdot\,) + \tau S(\,\cdot\,; u^p)$ comes down to a series of convex optimisation tasks. $R(\,\cdot\,; \hat{u}^{(l)})$ is smooth and convex, although not necessarily strictly, nor strongly convex. On the other hand, our smoothness term $S_2(\,\cdot\,; u^p)$ is convex and smooth, whereas $S_1(\,\cdot\,; u^p)$ and $S_\ell(\,\cdot\,; u^p)$ are convex but continuous at the very best. Our convex energies $R(\,\cdot\,; \hat{u}^{(l)}) + \tau S(\,\cdot\,; u^p)$ also suggest a natural splitting into a sum of two terms, here $R(\,\cdot\,; \hat{u}^{(l)})$ and $\tau S(\,\cdot\,; u^p)$. One of our goals is to present a single concise but modular framework that is flexible enough to handle all presented setups efficiently. To this end we propose to use the inertial proximal algorithm for nonconvex optimization (iPiano) by Ochs et al. [14]. In its most generic form it considers the minimisation of cost functions $f + g$, where g is a convex (possibly non-smooth) and f a smooth (possibly non-convex) function. Thus, it suits our requirements well. The algorithm, inspired by the Heavy-ball method of Polyak, combines a forward-backward splitting scheme with an additional inertial force to improve the convergence properties. In its generic form it iterates

$$x^{(k+1)} = \mathrm{prox}_{\alpha_k g}\left(x^{(k)} - \alpha_k \nabla f\left(x^{(k)}\right) + \beta_k\left(x^{(k)} - x^{(k-1)}\right)\right), \qquad (16)$$

where α_k and β_k are parameters that stem from the numerical scheme. Here, the function $\mathrm{prox}_{\alpha g}$ denotes the proximal operator given by

$$\mathrm{prox}_{\alpha g}(y) := \arg\min_x \left(\frac{1}{2}\|x - y\|^2 + \alpha g(x)\right). \qquad\qquad (17)$$

Algorithm 1. Poisson Image denoising via iPiano

Choose $\sigma = \frac{3}{4}$, a diffusivity $g(\cdot)$, a diffusivity parameter $\lambda > 0$, a data fidelity $S(\,\cdot\,;\,u^p)$, $\tau > 0$, $c = \frac{1}{1000}$, $\beta = \frac{4}{5}$

Initialise $\hat{u}^{(1)} = u^p$

repeat

$\quad$ Compute D for $\hat{u}^{(l)}$ and the matrix $A(\hat{u}^{(l)}) = A_l$ that realises (9)

$\quad$ Estimate Lipschitz constant L_l by Geršgorin circle theorem

$\quad$ Compute $\alpha_l = \frac{2(1-\beta)}{L_l+c}$

$\quad$ Initialise $u^{(0)} = u^{(1)} = \hat{u}^{(l)}$

$\quad$ **repeat**

$$u^{(k+1)} = \text{prox}_{\alpha_l \tau S(\,\cdot\,,u^p)} \left(u^{(k)} + \alpha_l A_l u^{(k)} + \beta_k \left(u^{(k)} - u^{(k-1)} \right) \right)$$

$\quad$ **until** *convergence of the iPiano scheme towards $u^{(\infty)}$*

$\quad$ Set $\hat{u}^{(l+1)} = u^{(\infty)}$

until *convergence of the lagged diffusivity scheme*

Applying the iPiano formalism to our setup in (10) yields the following iterative strategy:

$$u^{(k+1)} = \text{prox}_{\alpha_k \tau S(\,\cdot\,;u^p)} \left(u^{(k)} - \alpha_k \nabla R \left(u^{(k)}; \hat{u}^{(l)} \right) + \beta_k \left(u^{(k)} - u^{(k-1)} \right) \right) \quad (18)$$

with initial values $u^{(0)} = u^{(1)}$ and certain parameters α_k and β_k.

Whether the iPiano algorithm is fast hinges on an effective evaluation of the proximal mapping. For our choices of S_1, S_2, and S_ℓ the corresponding proximal mappings are well known and can be expressed in closed form. The proximal mapping of S_1 is given by the soft shrinkage operation, while it corresponds to a simple averaging for S_2. Finally, the proximal mapping for S_ℓ is easily found by setting the first derivative to 0. We also remark in this context that it suffices in each case to consider the 1D formulation. The computation of the proximal mapping decouples for all our choices of S.

Besides the computation of the proximal mapping we must also specify an update strategy for the free parameters α_k and β_k in (18) (resp. (17)). Here, we follow the recommendations found in [14] and set $\beta_k = \frac{4}{5}$ and $\alpha_k = \frac{2(1-\beta_k)}{L+c}$, where $c = \frac{1}{1000}$ and L being an upper bound of the spectral norm of $A(\hat{u}^{(l)})$, that is, the Lipschitz constant of $\nabla R(\,\cdot\,;\,\hat{u}^{(l)})$. For performance reasons we opt to estimate L by means of the Geršgorin circle theorem. A detailed listing of our approach is given in Algorithm 1.

We also use the following stopping criteria in our implementation. The outer iteration stops when $\hat{u}^{(l+1)}$ fulfils either

$$\frac{|E_{l+1}(\hat{u}^{(l+1)}) - E_l(\hat{u}^{(l)})|}{|E_{l+1}(\hat{u}^{(l+1)}) + E_l(\hat{u}^{(l)})|} < 5 \cdot 10^{-7} \quad (19)$$

or

$$\frac{1}{MN} \left\| \hat{u}^{(l+1)} - \hat{u}^{(l)} \right\|_1 < 10^{-6} \left\| u^p \right\|_\infty, \quad (20)$$

where E_l denotes the function E with the diffusion tensor fixed with the iterate $\hat{u}^{(l)}$. For the iPiano algorithm we use nearly the same break criteria, we substitute $\hat{u}^{(l+1)}$ with $u^{(k+1)}$, $\hat{u}^{(l)}$ with $u^{(k)}$ and E_{l+1} with E_l.

6   Additional Model Improvements

Even though our approach already delivers convincing results, we investigate further means to improve the quality of our findings. Our real world data is assumed to have hard edges with almost constant regions in between in the ideal case. Such results can, for example, be obtained by adding a second regulariser in form of a TV term. Thus, we must minimise the following energy.

$$F(u) := R(u) + \tau S(u \; ; \; u^p) + \kappa TV(u) \tag{21}$$

Algorithm 2. Poisson Image denoising via three-operator splitting

---

Choose $\sigma = \frac{3}{4}$, a diffusivity $g(\cdot)$, a diffusivity parameter $\lambda > 0$, a data fidelity $S(\cdot \; ; \; u^p)$, $\tau > 0$, $c = \frac{1}{1000}$, $\kappa > 0$

Initialise $\hat{u}^{(1)} = u^p$

repeat

    Compute D for $\hat{u}^{(l)}$ and the matrix $A(\hat{u}^{(l)}) = A_l$ that realises (9)

    Estimate Lipschitz constant L_l by Geršgorin circle theorem

    Compute $\alpha_l = \frac{2}{L_l+c}$

    Initialise $u^{(1)} = \hat{u}^{(l)}$

    repeat

$$x^{(k)} = \text{prox}_{\alpha_l \kappa T}\left(u^{(k)}\right)$$

$$y^{(k)} = \text{prox}_{\alpha_l \tau S(\,\cdot\,;\,u^p)}\left(2x^{(k)} - u^{(k)} + \alpha_l A_l x^{(k)}\right)$$

$$u^{(k+1)} = u^{(k)} + y^{(k)} - x^{(k)}$$

    until *convergence of the three-operator splitting scheme towards $y^{(\infty)}$*

    Set $\hat{u}^{(l+1)} = y^{(\infty)}$

until *convergence of the lagged diffusivity scheme*

Here, $TV(u)$ defines the discretised form of the total variation of a smooth function.

$$TV(u) := \sum_{(i,j) \in I} \left\| \begin{pmatrix} d_x u_{i,j} \\ d_y u_{i,j} \end{pmatrix} \right\|_2 , \tag{22}$$

where d_x and d_y denote standard forward differences. To find a minimiser u^* we proceed similarly as before. We use a lagged diffusivity fixed-point iteration to overcome the nonlinearities in the regulariser R. Similarly as before, we obtain a series of convex problems. The difference lies in the fact that we must minimise a sum of three terms now. Even though one could apply iPiano, we opt for a better adapted algorithmic approach. We use the three-operator splitting scheme presented in [6] to handle the additional TV term. It requires us to evaluate the proximal mapping of the TV term

$$\text{prox}_{\alpha TV}(\tilde{u}) = \arg\min_{u} \left(\frac{\|u - \tilde{u}\|_2^2}{2} + \alpha \sum_{(i,j) \in I} \left\| \begin{pmatrix} d_x u_{i,j} \\ d_y u_{i,j} \end{pmatrix} \right\|_2 \right). \qquad (23)$$

In this work, we use the algorithm of Chambolle and Pock [3]. It is appealing due to its simple form and decent efficiency. The complete algorithm, including the numerical strategy with the three-operator splitting is listed in Algorithm 2.

Here, we use the same stopping criteria as for Algorithm 1, see (19) and (20). In addition, we abort the iterative algorithm for computing $\text{prox}_{\alpha TV}(\tilde{u})$ when the following condition is met.

$$\frac{1}{MN} \left\| u^{(m+1)} - u^{(m)} \right\|_1 < 10^{-6} \left(\frac{9}{10} \right)^l \|u^p\|_\infty \qquad (24)$$

Here, l corresponds to the loop counter for the lagged diffusivity scheme. By tightening the convergence requirements while progressing through the outer iteration of Algorithm 2, we aim to shorten computing time at the start and get a better final result. It seems plausible that the first few iterates can be rather rough, whereas accurate estimates are much more important towards the end.

7 Numerical Results

We use three different synthetic test images for a qualitative analysis of our methods. These images are chosen in such a way that they contain features commonly occurring in microscopic images. The considered samples are depicted in Fig. 2. For our evaluation we corrupt each image three times with Poisson noise in three different strengths, yielding a total of nine noisy images. One of them is shown

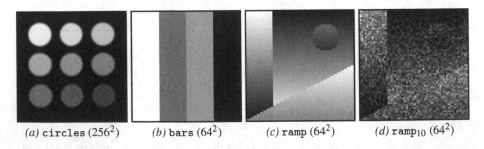

(a) circles (256^2)   (b) bars (64^2)   (c) ramp (64^2)   (d) ramp$_{10}$ (64^2)

Fig. 2. (a–c) Noise-free original versions of our considered test images. The numbers in brackets indicate the size of the square images. (d) Exemplary noisy version of ramp. The lowercase index states the peak value, that the original image was scaled to before computing the Poisson noisy image. The noisy image has a range going from 0 to 19.

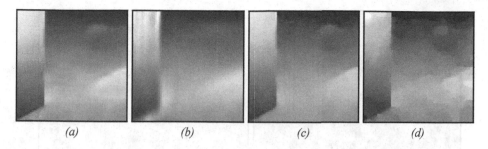

<table>
<tr><td>(a)</td><td>(b)</td><td>(c)</td><td>(d)</td></tr>
</table>

Fig. 3. *(a)* denoised image obtained by using the method from [2] for $\mathtt{ramp}_{10}$ (SSIM $=$ 0.7845) *(b)* denoised image obtained by using the method from [9] for $\mathtt{ramp}_{10}$ (SSIM $=$ 0.7437) *(c)* result of Algorithm 1 optimised for maximal SSIM with S_ℓ, g_{ePM}, $\lambda \approx$ 0.2446, $\tau \approx 0.2443$ for $\mathtt{ramp}_{10}$ (SSIM $=$ 0.7817) *(d)* result of Algorithm 2 optimised for maximal SSIM with S_1, g_W, $\lambda \approx 10.39$, $\tau \approx 96.74$, $\kappa \approx 226.2$ for $\mathtt{ramp}_{10}$ (SSIM $=$ 0.6552)

Table 1. MSE and SSIM for all tested images. Numbers in bold show the best result of all considered methods. The index after the image name denotes the peak value to which the image was scaled for the computation of the noisy image. Our approach outperforms the reference algorithms in terms of MSE for **bars** and **ramp**. Also, we obtain a higher SSIM for **bars** and **circles**.

Image	MSE			SSIM		
	[2]	[9]	Algorithm 1	[2]	[9]	Algorithm 1
$\mathtt{bars}_{255}$	**1.3295**	21.0788	1.6132	0.5381	0.4535	**0.8264**
$\mathtt{bars}_{10}$	23.9553	25.3805	**17.0658**	0.4448	0.4156	**0.5685**
$\mathtt{bars}_{1}$	428.8934	296.7009	**292.8159**	0.2744	0.3033	**0.3122**
$\mathtt{ramp}_{255}$	**9.4805**	160.4250	18.4692	**0.9618**	0.8086	0.9375
$\mathtt{ramp}_{10}$	114.0484	186.8883	**111.2820**	**0.7845**	0.7437	0.7817
$\mathtt{ramp}_{1}$	544.2952	702.1416	**459.8652**	0.4758	0.4836	**0.5196**
$\mathtt{circles}_{239}$	1.6322	**0.8585**	0.8934	0.8284	0.8176	**0.9394**
$\mathtt{circles}_{10}$	20.9328	9.7073	**9.4121**	0.6912	0.6563	**0.8226**
$\mathtt{circles}_{1}$	103.5318	**56.7450**	85.3593	0.2714	0.4334	**0.6910**

in Fig. 2, in an exemplary manner. In order to simulate different noise levels we rescale the images such that their peak value equals 255, 10, and 1 respectively before computing the sample of the probability distribution (2). In the last case the strength of noise is fairly high and the features of the original image are barely visible. We also use the mean squared error (MSE) and the structural similarity (SSIM) as a means to measure the quality of our reconstructions (see [19] for a reference on the SSIM). In addition, we consider the methods proposed in [2] and [9] for comparison. The authors of these works provide reference implementations of their algorithms, allowing us to do an objective evaluation.

512   G. Radow et al.

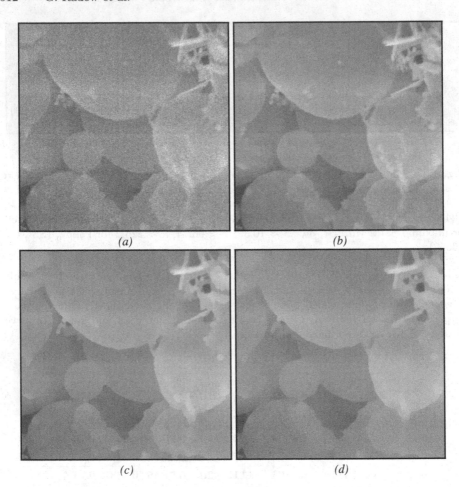

(a)   *(b)*

(c)   *(d)*

Fig. 4. *(a)* Original noisy image, cut out from Fig. 1 *(b)* result of VST from [2] *(c)* result of Algorithm 1 with S_ℓ, g_W $\lambda = 1.25$ and $\tau = 2.5$ *(d)* Result of Algorithm 2 with S_ℓ, g_W $\lambda = 1.25$ and $\tau = 2.5$ and $\kappa = 0.5$

Table 1 presents our findings for all considered images. Figure 3 shows one of the denoised images for every considered algorithm. The results of Algorithms 1 and 2 were achieved by thoroughly tuning all the model parameters to minimise the MSE respectively maximise the SSIM. The findings from Table 1 indicate that we achieve competitive results in most cases. If we focus solely on the SSIM, then our results are notably better (0.8226 vs. 0.6563 resp. 0.6912 for circles$_{10}$).

Finally, Fig. 4 presents the results of our algorithms on a real world example. The images depict gold sputtered PTFE particles recorded by a secondary electron detector of a ZEISS DSM 962 scanning electron microscope at an accelerating voltage of 20 kV. We have used a working distance of 6 mm and a magnification factor of 5000. Since the ground truth is unknown, the parameters have been tuned to obtain the most visually appealing result. Clearly, we are able to remove noise and other artefacts while preserving the contrast and sharp edges.

8 Conclusion

We have shown that modern optimisation methods and discretisation schemes can be applied with benefit in electron microscopy. Although our basic anisotropic model is relatively simple, it already yields visually convincing results in our test application. The flexible variational framework enables many possible extensions that can be used for specific applications. For future work we may especially include a thorough study of possible optimisation methods for optimal balancing of data term and regularisers.

References

1. Anscombe, F.J.: The transformation of Poisson, binomial and negative-binomial data. Biometrika **35**(3/4), 246–254 (1948)
2. Azzari, L., Foi, A.: Variance stabilization for noisy+estimate combination in iterative Poisson denoising. IEEE Signal Process. Lett. **23**(8), 1086–1090 (2016)
3. Chambolle, A., Pock, T.: A first-order primal-dual algorithm for convex problems with applications to imaging. J. Math. Imaging Vis. **40**(1), 120–145 (2010)
4. Chan, T.F., Shen, J.: Image Processing and Analysis: Variational, PDE, Wavelet, and Stochastic Methods. Society for Industrial and Applied Mathematic, Philadelphia (2005)
5. Charbonnier, P., Blanc-Feraud, L., Aubert, G., Barlaud, M.: Deterministic edge-preserving regularization in computed imaging. IEEE Trans. Image Process. **6**(2), 298–311 (1997)
6. Davis, D., Yin, W.: A three-operator splitting scheme and its optimization applications (2015). https://arxiv.org/abs/1504.01032
7. Feng, W., Qiao, H., Chen, Y.: Poisson noise reduction with higher-order natural image prior model. SIAM J. Imaging Sci. **9**(3), 1502–1524 (2016)
8. Figueiredo, M.A.T., Bioucas-Dias, J.M.: Restoration of Poissonian images using alternating direction optimization. IEEE Trans. Image Process. **19**(12), 3133–3145 (2010)
9. Giryes, R., Elad, M.: Sparsity-based Poisson denoising with dictionary learning. IEEE Trans. Image Process. **23**(12), 5057–5069 (2014)
10. Le, T., Chartrand, R., Asaki, T.J.: A variational approach to reconstructing images corrupted by Poisson noise. J. Math. Imaging Vis. **27**(3), 257–263 (2007)
11. Lefkimmiatis, S., Maragos, P., Papandreou, G.: Bayesian inference on multiscale models for Poisson intensity estimation: applications to photon-limited image denoising. IEEE Trans. Image Process. **18**(8), 1724–1741 (2009)
12. Luisier, F., Blu, T., Unser, M.: Image denoising in mixed Poisson-Gaussian noise. IEEE Trans. Image Process. **20**(3), 696–708 (2011)
13. Mäkitalo, M., Foi, A.: Optimal inversion of the anscombe transformation in low-count poisson image denoising. IEEE Trans. Image Process. **20**(1), 99–109 (2011)
14. Ochs, P., Chen, Y., Brox, T., Pock, T.: iPiano: Inertial proximal algorithm for nonconvex optimization. SIAM J. Imaging Sci. **7**(2), 1388–1419 (2014)

15. Perona, P., Malik, J.: Scale-space and edge detection using anisotropic diffusion. IEEE Trans. Pattern Anal. Mach. Intell. **12**(7), 629–639 (1990)
16. Rudin, L.I., Osher, S., Fatemi, E.: Nonlinear total variation based noise removal algorithms. Phys. D **60**(1–4), 259–268 (1992)
17. Salmon, J., Harmany, Z., Deledalle, C.A., Willett, R.: Poisson noise reduction with non-local PCA. J. Math. Imaging Vis. **48**(2), 279–294 (2013)
18. Timischl, F., Date, M., Nemoto, S.: A statistical model of signal-noise in scanning electron microscopy. Scanning **34**(3), 137–144 (2011)
19. Wang, Z., Bovik, A., Sheikh, H., Simoncelli, E.: Image quality assessment: from error visibility to structural similarity. IEEE Trans. Image Process. **13**(4), 600–612 (2004)
20. Weickert, J.: Anisotropic diffusion in image processing. Teubner, Stuttgart (1998)
21. Weickert, J., Welk, M., Wickert, M.: L^2-stable nonstandard finite differences for anisotropic diffusion. In: Kuijper, A., Bredies, K., Pock, T., Bischof, H. (eds.) SSVM 2013. LNCS, vol. 7893, pp. 380–391. Springer, Heidelberg (2013). doi:10.1007/978-3-642-38267-3_32
22. Zhang, B., Fadili, J., Starck, J.: Wavelets, ridgelets, and curvelets for Poisson noise removal. IEEE Trans. Image Process. **17**(7), 1093–1108 (2008)

Augmented Reality Interfaces
for Additive Manufacturing

Eythor R. Eiriksson[1]([✉]), David B. Pedersen[1], Jeppe R. Frisvad[1],
Linda Skovmand[1], Valentin Heun[2], Pattie Maes[2], and Henrik Aanæs[1]

[1] Technical University of Denmark, Kgs., Lyngby, Denmark
eruei@dtu.dk
[2] MIT Media Lab, Cambridge, MA, USA

Abstract. This paper explores potential use cases for using augmented
reality (AR) as a tool to operate industrial machines. As a baseline
we use an additive manufacturing system, more commonly known as
a 3D printer. We implement novel augmented interfaces and controls
using readily available open source frameworks and low cost hardware.
Our results show that the technology enables richer and more intuitive
printer control and performance monitoring than currently available on
the market. Therefore, there is a great deal of potential for these types
of technologies in future digital factories.

Keywords: 3D Printing · Additive manufacturing · Augmented reality

1 Introduction

With recent efforts in industrial digitalization, commonly referred to as 'industry 4.0', the core aim is to realize the factory of the future. These factories are
envisioned to be agile and flexible using complex autonomous manufacturing
technologies combined with the human skills of reasoning. As the manufacturing technology becomes more and more autonomous the need for intuitive and
fluid human-machine interaction becomes necessary. This is in part due to the
massive increase in data available to the users. Additionally, it is due to increasing abstraction where complex inter-connectivity between factory elements has
become invisible to the operators.

A promising, highly digitized and automated manufacturing technology that
frequently comes up in relation to the factory of the future, is additive manufacturing (AM), commonly known as 3D printing. For large scale manufacturing using not only several AM systems but also other highly automated
machine tools, the need for a larger control framework is required. Managing
multiple machinery as well as monitoring their performance often cannot easily
be achieved through their conventional physical user interfaces. However, these
interfaces can be dynamically scaled in an augmented reality control interface.
With such an interface, it can be evolved and iteratively tailored towards the

P. Sharma and F.M. Bianchi (Eds.): SCIA 2017, Part I, LNCS 10269, pp. 515–525, 2017.
DOI: 10.1007/978-3-319-59126-1_43

specific use cases at any given time. As an example, multiple machines can be controlled using a single AR user interface.

In this paper, we focus on the integration of an augmented reality based human-machine interface that substitutes a conventional user interface on an additive manufacturing (AM) machine tool. This is achieved using readily available open source frameworks, web technologies and low cost mobile hardware.

2 Related Work

Mobile embedded systems such as smartphones, tablets and head mounted displays (HMD) [1,2] now have enough computing power to enable augmented reality applications at a low cost. In an industrial scenario, applications such as remote technical support [3], worker training [4] and design [5], can be achieved.

In the context of AR industrial machine control and visualization, Zhang et al. [6] performed augmented cutting simulations on a industrial milling machine, as well as provided an interface control panel using a camera and high end processing computer. Olwal et al. demonstrated an industrial AR projection system that augments machine information onto a transparent display [7]. Both examples above, classify as tethered solutions that do not offer the flexibility achievable with a mobile device. Kollatsch et al. [8] presented a mobile implementation using a Windows tablet PC that displayed machine status parameters when presented with an AR marker, however no machine control was possible.

Our method takes advantage of recent advancement in high performance mobile technology that enables use of both consumer tablets and smartphones. Using open source systems we have implemented bi-directional AR interfaces that both monitor and control a 3D printer, in such a way that any changes made on the printer from external sources are reflected in the AR interface and vice versa.

3 Method

As consumer mobile tablet devices have already made their way into the industrial environment, we use a traditional iPad Mini 3 for this study. It is responsible for both displaying augmented interfaces as well as performing the computer vision necessary to identify and pose estimate unique optical markers that are placed on relevant locations on the machine. For this study, we employ the Ultimaker Original [9] desktop 3D printer to represent an industrial AM machine tool on a factory floor. It is connected to a laptop via a USB-serial interface, where the control and readout takes place. For this implementation, any device capable of running the node.js JavaScript runtime environment [10] could be used to handle the printer communication. This includes popular devices such as the Raspberry Pi, Arduino Yún as well as many industry standard embedded systems built on ARM technology. Thus it is entirely feasible for several machines to be equipped with AR functionality at a very low cost. Both the tablet, and laptop, are connected to a common wireless network (WIFI). The laptop runs a node.js web server that communicates to both tablet and printer. See Fig. 1.

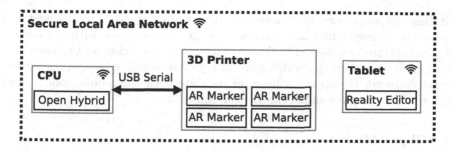

Fig. 1. Diagram illustrating the hardware and software components used.

3.1 Additive Manufacturing Machine Tool

The Ultimaker Original is entirely open source and belongs to the class of extrusion based printers. Its construction is cartesian with a motorized horizontal X-Y carriage system. It is used to trace 2D cross-sectional slices of 3D objects, whilst simultaneously extruding melted thermoplastic material through a heated extruder nozzle. The vertical Z-axis, that translates the build plate, is then moved downward. Typically in sub-mm increments for each new layer to be produced. The printer belongs to a large family of numerically controlled (NC) devices following the G-code standard [11]. It is possible to control the machine tool through its USB hardware interfaces which presents itself as a standard serial port when connected to a computer. The printer can then be operated using G-code commands which are conventionally used in computer-aided manufacturing to control automated technologies (CAx). It is thus possible to set and query all features of the printer ranging from motion control to tooling settings.

3.2 Open Hybrid

Open hybrid [12] is a highly customizable open source platform from the MIT Media Lab, that enables flexible interactions between physical objects through augmented user interfaces. Objects or systems powered by Open Hybrid enable the possibility of inter-connectivity between each one. Such that the underlying input- and output connectivity routing between objects can be dynamically altered though simple drag and drop gestures. The augmentation system is enabled by web standards, and as such each object is set up solely in standardized protocols through a simple Application Programming Interface (API).

3.3 Reality Editor

The Reality Editor [13] is a publicly available mobile iOS app that functions as a digital window and multi-tool for Open Hybrid objects. Allowing the user to connect and manipulate the functional behavior of any physical object that is enabled by the Open Hybrid platform. As a mobile device (tablet or phone) is pointed toward an object AR marker, the Reality Editor recognizes it and performs camera pose estimation in real time. It then proceeds to augment and display the corresponding user interface on the mobile device with the correct perspective transformation. Physical buttons, indicator lights and sensors may be shown as nodes between which relationships can be defined through gestures. Once configured, the object will retain its state until the mobile tool is used again to alter the previously defined relations.

3.4 Object Markers

Throughout this study we use a novel type of quick response (QR) codes called Human Readable Quick Response Codes [14] (HRQR). The marker consist of high contrast visual features that allow for robust marker detection and pose estimation. Furthermore, as the marker is readable by humans, its text will give an indication to what type of interface is to be expected.

4 Use Cases

The following sections describe use cases implemented in this study. We identified these cases based on common interactions for this type of printer and how they might be improved using AR.

4.1 Print Visualization

In a manufacturing setting, the ability to see what object is being produced per machine can be beneficial. Giving the user a overview on the current status on the production floor. Such functionality can be provided by AR interface. Currently, there are two ways of identifying the object under print. One is by visual inspection of the actual part and secondly by its filename, which is usually displayed on the printers graphical display. For the latter, special care must be taken in naming the files accordingly, especially when dealing with different versions of the same part. Throughout a print process it can be unclear what model is being printed. This also holds true when parts are being printed that vary only slightly in form. That being said, there could be cases where it is non-trivial what part is being printed. To address the above, we implemented a use case where the entire print bed consists of an AR marker. When the marker is observed using the Reality Editor, the 3D model being printed is augmented on the print platform itself. As the AR interface supports standard web technologies, we use the three.js Javascript 3D library for loading and rendering the mesh geometry. Camera pose estimation is obtained each frame by observing the AR marker, and the mesh is perspectively transformed accordingly. See Fig. 2.

In line with our previous work on pre-visualization of printed parts prior to printing [15,16], we here add the staircase effect often observed in layered manufacturing [17] to visualize for the user the surface structure that the printer will produce. We render the triangle mesh using a pixel shader with standard Phong lighting [18] and shadow mapping [19]. We customize our shader by adding

Fig. 2. Print visualization example. A 3D model (Bunny) is augmented whilst printing.

procedural normal mapping. Having the Z-axis as the printing direction, our procedure for normal mapping is to find the layer index i of the surface fragment observed in the pixel, calculate the distance t to the intersection of the eye ray with the next layer, and use this to modify the Z-coordinate of the normal. In mathematical terms,

$$i = \left\lfloor \frac{p_z}{\ell} \right\rfloor + H(\omega_z) \quad , \quad t = \frac{i\ell - p_z}{\omega_z} \quad , \quad \boldsymbol{n}^* = (n_x, n_y, H(\ell - t)\,\mathrm{sign}(n_z))$$

where $\boldsymbol{n}$ is the unit length surface normal, $\boldsymbol{p}$ is the world space fragment position, $\boldsymbol{\omega}$ is the unit length direction vector from the camera position toward $\boldsymbol{p}$, ℓ is the layer thickness, $H(x)$ is the Heaviside step function (which is 1 for $x \geq 0$ and 0 otherwise), and $\boldsymbol{n}^*$ is the modified normal that produces the staircase effect. See Fig. 3 for an example.

The sharp edges in our staircase normal mapping produce aliasing artifacts in excess. We anti-alias our renderings by sampling $N = 50$ positions $\boldsymbol{p}$ in an area of 2 by 2 pixels around the original fragment position. We use a linear congruential generator with the same seed in all pixels to generate a pseudo-random distribution of window space positions without expensive texture look-ups. Still in the pixel shader, we invert the graphics pipeline transformations to get the world space coordinates of the sampled positions. For every sampled position, we then compute the normal using the given procedure and evaluate the lighting. The final pixel value is the mean of these N lighting evaluations. By reducing the number of samples ($N = 25$), we can trade render quality for performance. Very thin layers (small ℓ) introduce Moiré patterns in the renderings.

Fig. 3. Reality Editor screenshot showing model augmentation using the staircase effect shader. Due to high performance AR marker tracking, obscured viewing angles such as this can be achieved. Note: Only the part of the model that is missing is augmented.

4.2 Nozzle Thermal Control

Thermal control of the nozzle that extrudes liquid plastic, is one of the most frequently operated settings on the Ultimaker. Currently, users have to perform five menu operations to reach this setting. We propose to tie the relevant parameters to a simple AR interface that is accesses through an AR marker on the print tool-head. There, the user can instantaneously monitor the current temperature, as well as easily change the set temperature through an intuitive slider. Figure 4 shows the interface under use.

4.3 3D Model Selection

The work flow for setting up a print job is conventionally by pre-processing a 3D geometry in a standalone software that generates a G-code print job. However, in a manufacturing setting, more often than not, the same set of models will be printed repeatedly. For most printers, these job files are stored on an on-board memory but cannot be pre-visualized or identified, other than by its filename. To improve upon this, we propose a print job selector, that allows the user to preview available models and select which should be printed. Allowing for very rapid job selections. The interface may be seen in Fig. 5.

Fig. 4. A minimal augmented interface showing the target and current temperatures of the heated print nozzle. Additionally a slider is included for easy manual operation.

Fig. 5. Our implemented selection interface for easy selection of parts to be printed. To select a job, the user simply presses on the responding augmented model.

4.4 Carriage Motion Control

Frequently, the X-Y carriage of the printer is moved for maintenance. To do this, only one axis can be jogged at a time through a cumbersome menu system on the

Fig. 6. Example interaction of intuitive carriage motion control.

printers current interface. For some systems, the implementation is non-direct. Then by pressing a button, the axis starts by moving slowly. However, the speed of the axis increases the longer the button is pressed. This non-direct implementation results in less controlled motion by the user. We have implemented direct augmented control of the carriage that allows it to be moved using the AR interface. Enabling a more precise motion control tool that follows the speed of the user. The implementation can be seen in Fig. 6. When a user touches the augmented joystick, we record the markers (u, v) screen coordinates on the display device. Then the user can proceed to pan the device, which updates the markers coordinates on the screen. This causes a disparity which is the used as a control signal for the X and Y motors. The greater the disparity, the faster the carriage will move. In practice, the absolute scale of the AR marker could be used as control for the Z-axis. This Z-axis control was however not implemented.

5   Future Work

For future interfaces we envision the following:

Extrusion control that enables users to change plastic feed-rates, as well as the initiation of extrusion.

Model placement on the build plate to be used to position one or more 3D objects in an intuitive fashion. Object collision or out of bounds events may be visualized by changing the objects color. Additionally, dimensional scaling of the geometries could be performed through common zoom pinching gestures.

Printer-to-printer interactions are possible as more printers are enabled by the Open Hybrid platform. This could allow for 'drag and drop' of 3D models between printers such that high demand geometry easily can be placed in a print queue. This allows for agile production ramp up of parts. An implementation could allow various types of printer technologies to interact, as the de-facto standard fileformat is based upon wireframe meshes.

Drag and dropping of models directly from community driven services such as thingiverse.com could be made possible. The back-end AR server would then be responsible for downloading the STL geometry, slicing the model and generating the machine code.

Layer by layer visualization could be implemented, where tool planning and support structures are shown. In addition, model areas lacking support structures could be highlighted.

Drawing shapes on the build plate, as well as simple sculpting operations can be supported. Drawing modes could for example include manual drawing of support structures.

6   Discussion and Conclusion

In this paper, we have explored and touched upon possible use cases for augmented reality systems in a manufacturing environment. This was done as an effort to explore how augmented reality can be used as a seamless interface for the factory of the future. In this study, we have successfully integrated state of the art and open source platforms that have been developed at the MIT Media Lab. In this spirit, the implementation has been realized on the next-gen manufacturing method of 3D printing. We demonstrate novel use cases in which augmented reality may outperform conventional control mechanism in such systems. However, it is important to note the presented augmented reality interfaces are not limited to this specific family of manufacturing technologies. In fact, similar augmented interfaces can be implemented on any kind of manufacturing tools and machinery such as CNC machine tools, water jet- and laser cutters, injection moulding machines, forging presses etc. Any modern conventional manufacturing process chain can readily be set up as an augmented object and monitored and/or controlled through means presented in this paper.

With the rapidly increasing amounts of data produced by modern machine tools, AR can serve as a medium that provides operators with context related data on demand. Thus bridging a gap that may be created as content complexity increases. The scalability enabled by AR interfaces has the potential to be an enabling technology for modular production platforms, rapid product development and hyperflexible automation. By using constantly evolving bi-directional interfaces and the ability to instantaneously switch between them, it is possible to rapidly adapt to changes in production, bring products to market faster and small series production platforms can be set up quickly.

With few limits and near endless possibilities we have demonstrated the importance of embracing augmented reality in manufacturing engineering such that advanced manufacturing processes and process chains can be interfaced in a simple and intuitive manner.

References

1. Microsoft HoloLens. https://www.microsoft.com/microsoft-hololens/en-us. Accessed 05 Jan 2017
2. DAQRI Smart Helmet. https://daqri.com/products/smart-helmet/. Accessed 05 Jan 2017
3. Billinghurst, M., Kato, H.: Collaborative augmented reality. Commun. ACM **45**(7), 64–70 (2002)
4. Webel, S., Bockholt, U., Engelke, T., Gavish, N., Olbrich, M., Preusche, C.: An augmented reality training platform for assembly and maintenance skills. Robot. Auton. Syst. **61**(4), 398–403 (2013)
5. Nee, A., Ong, S., Chryssolouris, G., Mourtzis, D.: Augmented reality applications in design and manufacturing. CIRP Ann.-Manuf. Technol. **61**(2), 657–679 (2012)
6. Zhang, J., Ong, S., Nee, A.: A volumetric model-based CNC simulation and monitoring system in augmented environments. In: International Conference on Cyberworlds 2006, CW 2006, pp. 33–42. IEEE (2006)
7. Olwal, A., Gustafsson, J., Lindfors, C.: Spatial augmented reality on industrial CNC-machines. In: Electronic Imaging 2008. International Society for Optics and Photonics (2008)
8. Kollatsch, C., Schumann, M., Klimant, P., Wittstock, V., Putz, M.: Mobile augmented reality based monitoring of assembly lines. Procedia CIRP **23**, 246–251 (2014)
9. Ultimaker Original, BOM and parts. http://github.com/Ultimaker/UltimakerOriginal. Accessed 05 Jan 2017
10. Node.js. https://nodejs.org/en/. Accessed 09 Jan 2017
11. ISO 6983-1:2009. Automation systems and integration - numerical control of machines - program format and definitions of address words - Part 1: Data format for positioning, line motion and contouring control systems. Standard, International Organization for Standardization, Geneva, CH, December 2009
12. Open Hybrid: Platform for interaction with everyday objects. http://openhybrid. org/. Accessed 05 Jan 2017
13. Heun, V., Hobin, J., Maes, P.: Reality editor: programming smarter objects. In: Proceedings of the 2013 ACM Conference on Pervasive and Ubiquitous Computing Adjunct Publication, pp. 307–310. ACM (2013)
14. HRQR: Human Readable Quick Response Code. http://hrqr.org/. Accessed 05 Jan 2017
15. Eiríksson, E.R., Pedersen, D.B., Aanaes, H.: Predicting color output of additive manufactured parts. In: ASPE 2015 Spring Topical Meeting, pp. 95–99 (2015)
16. Eiriksson, E.R., Luongo, A., Frisvad, J.R., Pedersen, D.B., Aanaes, H.: Designing for color in additive manufacturing. In: Proceedings of the ASPE/EUSPEN 2016 Summer Topical Meeting on Dimensional Accuracy and Surface Finish in Additive Manufacturing. ASPE-The American Society for Precision Engineering (2016)

17. He, Y., huai Xue, G., zhong Fu, J.: Fabrication of low cost soft tissue prostheses with the desktop. 3D printer, Scientific reports, vol. 4, no. 6973 (2014)
18. Phong, B.T.: Illumination for computer generated pictures. Commun. ACM **18**, 311–317 (1975)
19. Williams, L.: Casting curved shadows on curved surfaces. In: Proceedings of SIG-GRAPH 1978 Computer Graphics, vol. 12, pp. 270–274, August 1978

General Cramér-von Mises, a Helpful Ally for Transparent Object Inspection Using Deflection Maps?

Johannes Meyer[1](✉), Thomas Längle[2], and Jürgen Beyerer[2]

[1] Karlsruhe Institute of Technology, Karlsruhe, Germany
johannes.meyer@kit.edu
[2] Fraunhofer-Institute of Optronics,
System Technologies and Image Exploitation IOSB,
Karlsruhe, Germany
http://www.meyer-research.de

Abstract. Transparent materials are utilized in different products and have to meet high quality requirements, i.a., they have to be free from scattering defects. Such material defects are mainly manifested in changes of the direction of light transmitted through the object. Laser deflection scanners can acquire so-called four-dimensional light deflection maps conveying both, the spatial and angular information about captured light rays. In order to detect scattering defects, spatial discontinuities of the angular deflection distribution have to be extracted out of the deflection maps. This is necessary since the transparent object itself and possibly present scattering defects can deflect incident light rays into other directions. This contribution introduces a novel distance measure based on the generalized Cramér-von Mises distance that is suitable for comparing spatially adjacent deflection maps. The approach is evaluated by conducting experiments using both, simulated data and existing deflection maps acquired with a prototype of a deflection scanner. The results show, that the method is not as sensitive as the recently proposed earth mover's distance but might be able to yield spatially more accurate visualizations of scattering material defects.

Keywords: Transparent object inspection · Deflection maps · Light fields · Image processing · Histogram comparison

1 Introduction

For humans' every day life, objects made out of transparent materials are of utter importance. Such objects are important components in diverse products and applications and have to meet high quality requirements. For example, the windshields of automobiles and aircrafts have to protect the driver or pilot from the surroundings but also have to enable clear sight. Spectacle glasses or contact lenses are another example for which high requirements apply—they have

© Springer International Publishing AG 2017
P. Sharma and F.M. Bianchi (Eds.): SCIA 2017, Part I, LNCS 10269, pp. 526–537, 2017.
DOI: 10.1007/978-3-319-59126-1_44

to be free from even the smallest inhomogeneities. Apparently, the employed materials are required to have no scratches, cracks or enclosed air bubbles that might impair the mechanical stability. Furthermore, there should be no opaque contaminants in or on the windshield, respectively, the glasses, so that the user has clear sight.

Considering these two examples it already gets clear, that there has to be some kind of visual inspection of transparent materials after their production. On the one hand, workers can be employed for performing the inspection. However, since this is a fatiguing task for humans, it is prone to missing defects, what could have disastrous consequences. On the other hand, automated visual inspection systems can carry out the quality control.

Existing machine vision systems based on cameras or laser scanners are able to reliably inspect certain kinds of transparent test objects for being free from defects [1,8]. Opaque contaminants, which mainly absorb incident light, can usually be found by looking for regions of the test object where the exiting light has lower intensities [4]. In contrast, scattering defects (e.g., enclosed air bubbles) are harder to detect since they mainly cause incident light rays to be deflected into multiple directions [2,5]. Therefore, scattering defects are manifested in changes of the distribution of the light's propagation direction.

Indeed, the camera-based systems have a limited depth of field that prevents them from being used to inspect large objects like windshields. Although laser scanner-based systems usually have a larger depth of field, they cannot be used to inspect test objects having a strongly curved surface (e.g., large lenses or cover glasses of automotive headlamps) since they require the emitted laser beam to propagate approximately straightly through the measurement field.

Recently, laser deflection scanners were introduced which are able to capture the angles by which laser beams are deflected when propagating through a transparent test object. Since the acquired so-called deflection maps also contain information about the direction of the observed light, they can be used for finding scattering defects. By means of adequately processing the deflection maps, it is possible to image scattering defects even for test objects having a strongly curved 3D shape. However, the necessary image processing methods are still in an early stage and further approaches have to be evaluated.

This article proposes a novel approach for processing acquired light deflection maps in order to visualize scattering material defects. The introduced method is based on localized cumulative distributions and a generalized formulation of the Cramér-von Mises distance. In Sect. 2, the most relevant work performed by other researchers is presented and summarized. Section 3 explains the mentioned Cramér-von Mises distance in detail and introduces the proposed novel processing approach. The performed experiments are covered and discussed in Sect. 4. Section 5 closes the paper with a short summary and some ideas concerning further research.

This contribution can be seen as an extension of [7,9]. The previously introduced processing methods are extended by a novel approach using the generalized

Cramér-von Mises distance. Besides, synthetic experiments are performed via an appropriately adapted rendering framework.

2   Related Work

Light deflection maps first got mentioned in [11]. Here, the authors introduce a novel optical system which they call a Schlieren deflectometer. They illuminate transparent test objects with parallel light beams whose orientation is successively tilted with respect to the optical axis. For every tilt angle of the parallel light beams, they observe the backside of the test object with a camera system. By using a telecentric lens, they only capture those rays that propagate approximately parallel to the optical axis. As for every acquired image the tilt angle of the illumination is known, the method allows to determine the deflection angles for all rays exiting the test object. Since rays can also be scattered or spread inside the test object, not only one angle is acquired per pixel but a two-dimensional distribution which is called a deflection map. The authors did not propose to further evaluate the acquired deflection maps in order to check the test object for defects or impurities.

In [9], Meyer et al. introduce a novel optical inspection setup consisting of a collimated illumination source and a so-called $4f$ light field camera. This camera allows to capture spatially resolved measurements of the deflection angles of light rays propagating through the measurement field. By this means, they can acquire deflection maps of transparent test objects. In order to image scattering defects present in the test objects, the authors propose two methods for calculating image gradients based on the earth mover's distance and on vector fields of the mean local deflection direction. These two approaches allow to visualize spatial discontinuities of the deflection direction distribution. The authors positively evaluated their method using simulated data—however, the achievable spatial resolution is limited by the trade-off between angular and spatial resolution which is characteristic for optical setups based on light field cameras.

The first prototype of a laser scanner setup capable of acquiring deflection maps has been described in [7]. Figure 1 shows the principle optical setup of the deflection laser scanner proposed by the authors. Sequentially, parallel laser beams are emitted into the measurement field. The receiving part of the system consists of a convex lens or a parabolic mirror having a two-dimensional detector array in its focal plane. By this means, the angular information about the propagation direction of the light rays is transformed into a spatial information. Any light ray that is deflected inside the measurement field by an angle α will reach the detector array with a spatial displacement of $\delta = \tan(\alpha)f$ with respect to the intersection of the optical axis and the detector plane. So for every emitted laser beam a deflection map is acquired. The authors successfully evaluated their approach by setting up an early prototype of such a laser deflection scanner and by employing the processing methods introduced in [9] to clearly visualize scattering defects on two different transparent test objects.

The group of Hanebeck et al. introduced a new type of characterizing multivariate probability distributions, the so-called localized cumulative distribution

(LCD) [3]. The LCD can be summarized as a cumulative rectangular kernel transform of the respective probability density function. The authors further introduce a generalized form of the Cramér-von Mises distance (CVM-distance) that is based on the LCDs of two probability density functions and also suitable for the multivariate case. In probability theory, this generalized CVM-distance can be employed, e.g., to measure the similarity between two probability distributions, of which one is of a discrete type. In Sect. 3, this paper shows how the CVM-distance can be employed to analyze deflection maps in order to find spatial discontinuities of the angular deflection distribution of captured light rays.

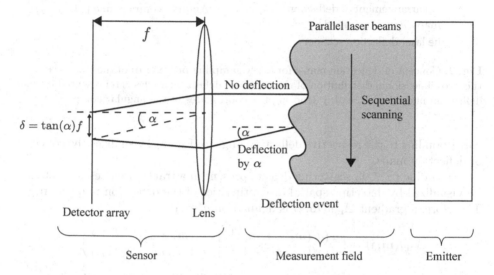

Fig. 1. Principle optical setup of a laser deflection scanner: the main components are an emitter sequentially emitting parallel laser beams and the receiving part consisting of a lens or a parabolic mirror with focal length f having a two-dimensional detector array in its focal plane. If a beam is deflected by an angle α inside the measurement field and still reaches the lens of the receiver, it will hit the detector array with a spatial displacement of $\delta = \tan(\alpha)f$ with respect to the intersection of the optical axis and the detector plane.

3 Deflection Map Processing Using the Generalized Cramér-von Mises Distance

A laser deflection scanner as proposed in [7] captures light intensities, that can be represented as a four-dimensional function $a(\mathbf{m}, \mathbf{j})$, with $\mathbf{m} = (m_1, m_2)^T$ corresponding to the respective spatial position of the laser source and $\mathbf{j} = (j_1, j_2)^T$ denoting the position on the detector array, i.e., the angular deflection component. It has to be noted that $\mathbf{m}, \mathbf{j}$ represent discretized quantities and that—especially for the angular coordinates $(j_1, j_2)^T$—the discretization involves non-linear operations, e.g., the calculation of the spatial displacement $\delta = f \tan(\alpha)$

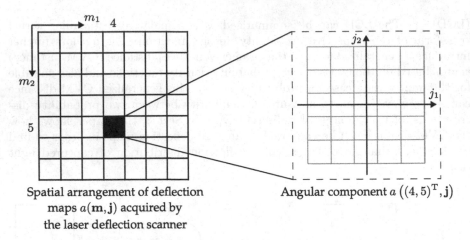

Spatial arrangement of deflection
maps $a(\mathbf{m},\mathbf{j})$ acquired by
the laser deflection scanner

Angular component $a\left((4,5)^{\mathrm{T}},\mathbf{j}\right)$

Fig. 2. Concept of deflection maps: for every sampling position $\mathbf{m}$ of the laser scanner, the two-dimensional distribution of the captured deflection angles is represented by the deflection map $a(\mathbf{m},\mathbf{j})$, with $\mathbf{j}$ denoting the coordinates of the angular component.

corresponding to the respective deflection angle α. Figure 2 visualizes the concept of deflection maps.

According to [7,9], scattering defects present in a transparent test object can be visualized by detecting spatial discontinuities of the deflection maps $a(\mathbf{m},\mathbf{j})$. Therefore, a gradient $\Delta_{\mathbf{m}}a(\mathbf{m},\mathbf{j})$ is defined on $a(\mathbf{m},\mathbf{j})$:

$$\Delta_{\mathbf{m}}a(\mathbf{m},\mathbf{j}) = \begin{pmatrix} d\left(a((m_1-1,m_2)^{\mathrm{T}},\mathbf{j}),a((m_1+1,m_2)^{\mathrm{T}},\mathbf{j})\right) \\ d\left(a((m_1,m_2-1)^{\mathrm{T}},\mathbf{j}),a((m_1,m_2+1)^{\mathrm{T}},\mathbf{j})\right) \end{pmatrix}, \tag{1}$$

with $d(\cdot,\cdot)$ denoting a distance function suitable for calculating the distance between two deflection maps. The gradient $\Delta_{\mathbf{m}}a(\mathbf{m},\mathbf{j})$ can be used for visualizing material defects or it can be used as a feature for an automated classification. As mentioned in the introduction, a generalized version of the Cramér-von Mises distance will be used for $d(\cdot,\cdot)$ in this paper. Therefore, the localized cumulative distribution has to be introduced first.

3.1   Localized Cumulative Distribution

For a random vector $\tilde{\mathbf{x}} \in \mathbb{R}^N, N \in \mathbb{N}$ that is characterized by an N-dimensional probability density function $f \colon \mathbb{R}^N \to \mathbb{R}_+$ there is no unique cumulative distribution function for $N > 1$. As mentioned before, [3] propose a localized cumulative distribution (LCD) that is well-defined for the multivariate case. The LCD $F(\mathbf{x},\mathbf{b})$ of f is defined as

$$F(\mathbf{x},\mathbf{b}) := P\left(|\tilde{\mathbf{x}} - \mathbf{x}| \le \frac{1}{2}\mathbf{b}\right), \tag{2}$$

$$F(\cdot,\cdot) \colon \Omega \to [0,1], \Omega \subset \mathbb{R}_+^N \times \mathbb{R}_+^N, \tag{3}$$

with $\mathbf{b} \in \mathbb{R}_+^N$ denoting the sizes of the integration kernel in the individual dimensions and the relation $\mathbf{x} \leq \mathbf{y}, \mathbf{x}, \mathbf{y} \in \mathbb{R}_+^N$ only holding if $\forall i \in [1, \dots, N] : x_i \leq y_i$. From a probability density function $f(\mathbf{x})$ the respective LCD $F(\mathbf{x}, \mathbf{b})$ is calculated by

$$F(\mathbf{x}, \mathbf{b}) = \int_{\mathbf{x} - \frac{1}{2}\mathbf{b}}^{\mathbf{x} + \frac{1}{2}\mathbf{b}} f(\mathbf{t}) d\mathbf{t}. \tag{4}$$

In order to apply this formulation to the deflection maps $a(\mathbf{m}, \mathbf{j})$, they first have to be normalized so that their values are in the range of $[0, 1]$ and their discrete nature has to be taken into account. Therefore, normalized pendants $\tilde{a}(\mathbf{m}, \mathbf{j})$ are defined as

$$\tilde{a}(\mathbf{m}, \mathbf{j}) = \frac{a(\mathbf{m}, \mathbf{j})}{\sum\limits_{(k,l)^T \in \Omega_j} a(\mathbf{m}, (k,l)^T)}, \tag{5}$$

with Ω_j denoting the angular support of the deflection maps, i.e., the set of valid coordinates with respect to the detector array of the employed laser deflection scanner. By means of Eq. (5), the formulation (4) of the LCD can be adapted, so that the LCD of $a(\mathbf{m}, \mathbf{j})$ is given by

$$A(\mathbf{m}, \mathbf{j}, \mathbf{b}) = \sum_{i = \max\{1, j - b\}}^{\min\{j_{\max}, j + b\}} \tilde{a}(\mathbf{m}, \mathbf{i}), \tag{6}$$

with $\mathbf{j}_{\max}$ denoting the maximum deflection captured by the system and $\max\{\cdot\}$ and $\min\{\cdot\}$ representing component wise vector operations. Based on these formulations, the generalized Cramér-von Mises distance can now be introduced.

3.2   Generalized Cramér-von Mises Distance

The generalized CVM-distance is a distance measure between two multivariate probability distributions based on their LCDs. Given the LCDs $F(\mathbf{x}, \mathbf{b})$ and $G(\mathbf{x}, \mathbf{b})$ of two probability density functions $f(\mathbf{x})$ and $g(\mathbf{x})$, their generalized CVM-distance is given by

$$D(f, g) := \int_{\mathbb{R}^N} \int_{\mathbb{R}_+^N} \left(F(\mathbf{x}, \mathbf{b}) - G(\mathbf{x}, \mathbf{b}) \right)^2 d\mathbf{b} d\mathbf{x}. \tag{7}$$

Equation (7) can now be employed in concert with (6)—the adapted formulation of the LCD for the deflection maps—to obtain a formulation of the generalized CVM-distance $d_{\text{CVM}}(a_1, a_2)$ that can be used as a distance measure in (1) between two deflection maps a_1 and a_2:

$$d_{\text{CVM}}(a_1, a_2) = \sum_{\mathbf{j} \in \Omega_j} \sum_{\mathbf{b} \in \Omega_j} \left(A_1(\mathbf{m}, \mathbf{j}, \mathbf{b}) - A_2(\mathbf{m}, \mathbf{j}, \mathbf{b}) \right)^2. \tag{8}$$

Furthermore, Eq. (8) can be extended by weights $w(\mathbf{b})$ corresponding to the individual kernel sizes in $\mathbf{b}$:

$$d_{\text{CVM}}(a_1, a_2) = \sum_{j \in \Omega_j} \sum_{\mathbf{b} \in \Omega_j} w(\mathbf{b}) \left(A_1(\mathbf{m}, \mathbf{j}, \mathbf{b}) - A_2(\mathbf{m}, \mathbf{j}, \mathbf{b}) \right)^2, \qquad (9)$$

with

$$\sum_{\mathbf{b} \in \Omega_j} w(\mathbf{b}) = 1. \qquad (10)$$

4   Evaluation

Simulated and empirically obtained data has been used to obtain a first evaluation of the proposed approach. By means of the physically based rendering framework Mitsuba [6], sensor readings of a laser deflection scanner as mentioned in Sect. 2 have been simulated. Further evaluations were based on the sensor readings from [7], that have been acquired using an early prototype of a laser deflection scanner. For the interpretation of the results described in the following sections it has to be taken into account that only a limited amount of data had been available when conducting the experiments.

4.1   Simulated Experiments

For performing the simulations, the Mitsuba framework has been extended by several components and plugins[1]. A sensor plugin has been introduced that models the receiving part of the described laser deflection scanner (see Fig. 1). Further modifications were needed in order to obtain the time sequential sensor readings for the different spatial positions of the laser source. For this purpose, a collimated light source has been realized, that is attached to a rectangular shape with a virtual 2D-grid representing the discrete positions $(m_1, m_2)^{\mathrm{T}}$ of the emitted light beams. Additionally, the rendering's ray tracing engine has been adequately adapted. Whenever one of the sensors' ray of sights is traced to the introduced light source, the respective 2D-position on the mentioned grid is combined with the 2D-coordinate $(j_1, j_2)^{\mathrm{T}}$ of the sampled sensor element and added to the ray data structure. Together, these two 2D-vectors finally allow to assemble the two-dimensional array $a(\mathbf{m}, \mathbf{j})$ of 2D-deflection maps.

In the simulated experiments, a double-convex lens suits as the test object. In order to simulate scattering material defects (enclosed air bubbles), small spheres of different sizes with an index of refraction of $n = 1$ have been placed inside the test objects. The introduced CVM-distance is compared against an existing approach, the so-called earth mover's (EM) distance [7,9,10], for processing deflection maps in order to visualize material defects.

The simulated laser deflection scanner setup had a spatial resolution of 300×300, an angular resolution of 15×15 and a pitch between the laser sampling positions of $334\,\mu$m. This choice of the parameter values is motivated by

---

[1] The respective source files can be downloaded from the main author's homepage.

the parameters of the empirical experiments. Small spheres with an index of refraction of $n = 1$ (air) have been inserted into the test object in order to simulate scattering material defects. Sensor readings have been simulated for a defect-free test object instance and for test objects with enclosed air spheres with radii $r \in \{83\,\mu\text{m}, 167\,\mu\text{m}, 334\,\mu\text{m}\}$. The radii of the scattering defects were chosen in order to evaluate the ability of the investigated methods to visualize defects of sizes around the sampling pitch of the laser scanner. Of every defect size, three defect instances have been inserted into the test object, one in the lower left corner, one in the center and one in the upper right corner.

The results calculated out of the simulated sensor readings using the earth mover's distance and the generalized CVM-distance are shown in Fig. 3 as pseudo color images. The grid-like structures present in all images is a result of the discrete nature and the limited resolution of the detector array and of the simulation. Whenever deflection angles change so that the laser spot reaches another sensor element on the detector array there is a spatial discontinuity in the deflection map at the respective position. Furthermore, the rendering framework simulates infinitely thin rays and does not take divergence or other effects into account. Hence, for sensor readings from a real physical system, such artifacts would not occur.

The smallest defects ($r = 83\,\mu\text{m}$) are only visualized by the EM-distance but with low contrast. The defects with radius of $r = 167\,\mu\text{m}$ are visible in the images obtained using the EM-distance and the CVM-distance. The largest simulated defects are clearly visualized by both methods but the image resulting from the generalized CVM-distance shows the defects slightly clearer.

4.2 Empirical Experiments

Further experiments were performed using empirical data from [7]. The sensor readings have been obtained using a prototype of a laser deflection scanner system with a spatial resolution of 39×39, an angular resolution of 9×9 and a pitch between the laser sampling positions of approximately $334\,\mu\text{m}$. A cylindrical convex lens served as the test object. There are sensor readings for a defect-free test object and for one affected by two scattering surface defects.

Figure 4 shows the results calculated using the EM-distance and the generalized CVM-distance as pseudo color images. As for the simulated sensor readings, the images obtained using the generalized CVM-distance have a clearer appearance due to their higher contrast compared to the images resulting from the EM-distance. Still, in both images the defect-free test object instance can clearly be recognized as being free from defects and the two defects present on the second test object instance are clearly visualized by both approaches.

For a quantitative evaluation, the signal-to-noise ratio SNR and the peak signal-to-noise ratio $PSNR$ can be taken into account. The SNR is defined by

$$SNR := \frac{\mu_{\text{defect}}}{\sigma_{\text{defect-free}}}, \tag{11}$$

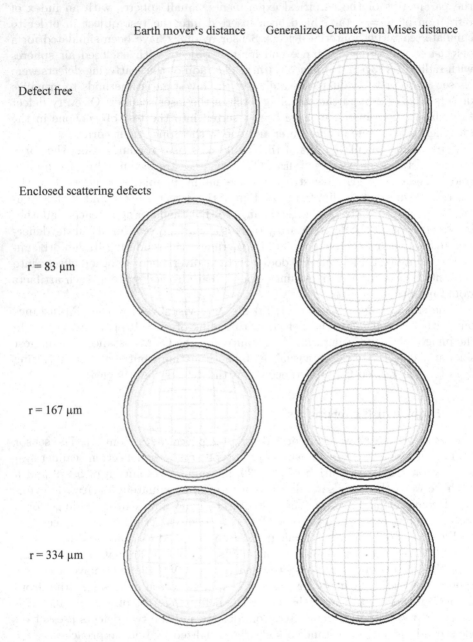

Fig. 3. Pseudo color images showing the results of the performed experiments. A double-convex lens has been used as the test object. The radii of the simulated scattering defects are denoted by r. Bright, respectively, dark regions correspond to low, respectively, high distance values.

with μ_{defect} denoting the mean signal value for pixels belonging to defective test object regions and $\sigma_{\text{defect-free}}$ representing the standard deviation of pixels belonging to defect-free regions of the test object. The PSNR is calculated via

$$PSNR := \frac{\hat{s}_{\text{defect}}}{\sigma_{\text{defect-free}}}, \tag{12}$$

with $\hat{s}_{\text{defect}}$ denoting the maximum signal intensity among the pixels corresponding to the defective test object region. Table 1 shows the SNR and $PSNR$ values resulting from the performed empirical experiments. The CVM-distance yields higher SNR and $PSNR$ values than the EM-distance but indeed, more experiments are needed to support these results.

Table 1. Comparison of the signal-to-noise ratio (SNR) and the peak signal-to-noise ratio ($PSNR$) for the two processing approaches.

	SNR	$PSNR$
EM-distance	15.3	28.3
CVM-distance	18.1	39.8

In the image obtained with the generalized CVM-distance, it might be possible to infer the size of the defect more accurately. The defects' actual sizes were manually measured and are approximately 2.8 mm × 2.8 mm for the upper left defect and 4 mm × 4 mm for the lower right defect. Table 2 shows rough estimations of the defects' sizes that can be obtained by counting the pixels clearly belonging to the defective regions. These results show that the EM-distance slightly tends to render the defects larger than they actually are—however more experiments are needed to support these findings.

Table 2. Comparison of the manually measured defect sizes and the rough estimations obtained by counting the pixels clearly belonging to the defective regions in the images resulting from the two processing approaches.

Defect	Upper left	Lower right
Manually measured size	2.8 mm × 2.8 mm	4mm × 4mm
EM-distance based estimation	4 mm × 4 mm	4.68 mm × 4.68 mm
CVM-distance based estimation	3 mm × 3 mm	4.34 mm × 4.34 mm

Earth mover's distance   Generalized Cramér-von Mises distance

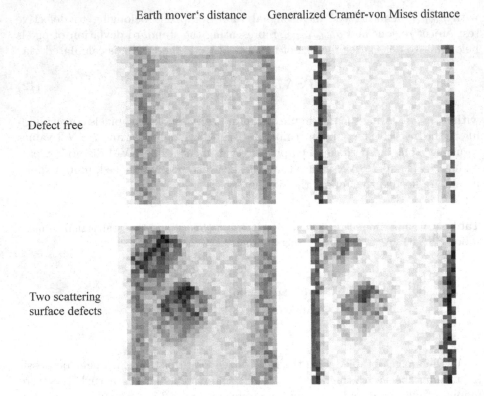

Defect free

Two scattering
surface defects

Fig. 4. Pseudo color images showing the results of the performed experiments based on empirical data. Bright, respectively, dark regions correspond to low, respectively, high distance values.

5   Summary and Outlook

This contribution proposed a novel method for processing light deflection maps in order to visualize scattering material defects present in transparent objects. Since these defects are mainly manifested in changes of the direction of transmitted light, spatial discontinuities in captured light deflection maps have to be found. Therefore, adequate distance methods between spatially adjacent deflection maps are needed. In this paper, the recently introduced concept of localized cumulative distributions and the generalized formulation of the Cramér-von Mises distance have been adapted so that they are applicable for comparing deflection maps. The proposed approach has been successfully validated by means of experiments based on both, simulated sensor readings and existing empirical data acquired using a prototype of a deflection laser scanner. The experiments showed that the introduced method is able to accurately visualize scattering defects. However, the earth mover's distance seems to be slightly superior to the CVM-distance regarding the visualization of smallest defects.

As further steps, the authors plan to conduct additional experiments based on an improved laser deflection scanner prototype in order to quantitatively evaluate and compare the capabilities and limits of different adequate distance measures. Regarding the optical setup, the dependence of the smallest detectable defect size on the system parameters (focal length, array size and resolution) will be investigated. Besides the suitability for visualizing material defects, other properties of the investigated methods, e.g., their computational complexity and processing time will be studied.

Furthermore, the authors plan to evaluate more physically motivated approaches for processing the deflection maps. On the one hand, small material defects could cause diffraction effects that lead to diffraction patterns on the sensor which can be used to infer the defects' sizes. On the other hand, also scattering theory, e.g., Mie scattering could be used to obtain more information about defects based on the characteristics of the observed intensity patterns.

References

1. Beyerer, J., León, F.P., Frese, C.: Machine Vision: Automated Visual Inspection: Theory, Practice and Applications. Springer, Heidelberg (2016)
2. Bohren, C.F., Huffman, D.R.: Absorption and Scattering of Light by Small Particles. Wiley-VCH Verlag GmbH, Berlin (2007)
3. Hanebeck, U.D., Klumpp, V.: Localized cumulative distributions and a multivariate generalization of the cramér-von mises distance. In: IEEE International Conference on Multisensor Fusion and Integration for Intelligent Systems (MFI 2008), pp. 33–39. IEEE (2008)
4. Hartrumpf, M., Heintz, R.: Device and method for the classification of transparent components in a material flow. Patent WO 2009/049594 A1 (2009)
5. van de Hulst, H.: Light Scattering by Small Particles. Dover Books on Physics. Dover Publications, New York (1957)
6. Jakob, W.: Mitsuba renderer (2010). http://www.mitsuba-renderer.org
7. Meyer, J., Längle, T., Beyerer, J.: Acquiring and processing light deflection maps for transparent object inspection. In: 2016 2nd International Conference on Frontiers of Signal Processing (ICFSP), pp. 104–109, October 2016
8. Meyer, J.: Overview on machine vision methods for finding defects in transparent objects. Technical report IES-2015-08, Karlsruhe Institute of Technology (2015)
9. Meyer, J., Längle, T., Beyerer, J.: About the acquisition and processing of ray deflection histograms for transparent object inspection. In: Irish Machine Vision & Image Processing Conference proceedings (2016)
10. Rubner, Y., Tomasi, C., Guibas, L.: A metric for distributions with applications to image databases. In: Sixth International Conference on Computer Vision, pp. 59–66. IEEE (1998)
11. Sudhakar, P., Jacques, L., Dubois, X., Antoine, P., Joannes, L.: Compressive imaging and characterization of sparse light deflection maps. SIAM J. Imag. Sci. 8(3), 1824–1856 (2015)

Dynamic Exploratory Search in Content-Based Image Retrieval

Joel Pyykkö$^{(\boxtimes)}$ and Dorota Głowacka

Department of Computer Science, University of Helsinki, Helsinki, Finland
{joel.pyykko,dorota.glowacka}@cs.helsinki.fi

Abstract. With the increase of digital media databases, the need for methods that can allow the user to efficiently peruse them has risen dramatically. This paper studies how to explore image datasets more efficiently in online content-based image retrieval (CBIR). We present a new approach for exploratory CBIR that is dynamic, robust and gives a good coverage of the search space, while maintaining a high retrieval precision. Our method uses deep similarity-based learning to find a new representation of the image space. With this metric, it finds the central point of interest and clusters its local region to present the user with representative images within the vicinity of their target search. This clustering provides a more varied training set for the next iteration, allowing the location of relevant features faster. Additionally, relearning a representation of the user's search interest in each round enables the system to find other non-local regions of interest in the search space, thus preventing the user from getting stuck in a context trap. We test our method in a simulated online setting, taking into consideration the accuracy, coverage and flexibility of adapting to changes in the user's interest.

Keywords: Content based image retrieval (CBIR) · Deep neural networks · Vector space models · Interactive information retrieval · Exploratory search

1 Introduction

Actively learning the user's search target is an important aspect of content-based image retrieval techniques (CBIR) [11]. Traditionally, image retrieval techniques were dependent on meta-data or otherwise restrictive features, which limited the possible directions of search. In recent years, there has been an interest in developing methods that dynamically react to changes in the semantic target search of the user. A related issue is system responsiveness both in terms of processing time and the amount of user feedback required for the system to converge to the user's search target.

There are many scenarios which would benefit from a highly dynamic and fast image retrieval systems, which can adjust to changes in what the user's ideal target image is. For example, an artist or journalist browsing stock photos. They might be looking for an image with a particular mood, which cannot be

© Springer International Publishing AG 2017
P. Sharma and F.M. Bianchi (Eds.): SCIA 2017, Part I, LNCS 10269, pp. 538–549, 2017.
DOI: 10.1007/978-3-319-59126-1_45

easily captured with tags, and they will only know they have found the right images once they see it. This would require the system to be able to learn specific features from the data throughout a given search session even though they may not have been in the original training set. Another case could be a doctor searching through a medical image database for images similar to an image of their patients condition.

In each scenario, the system has to be responsive and answer to the needs of the user immediately, providing relevant search results after a few training examples [32]. Furthermore, the user might not be sure what they are looking for at the beginning of the search session, but are refining their target as they see more images from a given dataset. Covering the various types of images present in a given dataset in a fair manner is important to avoid the context trap [17], where the user "gets stuck" in a single location in the search space. Sometimes the user may simply wish to browse the contents of the database, knowing their target only once they see it.

With these challenges in mind, we introduce a framework for exploratory CBIR that tackles all the above challenges. Our system flexibly relearns the relevant features from a simple binary feedback within a few search iterations, while balancing a good coverage of the database and maintaining high accuracy. Furthermore, the suggested framework allows the user to peruse images from unannotated databases. We also propose a metric to measure the coverage of the search space suitable for datasets of any size.

Our architecture utilizes pre-extracted features from deep neural networks, on top of which we learn a new distance representation of the images based on the user's relevance feedback. For the distance measure, we use the Siamese architecture, which was originally used for face verification [5]. The method then uses these distances to find interesting regions in the data set, identifying an efficient ranking for each image to help the user to learn what is available. These regions are found by clustering the images in the new representation and then choosing central images from each cluster to represent all the images in a given region. When the user chooses an image from one of these regions, the search may proceed faster towards the images most relevant to the task at hand.

2   Related Work

The first CBIR experiments date back to 1992 [16]. Since then, a variety of feature descriptors or local representations have been used for image representation, such as color, edge, texture and GIST [21], as well as local feature representations, such as the bag-of-words models [31] in conjunction with local feature descriptors (e.g. SIFT [20]). However, using such low-level feature representation may not be always optimal for more complex image retrieval tasks due to the semantic gap between such features and high-level human perception. Hence, in recent years there has been an increased interest in developing similarity measures specifically for such low-level feature image representation [3] as well as enhancing the feature representation in distance metric learning [28].

Over the past decade deep neural networks have been successfully utilized in CBIR tasks, bringing strong feature representations to the field [22]. For example, learning deep hierarchies that formulate short binary codes for image data sets that allow fast retrieval was considered on two occasions: by using autoencoders to automatically formulate a structure for the data [19], or creating hash codes based on deep semantic ranking [33]. While both methods are fast, neither is flexible enough to learn the image target based on the small amount of user's relevance feedback. Wan et al. [28] is the first study to apply deep learning to learn a similarity measure between images in a CBIR setting. The method relies on metric learning methods to learn the similarity between images. Their initial results show a potential for using deep similarity measures in image retrieval tasks.

Unfortunately, no consideration was given to the time requirements of the learning task, which is an important aspect of online interactive retrieval systems. A comprehensive case-study of deep learning in CBIR was conducted in 2014 [29]. Many state-of-the-art techniques were utilized in four different image retrieval tasks, showing the usability of convolutional networks in CBIR. These methodologies are similar to our baseline setups, utilizing similar feature representations and ranking principles.

Exploratory search for image retrieval is a field of research that has seen more activity in recent years [8,10,12–14,18,27]. There are a number of approaches comparable to our work. AIDE [6] is a framework for generic information retrieval settings in metric environments, which is also applicable for image retrieval. The method employs several exploration phases for even sampling of the search space, giving a good overview of the present items. In iconic images [1] a framework was proposed that shows clusters of images to the user. This setup finds images that represent existing categories in the search space by finding most salient and complete examples of possible targets. In [2] a kernel based method was suggested for image set exploration, which allowed users to utilize image representation, summarization and visualization. Although relevant to our work, their method focused solely on browsing the system. CLUE [4] is a framework for CBIR that utilizes graph-theoretic clustering to find similarities between images, providing a good comparison for images.

Previous work focused mainly on non-exploratory methodologies, which work best in simple look-up scenarios, or exploratory methodologies that use existing or rigid structures to measure the search space. The aim of our work is to create an exploratory CBIR system, which requires little preprocessing to start with, and is not constrained by predefined structures. Instead, it learns the user's interests dynamically during the search. Due to this, we assess the quality of exploration by comparing jointly both the precision and the coverage of the retrieved images. This ensures that the user does not only find a local cluster of good images within the search space, but also obtains a more comprehensive view of the search space.

3   System Overview

Our system assists the user in finding relevant images from databases that have minimal preprocessing done to them before hand. The user may feed an example image into the system at the beginning of the search to speed the process of finding relevant images. Next, at each search iteration the user is presented with k images and they indicate the relevant ones for their search. The remaining images in the set of k images that did not receive any feedback are treated as irrelevant. Based on this feedback, all the images in a dataset are re-ranked, providing a more refined representation on what is relevant for the user. The system aims to identify the subspace from the dataset that is relevant for the user in as few iterations as possible. We do this by exploring regions near the images already tagged as relevant.

3.1   Feature Extraction

In order to obtain a good base representation, we use features extracted with OverFeat [26] and relearn the last fully connected layers as the target representation. OverFeat is a publicly available convolutional neural network, trained with the ILSVRC13 data set [24], on which it achieved an error rate of 14.2%. ILVSRC13 contains 1000 object classes from a total of 1.2 million images. Over-Feat has been shown to be successful in various image recognition tasks from fine-grained classification to generic visual instance recognition tasks [23]. The selected features were a set of hidden nodes as the fully connected graph begins from layer 7 (19 within the architecture), totalling 4096 features. The images were shrunk and then cropped from all sides to produce images of equal size of 231×231 pixels.

3.2   Siamese Architecture

Our system employs the Siamese architecture [5], which is used to learn a similarity metric between images. This is done by maximizing the pair-wise distance of dissimilar images, and minimizing the same for similar images. The resulting representation maps images into a euclidean space where images from one class are grouped together, while being separated from images from other classes. We employ user relevance feedback to divide the presented images into the two classes, i.e. images with positive feedback (relevant class) and images with negative feedback (non-relevant class).

Siamese architecture D consists of two networks, both of which take as input an image X_1 and X_2, respectively. The networks share their weights W, which are trained to learn a location in the new metric representation. The system uses a contrastive loss function:

$$L(W, Y, X_1, X_2) = (1 - Y)\frac{1}{2}(D_W)^2 + Y\frac{1}{2}\{max(0, m - D_W)\}^2,$$

where $Y = 1$ if the two images are from the same class, and $Y = 0$ if they are from different classes, D_W is the network's predicted distance between X_1

542   J. Pyykkö and D. Głowacka

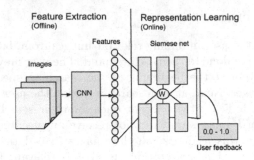

Fig. 1. The Siamese architecture and the image feature preprocessing step. The online component accepts two feature vectors, one per image, and user feedback as the label.

and X_2 given W, and m is a margin we wish to keep between different classes of images. This metric aims to maximize the intra-class similarity, in the case where X_1 and X_2 belong to the same class, and to minimize the inter-class similarity if they belong to different classes.

The Siamese architecture (Right side in Fig. 1) is able to find a new representation in the feature space that helps to distinguish between different aspects of the image, making it an ideal choice for our application. An important aspect of this architecture is that it generates a distance metric, which may be used to rank or generate relevance scores for all the images in a dataset.

3.3   Exploratory Search

As the Siamese neural network produces a metric representation for the images, our exploratory methodologies are able to separate data points spatially into regions of interest. The exploratory methodologies we present here affect two variables in the framework: first, what the primary point of interest in the search space is, and second, how to explore the region around this point. For our primary location, we evaluate a focal point that is far from irrelevant images and close to the center of the cluster of relevant images. From here we explore nearby groups of images to find images that cover as large a portion of images as possible.

Central target γ is the starting point of the exploration in our method. It is chosen to be the center of the cluster of images rated as relevant, which should be close to the highest estimated relevance according to the Siamese network. This point is found by minimizing the distance for each positively ranked image $x_+ \in X$, while maximizing the distance to all the negatively ranked images $x_- \in X$. Central target is thus selected by the following function:

$$\gamma = argmin\big(\Sigma(\|x, x_+\|) + \Sigma(C - \|x, x_-\|) \mid x \in X\big),$$

where x is an image that may or may not have yet been ranked, $\|x, x_+\|$ is the distance measure between x and x_+, and C is a suitably selected constant. The purpose of the constant is to work as the exploratory term – moving more aggressively away from the edges of positive clusters allows the method to find the center of each relevant region faster.

After the central target has been located, we define the local region as the nearest n images. This neighbourhood is then clustered to locate images that describe the content best. Defining a range parameter such as this one allows our method to utilize the relearned metric to increase the relevance. The parameter may be chosen as a proportion of the total size of the dataset balanced with the processing capabilities of the whole system.

In our system we used DBScan [7] for clustering as it generates clusters based on the form of the data itself, creating as many clusters as there are concentrated regions. This phase handles the exploratory phase of the search. As DBScan does not generate centroids, we use the image closest to the center as the centroid. This image is presented to the user if there are enough exploratory slots left. Depending on the precision rate of the previous iteration, we explore more or fewer items close to these centroids. If the previous iteration resulted in only relevant images, no exploration is done but rather the primary central region is exploited until it is exhausted. If, on the other hand, the precision is low, we look for more images from the nearby clusters.

Our algorithm moves around the image space due to the changes the neural net imposes on the representation. It reconfigures the center of interest at each iteration, assuming that interesting images were successfully separated from the rest.

4   Experiments

We conducted two sets of experiments with three different datasets to evaluate the applicability of the proposed systems in interactive CBIR. For this, we identified the following aspects of the system's performance to be crucial. The system needs to work with relatively few training examples, i.e. at each search iteration, the user is presented with only a small number of images and often provides feedback only to a subset of these. The retrieval system needs to be able to "learn" what the user is looking for based on this limited feedback. The search target may be something very concrete, e.g. "red rose", or very abstract, e.g. "happiness", and the system needs to support all types of searches with varying degree of abstractness. Furthermore, the training time has to be very short for the system to be interactive from the user's perspective.

We first test the overall precision and running time of our method. In the second part of our experiments, we test how well our method compares with other algorithms in a simulated setting, measuring both cumulative precision and the coverage of the search space as the user gives more feedback.

4.1   Experimental Setup

In our experiments we used three different datasets (Fig. 2), where the class labels represent the targets of the search. The first dataset consisted of 1096

Fig. 2. Sample images from the datasets used in our experiments.

images from the MIRFlickr dataset [15], which contains images of various mammals, birds, insects and vehicles. The original labels were combined into larger classes, e.g. images of hens, finches and parrots were labeled as 'birds'. The aim of this dataset is to show that our method is able to transfer learn an abstract, combinatorial concept, that the original features were not meant for. Here the features for example, for hens and parrots are distinct, and no semantic relation has been separately taught to the original feature representation – the new representation has to be able to learn it.

The next dataset was our own collection of 294 images of 6 different dog breeds, of which only four are included in the OverFeat's classification list. This dataset allows us to test whether the model is able to transfer learn the target in the presence of semantically related images, some of which are not included in original scope of features used for training.

Lastly, we used a classical dataset to test how well the method scales to larger datasets: 100 classes from the ILSVRC2013 dataset [24], totalling 128894 images.

We tested our method against three other CBIR setups that work in comparable settings. Each of them is based on similarity measures and assists the user in exploring a given dataset. First, Rocchio's algorithm [25], which is a widely used ranking method for vector space settings. It finds a vector from around which documents are shown to the user. The relevance score given to the method directs this vector towards a space with more related documents. We also paired Rocchio's algorithm with a classical exploratory method from multi-armed bandit literature: ϵ-greedy exploration [30]. Here, a certain number of actions are randomized to avoid policy stagnation. More precisely, the estimated optimal action is taken with a chance $1 - \epsilon$, and a random choice with chance ϵ. The initial ϵ was set to 0.5, which was annealed linearly to 0 after 10 simulation iterations with steps of 0.05. Finally, AIDE [6] is a recent exploratory framework for information retrieval that attempts to provide a good coverage of the whole dataset. It partitions the search space into subspaces, from where it attempts to find all the relevant regions by presenting to the user samples from each of them. We conducted one initial test for our system with small training set sizes ranging from 10 up to 150 presented images. This is the average number of images in a typical CBIR search session [9], when the user is presented with

10 images in a single iteration. We measured the precision and running time of our method, with results presented in Tables 1 and 2.

Next, we conducted exploratory tests for all the methodologies outlined above. In this setup we simulated an image retrieval task, where the system presents 10 images to the simulated user over 20 iterations, yielding a total of 200 images. In the simulations, the user feedback is 1 for images with a relevant label and 0 for the remaining images in the presented set. The search starts with a random selection of 9 irrelevant images, plus one relevant image "chosen by the user" – this setting allows us to ensure that all the simulation experiments have a comparable starting point. For the ILSVRC2013 dataset we sampled evenly 20000 images for the testing for each simulation due to space and time constraints set by some of the baseline methodologies. This setting allowed us to test if our system scales well early in the search task, reaching good enough performance in precision and running speed. All the reported results are averaged over 5 training runs for each of the existing classes in the datasets.

To test the quality of exploration, we measured the precision of the retrieved images as well as the coverage of the search space. We measured coverage C as the average of distances between all retrieved items compared to the dataset size:

$$ C = \sum_{i,j} \left(\|x_i^s, x_j^s\| / maxDist(X) / \mid x^s \mid \right), $$

where x^s is the set of retrieved images, $\mid x^s \mid$ its size, $\|x_i^s, x_j^s\|$ is the distance between i:th and j:th member in the set averaged over the number of retrieved images. The term $maxDist(X)$ is the maximum distance between two points within the dataset, scaling the sum to be between 0 and 1. The greater the average sum of these distances, the further apart the data points are in the similarity space, and thus the larger the view over the data set is.

4.2 Experimental Results

The initial precision results are shown in Table 1. As can be seen, our system is able to retrieve relevant images with high accuracy even within the first few iterations. As the training set is increased to 150 images, the precisions become comparable to modern ranking methodologies.

Table 1. Average precision with the three datasets.

Initial test: Precision				
Data set/#Images	10	50	100	150
ILSVRC2013	0.709	0.810	0.835	0.846
Dogs	0.691	0.921	0.953	0.958
MIRFlickr	0.482	0.611	0.690	0.722

Table 2. Training time in seconds for the three datasets as well as sizes of the training datasets.

Initial test: Time taken (s)				
Data set/#Images	10	50	100	150
ILSVRC2013	2.911	3.042	3.208	3.315
Dogs	2.861	2.903	3.181	3.302
MIRFlickr	2.872	2.997	3.196	3.414

In Table 2 we show the average training time for each training set size. For each dataset, the average duration for each search iteration is below 4 s. This makes the system interactive from the usability perspective, and grows linearly even as the number of the training data points grows larger.

In the second set of experiments we look at the performance of the various exploratory methods (Fig. 3). We report the cumulative precision until a given point with the previous iterations acting as the context for the user throughout the search session.

Our centroid-based method gains clear advantage after approximately 5 iterations as the system learns the target representation. Due to the small number of images present in the dog dataset, the curves for this dataset turn downwards for most methods as the search progresses as all the relevant images have been exhausted early on in the search. Still, our method finds the relevant images sooner and finds a larger portion of them at the end of the search. For the MIR-Flickr dataset, we can see how Rocchio's exploits an early local cluster of good images but fails later on in the search as it is unable to break out of the initial context. Meanwhile our method sacrifices a number of attempts early on and gradually achieves a larger number of correctly retrieved images.

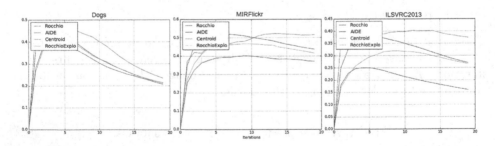

Fig. 3. Exploration tests for the three datasets, shown with cumulative precision for the following methods: the Rocchio's algorithm, central exploration, and AIDE.

The cumulative average coverage shows interesting trends with different methodologies. The baseline methods proceed steadily through the dataset adding relatively small gains throughout the search. Our method, on the other

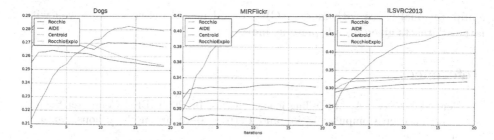

Fig. 4. Exploration tests for the three datasets, shown with cumulative average of coverage for the following methods: the Rocchio's algorithm, central exploration and AIDE.

hand, keeps finding new regions for a long time until slowing down after approximately 7 iterations. The overall coverage with our method is significantly larger in the various settings, showing how exploring the changing metric spaces helps us to find new regions of interest faster.

Finally, with the ILSVRC2013 dataset the effect of sparse targets highlights the efficiency of our method. With 100 target classes present, the local space of the initial target quickly exhausts valid images with centroid exploration. It is likely that if the first image is on the edge of the valid cluster of images, the nearby images will quickly present neighbouring classes.

In Fig. 4, we see the coverage for each method and dataset. The more exploratory methods keep covering a larger section of the datasets faster, while the greedy Rocchio's lags behind. With the MIRFlickr dataset, our method reaches the same coverage as AIDE after approximately 18 iterations. This suggests that the refined representation is able to find relevant locations beyond the immediate neighbourhood of the starting location.

5   Conclusions

We presented a deep exploratory search framework for online interactive CBIR settings, which reacts to the users feedback dynamically and covers a larger portion of the search space than conventional retrieval tools. The system allows users to conduct searches for concepts outside of the initially used features. We showed that this transfer learning is able to extend to abstract targets, learning concepts robustly that were not originally intended for the starting features.

The system is highly dependent on good initial image features. For cases where the dataset has not been annotated but presents natural images, a good object classification CNN is required. In the case of specialised image datasets, such as medical imaging, a separate neural network should be trained just for that purpose, after which the presented methodologies are able to transfer learn the various combinations required to identify the target. Furthermore, efficient sampling (or better hardware) is required to process more images. Fortunately, computational times with modern GPU-based neural networks scale well with larger datasets given an adequate memory.

References

1. Berg, T.L., Berg, A.C.: Finding iconic images. In: 2009 IEEE Computer Society Conference on Computer Vision and Pattern Recognition Workshops, pp. 1–8, June 2009
2. Camargo, J.E., Caicedo, J.C., Gonzalez, F.A.: A kernel-based framework for image collection exploration. J. Vis. Lang. Comput. **24**(1), 53–67 (2013)
3. Chechik, G., Sharma, V., Shalit, U., Bengio, S.: Large scale online learning of image similarity through ranking. J. Mach. Learn. Res. **11**, 1109–1135 (2010)
4. Chen, Y., Wang, J.Z., Krovetz, R.: Clue: cluster-based retrieval of images by unsupervised learning. IEEE Trans. Image Process. **14**(8), 1187–1201 (2005)
5. Chopra, S., Hadsell, R., LeCun, Y.: Learning a similarity metric discriminatively, with application to face verification. In: Proceedings of CVPR (2005)
6. Dimitriadou, K., Papaemmanouil, O., Diao, Y.: Explore-by-example: an automatic query steering framework for interactive data exploration. In: Proceedings of the 2014 ACM SIGMOD International Conference on Management of Data (SIGMOD 2014), pp. 517–528, New York, NY, USA. ACM (2014)
7. Ester, M., Kriegel, H.-P., Sander, J., Xiaowei, X.: A density-based algorithm for discovering clusters in large spatial databases with noise, pp. 226–231. AAAI Press (1996)
8. Głowacka, D., Shawe-Taylor, J.: Content-based image retrieval with multinomial relevance feedback. In: Proceedings of ACML, pp. 111–125 (2010)
9. Głowacka, D., Hore, S.: Balancing exploration-exploitation in image retrieval. In: Proceedings of UMAP (2014)
10. Glowacka, D., Teh, Y.W., Shawe-Taylor, J.: Image retrieval with a Bayesian model of relevance feedback. arXiv preprint (2016). arXiv:1603.09522
11. Heesch, D.: A survey of browsing models for content based image retrieval. Multimed. Tools Appl. **40**(2), 261–284 (2008)
12. Hoque, E., Hoeber, O., Gong, M.: CIDER: concept-based image diversification, exploration, and retrieval. Inf. Process. Manage. **49**(5), 1122–1138 (2013)
13. Hore, S., Glowacka, D., Kosunen, I., Athukorala, K., Jacucci, G.: Futureview: enhancing exploratory image search. In: IntRS@RecSys, pp. 37–40 (2015)
14. Hore, S., Tyrvainen, L., Pyykko, J., Glowacka, D.: A reinforcement learning approach to query-less image retrieval. In: Jacucci, G., Gamberini, L., Freeman, J., Spagnolli, A. (eds.) Symbiotic 2014. LNCS, vol. 8820, pp. 121–126. Springer, Cham (2014). doi:10.1007/978-3-319-13500-7_10
15. Huiskes, M.J., Lew, M.S.: The MIR flickr retrieval evaluation. In: Proceedings of MIR (2008)
16. Kato, T., Kurita, T., Otsu, N., Hirata, K.: A sketch retrieval method for full color image database-query by visual example. In: Pattern Recognition. Computer Vision and Applications, pp. 530–533 (1992)
17. Kelly, D., Xin, F.: Elicitation of term relevance feedback: an investigation of term source and context. In: Proceedings of the 29th Annual International ACM SIGIR Conference on Research and Development in Information Retrieval (SIGIR 2006), pp. 453–460, New York, NY, USA. ACM (2006)

18. Konyushkova, K., Glowacka, D.: Content-based image retrieval with hierarchical gaussian process bandits with self-organizing maps. In: ESANN (2013)
19. Krizhevsky, A., Hinton, G.E.: Using very deep autoencoders for content-based image retrieval. In: Proceedings of ESANN (2011)
20. Lowe, D.G.: Object recognition from local scale-invariant features. In: ICCV, pp. 1150–1157 (1999)
21. Oliva, A., Torralba, A.: Modeling the shape of the scene: a holistic representation of the spatial envelope. Int. J. Comput. Vis. **42**(3), 145–175 (2001)
22. Pyykko, J., Glowacka, D.: Interactive content-based image retrieval with deep neural networks. In: Gamberini, L., et al. (eds.) Symbiotic 2016. LNCS, vol. 9961. Springer, Heidelberg (2017)
23. Razavian, A.S., Azizpour, H., Sullivan, J., Carlsson, S.: CNN features off-the-shelf: an astounding baseline for recognition. CoRR, abs/1403.6382 (2014)
24. Russakovsky, O., Deng, J., Su, H., Krause, J., Satheesh, S., Ma, S., Huang, Z., Karpathy, A., Khosla, A., Bernstein, M., Berg, A.C., Fei-Fei, L.: Imagenet large scale visual recognition challenge. Int. J. Comput. Vis. **115**, 211–252 (2014)
25. Salton, G.: The SMART Retrieval System-Experiments in Automatic Document Processing. Prentice-Hall Inc., Upper Saddle River (1971)
26. Sermanet, P., Eigen, D., Zhang, X., Mathieu, M., Fergus, R., LeCun, Y.: Overfeat: integrated recognition, localization and detection using convolutional networks. In: Proceedings of ICLR (2014)
27. Suditu, N., Fleuret, F.: Iterative relevance feedback with adaptive exploration/exploitation trade-off. In: Proceedings of CIKM (2012)
28. Wan, J., Wang, D., Hoi, S.C.H., Wu, P., Zhu, J., Zhang, Y., Li, J.: Deep learning for content-based image retrieval: a comprehensive study. In: Proceedings of MM (2014)
29. Wan, J., Wang, D., Hoi, S.C., Wu, P., Zhu, J., Zhang, Y., Li, J.: Deep learning for content-based image retrieval: a comprehensive study. In: Proceedings of the 22nd ACM International Conference on Multimedia (MM 2014), pp. 157–166, New York, NY, USA. ACM (2014)
30. Watkins, C.J.C.H.: Learning from delayed rewards. Ph.D. thesis, King's College, Cambridge, UK, May 1989
31. Yang, J., Jiang, Y.-G., Hauptmann, A.G., Ngo, C.-W.: Evaluating bag-of-visual-words representations in scene classification. In: Multimedia, Information Retrieval, pp. 197–206 (2007)
32. Yee, K.-P., Swearingen, K., Li, K., Hearst, M.: Faceted metadata for image search and browsing. In: Proceedings of the SIGCHI Conference on Human Factors in Computing Systems (CHI 2003), pp. 401–408, New York, NY, USA. ACM (2003)
33. Zhao, F., Huang, Y., Wang, L., Tan, T.: Deep semantic ranking based hashing for multi-label image retrieval. ArXiv e-prints, January 2015

Robust Anomaly Detection Using Reflectance Transformation Imaging for Surface Quality Inspection

Gilles Pitard[1]([✉]), Gaëtan Le Goïc[2], Alamin Mansouri[2], Hugues Favrelière[3], Maurice Pillet[3], Sony George[1], and Jon Yngve Hardeberg[1]

[1] The Norwegian Colour and Visual Computing Laboratory,
Department of Computer Science, NTNU, Gjøvik, Norway
gilles.pitard@ntnu.no
[2] Laboratoire LE2I, FRE CNRS 2005, UBFC, Auxerre, France
[3] Laboratoire SYMME, EA 4144, USMB, Annecy, France

Abstract. We propose a novel methodology for the detection and analysis of visual anomalies on challenging surfaces (metallic). The method is based on a local assessment of the reflectance across the inspected surface, using Reflectance Transformation Imaging data: a set of luminance images captured by a fixed camera while varying light spatial positions. The reflectance, in each pixel, is modelled by means of a projection of the measured luminances onto a basis of geometric functions, in this case, the Discrete Modal Decomposition (DMD) basis. However, a robust detection and analysis of surface visual anomalies requires that the method must not be affected neither by the geometry (sensor and surface orientation) nor by the texture pattern orientation of the inspected surface. We therefore introduce a rotation-invariant representation on the DMD, from which we devise saliency maps representing the local differences on reflectances. The methodology is tested on different engineering metallic samples exhibiting several types of defects. Compared to other saliency assessments, the results of our methodology demonstrate the best performance regarding anomaly detection, localisation and analysis.

Keywords: Anomaly detection · Metallic surfaces · Reflectance · RTI

1 Introduction

The precise detection and analysis of visual anomalies are of primary importance in many fields involving the visual inspection of surfaces. This is particularly true in the case of metallic engineered surfaces where machining and finishing operations often produce some small and local defects on surfaces. Moreover, a solid trend in the industry of high added-value products consists in the automation of the visual inspection process. Indeed, various automated visual inspection systems for the detection of anomalies on metallic surfaces have been recently implemented [1], based on different set-up principles. In a non-exhaustive way,

© Springer International Publishing AG 2017
P. Sharma and F.M. Bianchi (Eds.): SCIA 2017, Part I, LNCS 10269, pp. 550–561, 2017.
DOI: 10.1007/978-3-319-59126-1_46

the *dark field* imaging technique consists of illuminating the surface by a directional light source at a grazing angle of incidence, whereas *bright field* imaging [2] uses a sensor placed along the direction of the specular reflection. As the curvature is known to correlate well with the visual perception, the D-Sight system [3,4] focused on the detection of curvature variations by observing the reflected light through a retro-reflective screen. However, since the variety of potential visual anomalies is very large these systems are often designed for a specific kind of anomaly, finishing process, surface model, and/or object. Consequently, this leads to non detection of unexpected (out of the considered model) visual anomalies. Furthermore, these methods do not allow an accurate description of the visual anomalies, which are often compressed in a single descriptor (e.g. D-sight curvature maps), or reduced to an intensity-based variation in the used imaging modality.

We deal with this issue by proposing a methodology based on the detection and the analysis of the changes on the local "visual behaviour" of the surface using reflectance information. The detection is indeed eased by the utilization of reflectance features on the inspected surfaces, which are very discriminating when calculated locally (on each point of the surface) and made invariant to rotation. To achieve the acquisition of reflectance characteristics, the method proposed in this work is placed within the framework of the techniques called Reflectance Transformation Imaging (RTI) which are widespread in the field of Cultural Heritage for the digitalization of the visual appearance of objects [5]. In the RTI technique, for each pixel of the inspected surface, a set of values representing the reflected light for each light direction is acquired allowing the modelling of what we call the angular reflectance (in contrast with spectral reflectance). The approximation of this discrete set of values by a surface is a least square regression for which we propose the method based on the Discrete Modal Decomposition (DMD) [6]. This latter provides indeed a fine description of the local angular reflectance particularly in the case of local sharp variations of angular reflectance, such as specularities, shadows, or sparkles, which can induce visual anomalies by disrupting the reflectance spatial homogeneity. We then introduce a new rotation-invariant representation derived from the DMD that enables the method to be insensitive to the object/sensor relative orientation during acquisition, or to the orientation of the intrinsic surface roughness pattern. Saliency maps are then estimated from this information as means of defects detection and characterisation. Experimental results are presented on different engineering metallic samples exhibiting several types of appearance defects and compared to other conventional saliency assessments.

2   Related Work

The most exhaustive description of the reflectance is the Bidirectional Reflectance Distribution Function [7] (BRDF). However, the assessment of this function is costly in terms of processing time and amount of data, and is therefore not adequate for the detection of visual anomalies. A simplification of the

BRDF consists in the acquisition of the angular component of the reflectance – the sensor is in a fixed position normal to the surface while only the spatial positions of lighting vary during acquisitions – leads to achieve a good balance between a fine description of local visual behaviour and the practicality. We propose thus to implement RTI technique along with its subsequent processing for aiding the visual inspection of surfaces by allowing the detection and description of visual appearance anomalies in terms of visual behaviour.

2.1   RTI Data Acquisition Requirements

As the high quality of the source images is essential, we first briefly recall the main requirements for RTI stereo-photometric data acquisitions, before detailing the next steps of our methodology. An RTI setup is used to acquire for each pixel of the scene the luminance values $\mathcal{L}$ under varying light positions defined in the coordinate system $(l_u, l_v, \mathcal{L})$, where (l_u, l_v) are the components associated with light positions (LP) projected in the horizontal plane, as shown in Fig. 1 (left). The sensor is generally fixed at the centre of a dome [8], orthogonal to the inspected surface. Other RTI setups based on this principle have been designed to adapt the capture of image sources from *in situ* archaeological material to the intrinsic constraints of the object and its environment [9,10].

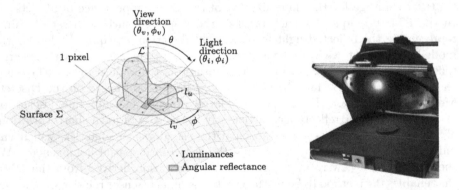

Fig. 1. RTI principle. (left) Angular reflectance approximated from the measured luminances expressed in the coordinate system $(l_u, l_v, \mathcal{L})$ and (right) our RTI setup

2.2   Angular Reflectance Modelling

Modelling of the angular reflectance from RTI acquisitions aims at approximating the set of the discrete points representing the luminance values at each pixel for all light positions (hemispheric shaped) by a continuous surface as shown in Fig. 2 (right). This step is called surface parametrization for which, to our knowledge, only three methods exist in literature: PTM [8,11], HSH [12] and DMD [6]. The objective of the parametrization is to obtain the best fitting of the information held by the discrete points while compacting it reliably. We chose to implement here the Discrete Modal Decomposition (DMD), which outperforms the others, especially in detecting and highlighting singularities on surfaces [6]. The DMD method is based on a projection on the Modal bases. The modal bases are composed of elementary forms that take their origin of a structural dynamic problem. We adapt these modal bases to the case of angular reflectance reconstruction by devising new modal shapes that we call Reflectance Modal Basis (RMB), solution of the structural dynamic problem whose reference surface is a hemisphere. This set of elementary forms noted Q_i (with $i = 1..m$ where m is the number of modes) is used to reconstruct the reflectance shape from discrete values (luminance) obtained during the acquisition stage [13]. A representation of first modes of the RMB is given in Fig. 3. Therefore, the local reflectance associated to each point/pixel of the inspected surface is then described by a modal spectrum representing the contributions λ_i of each mode of the RMB (Fig. 2 - left).

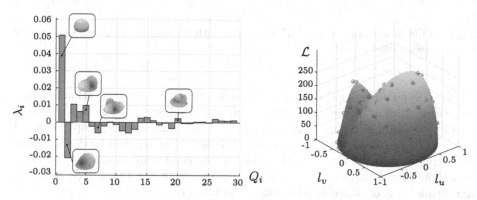

Fig. 2. Angular reflectance modelling at a pixel (right) approximated from measured luminances (blue dots) using DMD method and (left) the associated spectrum λ_i representing the contributions of each mode of the RMB (Color figure online)

3 Proposed Methodology

In the following we show how we extend DMD in order to automate visual inspection by introducing a rotation-invariant representation of the reflectance characteristics of any inspected surface.

3.1 Rotation-Invariant Representation

One interesting property of the modal shapes is that it can be separated in rotation-dependent and rotation invariant modes, named respectively *simple* and *congruent* modes. The shape of the *simple* modes presents a rotational symmetry which preserve their shape for any rotation around the vertical axis (modes Q_1 or Q_6 in Fig. 3). They are thereby originally rotation-invariants. Some pairs, namely the *congruent* modes, show the same shape but are oriented differently (for instance Q_{2-3} or Q_{4-5} in Fig. 3). As a consequence, in this original form, the modal spectrum composed of λ_i coefficients will then vary when the orientation (rotation for example) of the object under the camera is modified. The same principle applies to, for example, a sample with a regular pattern oriented differently across its surface: two points with the same visual appearance behaviour but different orientation will be described by completely different modal spectra. In order to use the modal information for the estimation of visual saliency maps, it is thus essential to transform the original modal shapes to make them rotation-invariant.

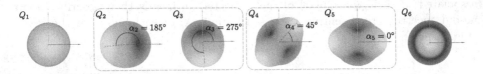

Fig. 3. First descriptors of the RMB noted Q_i, with the associated phase-angle α_i for the *congruent* modes

The proposed method to obtain a rotation-invariant representation is based on the separation of the phase and amplitude components. This separation is performed through a change applied on the congruent modes group presented below. The amplitude and the phase-angle (resp. λ'_j et α'_j) can be derived from the expression of the linear combination of two congruent modes Q_i and Q_{i+1} and their modal coefficients λ_i and λ_{i+1} as shown in the Fig. 4. The resulting amplitude λ'_j is obtained by computing the L2-norm: $\lambda'_j = (\lambda_i^2 + \lambda_{i+1}^2)^{1/2}$.

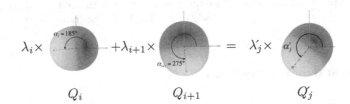

Fig. 4. Change of parametrization in the amplitude/phase-angle form (e.g. Q_{2-3} pair)

The L2-norm does not vary when the orientation of the reflectance is rotated around the vertical axis of $(l_u, l_v, \mathcal{L})$. The vector of resulting amplitudes λ'_j is therefore a subset of the initial coefficients λ_i which characterizes the shape of any reflectances independently of its spatial orientation. In other words, the angular reflectances which are identical in shape but differently oriented, i.e. which can be brought into coincidence by rotation, present the same modal spectra in this rotation-invariant reflectance description. The phase-angle α'_j of the associated Q'_j mode can be determined by the following Eq. 1, where the phase-angle α_i of each mode is known *a priori* and saved with the RMB.

$$\alpha'_j = \alpha_i + \arctan \frac{\lambda_{i+1} sin(\alpha_{i+1} - \alpha_i)}{\lambda_i + \lambda_{i+1} \cos(\alpha_{i+1} - \alpha_i)} \tag{1}$$

This phase information is not used for this methodology, but could be relevant for other applications, as for example to assess the orientation of a textural pattern, or to determine the appropriate light directions in order to find specific visual appearance features on surfaces.

3.2 Visual Saliency Estimation

In order to enable the identification of atypical (salient) behaviours in term of angular reflectance, we implement a multivariate statistical analysis on the local invariant modal spectra. The multivariate image is defined as a three-dimensional $n_1 \times n_2 \times m_B$ data matrix, where two spatial dimensions represent the $n_1 \times n_2$ image pixels and the third dimension represents the m_B modal rotation-invariant coefficients λ'_j. We can reshape the $n_1 \times n_2$−pixel images of the amplitude series into column vectors with dimension $n = n_1 n_2$ and express the multivariate image data as two-dimensional $n \times m_B$ matrix $\boldsymbol{\lambda}' = [\lambda'_1, \lambda'_2, ..., \lambda'_{m_B}]$. Then, we use the Mahalanobis distance D_{Mahal} [14] to assess the distance between the reflectance modal spectrum of each element (pixel) and the average modal spectrum of reflectance estimated on the whole image, i.e. from all vectors $\boldsymbol{\lambda}'$. The expression of D_{Mahal} is given by:

$$D_{\text{Mahal}} = \sqrt{(\boldsymbol{\lambda}' - \boldsymbol{\mu})^T \boldsymbol{\Sigma}^{-1} (\boldsymbol{\lambda}' - \boldsymbol{\mu})} \tag{2}$$

where $\boldsymbol{\Sigma}$ and $\boldsymbol{\mu} = (\mu_1, \mu_2, \mu_3, ..., \mu_p)^T$ are respectively the covariance matrix and the average vector describing the mean reflectance shape across the surface. The visual saliency is finally computed from this distance estimation on each point of the inspected surface and plotted as saliency maps. Figure 5 summarizes this methodology.

4 Results and Discussion

In order to evaluate the proposed method, we tested it on different metallic objects (samples). We compared our results with both human observers assessment and a state-of-the art method for saliency maps computation. Since the state-of-the-art method was not initially adapted to RTI data and objects we studied, we introduced an adaptation on it. Then the rotational invariance of our methodology is demonstrated on a real case. The samples described in Sect. 4.1 were acquired with a dome with 96 light sources evenly distributed over the hemisphere presented in Fig. 1 (right).

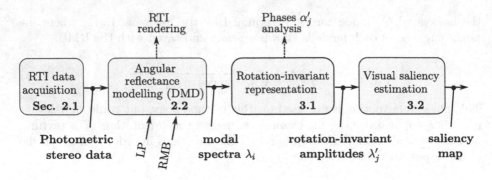

Fig. 5. Methodology flow chart

4.1   Datasets

Dataset D_1. The first dataset we used presents the interest that the scratchs introduced on it are controlled (well known features and locations). The samples are polished metallic surfaces (Fig. 6 on top) which were indented with a Berkovish nanoindenter. The indented scratches of the three tested samples are described by the geometric parameters presented in Table 1. These samples were assessed through a sensorial analysis performed by trained quality controllers, as described in [15]. The adaptation phenomenon [16] during this sensory evaluation was notably reduced by positioning the scratches at random locations on each sample (Fig. 6). A *visibility* ratio is thereby associated to each sample (Table 1). As other uncontrolled degradations are observed on the samples, these defects present a good challenge for our approach.

Dataset D_2. The second dataset corresponds to a gauge block that is used as a standard reference to check the calibration of measurement tools such as micrometers. The manufacture of gauge blocks implies lapping and polishing processes with high precision for achieving a fine surface finish. These gauges have to be of very high quality (precision and location of patterns) which requires a strong quality control (anomaly detection) stage. One side of the gauge block of the dataset D_2 has an unidirectional brushed pattern. For an identical scene of observation (light position and view directions) given in Fig. 7a, the surface appearance of the object changes when its orientation

Table 1. Geometric parameters of the indented scratch and results of the sensory analysis on S_{1-3} samples [15]

Sample	a (μm)	b (μm)	c (μm)	Visibility (%)
S_1	1.25	0.75	58.32	64.6
S_2	0.98	0.61	43.57	14.5
S_3	0.68	0.44	26.1	5

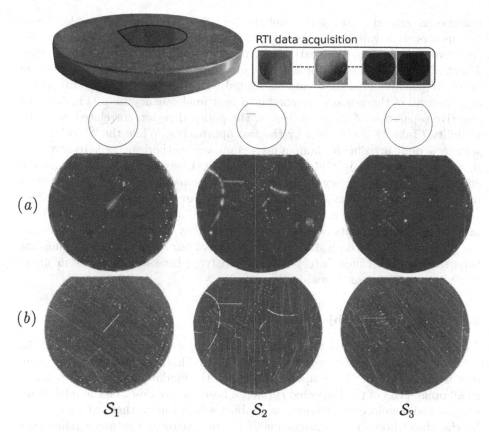

Fig. 6. Saliency maps S_{1-3} obtained, from RTI data, with (a) the Itti and Koch's saliency model and (b) our method

(rotation) varies. The rotation of the object is thus not equivalent to a rotation of the image. However, the angular reflectance associated to a point of the object in any spatial orientation has to be identical, only phase shifted.

4.2   Anomaly Detection Performances

The resulting saliency maps obtained with our method (Sect. 3) are given in Fig. 6b for each sample of the first dataset D_1. Beside, in a purpose of comparison, we use a well-known implementation of Itti and Koch's saliency model (Matlab Saliency Toolbox [17,18]) based on three visual features - luminance contrast, orientation and colour - with each feature computed over a multi-scale Gaussian pyramid. Itti's model was largely extended to the saliency estimation in a dynamic scene (video) by integrating motion estimators [19], applications on 3D meshes [20], or multispectral content [21]. As this model was optimized for a single image scene with macroscopic objects, a change is thus applied on the parameters related to the scale levels ($c \in \{1, 2, 3, 4\}$ and $\delta \in \{1, 2, 3\}$). These parameters were validated on synthetic rough surfaces [22,23] in order to be sensitive to high frequency spatial information contained in the surface images. The Itti's method is then applied on each input image of the RTI dataset in order to generate a saliency map associated with the different light directions, and then all of those maps are combined in a final saliency map (Fig. 6a). An effective separation of the scratches on the polished region correlated with the visibility (Table 1) is obtained by the two approaches. While the S_3 calibrated scratch is quite invisible for human inspection, our method succeeds to reveal its proper signature (straight line). The results on these samples demonstrate the effectiveness of our saliency maps as means of anomaly detection. With regards to Itti's methods, the saliency maps generated are rather rotation-dependent since the small scratches are not detected. In addition, major limitations of Itti's method are related to its computational complexity and time costs of processing all images of RTI data structure. In contrast, our method allows significant time savings (150 times faster) thereby satisfying time requirements in many inspection applications of real-world surfaces.

4.3   Invariance to Object Rotation

Our approach is applied on the second dataset D_2 for illustration purposes of the rotation invariance property of our methodology. This rotation-invariance property is important for the comparison between the modal spectra corresponding to all point/pixel of the inspected surface. Those spectra describe the reflectance shapes, and therefore allow to generate robust saliency maps that are not affected by the object pose or the orientation of texture pattern. The mean reflectance shape (Fig. 7b) characterizing the normal behaviour over the whole surface of the gauge block corresponds to the linear combination of the mean vector of coefficients λ_i with the associated modes Q_i of the RMB. In this original space of representation, we can easily observe that the spectra λ_i vary while the surface orientation of the gauge block is rotated, as shown in the first chart of Fig. 8. The change of parametrization consists in replacing the set of coefficients λ_i by a subset of amplitudes λ'_j (second chart of Fig. 8), after separating the component linked to the orientation as described in Sect. 3.1. The resulting spectra of amplitudes λ'_j constructed from the initial set of coefficients λ_i, are quite similar (invariant) for the three different orientations (0, 60 and 90° rotation) of the gauge block. The small differences between these spectra can be explained by the light positions that can not be strictly identical in practice when the surface is rotated inside the RTI dome. However these deviations are not critical for the visual saliency estimation. The resulting saliency maps given in Fig. 7c obtained from these data show that our approach is robust to a change of orientation of the object. The salient features detected over the metal surface of the gauge block corresponds to many scratches, marks, and regions where the brushed pattern displays more pronounced (deeper) lines than normal visual appearance.

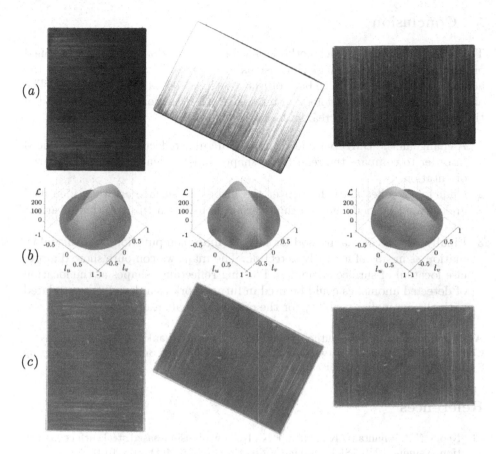

Fig. 7. Effects of the sample orientation change. (a) The same sample acquired for the same light position for three orientations (0, 60 and 90° rotation): (b) the corresponding estimated angular reflectance, (c) the saliency maps using our methodology

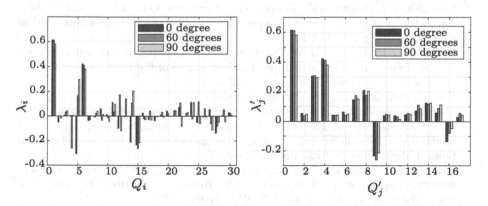

Fig. 8. Change of parametrization from the original spectra λ_i to its rotation-invariant spectra λ'_j, characterizing the reflectance shape independently of the surface orientation

5   Conclusion

This paper presented a novel methodology for the detection of anomalies applied on metallic surfaces from high dimensional RTI data. The method presents the advantage to be insensitive to both object rotations and pattern orientation. The effectiveness of our methodology in terms of both detection performance and of time-saving was shown on real surfaces. We claim:

1. A change in the DMD parametrization of angular reflectances was introduced in order to compare the reflectance shapes independently from their spatial orientation.
2. Changes in reflectance shape over the inspected surface are then identified and located by performing a multivariate analysis in this rotation-invariant space.
3. Finally, we provided a method for saliency maps computation from this information as means of anomaly detection. The maps we compute showed a precise location of surface anomalies. Plotting reflectance shapes at the location of detected anomalies could be used in future work as a descriptor correlated with the anomalies visibility for the criticality evaluation.

Acknowledgments. The authors would like to warmly thank the Regional Research Council (RFF-Innlandet, Norway) and the partners of the MeSurA project (Measuring Surface Appearance) for their support.

References

1. Neogi, N., Mohanta, D.K., Dutta, P.K.: Review of vision-based steel surface inspection systems. EURASIP J. Image Video Process. **2014**(1), 50 (2014)
2. Wu, G., Kwak, H., Jang, S., Xu, K., Xu, J.: Design of online surface inspection system of hot rolled strips. In: 2008 IEEE International Conference on Automation and Logistics (ICAL), pp. 2291–2295 (2008)
3. Reynolds, R.L., Karpala, F., Clarke, D.A., Hageniers, O.L.: Theory and applications of a surface inspection technique using double-pass retroreflection. Optical Eng. **32**(9), 2122–2129 (1993)
4. Heida, J.H., Bruinsma, A.J.A.: D-sight technique for rapid impact damage detection on composite aircraft structures. In: Proceedings of the 7th European Conference on Non-Destructive Testing, pp. 1–12 (1998)
5. Earl, G., Martinez, K.: Archaeological applications of polynomial texture mapping: analysis, conservation and representation. J. Archaeol. Sci. **37**(8), 2040–2050 (2010)
6. Pitard, G., Le Goïc, G., Favreliere, H., Samper, S., Desage, S., Pillet, M.: Discrete modal decomposition for surface appearance modelling and rendering. In: SPIE Optical Metrology, vol. 9525, pp. 952523–952523-10 (2015)
7. Nicodemus, F.E., Richmond, J.C., Hsia, J.J., Ginsberg, I.W., Limperis, T.: Geometrical considerations and nomenclature for Reflectance. Institute for Basic Standards, National Bureau of Standards, Washington (1977)
8. Malzbender, T., Gelb, D., Wolters, H.: Polynomial texture maps. In: Proceedings of the 28th Annual Conference on Computer Graphics and Interactive Techniques (2001)

9. Dellepiane, M., Corsini, M., Callieri, M., Scopigno, R.: High quality PTM acquisition: reflection transformation imaging for large objects. In: VAST (2006)
10. Selmo, D., Sturt, F., Miles, J., Basford, P., Malzbender, T., Martinez, K., Thompson, C., Earl, G., Bevan, G.: Underwater reflectance transformation imaging: a technology for in situ underwater cultural heritage object-level recording. J. Electron. Imag. **26**, 011029 (2017)
11. Drew, M.S., Hel-Or, Y., Malzbender, T., Hajari, N.: Robust estimation of surface properties and interpolation of shadow/specularity components. Image Vis. Comput. **30**(4–5), 317–331 (2012)
12. Zhang, M., Drew, M.S.: Efficient robust image interpolation and surface properties using polynomial texture mapping. EURASIP J. Image Video Process. **2014**(1), 1–19 (2014)
13. Pitard, G.: Surface appearance metrology and modeling for industrial quality inspection. Ph.D. thesis, Université Grenoble Alpes (2016)
14. Mahalanobis, P.C.: On the generalized distance in statistics. In: Proceedings of the National Institute of Sciences (Calcutta) (1936)
15. Puntous, T., Pavan, S., Delafosse, D., Jourlin, M., Rech, J.: Ability of quality controllers to detect standard scratches on polished surfaces. Precis. Eng. **37**(4), 924–928 (2013)
16. Fecteau, J.H., Munoz, D.P.: Exploring the consequences of the previous trial. Nat. Rev. Neurosci. **4**(6), 435–443 (2003)
17. Itti, L., Koch, C.: A saliency-based search mechanism for overt and covert shifts of visual attention. Vis. Res. **40**(10–12), 1489–1506 (2000)
18. Walther, D., Koch, C.: Saliency Toolbox 2.3 (2006). http://www.saliencytoolbox.net
19. Itti, L., Dhavale, N., Pighin, F.: Realistic avatar eye and head animation using a neurobiological model of visual attention. In: SPIE's 48th Annual Meeting on Optical Science and Technology, vol. 5200, pp. 64–78, January 2004
20. Lee, C.H., Varshney, A., Jacobs, D.W., Lee, C.H., Varshney, A., Jacobs, D.W.: Mesh saliency. ACM Trans. Graph. (TOG) **24**, 659–666 (2005)
21. Le Moan, S., Mansouri, A., Hardeberg, J.Y., Voisin, Y.: Saliency for spectral image analysis. IEEE J. Sel. Topics Appl. Earth Obs. Remote Sens. **6**(6), 2472–2479 (2013)
22. Clarke, A.D.F., Green, P.R., Chantler, M.J., Emrith, K.: Visual search for a target against a $1/f\beta$ continuous textured background. Vis. Res. **48**(21), 2193–2203 (2008)
23. Clarke, A.D.F., Chantler, M.J., Green, P.R.: Modeling visual search on a rough surface. J. Vis. **9**(4), 11.1–11.12 (2009)

Block-Permutation-Based Encryption Scheme with Enhanced Color Scrambling

Shoko Imaizumi[1(✉)], Takeshi Ogasawara[2], and Hitoshi Kiya[3]

[1] Graduate School of Engineering,
Chiba University, 1–33 Yayoicho, Inage-ku, Chiba 263-8522, Japan
imaizumi@chiba-u.jp
[2] Graduate School of Advanced Integration Science,
Chiba University, 1–33 Yayoicho, Inage-ku, Chiba 263-8522, Japan
[3] Department of Information and Communication Systems,
Tokyo Metropolitan University, 6-6 Asahigaoka, Hino-shi, Tokyo, Japan
kiya@sd.tmu.ac.jp

Abstract. This paper proposes an extension of block-permutation-based encryption (BPBE) for the encryption-then-compression (ETC) system, which is more robust against some possible attacks compared to the conventional BPBE schemes. After dividing the original image into multiple blocks, the conventional schemes generate an encrypted image through four processes: positional scrambling, block rotation/inversion, negative-positive transformation, and color component shuffling. The proposed scheme achieves enhanced color scrambling by extending three of the four processes. The resilience against jigsaw puzzle solving problems can be consequently increased. The key space against brute-force attacks has also been expanded exponentially. Our scheme can maintain approximately the same compression efficiency compared with that of the conventional schemes.

Keywords: Block-permutation-based encryption · Lossless image compression · Color scrambling · Jigsaw puzzle solver · Brute-force attack

1 Introduction

Due to the rapid progress of information and communications technology, privacy and copyright protection for digital images has been a serious concern in cloud services, social networking services, and so forth. One of the traditional techniques to securely transmit images is a compression-then-encryption (CTE) system, which performs compression before encryption. However, the image owner has to disclose the image content to a network provider in the CTE system. For this reason, another technique for secure image transmission, that is, an encryption-then-compression (ETC) system, has been studied as the framework where encryption is performed by the image owner before compression/transmission [1–3]. The symmetric key cryptosystems, such as AES and

© Springer International Publishing AG 2017
P. Sharma and F.M. Bianchi (Eds.): SCIA 2017, Part I, LNCS 10269, pp. 562–573, 2017.
DOI: 10.1007/978-3-319-59126-1_47

Triple DES, are frequently used for image protection. However, there is a trade-off between security and additional signal processing in the encryption domain for image transmission systems. Because of this, soft encryption schemes have also been studied.

Block-permutation-based encryption (BPBE) [4–6] first divides the original image into definite size blocks and performs four processes: positional scrambling, block rotation/inversion, negative-positive transformation, and color component shuffling. However, it processes the R, G, and B components identically in each process, and the color distribution of the original image strongly affects that of its encrypted image.

Jigsaw puzzle solvers (JPSs) [7–10] are attacks that aim to retrieve the original image from a large number of pieces by utilizing the correlation among them. Because BPBE images consist of multiple blocks, those blocks can be assumed to be the puzzle pieces, and JPSs should be considered as one type of attacks for BPBE schemes. It has been reported that a jigsaw puzzle consisting of 30,745 pieces can be solved completely by using the conventional JPS [9]. Another JPS has successfully solved a puzzle where the directional information of each piece was not known [10]. It has been confirmed that some encryption schemes, where the key spaces are large enough to protect against brute-force attacks, are still vulnerable to JPSs [11]. On the other hand, it has been confirmed that conventional JPSs can hardly ever solve a jigsaw puzzle where the color distribution is modified.

In this paper, we propose an extended BPBE algorithm to deal with the above problem. The proposed scheme processes the three color components independently in each process. As a consequence, the color distribution of the original image does not severely affect that of its encrypted image in our scheme. Owing to the proposed algorithm, the security against some possible attacks can be improved. We confirm that the compression efficiency of JPEG-LS [12] in the proposed scheme is approximately the same as that of the conventional schemes.

2 Block-Permutation-Based Encryption [4–6]

By encrypting the original images before sending them to the provider, the image owner does not need to disclose the image content to a network provider. As shown in Fig. 1, the conventional BPBE algorithms [4–6] first divide the original image into definite size blocks and then execute four processes: positional scrambling, block rotation/inversion, negative-positive transformation, and color component shuffling. Finally, they integrate the blocks into one encrypted image. The noteworthy advantage of BPBE is that the compression efficiency of the encrypted images can be maintained high compared to that of the original images. The BPBE schemes can also control the quality of the encrypted image and the encryption strength by changing the block size. The encryption procedure is as follows.

Step 1: Divide an original image $I = \{I_R, I_G, I_B\}$ with $M \times N$ pixels into multiple blocks with $B_x \times B_y$ pixels.

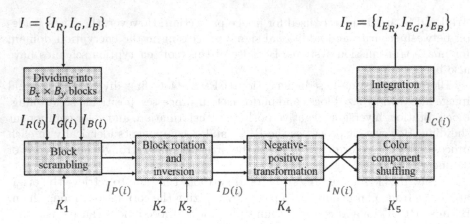

Fig. 1. Block-permutation-based encryption procedure.

Step 2: Scramble the position of each block using a random number generated by key K_1.

Step 3: Rotate and invert each block using random numbers generated by keys K_2 and K_3.

Step 4: Perform negative-positive transformation on each block using a random number generated by key K_4.

Step 5: Shuffle the R, G, and B components in each block using a random number generated by key K_5.

Step 6: Integrate all the blocks and generate the encrypted image.

Note that the keys K_1, K_2, K_3, and K_4 are commonly used for the three color components in the conventional schemes. We explain the main four processes in detail below.

2.1   Positional Scrambling

According to a random number generated by key K_1, the positions of the divided blocks are shuffled. Note that the R, G, and B components in each block are identically shuffled by commonly using K_1.

2.2   Block Rotation and Inversion

As shown in Fig. 2(a), each block, where $B_x = B_y$ is supposed, is rotated 0, 90, 180, or 270 degrees using a random number generated by key K_2. Figure 2(b) shows the block inversion process. Each block is decided to be inverted horizontally and/or vertically or is not inverted according to a random number generated by key K_3. Note that the R, G, and B components in each block are identically rotated and inverted by commonly using K_2 and K_3.

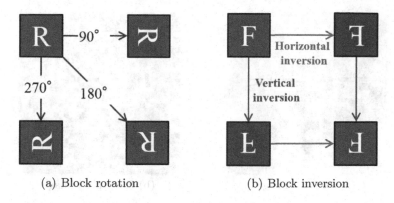

(a) Block rotation   (b) Block inversion

Fig. 2. Block rotation and inversion.

2.3 Negative-Positive Transformation

The negative-positive transformation reverses all of the pixel values in a block according to a random number of either zero or one, which is generated by key K_4. The transformed pixel value p' is obtained by

$$p' = \begin{cases} p & (r(i) = 0) \\ 255 - p & (r(i) = 1), \end{cases} \tag{1}$$

where p is the original pixel value and $r(i)$ is a random integer given for the i-th block by using K_4. Note that the R, G, and B components in each block are identically transformed by commonly using K_4.

2.4 Color Component Shuffling

The color component shuffling is the operation that changes the R, G, and B components in each block according to a random number in the range of zero to five, which is generated by key K_5. It is operated as shown in Table 1.

Table 1. Color component shuffling.

Random number	R	G	B
0	R	B	G
1	B	R	G
2	B	G	R
3	G	R	B
4	G	B	R
5	R	G	B

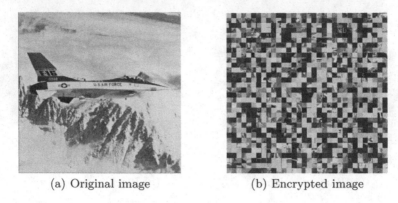

(a) Original image   (b) Encrypted image

Fig. 3. Encrypted image obtained by conventional scheme [5].

Figure 3 shows the original image and its encrypted image where all of the operated blocks have been integrated. The conventional schemes not only change the spatial positions and directions but also reverse the pixel values and changes the three color components in each block. However, the color distribution of the original image directly affects that of the encrypted image. It is assumed to be vulnerable to jigsaw puzzle solvers (JPSs). In the next section, we propose an extended BPBE approach to decrease the effect of the color distribution of the original image.

3   Proposed Scheme

We propose an extended BPBE scheme to improve security against some possible attacks. In the conventional schemes [4–6], the color distribution of the original image strongly affects that of its encrypted image. The proposed scheme aims to solve the above problem by expanding the number of random numbers used in three of the four processes, namely, positional scrambling, block rotation/inversion, and negative-positive transformation, without severe degradation of the compression efficiency. Note that the procedure of the BPBE is the same as that of the conventional schemes, which is shown in Fig. 1.

As described in Sect. 2, the keys K_1, K_2, K_3, and K_4 for the three processes are commonly used for the R, G, and B components in the conventional schemes. In other words, the three color components in each block are identically operated. The proposed scheme prepares three keys for each process, e.g., $K_{1,R}$, $K_{1,G}$, and $K_{1,B}$ for positional scrambling and independently operates the three color components. We give a detailed account using the negative-positive transformation as follows.

Here, we make a concrete example of the negative-positive transformation. In the conventional schemes, this transformation is identically operated for the three color components by using a random number of either zero or one, which

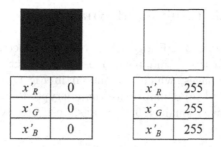

Fig. 4. Negative-positive transformation in conventional schemes [4–6] where $\{x_R, x_G, x_B\} = \{255, 255, 255\}$.

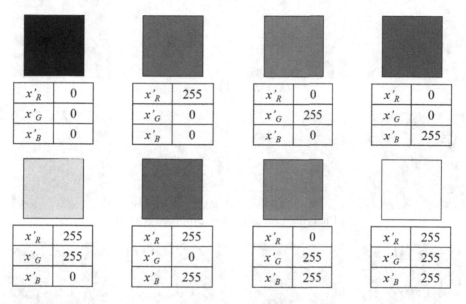

Fig. 5. Negative-positive transformation in proposed scheme where $\{x_R, x_G, x_B\} = \{255, 255, 255\}$.

is generated by key K_4. In the case when pixel x has the values $\{x_R, x_G, x_B\} = \{255, 255, 255\}$, they are changed to $\{x'_R, x'_G, x'_B\} = \{0, 0, 0\}$ or are not changed, that is, $\{x'_R, x'_G, x'_B\} = \{255, 255, 255\}$, as shown in Fig. 4. On the other hand, the proposed scheme prepares three random numbers of either zero or one, which are generated by three keys $K_{4,R}$, $K_{4,G}$, and $K_{4,B}$. The three color components are operated independently according to the random numbers. In the case of the above example where $\{x_R, x_G, x_B\} = \{255, 255, 255\}$, they could be changed to $\{x'_R, x'_G, x'_B\} = \{0, 0, 0\}$, $\{255, 0, 0\}$, $\{0, 255, 0\}$, $\{0, 0, 255\}$, $\{255, 255, 0\}$, $\{255, 0, 255\}$, $\{0, 255, 255\}$, or $\{255, 255, 255\}$, as shown in Fig. 5.

In the proposed scheme, the color information is consequently more scrambled than in the conventional schemes. In the next section, we will demonstrate and evaluate the effectiveness of our algorithm.

4   Experimental Results and Analysis

We evaluate the proposed BPBE algorithm from the aspects of the color scrambling, compression efficiency, and resilience against brute-force attacks/JPSs. The four 512×512 images, which are shown in Fig. 6(a), were used as test images. Figures 6(b) and (c) demonstrate the encrypted images obtained by the proposed and the conventional [4–6] schemes, respectively, where the divided block size is 16×16 pixels.

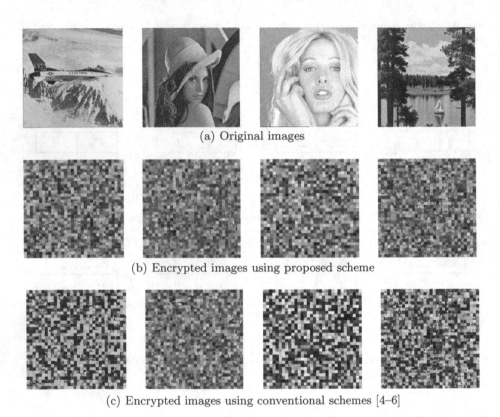

(a) Original images

(b) Encrypted images using proposed scheme

(c) Encrypted images using conventional schemes [4–6]

Fig. 6. Test images and their encrypted images obtained by proposed and conventional [4–6] schemes.

4.1   Color Scrambling

Table 2 indicates the entropies of the original and the encrypted images shown in Fig. 6. The entropy $H(A)$ is defined as

$$H(A) = -\sum_{i=1}^{2^{24}} p(a_i) \log_2 p(a_i), \tag{2}$$

Table 2. Entropies of original and encrypted images shown in Fig. 6.

	Airplane	Lena	Girl	Sailboat
Original	13.96	16.84	14.77	16.88
Proposed	17.78	17.95	17.56	17.96
Conventional	15.62	17.73	16.69	17.63

where A represents the finite set of 24-bit colors a_i ($i = 1, 2, ..., 2_{24}$), that is, $A = \{a_1, a_2, ..., a_{2^{24}}\}$, and $p(a_i)$ is the occurrence probability of a_i. The entropies of the encrypted images obtained by the proposed scheme are higher than those of the original images and the encrypted images obtained by the conventional schemes.

We also compare the color distributions in the original image and its encrypted images obtained by the proposed and the conventional schemes. Figure 7 shows the histograms of the test image 'Lena', where the vertical/horizontal axes represent the saturation/hue values, respectively. The encrypted image produced by the proposed scheme exhibits a wider distribution than the original image and the encrypted image produced by the conventional schemes.

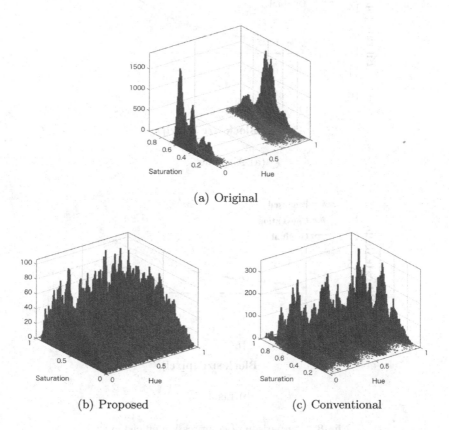

(a) Original

(b) Proposed   (c) Conventional

Fig. 7. Comparison of color distribution (Lena).

From those results, it is verified that the color information of the encrypted images produced by the proposed scheme could become more scrambled relative to that produced by the conventional schemes.

4.2 Compression Efficiency

We compare the compression efficiency of JPEG-LS [12] among the original images and their encrypted images produced by the proposed and the conventional schemes. The compression efficiency is compared by calculating the bit rates.

$$Bit\ rate(\text{bpp}) = \frac{Size\ of\ image\ file}{Number\ of\ pixels\ in\ image(M \times N)} \tag{3}$$

where M and N are the vertical/horizontal sizes of the image.

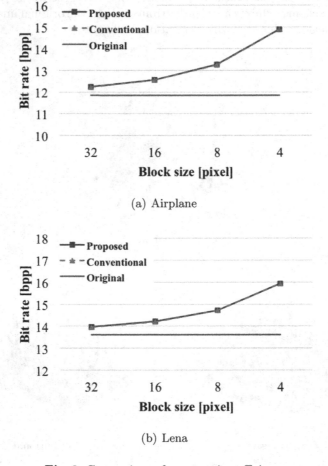

(a) Airplane

(b) Lena

Fig. 8. Comparison of compression efficiency.

Figure 8 shows the comparative results for two of the test images, that is, Airplane and Lena. The sizes of the divided blocks are 4×4, 8×8, 16×16, and 32×32 pixels in the figure. The results for the other two test images have shown analogous lines. It is certified that the proposed scheme can maintain approximately the same compression efficiency as that of the conventional schemes.

4.3   Resilience Against Brute-Force Attacks

We discuss the key space for resilience against brute-force attacks in this section. In the proposed scheme, the four encryption processes are carried out independently from each other. The total key space can be obtained by multiplying the key spaces for the four processes. In the case of dividing a $M \times N$ image into $B_x \times B_y$ blocks, the number of divided blocks L is given as

$$L = \lfloor \frac{M}{B_x} \rfloor \times \lfloor \frac{N}{B_y} \rfloor. \tag{4}$$

The key space N_P of the positional scrambling, which is the number of all the scrambling patterns of L blocks, is calculated by

$$N_P = (_L P_L)^3 = (L!)^3. \tag{5}$$

As shown in Fig. 2, the number of all patterns for both the block rotation and the block inversion is four. When combining those two processes, some combinations correspond to other combinations. Therefore, the number of total patterns for the block rotation/inversion becomes eight. The combined key space of the block rotation and inversion N_D is obtained by

$$N_D = (8^L)^3 = 512^L. \tag{6}$$

In the negative-positive transformation and the color component shuffling, the key spaces N_N and N_C are given by

$$N_N = (2^L)^3 = 8^L, \tag{7}$$

$$N_C = 6^L. \tag{8}$$

Consequently, the total key space N_A in the proposed scheme can be represented by

$$N_A = N_P \times N_D \times N_N \times N_C$$
$$= (L!)^3 \times 512^L \times 8^L \times 6^L, \tag{9}$$

while the total key space in the conventional schemes $N_{A,Conv}$ is given as

$$N_{A,Conv} = N_{P,Conv} \times N_{D,Conv} \times N_{N,Conv} \times N_{C,Conv}$$
$$= L! \times 8^L \times 2^L \times 6^L, \tag{10}$$

where $N_{P,Conv}$, $N_{D,Conv}$, $N_{N,Conv}$, and $N_{C,Conv}$ are the key spaces for the four encryption processes in the conventional schemes.

4.4   Resilience Against Jigsaw Puzzle Solvers [7–10]

Jigsaw puzzle solvers (JPSs) [7–10] are a kind of possible attack that attempts to retrieve the original image from a large number of pieces by using the correlation among them. JPSs have been actively studied in the area of computer vision and pattern recognition. They could be considered as possible attacks for the BPBE schemes because a BPBE image consists of multiple blocks. According to [11], the encrypted images obtained by the conventional schemes [4–6] have a great risk of being maliciously decrypted, even though they keep a sufficiently large key space, as described in Sect. 4.3.

On the other hand, it has also been demonstrated that the restorability of images attacked by JPSs can be greatly decreased in the case when the color information of the encrypted image has been modified. This is because it would be difficult for the conventional JPSs to solve the puzzle when the color correlation among the pieces becomes low. Therefore, it is assumed that the proposed algorithm, which can fully scramble the color information of the encrypted images, would be effective against JPSs.

5   Conclusion

We proposed an extended BPBE algorithm, where the color scrambling of the encrypted images has been enhanced relative to that of the conventional schemes. Consequently, it is effective for increasing the resilience against both brute-force attacks and JPSs. Furthermore, the compression efficiency using JPEG-LS can be maintained to be approximately the same as that of the conventional schemes.

References

1. Zhou, J., Liu, X., Au, O.C., Tang, Y.Y.: Designing an efficient image encryption-then-compression system via prediction error clustering and random permutation. IEEE Trans. Inf. Forensics Secur. **9**(1), 39–50 (2014)
2. Liu, W., Zeng, W., Dong, L., Yao, Q.: Efficient compression of encrypted grayscale images. IEEE Trans. Image Process. **19**(4), 1097–1102 (2010)
3. Johnson, M., Ishwar, P., Prabhakaran, V., Schinberg, D., Ramchandran, K.: On compressing encrypted data. IEEE Trans. Signal Process. **52**(10), 2992–3006 (2004)
4. Kurihara, K., Imaizumi, S., Shiota, S., Kiya, H.: An encryption-then-compression system for lossless image compression standards. IEICE Trans. Inf. Syst. **E100-D**(1), 52–56 (2017)
5. Kurihara, K., Kikuchi, M., Imaizumi, S., Shiota, S., Kiya, H.: An encryption-then-compression system for JPEG/motion JPEG standard. IEICE Trans. Fundam. **E98-A**(11), 2238–2245 (2015)
6. Watanabe, O., Uchida, A., Fukuhara, T., Kiya, H.: An encryption-then-compression system for JPEG 2000 standard. In: Proceedings on IEEE ICASSP, pp. 1226–1230 (2015)
7. Paikin, G., Tal, A.: Solving multiple square jigsaw puzzles with missing pieces. In: Proceedings on CVPR, pp. 4832–4839 (2015)

8. Son, K., Moreno, D., Hays, J., Cooper, D.B.: Solving small-piece jigsaw puzzles by growing consensus. In: Proceedings on CVPR, pp. 1193–1201 (2016)
9. Sholomon, D., David, O.E., Netanyahu, N.S.: An automatic solver for very large jigsaw puzzles using genetic algorithms. Genet. Program. Evolvable Mach. **17**(3), 291–313 (2016)
10. Gallagher, A.C.: Jigsaw puzzles with pieces of unknown orientation. In: Proceedings on CVPR, pp. 382–389 (2012)
11. Chuman, T., Kurihara, K., Kiya, H.: On the security of block scrambling-based ETC systems against jigsaw puzzle solver attacks. In: Proceedings on IEEE ICASSP, pp. 2157–2161 (2017)
12. Weinberger, M.J., Seroussi, G., Sapiro, G.: The LOCO-I lossless image compression algorithm: principles and standardization into JPEG-LS. IEEE Trans. Image Process. **9**(8), 1309–1324 (2000)

Author Index

Printed in the United States
By Bookmasters